Radio Systems: Engineering and Applications

Radio Systems: Engineering and Applications

Edited by
Francis Schmidt

WILLFORD PRESS

www.willfordpress.com

Published by Willford Press,
118-35 Queens Blvd., Suite 400,
Forest Hills, NY 11375, USA

ISBN: 978-1-68285-635-2

Cataloging-in-Publication Data

Radio systems : engineering and applications / edited by Francis Schmidt.
    p. cm.
Includes bibliographical references and index.
ISBN 978-1-68285-635-2
1. Radio. 2. Radio--Transmitters and transmission--Design and construction.
3. Radio--Equipment and supplies. I. Schmidt, Francis.
TK6550 .R33 2019
621.384--dc23

For information on all Willford Press publications
visit our website at www.willfordpress.com

WILLFORD PRESS

Contents

Preface

Radio engineering is a subset of electrical engineering. It deals with the engineering of devices in the radio frequency spectrum, between 3 kHz and 300 GHz. Some of the devices that use radio waves include mobile phones, radios, wifi, etc. A radio communication system comprises of a transmitter and a receiver. It also requires an antenna for converting electric currents into radio waves and vice versa. The principles of transmission, modulation, wave propagation, resonance and demodulation are important aspects of radio systems. An understanding of electronic hardware design, antenna radiation and circuit board material are also crucial in radio engineering. The book aims to shed light on some of the unexplored aspects of radio systems engineering and the recent researches in this field. The various studies that are constantly contributing towards advancing technologies and evolution of this field are examined in detail. It includes contributions of experts and scientists, which will provide innovative insights into this field. This book is an essential guide for both academicians and those who wish to pursue this discipline further.

This book is a result of research of several months to collate the most relevant data in the field.

When I was approached with the idea of this book and the proposal to edit it, I was overwhelmed. It gave me an opportunity to reach out to all those who share a common interest with me in this field. I had 3 main parameters for editing this text:

1. Accuracy – The data and information provided in this book should be up-to-date and valuable to the readers.

2. Structure – The data must be presented in a structured format for easy understanding and better grasping of the readers.

3. Universal Approach – This book not only targets students but also experts and innovators in the field, thus my aim was to present topics which are of use to all.

Thus, it took me a couple of months to finish the editing of this book.

I would like to make a special mention of my publisher who considered me worthy of this opportunity and also supported me throughout the editing process. I would also like to thank the editing team at the back-end who extended their help whenever required.

Editor

Properties of Multimode Optical Epoxy Polymer Waveguides Deposited on Silicon and TOPAS Substrate

Vaclav PRAJZLER [1], Milos NERUDA [1], Pavla NEKVINDOVA [2], Petr MIKULIK [3]

[1] Dept. of Microelectronics, Faculty of Electrical Engineering, Czech Technical University, Technická 2, 166 27 Prague, Czech Republic
[2] Institute of Chemical Technology, Technická 5, 166 27 Prague, Czech Republic
[3] Dept. of Condensed Matter Physics, Masaryk University, Kotlářska 2, 611 37 Brno, Czech Republic

xprajzlv@feld.cvut.cz

Abstract. *The paper reports on the fabrication and characterization of multimode polymer optical waveguides. Epoxy polymer EpoCore was used as the waveguide core material and EpoClad was used as a cladding and cover protection layer. The design of the waveguides was schemed for geometric dimensions of 50 μm core and for 850 nm and 1310 nm wavelengths. Proposed shapes of the waveguides were fabricated by standard photolithography process. Optical losses of the planar waveguides were measured by the fibre probe technique at 632.8 nm and 964 nm. Propagation optical loss measurements for rectangular waveguides were done by using the cut-back method and the best samples had optical losses lower than 0.53 dB/cm at 650 nm, 850 nm and 1310 nm.*

Keywords

Optical planar waveguides, optical rectangular waveguides, multimode waveguides, polymer

1. Introduction

Optical data transmission has become the obvious choice for communication over longer distances, but copper based electrical interconnects have dominated short-range communication links for over decades. The increasing demand in recent years for interconnection bandwidth in data centers and supercomputers, in conjunction with the inherent disadvantages of copper interconnects when operating at high data rates (>10 Gb/s), has led to the consideration of the use of optical technologies in very short communication links [1–3]. This new trends force designers to use optical interconnections also to bridge short distances including optical communications like Short Reach (SR) up to 300 m, Extra Short Reach (ESR) up to 5 cm and also Ultra Short Reach (USR) or even 1 cm. For such new optical communications systems, it is highly desirable to develop new technology techniques with new optical materials, which will allow high operating data rates and would make fabrication process simple and cheap.

The key waveguide components for formation of such complex on-board optical layouts include multimode ridge waveguides. Therefore, many research groups are looking for fabrication of suitable elements that would combine new materials having comparable properties and the same time a possibility to be prepared by easy fabrication process in order to make production of those structures possible in a large scale, low cost and also environmentally friendly.

Multimode polymer waveguides are a particularly attractive option for the ESR and USR communications because they exhibit favorable optical, mechanical and thermal properties allowing their direct integration onto low cost printed circuit boards with standard methods of conventional electronics manufacturing. Optical interconnects via polymer waveguides have attracted considerable attention during the last decade also because they enable fabrication of multimode optical waveguide structures on printed circuit boards [4–8] and flexible foil substrate using low-cost assembly and packaging methods [9], [10].

This paper presents a new approach to realize multimode polymer ridge waveguides for optical interconnection applications. In developing the new approach, epoxy polymer was selected as a core waveguide material while EpoClad polymer material was selected for cladding; both supported by Micro resist technology GmbH. These materials possess excellent properties such as high heat and pressure resistance, low optical losses (< 0.49 dB·cm⁻¹ at 633 nm [11], [12], 0.2 dB·cm⁻¹ at 850 nm, refractive index EpoCore 1.58, EpoClad 1.57, λ = 830 nm [13]), easy fabrication process, etc. For the substrate we used silicon and flexible TOPAS polymer foil 8007X4. TOPAS is Cyclic Olefin Copolymer and it is a glass-clear amorphous polymer with outstanding moisture barrier, chemical resistance, high purity and non-reactive surface making it an excellent choice for optical interconnections [14].

2. Design of Planar and Ridge Waveguides

To design and develop an optical rib waveguide one usually needs to start with a slab waveguide. The slab waveguide is a planar dielectric thin film (core) with refractive index n_f which is sandwiched between a substrate and cover (or superstrate, or cladding) with lower refractive indices n_s and n_c, respectively (see Fig. 1a).

The guiding mechanism is provided by total reflections along the vertical dimension. In the simplest case the core is deposited on the same substrate as a cover layer; than the slab is called symmetric ($n_s = n_c$). The ridge waveguides are ones of the most commonly used optical waveguides in integrated optics and photonics. They have shape of a rectangular as shown in Fig. 1b; in our case we used cladding and cover layer made of the same polymer materials (EpoClad) and therefore it may be called a symmetric waveguide.

Suitability of the polymer EpoCore and EpoClad materials proposed for the cores and cladding of the planar waveguides was checked by Spectrometer Shimadzu (UV-3600). The measurements showed that the materials were transparent within the range from 400 to 1600 nm (see Fig. 2) and therefore fulfill the requirements for their suitability stated above.

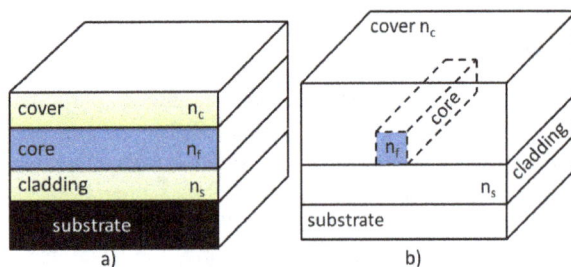

Fig. 1. Cross-sectional view: a) optical planar waveguide, b) ridge waveguide.

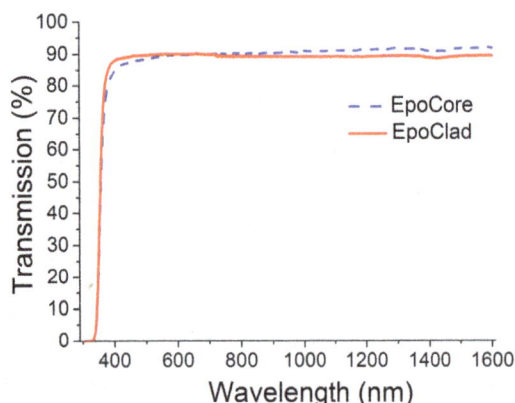

Fig. 2. Transmission spectra of EpoClad and EpoCore polymers.

3. Fabrication

The waveguides were prepared using epoxy negative tone photoresists EpoCore (core layer) and EpoClad polymer (cladding and cover layers) supplied by Micro resist technology GmbH. The waveguiding layers were deposited onto silicon and flexible TOPAS 8007X4 (thickness 0.125 and 0.3 mm) substrates.

The optimized fabrication process of the samples is illustrated step by step in Fig. 3 a-f and consisted of the following steps:

- Cleaning by plasma etching (Fig. 3a).
- Deposition of the EpoClad layers by spin coating (Fig. 3b) followed by soft bake process done for 10 minutes on a hot plate at 50°C after which the temperature was gradually increased to 90°C (at 10°C/min). Finally, a UV-curing process was applied followed by a bake process. This bake process was the same as before the UV curing.
- Deposition of an EpoCore layer on it by spin coating (Fig. 3c) and again, the soft-bake process was applied on a hotplate at 50°C for 10 min. Afterwards, the temperature was once again gradually increased to 90°C (at 10°C/min).
- Photolithography process using PERKIN-ELMER 300 HT Micralign was performed (Fig. 3d). Then again bake process was applied at 50°C for 10 min on hotplate. After that the temperature was gradually increased to 90°C (10°C/min) and after cooling down it was followed with dipping into mr-Dev 600 developer (Fig. 3e). The developing process was stopped by dipping the samples into isopropanol.
- Deposition of EpoClad layer as a cover protection layer by spin coating (Fig. 3f) was the last step.

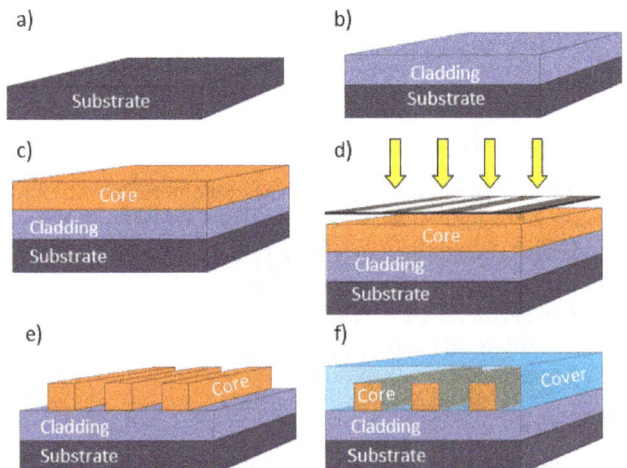

Fig. 3. Fabrication process for EpoCore optical planar waveguides a) substrate cleaning, b) deposition of EpoClad cladding layer, c) deposition of EpoCore waveguide layer, d) UV curing process, e) wet etching, f) deposition of EpoClad cover layer.

4. Measurements and Results

The thicknesses of the fabricated polymer layers were measured by profile-meters Talystep Hommel Tester 1000. The experimentally found thickness of the cladding layer was around 20 μm and that of the core waveguide layers varied from 20 up to 80 μm depending on the rate of spinning of the coater during the deposition.

Properties of the ridge waveguides were checked using optical digital camera ARTCAMI which is equipped with optical head ZOOM Optics (Olympus Czech Group Ltd.). The camera was controlled by QUICKFOTO software. The measurement revealed that the ridge waveguides had good optical quality and dimension of the fabricated structure corresponded well with the size of the proposed waveguides. The images of the ridge waveguides are shown in Fig. 4 where Fig. 4a shows top view image and Fig. 4b shows the edge view.

Waveguiding properties of the planar waveguides were examined by dark mode spectroscopy using Metricon 2010 prism-coupler system [15] at five wavelengths 473, 632.8, 964, 1311 and 1552 nm. Index refraction of the planar waveguide can be determined by measuring of the critical angle of the incidence at the interface between the prism and the material in contact with the coupling prism. A typical example of the result in a form of mode spectra is given in Fig. 5.

Critical angles of incidence determine refractive indices of the waveguiding (EpoCore, n_f) and cladding

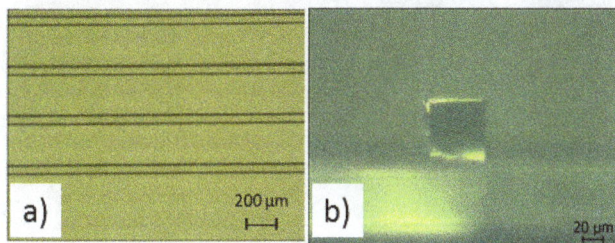

Fig. 4. Images of EpoClad/EpoCore ridge waveguides: a) top view, b) edge view.

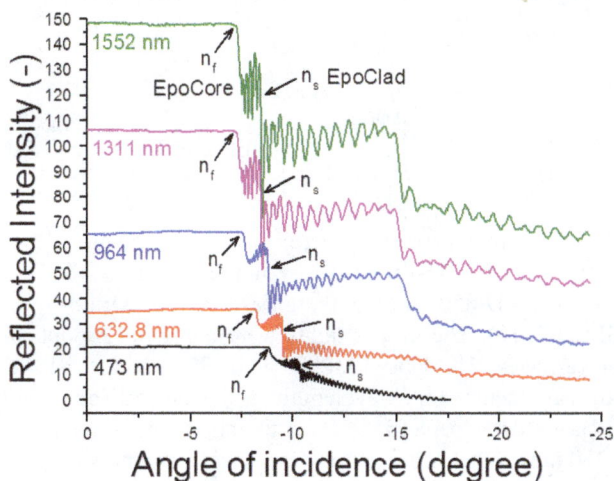

Fig. 5. Mode pattern of EpoCore/EpoClad planar waveguides.

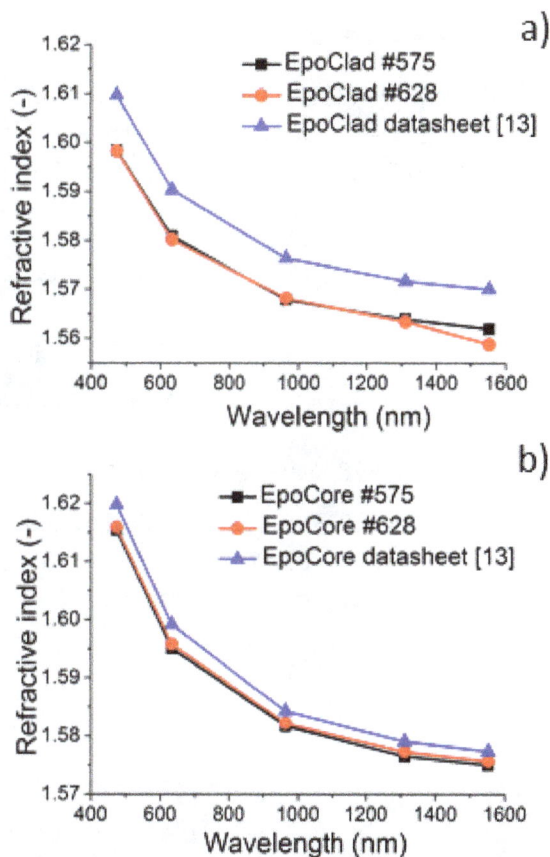

Fig. 6. Refractive indices measured by Metricon 2010 prism-coupler system and compared with datasheet values: a) EpoClad, b) EpoCore.

(EpoClad, n_s) layers. For more details of such measurement see [11, 16].

Refractive indices for the EpoClad cladding and EpoCore core waveguides determined for the mode patterns are shown in Fig. 6.

In Fig. 6, measured refractive indices are compared to the data sheet [13] values for EpoClad and EpoCore epoxy polymer and it was found that value of refractive index for EpoClad were very similar to the tabular values. The values for the EpoCore were only marginally lower than those of the declared ones, in contrast with the measured refractive index values of EpoClad, that were found lower than tabular values. The reason for that was probably lower temperature of the final hardening of our samples, which was used to prevent deterioration of flexibility of our EpoClad foils.

Optical losses of our planar waveguides were measured by fiber probe technique. The principle of the measurement involves measurement of transmitted and scattering light intensity as a function of propagation distance along the waveguide [12, 17]. The light is coupled into the planar waveguides through optical coupling prism and the outgoing scattered light intensity was detected by optical fiber connected to Si detector.

The results of optical loss measurements of the EpoCore planar waveguides are demonstrated in Fig. 7.

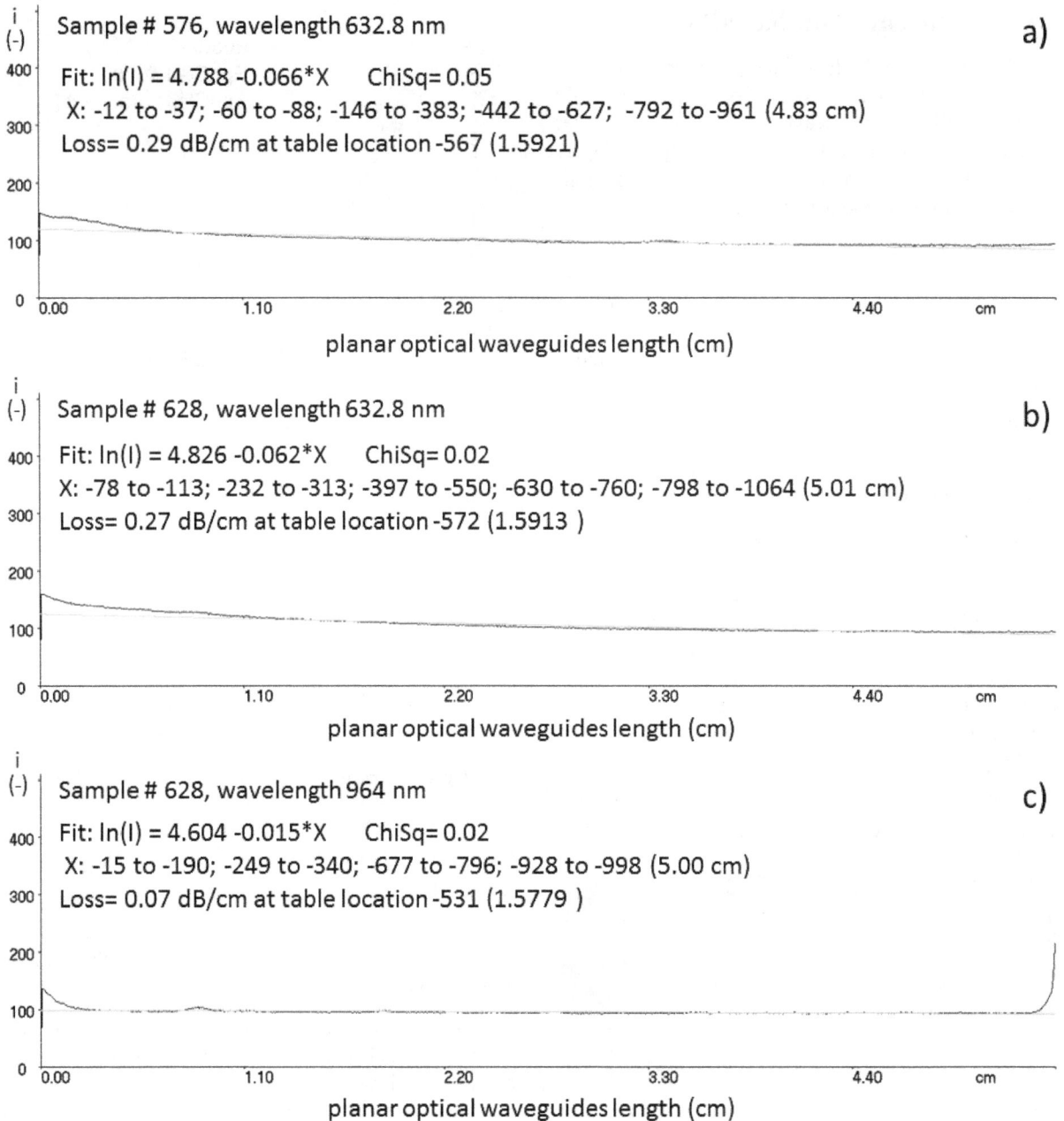

Fig. 7. Optical losses of the EpoCore/EpoClad planar waveguides on: a) Si substrate (632.8 nm), b) TOPAS substrate (632.8 nm), c) TOPAS substrate (964 nm).

Figure 7a shows the results for EpoCore waveguides deposited on silicon substrate and Figures 7b and c give the results for the waveguides deposited on TOPAS substrate. Figures 7a and b demonstrate the measured values for the wavelength 632.8 nm while Figure 7c shows it for the wavelength 964 nm. Our optical planar waveguides had optical losses lower than 0.3 dB·cm^{-1} with the best sample having optical losses as low as 0.27 dB·cm^{-1} at 632.8 nm and 0.07 dB·cm^{-1} at 964 nm.

Optical loss measurements for ridge waveguides were done using the cut-back method. The principle of the method is shown in Fig. 8a while Figure 8b shows a photo of the experimental set-up of the measurement. Optical losses were calculated using equation (1) [18]:

$$\alpha = \frac{10 \cdot \log \dfrac{P_1[\text{W}]}{P_2[\text{W}]}}{(l_1 - l_2)[\text{cm}]} \ [\text{dB·cm}^{-1}]. \qquad (1)$$

The measurements were done at: 650 nm (laser Safibra OFLS-5-FP-650), 850 nm (laser Safibra OFLS-6- LD-850) and 1310 nm (laser Safibra OFLS-6CH, SLED-1310). The output lights were measured by optical powermeter Thorlabs PM200 with Si detector S151C (measurements for wavelength 650 and 850 nm) and InGaAs detector S155C (measurements for wavelength 1310 nm). The accuracy of the measurement set-up is estimated to be ±5 %.

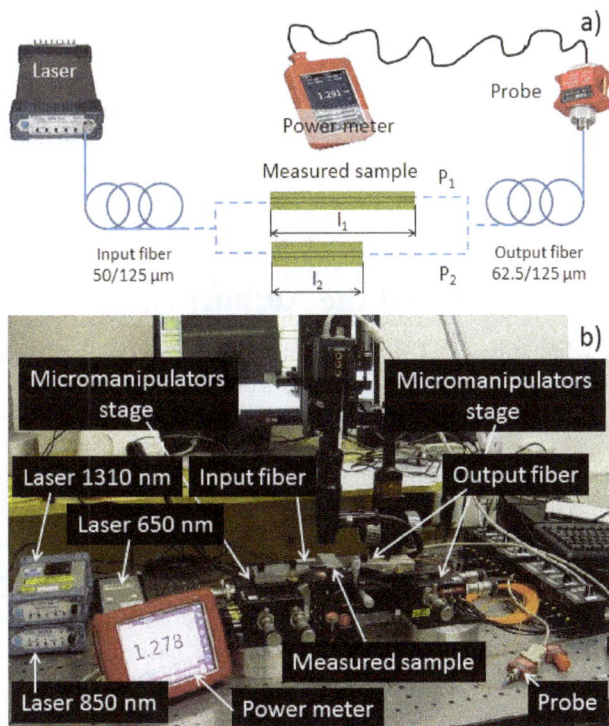

Fig. 8. Principle for insertion optical loss measurement: a) schematic view of the setup, b) image of the setup.

The measurement started with determination of the optical power (P_1) emitted from the source and passing through the whole length l_1 of the ridge waveguide and proceeding through to the detector via output fiber waveguides. P_2 is the output optical power obtained after breaking the waveguide, where l_2 is the length of the broken part of the optical waveguide.

Detailed image of the ridge waveguide transmitting optical light 650 nm for optical loss measurements using the cut-back method is shown in Fig. 9.

Our fabricated optical channel waveguides had insertion optical losses lower than $0.60\ dB \cdot cm^{-1}$. The average value of the insertion optical losses were $0.36\ dB \cdot cm^{-1}$ at 650 nm, $0.32\ dB \cdot cm^{-1}$ at 850 nm and $0.53\ dB \cdot cm^{-1}$ at 1310 nm. The best sample had optical losses $0.27\ dB \cdot cm^{-1}$ at 650 nm, $0.16\ dB\ cm^{-1}$ at 850 nm and $0.30\ dB \cdot cm^{-1}$ at 1310 nm.

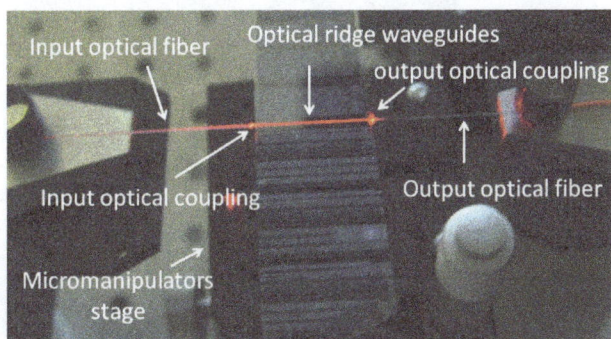

Fig. 9. Detailed image of the ridge waveguide transmitting optical light 650 nm for optical loss measurements using the cut-back method.

5. Conclusion

We report about properties of EpoClad/EpoCore polymer ridge waveguides fabricated on silicon and TOPAS 8007X4 substrate. Ridge waveguides were deposited by using spin coating and photolithography process.

Optical waveguiding properties of our planar waveguides samples were characterized by Metricon 2010 prism-coupler system for five wavelengths (473, 633, 964, 1311 and 1552 nm) and optical losses were measured by collecting the scattered light using fiber scanning along the waveguide read by the Si photodetector at 632.8 and 964 nm. The samples had optical losses less than $0.5\ dB \cdot cm^{-1}$ and the best sample has optical losses around $0.27\ dB \cdot cm^{-1}$ at 632.8 nm and $0.07\ dB \cdot cm^{-1}$ at 964 nm. Insertion losses of the ridge waveguides were measured by cut-back method and the samples had optical losses lower than $0.6\ dB \cdot cm^{-1}$. The best samples have optical losses $0.27\ dB \cdot cm^{-1}$ at 650 nm, $0.16\ dB \cdot cm^{-1}$ at 850 nm and $0.30\ dB \cdot cm^{-1}$ at 1310 nm.

Acknowledgments

The authors thank Topas Advanced Polymers company for providing TOPAS 8007X4 foils and Mr. M. Kucera from Masaryk University, Dept. of Condensed Matter Physics for technical support. This work was supported by the Epsilon Programme of the Technology Agency of the Czech Republic, project no. TH01020276, and by the CTU grant no. SGS17/188/OHK3/3T/13.

References

[1] BAMIEDAKIS, N., CHEN, J., PENTY, R.V., WHITE I. H. Bandwidth studies on multimode polymer waveguides for 25 Gb/s optical interconnects. *IEEE Photonics Technology Letters*, 2014, vol. 26, no. 20, p. 2004–2007. DOI 10.1109/LPT.2014.2342881

[2] BOSMAN, E., Van STEENBERGE, G., MILENKOV, I., PANAJOTOV, K., THIENPONT, H., BAUWELINCK, J., Van DAELE, P. Fully flexible optoelectronic foil. *IEEE Journal on Selected Topics in Quantum Electronics*, 2010, vol. 16, no. 5, p. 1355–1362. DOI 10.1109/JSTQE.2009.2039466

[3] DANGEL, R., BERGER, C., BEYELER, R., DELLMANN, L., GMUR, M., HAMELIN, R., HORST, F., LAMPRECHT, T., MORF, T., OGGIONI, S., SPREAFICO, M., OFFREIN, B.J. Polymer-waveguide-based board-level optical interconnect technology for datacom applications. *IEEE Transactions on Advanced Packaging*, 2008, vol. 31, no. 4, p. 759–767. DOI: 10.1109/TADVP.2008.2005996

[4] CHOI, C., LIN, L., LIU, Y., CHOI, J., WANG, L., HAAS, D., MAGERA, J., CHEN, R.T. Flexible optical waveguide film fabrications and optoelectronic devices integration for fully embedded board-level optical interconnects. *Journal of Lightwave Technology*, 2004, vol. 22, no. 9, p. 2168–2176. DOI: 10.1109/JLT.2004.833815

[5] BRUCK, R., MUELLNER, P., KATAEVA, N., KOECK, A., TRASSL, S., RINNERBAUER, V., SCHMIDEGG, K., HAINBERGER, R. Flexible thin-film polymer waveguides fabricated in an industrial roll-to-roll process. *Applied Optics*, 2013, vol. 52, no. 19, p. 4510–4514. DOI: 10.1364/AO.52.004510

[6] HWANG, S.H., LEE, W.J., KIM, M.J., JUNG, E.J., KIM, G.W., AN, J.B., JUNG, K.Y., CHA, K.S., RHO, B.S. Ultra-thin and low-power optical interconnect module based on a flexible optical printed circuit board. *Optical Engineering*, 2012, vol. 51, no. 7, Article Number: 075402. DOI: 10.1117/1.OE.51.7.075402

[7] IMMONEM, M., WU, J., YAN, H.J., ZHU, L.X., CHEN, P., RAPALA-VIRTANEN, T. Long distance optical printed circuit board for 10 Gbps optical interconnection. *Proceedings of SPIE*, 2012, vol. 8555, Article Number: UNSP 85551M. DOI: 10.1117/12.999969

[8] BAMIEDAKIS, N., CHEN, J., WESTBERGH, P., GUSTAVS-SON, J. S., LARSSON, A., PENTY, R. V., WHITE, I. H. 40 Gb/s data transmission over a 1-m-long multimode polymer spiral waveguide for board-level optical interconnects. *Journal of Lightwave Technology*, 2015, vol. 33, p. 882–888. DOI: 10.1109/JLT.2014.2371491

[9] KOBAYASHI, J., YAGI, S., HATAKEYAMA, Y., KAWAKAMI, N. Low loss polymer optical waveguide replicated from flexible film stamp made of polymeric material. *Japanese Journal of Applied Physics*, 2013, vol. 52, Article Number: UNSP 072501. DOI: 10.7567/JJAP.52.072501

[10] PRAJZLER, V., NEKVINDOVÁ, P., HYPŠ, P., LYUTAKOV, O., JERABEK, V. Flexible polymer planar optical waveguides. *Radioengineering*, 2014, vol. 23, no. 3, p. 776–782. ISSN: 1210-2512

[11] PRAJZLER, V., NEKVINDOVÁ, P., HYPŠ, P., JERABEK, V. Properties of the optical planar polymer waveguides deposited on printed circuit boards. *Radioengineering*, 2015, vol. 24, no. 2, p. 442-448. DOI: 10.13164/re.2015.0442

[12] PRAJZLER, V., NEKVINDOVÁ, P., HYPŠ, P., JERABEK, V. Optical properties of polymer planar waveguides deposited on flexible foils. *Journal of Optoelectronics and Advanced Materials*, 2015, vol. 17, no. 11-12, p. 1597–1602.

[13] Micro resist technology GmbH: Datasheet. Available at: http://www.microresist.de

[14] TOPAS Advanced Polymers: Datasheet. Available at: http://www.topas.com/tech-center

[15] web Metricon Corporation. Available at: www.metricon.com

[16] ULRICH, R., TORGE, R. Measurement of thin film parameters with a prism coupler. *Applied Optics*, 1973, vol. 12, no. 12, p. 2901–2908. DOI: 10.1364/AO.12.002901

[17] NOURSHARGH, N., STARR, E. M., FOX, N. I., JONES, S. G. Simple technique for measuring attenuation of integrated optical waveguides. *Electronics Letters*, 1985, vol. 21, no. 18, p. 818–820. DOI: 10.1049/el:19850577

[18] ZIEMANN, O., KRAUSER, J., ZAMZOV, P. E., DAUM, W. *POF Handbook, Optical Short Range Transmission Systems*. 2nd ed. Berlin (Germany): Springer-Verlag Berlin Heidelberg, 2008. ISBN 978-3-540-76628-5

About the Authors ...

Václav PRAJZLER was born in 1976 in Prague, Czech Republic. In 2001 he graduated from the Dept. of Microelectronics, Faculty of Electrical Engineering, Czech Technical University in Prague. In 2007 he obtained the Ph.D. degree from the same university. Since 2014 he has been an Associate Professor of Electronics with the Department of Microelectronic, Czech Technical University in Prague. His current research is focused on design, fabrication and investigation of properties of photonics structures.

Miloš NERUDA was born in 1975. His bachelor program was reached at the Department of Microelectronics at CTU in Prague in 2012 and his bachelor thesis was focused on the design, fabrication and investigated properties of multimode polymer planar optical power splitters. His current research is focused to design photonics structures.

Pavla NEKVINDOVÁ was born in 1972 in Kolín, Czech Republic. She graduated from the Institute of Chemical Technology, Prague (ICTP) in 1999. Now she is an Assistant Professor at the ICTP giving lectures on general and inorganic chemistry. She has worked there continuously in materials chemistry research. She has a long-term experience with fabrication and characterization of optical waveguiding structures in single-crystalline and glass materials.

Petr MIKULÍK was born in 1969 in Brno, Czech Republic. In 1992 he graduated from the Masaryk University, Brno, Czech Republic in Solid State Physics. In 1997 he obtained the Ph.D. degree under double leadership from the Masaryk University and from Université Joseph Fourier, Grenoble, France. Since 2009 he has been an Associate Professor Masaryk University Brno and now he is working at the Department of Condensed Matter Physics from the same university.

2

Versatile Controllability over Cell Switching for Speedy Users in LTE HetNets

Mohammad T. KAWSER, Mohammad R. ISLAM, Muhammad R. RAHIM, Muhammad A. MASUD

Dept. of Electrical and Electronic Engineering, Islamic University of Technology, Board Bazar, Gazipur-1704, Bangladesh

mkawser@hotmail.com, rakibultowhid@yahoo.com, ridwan351990@gmail.com, atique76i@gmail.com

Abstract. *The heterogeneous networks (HetNets) are regarded as a promising solution in LTE-Advanced for ubiquitous and cost effective broadband user experience. But there are challenges to support seamless mobility in HetNets, especially, when the user speed is high. In this paper, we investigate these challenges and study the scopes to address them for the improvement of cell edge performance. The study indicates the requirement of enhanced and versatile controllability over adaptation of cell switching parameters that simultaneously depends on variation in user speeds, traffic loads, street patterns, types of cells involved in switching, and so forth. We propose a scheme to scale cell switching parameters that incorporates Doppler spread estimation and adapts smoothly to various changes. Both the eNodeB and the UE participate in a versatile control over the scaling. Limited simulations have been performed to partially reflect the outcome of the proposed scheme.*

Keywords

LTE, HetNets, cell reselection, handover, speedy users

1. Introduction

Long term evolution (LTE), and its later version LTE-Advanced, are the latest steps in the advancing series of cellular communication technologies. The heterogeneous networks (HetNets) are introduced in LTE-Advanced to enhance coverage and capacity greatly. A HetNet consists of regular macro cells typically transmitting at high power level, overlaid with low power small cells such as pico cell, femto cell, remote radio head (RRH), and relay node (RN) [1]. In this paper, in addition to the macro cell, the HetNet is considered to have only pico cells and femto cells and they both are referred to as small cells. The small cells offload traffic from the macro cell and offer extension of the reach of coverage. They improve the conditions in coverage holes providing higher data rates at cell edge or in hotspots. The small cells have smaller base stations with lower antenna gain compared to macro cells and so, their site acquisition can be simpler. However, there are some challenges that need to be addressed in HetNets.

The serving cell is updated automatically with the movement of the user to support user mobility in LTE. In the RRC_IDLE state, the cell reselection procedure is performed to change the cell on which the user equipment (UE) is camped. In the RRC_CONNECTED state, handover is performed to change the serving cell. The decision for the change of cell depends on the relative radio link quality between what are experienced by the UE from the currently serving cell and from the neighbor cells. Because of the time varying nature of the radio signals, it is possible that what appears to be an increase or decrease of the received signal strength is actually a transient fluctuation. Thus, the decision for cell switching must allow the target cell to remain better than the serving cell by a sufficient margin continuously for a sufficient period. Otherwise, there is a possibility of switching back and forth between cells unnecessarily, which is known as ping-pong events. The ping-pong events degrade the system performance using resources unnecessarily.

There are a lot of challenges in proper triggering of cell switching in HetNets, particularly with high speed users. These challenges must be addressed carefully as they have direct bearing on the cell edge performance. It may be argued that HetNets are mostly expected in urban areas and high speed users do not exist in urban areas due to frequent red light stops. But in fact, in modern cities, highways and bullet trains can go through cities and a good number of users can move at high speed without making stops [2].

In a traditional macro cellular network, the radio link failures (RLFs) and ping-pong events can typically be avoided to a good extent [3]. However, the use of the set of cell switching procedures and parameters designed for a macro cellular network can degrade the mobility performance in HetNets [4], albeit the same set is used at present [5]. The poor mobility performance in HetNets with high speed users can be serious problems [6], [7], [8]. The authors of [4] show that the mobility performance strongly depends on the cell size and the user speed in HetNets. In [9], the authors show that the optimal value of cell switching parameters considering both the radio link failure rate and the ping-pong rate depends on the various cell sizes in HetNets. The authors of [10] show improvement in cell switching performance when the scaling of cell switching parameters considers both the user speed and the difference in cell sizes in HetNets. [10] also shows improvement in performance when the high speed users are primarily kept under only macro service, limiting the number of cell changes and the low speed users are offloaded to the small cells. The authors in [11] also proposed keeping the high speed users under the macro service but additionally consider the QoS of the application and so, three speed levels are used for the decision. The authors in [7] also proposed a context-aware mobility management procedure for HetNets, in which the resources are allocated to the users based on their velocities and historical data rates. The authors in [12] proposed network controlled macro cell mobility, while the small cell mobility is proposed to be UE autonomous. This offloads the network from having to perform frequent small cell change decisions, and reduces the signaling overhead. The authors in [13] proposed the adjustment of cell switching parameter values depending on whether the handover occurs from macro cell to small cell or from small cell to macro cell, since the conditions are different in these two cases. In [14], the authors attempt to determine appropriate cell switching parameter values considering handovers between macro and small cells and between macro and macro cells, separately. The authors in [15] proposed that the adjustment of cell switching parameters for a homogenous network is defined by a cost function that uses the rates of ping-pong, call drop and handover failure while assigning each of them weight factors. In [16] and [17], the authors also proposed similar cost functions but for HetNets and considering ping-pong, too late handover and too early handover. The authors of [18] show that if the adjustment of cell switching parameters with user velocity uses higher number of steps, then it improves the performance for a homogenous network.

It is thus already manifested that there are many different factors which should dictate the adjustment of cell switching parameters. Therefore, a versatile controllability should be used to accomplish such an adjustment. In this paper, we further investigate the challenges in mobility support for high speed users in HetNets and the investigation substantiates the necessity of versatile controllability. However, the existing specification or an existing proposal takes only one or two factors into account and therefore, the cell edge performance is at risk. We propose a scheme that attempts to overcome the current shortcomings.

The remainder of this paper is organized as follows. Interference issues in HetNets are investigated in Sec. 2. In Sec. 3, the existing cell switching conditions and the scaling of cell switching parameters for speedy users are presented. We delve into the challenges in mobility support for high speed users in HetNets in Sec. 4. In Sec. 5, we propose a scheme for adaptive scaling of cell switching parameters that uses enhanced and versatile controllability. Simulation results are given in Sec. 6 and the whole paper is concluded in Sec. 7.

2. Interference Issues in HetNets

We assume that the HetNets are using co-channel de-

ployment, in which the small cells use the same spectrum as do the macro cells. Thus, the spectral efficiency is increased via spatial reuse but causing significant co-channel interference (CCI). Especially, the users in edge area of low power small cells are mostly vulnerable to strong interference signals from the high power macro cell. There are, in fact, two types of interference concerns: co-tier interference, which occurs between neighboring small cells, and cross-tier interference, which occurs between small cells and macro cells. The interference is a serious problem because of the following reasons.

1. A large number of cell boundaries are created in HetNets and there is a large difference between the transmit power levels of eNodeBs of macro cells and small cells, which makes the interference problem very complex. These interferences can arise in various scenarios as shown below [19].

i. The small cell users can be affected in downlink by the dominant macro eNodeBs.

ii. The small cell users can be affected in uplink by macro users. Also, the macro users can be affected in downlink by the small cell eNodeB. These interferences can be substantial when the macro users stay close to the small cells.

iii. The macro users can be affected in uplink by the small cell users. These interferences can be substantial when there are a large number of small cells deployed making the aggregate uplink interference from the large number of small cell users high.

iv. The small cell users can be affected both in downlink and in uplink by another small cell. These interferences can be substantial when the small cells are randomly located, which is more likely in residential deployments or in hot-spots. In enterprise deployments, small cells may be carefully located at predesigned places limiting the interference.

2. Since the eNodeB in the macro cell transmits much higher power than that in the small cell, the UE tends to remain connected to the macro cell even when the path loss between the UE and the small cell is smaller. This makes the load among tiers unbalanced. Therefore, 3GPP has standardized the provision for cell range expansion (CRE), which virtually increases the coverage area of small cells. In the case of CRE, by biasing handover decision toward a small cell, the users are handed over to the small cell earlier than usual and thus, some of the load is shifted from the macro cell to the small cell. Similarly, cell reselection can also be biased for users in RRC_IDLE state so that some additional users request services to the small cell when they need to. However, with CRE, the users, switched to small cells earlier than usual, receive low power from the serving small cell and high interference from the macro cell and thus, suffer from low SINR at the cell edge.

3. Many small cells, especially, femto cells are typically deployed in an ad hoc manner by users. They can

even be moved or switched on or off at any time. Hence, traditional network planning and optimization becomes inefficient because operators do not control either the number or the location of these cells. The unplanned deployment aggravates the interference problem.

4. Some femto cells may operate in closed subscriber group (CSG) mode, in which cell access is restricted and only the respective subscribers are allowed to access the femto cells. The users, who are close to these femto cells but disallowed to access them, can be exposed to severe interference from them.

3. Cell Switching Conditions

In the current 3GPP specifications, the cell switching procedure is the same for a macro cellular network and for HetNets. The conditions for triggering cell reselection or handover are explained below and they are required to be fulfilled neither too early nor too late. The neighbor cells are considered to operate at intra-frequency or at inter-frequency with equal priority.

3.1 Conditions for Cell Reselection

The UE evaluates the rank of the serving cell and the target neighbor cell frequently as

$$R_S = Q_{meas'S} + Q_{hyst} \text{ and } R_T = Q_{meas'T} - Q_{offset} \quad (1)$$

where $Q_{meas'S}$ and $Q_{meas'T}$ are the signal strength of the serving cell and the target neighbor cell, respectively, measured as reference signal received power (RSRP) in dBm. Q_{hyst} is called the hysteresis margin. Q_{offset} is a cell specific offset, which allows biasing the cell switching decision toward particular cells. If the UE finds that the target cell has been continuously better ranked for $T_{reselectionRAT}$ period, it reselects to the target cell. A separate $T_{reselectionRAT}$ timer is started for each cell that becomes better ranked than the serving cell. The UE reselects to the highest ranked cell when the corresponding timer expires. The eNodeB configures the values of Q_{hyst}, Q_{offset} and $T_{reselectionRAT}$ using system information messages.

3.2 Conditions for Handover

The UE sends measurement reports frequently to the eNodeB, which convey the signal strength of the serving cell and the neighbor cells. Typically, the serving eNodeB issues a handover command after a certain number of measurement reports are received from the UE, indicating the target cell is better than the serving cell by a sufficient margin. The measurement reporting can be either event-triggered or periodic as shown below.

- *Event-Triggered:* The UE is considered to enter and to leave a particular event when certain conditions are met. The UE begins sending measurement reports when the entering condition is met for *TimeToTrigger*

period. As long as the event remains activated, the UE keeps on sending measurement reports at a certain interval until it reaches a maximum number of reports. There are five types of events indicated as A1 through A5. The whole entering condition for event A1 and event A3 are shown below for exemplification. The entering condition includes hysteresis for all five events.

Event A1:
Serving cell measurement – Hysteresis > Threshold

Event A3:
Neighbor cell measurement
+ OffsetFreq for neighbor cell
+ CellIndividualOffset for neighbor cell
– Hysteresis > Serving cell measurement
+ OffsetFreq for serving cell
+ CellIndividualOffset for serving cell
+ A3_Offset

- *Periodical:* The UE keeps on sending the measurement reports one after another at a certain interval until it reaches a maximum number of reports.

As the UE moves, if the handover occurs too late, the received power from the source cell can fall too low. This can cause RLF. The RLF can lead to a handover failure (HF). When the HF occurs due to RLF, usually, the UE fails to receive RRCConnectionReconfiguration message from the source cell, which carries instruction to switch to the target cell. After the HF, the UE can attempt to reestablish RRC connection with the strongest cell. A call or session drop occurs if this reestablishment fails. If the RRC connection reestablishment becomes successful, the target cell can send a RLF indication message to the source cell.

As the UE moves, if the handover occurs too early, the radio link with the target cell may not have become sufficiently strong yet. In this case, during or after the handover procedure, RLF can occur at the target cell and then the UE tries to reestablish RRC connection with the old source cell. After the RRC connection reestablishment, the source cell can send a RLF indication message to the target cell. To inform the source cell about the reason for failure, the target cell can send a handover report message indicating 'handover too early'.

3.3 Scaling of Parameters for Speedy Users

The high speed users may be subject to very poor SINR due to what is often called a dragging effect. This is because the cell switching is not triggered until the required conditions have remained fulfilled for a certain period. This delay would let the speedy users move too far away from the old serving cell before switching to the target cell. This results in poor signal strength from the serving cell and strong interference from the target cell. Thus, the dragging effect causes poor cell edge SINR and extremely low data rate and in the worst case, it can cause RLF and HF. As a remedy to this problem, 3GPP standardizes a scaling of cell switching parameters for speedy users. Accordingly, apart from the normal-mobility state, the medium and high mobility states are defined as shown below and scaling is applied only in these two states.

- *High-Mobility State:* The UE enters this state if there are more than N_{CR_H} numbers of cell reselections or handovers in the last T_{CRmax} duration indicating high user speed.

- *Medium-Mobility State:* The UE enters this state if there are more than N_{CR_M} number but equal to or less than N_{CR_H} number of cell reselections or handovers in the last T_{CRmax} duration indicating medium user speed.

- *Normal-Mobility State:* The UE enters this state from medium or high mobility state if neither medium nor high mobility state is detected during $T_{CRmaxHyst}$ period. No scaling is performed in this state.

The eNodeB configures the values of T_{CRmax}, N_{CR_H}, N_{CR_M} and $T_{CRmaxHyst}$. In the case of cell reselection, the UE scales Q_{hyst} and $T_{reselectionRAT}$ as follows.

- Q_{hyst}: In high and medium mobility states, SF-High and SF-Medium fields are added to Q_{hyst}, respectively. The eNodeB sends these two fields on Q-HystSF IE via system information message and their values can be –6, –4, –2 or 0 dB.

- $T_{reselectionRAT}$: In high and medium mobility states, SF-High and SF-Medium fields are multiplied with $T_{reselectionRAT}$, respectively. The eNodeB sends these two fields on SpeedStateScaleFactors (SSSF) IE via system information message and their values can be 0.25, 0.5, 0.75 or 1.

In the case of handover, the hysteresis is not scaled in the current specifications. In high and medium mobility states, SF-High and SF-Medium fields are multiplied with *TimeToTrigger*, respectively. The eNodeB sends these two fields on SSSF IE via RRCConnectionReconfiguration message.

4. Challenges in Mobility Support in HetNets

The challenges in mobility support for speedy users in HetNets are manifold and we derive them as follows.

1. Section 2 explains that SINR can be very poor at the cell edge of small cells even for immobile users in HetNets. On top of that the dragging effect is particularly severe in a HetNet scenario with high speed users [2], which further aggravates the cell edge SINR. Thus, the cell edge performance degrades severely and there can be even RLF and HF.

2. Typically, the lowest received signal power is experienced in the system right before the cell switching and this power is denoted as P_{R_Min}. Currently, a scaling cell switching parameters is performed for speedy users but only at two discrete steps as shown in Sec. 3.3.

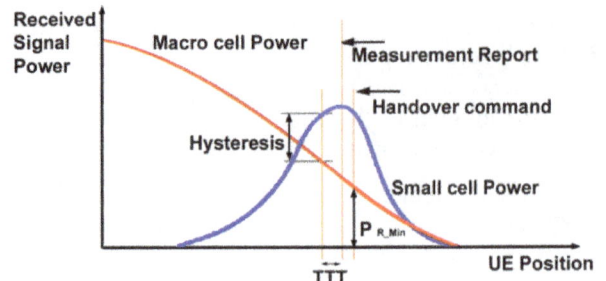

Fig. 1.   Reception of low P_{R_Min} for speedy users in HetNets.

Thus, for speedy users in HetNets, the cell switching may not be triggered as quickly as it is required and the UE can move far inside the new cell before the new cell takes over as illustrated in Fig. 1. Here, P_{R_Min} is not enough velocity independent and P_{R_Min} can fall very low. This can cause highly degraded data rate in the case of handover and there can be even RLF and HF. Similar problem can also occur in the case of cell reselection.

3. The receiver sensitivities for the eNodeB and the UE are required to support P_{R_Min} in uplink and downlink, respectively. But P_{R_Min} takes on widely varying values at different velocities with only two steps of scaling. This may pose difficulty to match between the receiver sensitivity and P_{R_Min} in the system.

4. In HetNets, the existence of high number of cells can lead to a lot of overlapping in cell coverage. Also, there can be multiple cell borders and different target cells at different times, due to the fast variation of SINR from each cell [2]. Thus, the cell boundaries and target cells can be far from clear and there can be difficulties in successful cell switching.

5. The existing method relies merely on the history of cell switching for the decision of scaling. But this history is not always a good indication of the user speed. For example, a user at a low velocity may take a tangled pathway within overlapped cells and undergo frequent cell switching. Then the existing method applies high scaling, which is not justified because a sufficient margin is required to avoid ping-pong effects. In this case, if the high scaling was not applied, the UE would not go far inside another cell. Thus, the possibility of ping-pong effects is unnecessarily increased.

6. In HetNets, the adjustment of cell switching parameters should consider the different characteristics of the neighbor cells. For switching from small cells to macro cells, the received power from the small cell drops much quicker compared to cell switching between macro cells. Thus, a late cell switching can lead to very poor SINR. For cell switching from macro cells to small cells, the received power from the small cell increases much quicker compared to cell switching between macro cells and the UE suffers from inter-cell interference from the small cell earlier. Thus, a late cell switching can again lead to very poor SINR.

7. The handover conditions should be adaptive to fit into the current loading conditions and this adaptability is not enough effective at present. The CRE can vary its extent depending on loading conditions. Moreover, the SINR on PDCCH depends on the traffic usage pattern. Since HARQ cannot be applied to PDCCH, if the BLER on PDCCH exceeds 10 % during handover, the UE can fail to receive response from the eNodeB. This causes HF. The impact of traffic pattern on SINR on PDCCH is explained below.

A small number of high data rate users, using applications like FTP, get PDCCH lightly loaded. On the other hand, PDCCH gets heavily loaded if there are a large number of low rate users in the cell, using applications like VoIP. A PDCCH instance uses a number of control channel elements (CCEs) where a CCE consists of nine resource element groups (REGs) or 36 resource elements. There are four different PDCCH formats that use different aggregation levels of CCEs. A higher aggregation level with a larger number of CCEs can be used to achieve stronger coding for cell edge users with poor SINR condition. But a high aggregation level may be difficult to be made available when the PDCCH loading is high. Besides, a light loading on PDCCH allows less co-channel interference from neighboring cells. This is because the whole control region is not used with the light loading and the symbol quadruplets of the CCEs from each cell are shifted to different sub-carrier frequencies or different symbols to randomize and reduce co-channel interference [1], [2].

8. In HetNets, the downlink coverage of a small cell is much smaller than that of a macro cell. But this is not the case for uplink; since all UEs have almost equal transmit power capabilities. This can create a mismatch between downlink and uplink handover boundaries compared to homogenous networks, in which those boundaries are more closely matched [20]. Since the handover decision is based on the downlink boundaries, a UE with ongoing data transfer only in uplink may not trigger handover at a very appropriate moment in HetNets. For example, a UE moving from small cell to macro cell and undergoing uplink data transfer will have to boost the uplink transmit power suddenly once the handover is triggered because the macro eNodeB is far away. This will cause additional interference, which could be avoided if the handover was triggered later. Similarly, if the UE was moving from macro cell to small cell, an earlier handover could preclude high uplink power transmission recently before the handover and its associated interference.

5. Proposed Scheme

It can be inferred from the discussion in Sec. 4 that a versatile controllability over cell switching is required to address the wide variety of challenges. We propose the

following changes in the scheme of scaling of cell switching parameters.

i. A versatile controllability over the adaptive scaling is used so that the scaling can better fit into the current status considering user speeds, traffic loads, street patterns and types of cells involved in switching.

ii. The user velocity v and the maximum Doppler shift f_d are related as $f_d\lambda$ where λ is the wavelength. In order to improve versatility in the controllability, instead of using only the recent history of cell switching for computation of the scaling factor, it is combined with the Doppler spread, which is also indicative of the user speed. The rationale behind this action is that one between the number of cell switching and the estimated Doppler spread can be found better indicative of the requirement of scaling than the other depending on the scenario in HetNets. Secondly, since the scaling with speed reduces the cell switching delay, it increases the possibility of ping-pong effects. The inclusion of Doppler spread ensures that the velocity is really high when high scaling is applied and the chances of ping-pong effects are thus mitigated.

iii. The scaling varies smoothly with the user speed instead of varying only at two discrete steps. The medium mobility state is removed and N_{CR_M} is not transmitted. A smooth scaling is performed in the high mobility state and the normal mobility uses no scaling. Because of the smooth scaling, P_{R_Min} becomes more independent of the user speed. Thus, P_{R_Min} does not fall too low providing better SINR and this may prevent RLF and HF. Also, due to the stability of P_{R_Min}, the receiver can be more easily designed with proper sensitivity.

iv. Instead of letting eNodeB gain the whole control over scaling, a part of the control is shifted to the UE. The eNodeB controlled scaling of the hysteresis margin is used to establish a set point. Then the UE performs adjustment of timer period as an overlay around the set point at the discretion of the UE itself. The adjustment by the UE does not require any feedback or any overhead and it can quickly adapt to the changes. Thus, it can function as a fine-tuning on top of the eNodeB controlled scaling.

Several methods have been suggested for the estimation of Doppler spread for OFDM signals [21], [22], [23] and the UE may use any reliable method in the proposed scheme. [21] uses the relationship

$$v = f_d\lambda = \frac{cf_d}{f_c} = \frac{2.405c}{2\pi f_c T_S \hat{l}_0} \quad (2)$$

where f_c is the signal frequency, c is the speed of light, T_S is OFDM symbol duration and $\hat{l}_0$ is the zero crossing point of the estimated covariance function, $\varphi_t(l)$ of the received signal at a certain carrier in the frequency domain with l representing the difference in time. The velocity or Doppler spread estimation is shown almost independent on SINR in

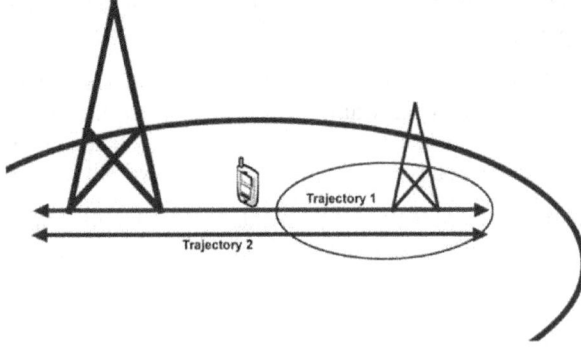

Fig. 2. The trajectory along which the user moves.

[21] and so, it can be used at the cell edge during cell switching.

We assume that the user moves along trajectory 1 in Fig. 2 in any direction at velocity v where trajectory 1 is a straight line connecting the eNodeBs of macro cell and small cell. D_S and D_T are the distances of the cell border from the serving eNodeB and the target eNodeB, respectively. The proposed scaling scheme is explained in Sec. 5.1 and 5.2 for cell reselection and handover, respectively.

5.1 Scheme for Cell Reselection

The proposed scaling of hysteresis and timer period for cell reselection is depicted in Fig. 3. We assume that the condition for the triggering of $T_{reselectionRAT}$ timer is met at a distance d from the cell border of the target cell. According to the log distance path loss model, the received power at d from the serving cell and the target cell can be expressed in dB as

$$\overline{Pr_S}(d_0) - 10n\log\left(\frac{D_S+d}{d_0}\right) \text{ and } \overline{Pr_T}(d_0) - 10n\log\left(\frac{D_T-d}{d_0}\right)$$

respectively, where n is the path loss exponent for the multipath environment. Ignoring the quantization in power level calculation, (1) can be shown as

$$10n\log\left(\frac{D_S+d}{d_0}\right) - 10n\log\left(\frac{D_T-d}{d_0}\right)$$
$$= Q_{hyst} + \overline{Pr_S}(d_0) - \overline{Pr_T}(d_0) \quad (3)$$

where Q_{offset} has been set to zero assuming zero bias to the target cell. Equation (3) can be rearranged as

$$4.343n\left[\ln\left(1+\frac{d}{D_S}\right) - \ln\left(1-\frac{d}{D_T}\right) + \ln\left(\frac{D_S}{D_T}\right)\right] - \Delta_{Pr}(d_0) = Q_{hyst} \quad (4)$$

where $\Delta_{Pr}(d_0) = \overline{Pr_S}(d_0) - \overline{Pr_T}(d_0)$. $\Delta_{Pr}(d_0)$ will be positive when the UE is moving from macro cell to small cell and it will be negative when the UE is moving from small cell to macro cell. Using Taylor series expansion, assuming $d \ll D_S$ and $d \ll D_T$ and ignoring higher order terms

$$4.343n\left(\frac{1}{D_S}+\frac{1}{D_T}\right)d + 4.343n\ln\left(\frac{D_S}{D_T}\right) - \Delta_{Pr}(d_0) = Q_{hyst} \quad (5)$$

$4.343\ n\ln(D_S/D_T) - \Delta_{Pr}(d_0)$ is a constant for a particular scenario. To attain smooth scaling, we proposed to be reduced linearly as the user velocity v increases and for this purpose, Q_{hyst} can be scaled as

$$4.343n\left(\frac{1}{D_S}+\frac{1}{D_T}\right)d + 4.343n\ln\left(\frac{D_S}{D_T}\right) - \Delta_{Pr}(d_0) = Q_{hyst} - k'v \quad (6)$$

k' in (6) would be configured by the eNodeB using the parameter $Q_{hyst}SF$ where $Q_{hyst}SF$ is negative in dB value as given by

$$4.343n\left(\frac{1}{D_S}+\frac{1}{D_T}\right)d + 4.343n\ln\left(\frac{D_S}{D_T}\right) - \Delta_{Pr}(d_0) \quad (7)$$
$$= Q_{hyst} + Q_{hyst}SF \cdot kv.$$

The UE determines a factor *Vel_factor_CR*, proportional to v, and uses it in place of kv as

$$4.343n\left(\frac{1}{D_S}+\frac{1}{D_T}\right)d + 4.343n\ln\left(\frac{D_S}{D_T}\right) - \Delta_{Pr}(d_0) \quad (8)$$
$$= Q_{hyst} + Q_{hyst}SF \cdot Vel\_factor\_CR.$$

Despite the deployment of HetNets, it may happen that the UE finds itself moving between two macro cells. Assuming all macro cells to have radius D and equal transmit power at eNodeBs, a similar derivation yields

$$\frac{8.686n}{D}d = Q_{hyst} + Q_{hyst}SF \cdot Vel\_factor\_CR \cdot \quad (9)$$

The *Vel_factor_CR* is computed from the number of cell reselections in the last T_{CRmax} period, N_{cr} and the estimated Doppler spread as

$$Vel\_factor\_CR = (1-\alpha)N_{cr_{filtered}} + \alpha\beta f_d \quad (10)$$

where $$N_{cr_{filtered}} = (1-\gamma)N_{cr} + \gamma N_{cr\_last} \quad (11)$$

and $$\gamma = 1/2^\eta. \quad (12)$$

Here, $N_{cr\ filtered}$ represents the filtered value of N_{cr} and it is updated in every T_{CRmax} period. $N_{cr\ last}$ gives the last filtered value of N_{cr}. η is the filter coefficient, which can vary between 0 and 6. It is expected that the impact of N_{cr} is more important than $N_{cr\ last}$ and η will generally be high. (11) and (12) allow better granularities for the impact of N_{cr} for higher values of η. B is a gain factor and the value of β should be chosen to yield the same order of values for $N_{cr\ filtered}$ and βf_d. β decreases with f_c and so, it will vary with the operating band. α is used to control the relative influence of the number of cell reselections and Doppler spread. The UE applies Q_{hyst_scaled} as the scaled hysteresis, which is computed as

$$Q_{hyst\_scaled} = Q_{hyst} + \left(Q_{hyst}SF \cdot Vel\_factor\_CR\right). \quad (13)$$

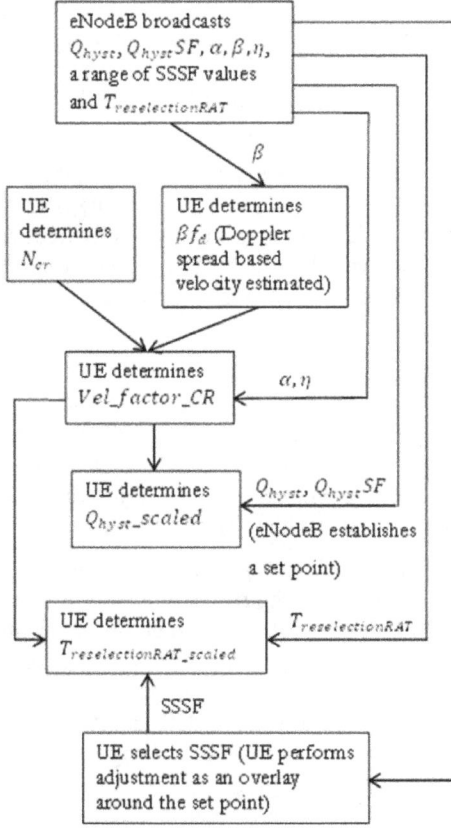

Fig. 3. A flowchart showing determination of scaled parameters for cell reselection.

Once the condition of (13) is fulfilled at a distance d, the timer $T_{\text{reselectionRAT}}$ is supposed to trigger and run over an additional distance d' before triggering cell reselection where

$$d' = v \cdot T_{\text{reselectionRAT}}. \qquad (14)$$

To keep d' reasonably constant, the UE uses a scaled timer period in place of $T_{\text{reselectionRAT}}$, which is given by

$$T_{\text{reselectionRAT\_scaled}} = T_{\text{reselectionRAT}} \frac{SSSF}{Vel\_factor\_CR}. \qquad (15)$$

The eNodeB transmits $Q_{\text{hyst}}SF$, α, β, η and a range of values for $SSSF$ using system information message. The UE selects $SSSF$ value from the given range.

5.2 Scheme for Handover

We propose a method of scaling for handover similar to what is explained for cell reselection in Sec. 5.1. We assume event-triggered measurement reporting. Vel_factor_HO, a factor proportional to the user velocity is computed from the number of handovers in last T_{CRmax} period, N_{ho} and the estimated Doppler spread as

$$Vel\_factor\_HO = (1-\alpha) N_{\text{ho\_filtered}} + \alpha\beta f_d \qquad (16)$$

where

$$N_{\text{ho\_filtered}} = (1-\gamma) N_{\text{ho}} + \gamma N_{\text{ho\_last}} \qquad (17)$$

and

$$\gamma = 1/2^\eta. \qquad (18)$$

The hysteresis in the entering conditions for event-triggered measurement reporting is not scaled with the user speed in the existing method. But we propose that the scaling of hysteresis is performed and thus, a set point for scaling is established. The UE applies $Hysteresis\_scaled$ as the scaled hysteresis, which is computed as

$$Hysteresis\_scaled = \\ Hysteresis + (Hyst\_SF \cdot Vel\_factor\_HO) \qquad (19)$$

where $Hyst\_SF$ is negative in dB value. The distance d'' over which the entering conditions for a particular event need to be met can be expressed as

$$d'' = v \cdot TimeToTrigger. \qquad (20)$$

To keep d'' reasonably constant, the UE uses scaled timer period given by

$$TimeToTrigger_{\text{scaled}} = TimeToTrigger \frac{SSSF}{Vel\_factor\_HO}. \qquad (21)$$

The eNodeB transmits $Hyst\_SF$, α, β, η, and a range of values for $SSSF$ using RRCConnectionReconfiguration message. The UE selects $SSSF$ value from the given range.

5.3 Control over Scaling

The proposed scheme provides versatile controllability over the scaling depending on the scenario. The scenarios in HetNets may vary widely as a result of differences in transmit power, coverage of macro and small cells, traffic loading in macro and small cells, position and traffic pattern of highways and other streets with respect to the coverage of macro and small cells, profile of the user speeds, operating frequency, multipath environment and fading conditions, QoS requirements of data services, and so forth. A dynamic control is proposed that uses $Q_{\text{hyst}}SF$ or $Hyst\_SF$, $SSSF$, α and η based on various factors of the particular scenario and for this purpose, some guidelines are given below. The operator may not necessarily limit to these guidelines but may further extend, if required.

1. The scaling of hysteresis establishes a set point using $Q_{\text{hyst}}SF$ or $Hyst\_SF$, configured by the eNodeB. The scaling of timer period provides an overlay of scaling around the set point using $SSSF$, configured by the UE. Increasing the absolute values of $Q_{\text{hyst}}SF$ or $Hyst\_SF$ reduces hysteresis and thus, yields a quicker cell switching. On the other hand, increasing the value of $SSSF$ increases timer period and thus, yields a delayed cell switching. Simultaneous control from both the eNodeB and the UE may often reduce the controllability. Therefore, $SSSF$ must be within a range of values, provided by the eNodeB.

In the case of handover, $Hyst\_SF$ and $SSSF$ values are adjusted based on too late handover, too early handover, and ping-pong events. Section 3.2 shows how too early handover and too late handover can be detected from different messages. Ping-pong events can be detected from the identity of the recently serving cells. Similar with the proposal in [17], the adjustment of *Hyst_SF* uses a cost function as

$$Adj_{\text{Hyst\_SF}} = w_{\text{l}} \frac{N_{\text{l}}}{N_{\text{T}}} - w_{\text{e}} \frac{N_{\text{e}}}{N_{\text{T}}} - w_{\text{p}} \frac{N_{\text{p}}}{N_{\text{T}}} \qquad (22)$$

where N_{l}, N_{e}, N_{p}, and N_{T} are the numbers of too late handovers, too early handovers, ping-pong events, and total handovers in last T_{ho} period and w_{l}, w_{e}, and w_{p}, are their respective weight factors. T_{ho} is configured by the eNodeB. *Hyst_SF* is updated as

$$\left| Hyst\_SF_{\text{new}} \right| = \left| Hyst\_SF + Adj_{\text{Hyst\_SF}} \right|. \qquad (23)$$

The above adaption of scaling of hysteresis will be slow in comparison with the fast changes for speedy users in HetNets. But the overlay adjustment by the UE is quick and so, it can take care of instant requirements. The UE will slightly decrease SSSF whenever it detects a too late handover and the UE will slightly increase SSSF whenever it detects a too early handover or a ping-pong event. In the case of cell reselection, the UE adjusts $Q_{\text{hyst}}SF$ and SSSF the way *Hyst_SF* and SSSF are adjusted in the case of handover. In the case of either handover or cell reselection, the UE detects a too late cell switching from RLF at the source cell and a too early cell switching from RLF at the target cell. The UE detects ping-pong events from the identity of the recently serving cells.

2. The eNodeBs are updated with the information of different cell sizes in HetNets. Then in RRC_CONNECTED state, the serving eNodeB can determine from the measurement report whether the UE is moving from macro cell to small cell, from small cell to macro cell or between two macro cells. A higher absolute value of *Hyst_SF* can be used if the UE is moving between a macro cell and a small cell compared to the movement between two macro cells. To adjust *Hyst_SF* for this purpose, the eNodeB can send a quick update to the UE.

3. When the UE performs data transfer only in the uplink, to limit increase in the uplink power transmission, the eNodeB can send a quick update to set a lower absolute value of *Hyst_SF* if the UE is moving from small cell to macro cell. Similarly, an update to set a higher value of *Hyst_SF* can be used if the UE is moving from macro cell to small cell. However, this can lead to poor received power in downlink but the strong modulation and coding scheme (MCS) used for PCFICH, PDCCH and PHICH may still enable their successful decoding.

4. α is used to control the relative influence of the number of cell switching and Doppler spread in the computation of the scaling factor. If the traffic pattern of the streets suggests that a user may keep varying his speed widely, a high α may be used. This is because the Doppler spread now better indicates the current state of the user. A low α may be used if there are frequent red light stops. In this case, although the users have low velocity, they have chances of ping-pong effects because they may stop or turn around [24]. Low α may also be used when the Doppler spread estimation in the physical layer is not very reliable.

5. When the UE finds that its own velocity, estimated from the Doppler spread, exceeds a threshold value v_{th}, it sets SSSF to a very high value $SSSF_{\text{high}}$. This will preclude switching to small cells frequently and the cell switching will rather occur only between macro cells. v_{th} and $SSSF_{\text{high}}$ are configured by the eNodeB.

6. When there is high PDCCH loading, a tighter control over handover timing can be used. And for this purpose, the required number of measurement reports with the information of a better neighbor cell, to issue a handover command, can be adjusted using a cost function similar with (22).

7. The filter coefficient η is used to control the relative influence between the recent and older number of cell switching and its value should be set accordingly. The filtering removes sudden errors in the estimation. An example of such error can be that a highway may take a turn around the eNodeB and the user at a high velocity may not undergo cell changes for a while.

6. Simulation

A MATLAB based simulation has been performed and the simulation environment is limited to partially implement the proposed features. Primarily, the simulation manifests the performance due to the smoothness in scaling in the proposed scheme and compares it with the two discrete steps of scaling in the existing scheme. A strong system level simulator is required for comprehensive validation of the proposals. But such a simulator cannot be made accessible at the moment and so, a proper simulation has been left as a future work. However, Section 5 logically explains that the proposed scheme has ample and versatile controllability and it can potentially have remarkable achievements. The simulation assumptions are shown in Tab. 1. The parameter values in Tab. 1 comply with 3GPP specifications.

In the limited simulation environment, the simulation results are not significantly different between the movements from macro cell to small cell and from small cell to macro cell. Therefore, the simulation results are shown only for the user moving from macro cell to small cell along the two straight line trajectories of Fig. 2. Trajectory 2 is a straight line parallel with Trajectory 1 at 150 m distance. The user moves along Trajectory 2 only in the simulation results of Fig. 6 while Trajectory 1 is used in all other figures in this section. The simulation was performed for a number of fixed user velocities, against which $P_{\text{R\_Min}}$ was plotted. $P_{\text{R\_Min}}$ was calculated as the received power from the source cell at the position where cell switching is triggered. The path loss was calculated using Okumura-Hata model. N_{cr} and N_{ho} were calculated as the average

Parameter	Value
Operating frequency (DL)	1700 MHz
Separation between trajectory 1 and 2	150 m
User velocity (constant)	10, 20, 30, ...,150 km/hr
User direction	Moving from macro cell to small cell
Path loss model	Okumura-Hata
EIRP at macro eNB	40 dBm
EIRP at smalleNB	23 dBm
Separation between macro eNB and small eNB	800 m
Antenna height at macro eNB	30 m
Antenna height at small eNB	20 m
User antenna height	1 m
T_{CRmax}	4 sec
$T_{reselectionRAT}$	6 sec
Q_{offset}	0 dB
Q_{hyst}	10 dB
SF-High (Q-$HystSF$)	4 dB
SF-Medium (Q-$HystSF$)	8 dB
SF-High ($SSSF$)	0.75
SF-Medium ($SSSF$)	0.25
$Q_{hyst}SF$ (proposed case)	−0.2, −0.4, −0.6 or −0.8 dB
Reporting method for HO	Event-triggered with event A3
CellIndividualOffset for neighbor cell	0 dB
CellIndividualOffset for serving cell	0 dB
A3_Offset	2 dB (IE value, 4)
Hysteresis	8 dB (IE value, 16)
ReportInterval	0.24 sec
No. of reports triggering HO	4
$Hyst\_SF$ (proposed case)	−0.4 or −0.7 dB
TimeToTrigger	2.56 sec
N_{CR_M} (existing case)	3
N_{CR_H} (existing case)	7
N_{CR_H} (proposed case)	2
$SSSF$ (proposed case)	0.5, 1, 1.5 or 2
α	0.4
β	4
η	3

Tab. 1. Simulation assumptions.

number of cell reselections or handovers in the last T_{CRmax} period for a particular user velocity. For simplicity, N_{cr_last} or N_{ho_last} were also calculated the same way but using immediate lower velocity from the velocity array. f_d was calculated as $f_d = v/\lambda$ and λ was calculated as c/f based on the operating downlink frequency. Fixed values of α, β and η were used in all simulations because the simplified simulation environment is unable to demonstrate discrepancies with variation in their values. The simulation results are shown in Sec. 6.1 and 6.2 for cell reselection and handover, respectively.

6.1 Results for Cell Reselection

Figures 4 to 6 show P_{R_Min} vs. user speed profile for different values of $SSSF$ comparing existing and proposed methods in the case of cell reselection. In Fig. 4, P_{R_Min} follows a smooth variation and it remains fairly velocity independent in the proposed scheme. Conversely, P_{R_Min}

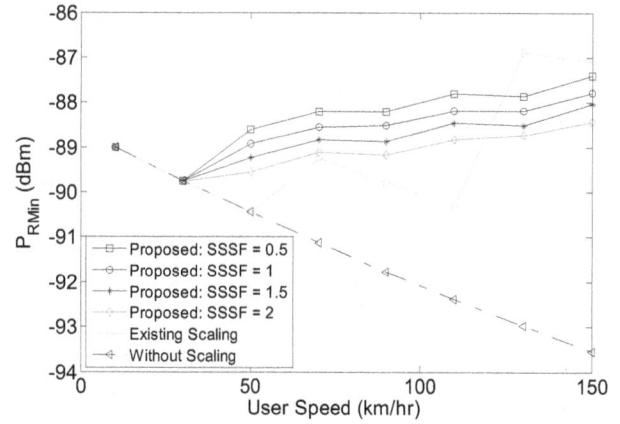

Fig. 4. P_{R_Min} vs. user speed for CR ($Q_{hyst}SF = -0.4$ dB).

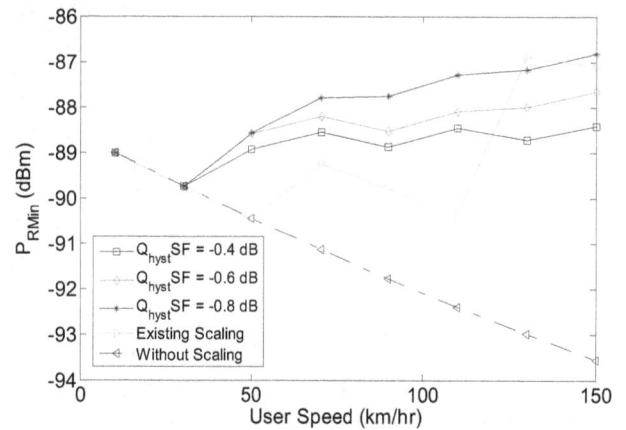

Fig. 5. P_{R_Min} vs. user speed for CR with adaptation of SSSF by the UE for $Q_{hyst}SF = -0.4$ dB, -0.6 dB and -0.8 dB.

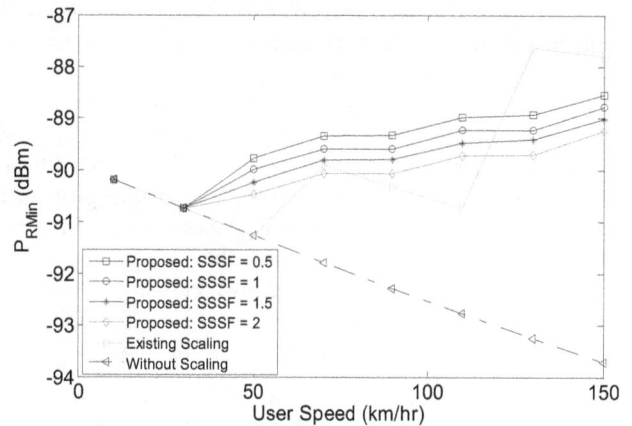

Fig. 6. P_{R_Min} vs. user speed for CR ($Q_{hyst}SF = -0.4$ dB, trajectory 2).

shoots within a wide range in the existing scheme. Here, P_{R_Min} shoots when there is a change in mobility state. Ordinarily, P_{R_Min} does not fall as low in the proposed scheme as it does in the existing scheme. However, as P_{R_Min} shoots widely in the existing scheme, it may often rise above the proposed scheme as occurs around user speed 140 km/hr. Thus, in the existing scheme, P_{R_Min} may get better, although rarely, but nevertheless, its instability remains as a problem. Figure 4 also shows that a decrease in $SSSF$ triggers cell reselections earlier boosting the level of P_{R_Min} and thus, it indicates how the UE can quickly perform necessary adjustment using SSSF.

Figure 5 demonstrates the actual proposal of establishment of a set point using a fixed $Q_{hyst}SF$ value and the use of instant adjustment using *SSSF* as an overlay. In Fig. 5, several fixed $Q_{hyst}SF$ values are used to establish different set points. On top of each of them, the UE gradually increases *SSSF* values as the user speed increases. It is evident from Fig. 5 that the profile of P_{R_Min} is under good control using $Q_{hyst}SF$ and *SSSF*, selected by the eNodeB and the UE, respectively. Using an appropriate control, P_{R_Min} values should be made as velocity independent as seems justified. The proposed scheme can achieve good stability of P_{R_Min}. The eNodeB and the UE, from the two ends, can employ the enhanced controllability considering the requirements of various factors of the particular scenario according to the guidelines shown in Sec. 5.3. Thus, the triggering of cell reselection will potentially adapt to various scenarios in a desired fashion. It may be noted that an attempt to raise P_{R_Min} too high using $Q_{hyst}SF$ and *SSSF* may lead to ping-pong effects. The user moves along Trajectory 2 in Fig. 6. The results of Fig. 4 and Fig. 6 are found to be very similar. Figure 6 basically shows that the proposed scheme is helpful in a different pathway too.

6.2 Results for Handover

Figures 7 to 9 show P_{R_Min} vs. user speed profile comparing existing and proposed schemes in the case of handover with a fixed value of *Hyst_SF* and different values of *SSSF*. The hysteresis is not scaled in the existing scheme as mentioned in Sec. 3.3 but it is scaled in the proposed scheme. To show the impact of this scaling, the proposed scheme avoids scaling of hysteresis in Fig. 7 whereas this scaling is applied in all other figures in this section. In Fig. 7, P_{R_Min} is found significantly higher in the proposed scheme compared to the existing scheme but its stability with the user speed is not a lot better. With scaling of hysteresis, in Fig. 8, P_{R_Min} stays far higher in the proposed scheme compared to the existing scheme and also, the stability of P_{R_Min} is way better in the proposed scheme. The decrease in *SSSF* triggers handover earlier boosting the level of P_{R_Min} and thus, it indicates how the UE can quickly perform necessary adjustment using *SSSF*.

In the existing scheme, the received power may remain very low for a period and in this period, the service is extremely degraded and it is at risk of disruption. Therefore, the length of this period is determined assuming that the service degrades extremely when the received power falls below –90 dBm. Table 2 shows the length of the period with such low received power. This period is not calculated for the proposed scheme as the received power is then controlled to stay above –90 dBm. Table 2 shows that the periods are not trivial. The period increases gradually with the user speed because the user goes further away from the old serving cell before handover. However, the period decreases between user speed 90 km/hr and 110 km/hr because of a change in mobility state.

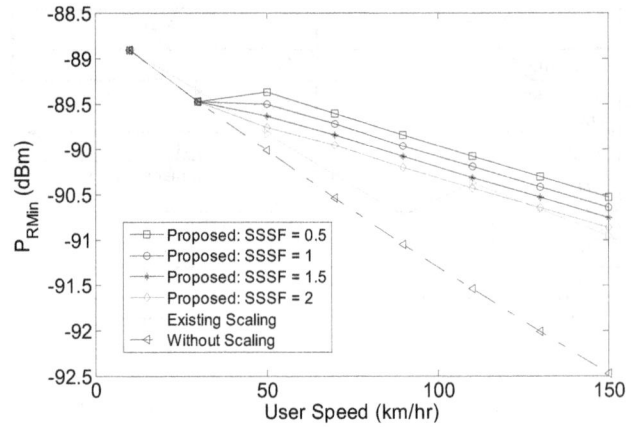

Fig. 7. P_{R_Min} vs. user speed for HO avoiding scaling of hysteresis in the proposed scheme (*Hyst_SF* = –0.4 dB).

Velocity (km/hr)	70	90	110	130	150
Period (ms)	696	1404	574	886	1114

Tab. 2. Period of power reception below –90 dBm in the existing scheme.

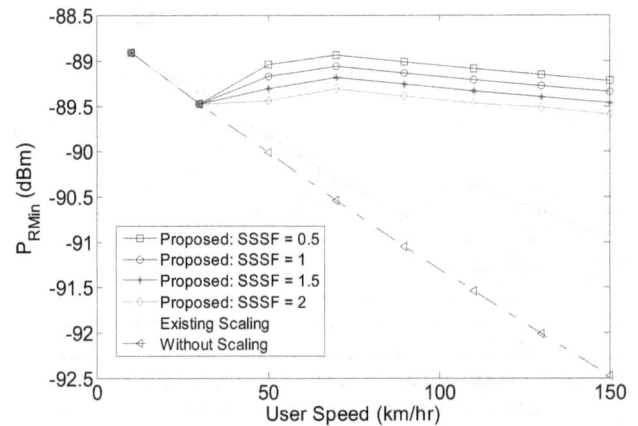

Fig. 8. P_{R_Min} vs. user speed for HO with scaling of hysteresis in the proposed scheme (*Hyst_SF*= –0.4 dB).

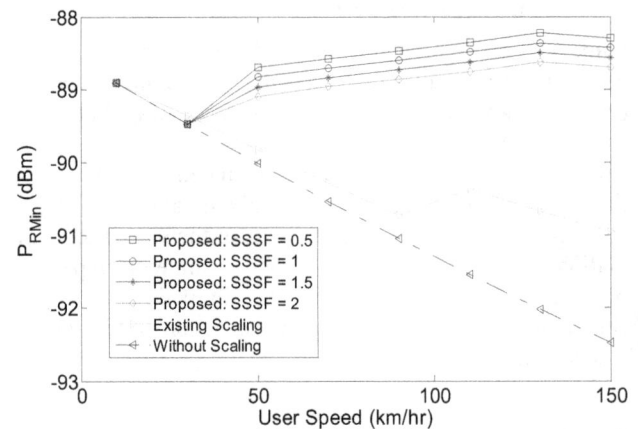

Fig. 9. P_{R_Min} vs. user speed for HO with scaling of hysteresis in the proposed scheme (*Hyst_SF*= –0.7 dB).

In Figure 9, the absolute value of *Hyst_SF* is higher compared to Fig. 8 and thus, Figure 9 has higher P_{R_Min} values for the same *SSSF* values. This shows that, in the

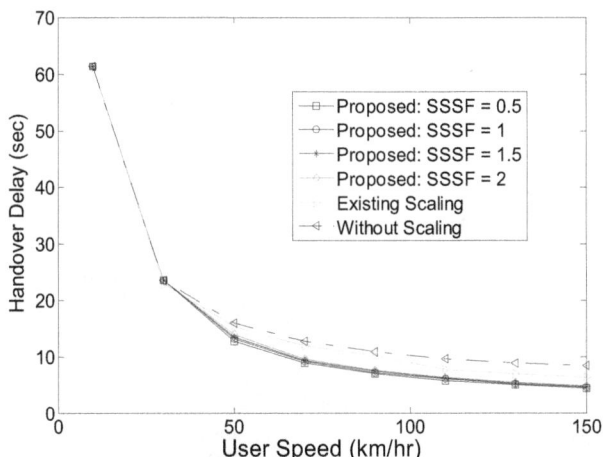

Fig. 10. Handover delay vs. user speed for HO with scaling of hysteresis in the proposed method ($Hyst\_SF$ = –0.4 dB).

proposed scheme, the eNodeB and the UE, from the two ends, can apply good control using $Hyst\_SF$ and $SSSF$, respectively. To ensure that the triggering of handover adapts to various scenarios in the desired fashion, the enhanced controllability should consider various factors, for example, user speeds, traffic loads, street patterns, types of cells, and so forth, according to the guidelines shown in Sec. 5.3.

Figure 10 shows the handover delay vs. user speed. The delay is calculated as the time that elapses between the moment of crossing the cell boundary and the moment of triggering handover. The lowest value of the delay is 4417 msec in the proposed scheme and it occurs at user velocity 150 km/hr with $SSSF$ 0.5. In a measurement period of 200 msec, the physical layer of the UE performs measurements of at least 8 identified intra-frequency cells and reports the results to layer 3 when no measurement gaps are activated [25]. Thus, in 4417 msec, assuming the presence of 16 intra-frequency cells, the layer 3 of the UE receives measurement results at least 11 times from the physical layer until it sends its last measurement report. This appears to be sufficient to ensure that the neighbor cell has really got better than the serving cell avoiding the possibility of ping-pong effects.

7. Conclusion

The challenges in the mobility support for speedy users in HetNets have been investigated and a wide variety of issues are derived. Addressing most of the issues, a scheme is proposed with versatile controllability over the scaling of cell switching parameters. Some of the features of the proposed scheme have been evaluated using limited simulations. The simulation results show that the lowest received power improves and also, it stabilizes better with the user speed in the proposed scheme. The illustrations of the proposed scheme bespeak the potential mitigation of various problems and the achievement of satisfactory cell edge performance.

References

[1] KAWSER, M. T. LTE *Air Interface Protocols*. Boston (USA): Artech House, 2011. ISBN: 978-1-60807-201-9

[2] Fujitsu Network Communications Inc. *Enhancing LTE Cell-Edge Performance via PDCCH ICIC*. 16 pages. [Online] Cited 2011. Available at: http://www.fujitsu.com/us/Images/Enhancing-LTE-Cell-Edge.pdf

[3] WILEY-GREEN, M. P., SVENSSON, T. Throughput, capacity, handover and latency performance in a 3GPP LTE FDD field trial. In *Proceedings of IEEE Global Telecommunications Conference (GLOBECOM 2010)*. Florida (USA), December 2010, p. 1–6. DOI: 10.1109/GLOCOM.2010.5683398

[4] 3GPP TR 36.839. *Evolved Universal Terrestrial Radio Access (E-UTRA); Mobility Enhancements in Heterogeneous Networks*. Release 11, 2012.

[5] 3GPP TS 36.331. *Evolved Universal Terrestrial Radio Access (E-UTRA); Radio Resource Control (RRC); Protocol specification*. Release 11, 2014.

[6] LOPEZ-PEREZ, D., GUVENC, I., CHU, X. Mobility enhancements for heterogeneous wireless networks through interference coordination. In *Proceedings of IEEE Wireless Communications and Networking Conference Workshops (WCNCW)*. Paris (France), April 2012, p. 69–74. DOI: 10.1109/WCNCW.2012.6215543

[7] SIMSEK, M., BENNIS, M., GUVENC, I. Mobility management in HetNets: a learning-based perspective. *EURASIP Journal on Wireless Communications and Networking*, February 2015, vol. 14. DOI 10.1186/s13638-015-0244-2

[8] PENG, Y., YANG, Y.Z.W., ZHU, Y. Mobility Performance enhancements for LTE-Advanced heterogeneous networks. In *Proceedings of IEEE 23rd International Symposium on Personal Indoor and Mobile Radio Communications (PIMRC)*. Sydney (Australia), September 2012, p. 413–418. DOI: 10.1109/PIMRC.2012.6362820

[9] BARBERA, S., MICHAELSEN, P., SAILY, M., PEDERSEN, K. Mobility performance of LTE co-channel deployment of macro and pico cells. In *Proceedings of IEEE Wireless Communications and Networking Conference (WCNC)*. Paris (France), April 2012, p. 2863–2868. DOI: 10.1109/WCNC.2012.6214290

[10] BARBERA, S., MICHAELSEN, P., SAILY, M., PEDERSEN, K. Improved mobility performance in LTE co-channel HetNets through speed differentiated enhancements. In *Proceedings of IEEE Globecom Workshops (GC Wkshps)*. Anaheim (CA, USA), Dec. 2012, p. 426–430. DOI: 10.1109/GLOCOMW.2012.6477610

[11] ZHANG, H., WEN, X., WANG, B., ZHENG, W., SUN, Y. A novel handover mechanism between femtocell and macrocell for LTE based networks. In *Proceedings of IEEE Second International Conference on Communication Software and Networks (IC-CSN)*. February 2010, p. 228–231. DOI: 10.1109/ICCSN.2010.91

[12] PEDERSEN, K.I., MICHAELSEN, P.H., ROSA, C., BARBERA, S. Mobility enhancements for LTE-advanced multilayer networks with inter-site carrier aggregation. *IEEE Communications Magazine*, May 2013, vol. 51, no. 5, p. 64–71. DOI: 10.1109/MCOM.2013.6515048

[13] PENG, Y., YANG, W., ZHANG, Y., ZHU, Y. Mobility performance enhancements for LTE-advanced heterogeneous networks. In *Proceedings of IEEE 23rd International Symposium on Personal Indoor and Mobile Radio Communications (PIMRC)*. Sydney (Australia), September 2012, p. 413–418. DOI: 10.1109/PIMRC.2012.6362820

[14] LEE, Y., SHIN, B., LIM, J., HONG, D. Effects of time-to-trigger parameter on handover performance in SON-based LTE systems.

In *Proceedings of IEEE 16th Asia-Pacific Conference on Communications (APCC)*. Auckland (New Zealand), October-November 2010, p. 492–496. DOI: 10.1109/APCC.2010.5680001

[15] JANSEN, T., BALAN, I., TURK, J., MOERMAN, I., KURNER, T. Handover parameter optimization in LTE self-organizing networks. In *Proceedings of IEEE 72nd Vehicular Technology Conference Fall (VTC 2010-Fall)*. Ottawa (Canada), September 2010, p. 1–5. DOI: 10.1109/VETECF.2010.5594245

[16] ZHENG, W., ZHANG, H., CHU, X., WEN, X. Mobility robustness optimization in self-organizing LTE femtocell networks. *EURASIP Journal on Wireless Communications and Networking*, 2013, 10 p. DOI: 10.1186/1687-1499-2013-27

[17] EMRAH TUNÇEL. *Tuning of Handover Parameters in LTE-A Heterogeneous Networks*. 94 pages. [Online] Cited 2014-09. Available at: http://etd.lib.metu.edu.tr/upload/12617911/index.pdf

[18] KIM, Y., LEE, K., CHIN, Y. Analysis of multi-level threshold handoff algorithm. In *Proceedings of IEEE Global Telecommunications Conference (GLOBECOM 96)*. London (UK), November 1996, p. 1141–1145. DOI: 10.1109/GLOCOM.1996.587613

[19] KAWSER, M. T., ISLAM, M. R., AHMED, K. I., et al. Efficient resource allocation and sectorization for fractional frequency reuse (FFR) in LTE femtocell systems. *Radioengineering*, December 2015, vol. 24, no. 4, p. 940–947. DOI: 10.13164/re.2015.0940

[20] AAMOD KHANDEKAR, NAGA BHUSHAN, JI TINGFANG, et al. LTE Advanced: Heterogeneous networks. In *Proceedings of IEEE European Wireless Conference (EW)*. Lucca (Italy), April 2010, p. 978–982. DOI: 10.1109/EW.2010.5483516

[21] SCHOBER, H., JONDRAL, F. Velocity estimation for OFDM based communication systems. In *Proceedings of IEEE 56th Vehicular Technology Conference (VTC 2002-Fall)*. Vancouver, (BC, Canada), September 2002, vol. 2, p. 715–718. DOI: 10.1109/VETECF.2002.1040692

[22] YUCEK, T., TANNIOUS, R. M. A., ARSLAN, H. Doppler spread estimation for wireless OFDM systems. In *Proceedings of IEEE/Sarnoff Symposium on Advances in Wired and Wireless Communication*. Princeton (NJ, USA), April 2005, p. 233–236. DOI: 10.1109/SARNOF.2005.1426552

[23] TEPEDELENLIOGLU, C., ABDI, A., GIANNAKIS, G. B., et al. Estimation of Doppler spread and signal strength in mobile communications with applications to handoff and adaptive transmission. *Wireless Communications and Mobile Computing*, April 2001, vol. 1, no. 2, p. 221–242. DOI: 10.1002/wcm.1

[24] BHATTACHARYA, P. P. A new environment dependent handoff technique for next generation mobile systems. *International Journal of Computer and Communications*, March 2011, vol. 1, no. 1, p. 15–24.

[25] 3GPP TS 36.133 V11.4.0. *Evolved Universal Terrestrial Radio Access (E-UTRA); Requirements for Support of Radio Resource Management*. Release 12, 2014.

About the Authors ...

Mohammad T. KAWSER received his Ph.D. degree from Islamic University of Technology, Bangladesh in 2016, MS degree from Virginia Tech, USA in 2005 and BS degree from Bangladesh University of Engineering and Technology, Bangladesh in 1999, all in Electrical and Electronic Engineering. He is serving as an assistant professor at Islamic University of Technology, Bangladesh. His research interests include layer 2 and layer 3 functions of cellular operation, LTE-Advanced features, etc.

Mohammad R. ISLAM received BS and MS degree in Electrical and Electronic Engineering from Bangladesh University of Engineering and Technology, Bangladesh in 1998 and 2004, respectively. He received MBA degree from the Institute of Business Administration, Bangladesh in 2006. He received his Ph.D. degree from Kyung Hee University, South Korea in 2010. He is serving as a professor at Islamic University of Technology, Bangladesh. His research interests include wireless sensor networks, LDPC and QC-LDPC codes, secrecy capacity and other wireless applications.

Muhammad R. RAHIM received his BS degree in Electrical and Electronic Engineering from Islamic University of Technology in 2012. Currently, he is enrolled in the MS program at RWTH Aachen University, Germany. His interest lies in mobile radio networking.

Muhammad A. MASUD received his BS degree in Electrical and Electronic Engineering from Islamic University of Technology in 2012. He has been serving as a lecturer at Uttara University, Bangladesh. His research interests include wireless communications, inter-networking and network security.

3

Mobile Signal Path Losses in Microcells behind Buildings

Saulius JAPERTAS, Vitas GRIMAILA

Dept. of Telecommunications, Kaunas University of Technology, K. Donelaičio str. 73, 44249 Kaunas, Lithuania

saulius.japertas@ktu.lt, vitas.grimaila@ktu.lt

Abstract. *The paper presents measurement results of the GSM (900 MHz band), UMTS (2100 MHz band), and LTE (1800 MHz band) propagation path loss (PL) in the urban area behind the buildings of ten different heights. The results were compared with the 7 most popular models. It was found that the existing models approximate the experimental results with relatively large errors. The new model, which evaluates the path loss variation nature behind the buildings, is proposed. This new model shows good agreement with measurements for all three mobile technologies. The average relative error is less than 6.5 %.*

Keywords

Cellular networks, radio wave propagation, mobile communications

1. Introduction

The number of mobile phone users is growing at high speed. This is related more to data services demand growth than to the needs of a voice service. Some of mobile phone manufacturers predict that the number of smart phone users will be more than 9.1 billion by 2020 [1]. The data transfer rate, compared with 2015, will increase 8 times and monthly global mobile data traffic will exceed 24.3 exabytes by 2019 [2]. Therefore, it will be necessary for such technologies, which can provide the high data transmission speed and quality. It will be done in the development of new technologies (5G and other) as well as improving the existing ones.

Most countries plan that 5G technology will be realized at frequencies exceeding 10 GHz [3]. However, as it is well known, the increase in frequency causes the decrease of the distance between the base station (hereinafter BS) and User Equipment (hereinafter UE). Therefore, it can be assumed that high speed data will be ensured at the relatively short distances: less than 500 m between the BS and the UE. The mobile coverage planning becomes very important in this case. The accuracy of this planning will be determined by the propagation model accuracy. In this way, the models assessing different effects in the microcells will require the 5G networks planning. Those effects are: reflection, diffraction, refraction, scattering, shadowing, and penetration. However, in order to investigate the 5G it is necessary to have deeper knowledge of the mobile signal propagation characteristics in the microcells for already existing technologies.

The main objectives of this work is to experimentally investigate the path losses in microcells for GSM, UMTS, LTE technologies, to compare these experimental data with the existing propagation prediction models and, if necessary, to propose a new path loss model. This work is continuation of [4] where the 2G technology signal propagation behind the buildings is analyzed.

2. Related Works

Currently, there are quite a lot of models (it is possible to charge more than 60), which allow to assess the propagation losses both in macrocells and microcells. All these models have certain limitations: according to frequency, to the distances between base station BS and user equipment UE, to BS and UE antenna heights, etc. Okumura-Hata and COST 231 Hata are the most popular propagation loss prediction models in macrocells. The path losses in the microcells are usually predictable using Walfish-Ikegami, LEE, ECC 33, Two slope, SUI, and other models. All these models are the functions of the distance between the BS and UE d_{BS}, frequency f, BS and UE antenna highs h_{BS}, and h_{UE} respectively, certain correction coefficients groups C_n:

$$PL = \phi\left(d_{BS}, h_{BS}, h_{UE}, f, C_n\right). \quad (1)$$

Some microcells models try to assess the diffraction and reflection effects using additional parameters, such as: the certain angles α, building heights h_b, certain spaces (street width) dimensions d_r (e.g., Walfish-Ikegami model) or terrain roughness h_t (Lee models). The path losses are described as the function in (2):

$$PL = \phi\left(d_{BS}, h_{BS}, h_{UE}, f, C_n, \begin{cases} \alpha, h_b, d_r \\ h_t \end{cases}\right). \quad (2)$$

Some of the works in order to approximate the path losses data in the microcells use models that formally have to be applied only in the macrocells (Okumura-Hata, COST 231 Hata, etc.) [4–9]. However, these macrocell models results do not differ from microcells models (Walfish-Ikegami, ECC 33, Two slope, etc.) results according to the accuracy. The above-mentioned works very clearly show this. The works' analysis shows that only certain models at the certain distance from the BS areas may coincide with the experimental results. But at the other distances, the same model results differ from experimental results rather significantly. In particular, the experimental and theoretical results show the clear mismatches in short distances from BS. The experimental results in [7] are well approximated using the SUI model but only over the distances of 300 m. The error between the experimental and model results can reach up to 30 % when the distances are shorter. The experimental results in [8] are compared with the Walfish-Ikegami model results; it is seen that at the distances from the BS shorter than 500 m the results mismatch is significantly higher than 10%. In addition, these experimental results show the clear results scattering similar to the slow fading influence. And it does not depend on the frequency and area (urban or suburban) where measurements are carried out [10]. However, none of models does not evaluate such results scattering. Sometimes there is an attempt to modify the known model using the obtained experimental results. But often such modified models provide not sufficiently good approximations even for the experimental results which were received at the same work. For example, in [11] the optimized Hata model is proposed, it gives good result coincidences in BS1 and BS3 stations. Meanwhile the error can reach about 10% for the BS4 case in the distances between 500 m and 1000 m. Furthermore, there is clearly seen that the errors between the experiment and modeling results are significant at less than 200 m distances.

The received results are spread out in the wide range of path losses values and seem as the certain „swarm" when the sufficient number of experimental measurements are carried out in the microcells [12–15]. Such "swarm" is also observed at short distances between BS and UE (less than 200 m) when the frequency is high enough (> 20 GHz) [16–18].

Thus, the works' analysis shows that there is still important to investigate the path loss changes in the microcells, especially for short distances between BS and UE.

3. Experimental Setup

The two base stations (hereinafter BS1 and BS2) near Kaunas University of Technology have been chosen for the experiments. The areas around BS1 and BS2 are densely populated. These base stations support GSM (EDGE), UMTS and LTE mobile technologies. Buildings of different heights, at the distances less than 50 m from each other, are in BS1 and BS2 coverage areas. The path losses variation of GSM, UMTS and LTE in BS1 coverage area and only GSM path losses variation in the BS2 coverage area were researched. BS1 frequency for the GSM is 956 MHz, for the UMTS is 2127 MHz, for the LTE is 1819 MHz. BS1 antennas height h_{BS1} is 43 m, ERP (equivalent radiated power) is 62.82 dBm for UMTS, and LTE for GSM ERP is 38.81 dBm. BS2 frequency (GSM) is 945 MHz, antennas height h_{BS2} is 32 m and ERP is 40.01 dBm. Receiver (UE) antennas height h_{UE} was constant at 1.3 m.

Table 1 summarizes the main experimental parameters. h_b means the height of the building, d_{BS} means the distance between the base station and the measurement point.

The signal strength measurements were carried out with a spectrum analyzer Anritsu Cell Master MT8212A; its frequency measurement range is from 10 MHz to 3.0 GHz. The results were processed using specialized software Handheld Software Tools (HHST 6.61). The measurements were carried out up to 500 m distance from the base station. UMTS and LTE signals were simultaneously measured and recorded in the spectrum analyzer's memory. GSM signal strength measurements were carried out at other times in similar weather conditions. The signal strength characteristics of behind the 17 buildings whose height varied from 6 to 41 m were measured. In some cases, buildings of the same height are at different distances from the BS. The measurements were carried out behind the building. The first measurement point was chosen at the distance of 1 m from the rear (with respect to the BS) wall of the building. The signal strength was measured every 5 m (UMTS and LTE) and 2 m (GSM), gradually moving away from the house. Ten measurements

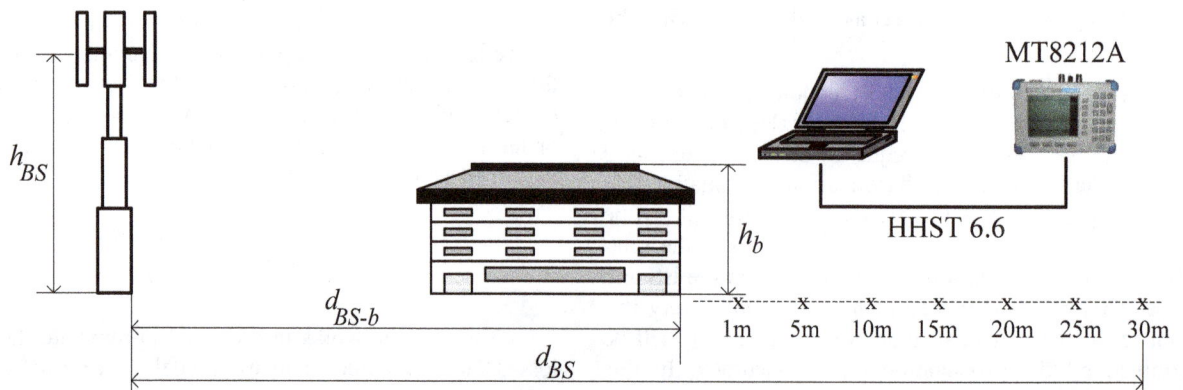

Fig. 1. Experiment scheme (sign x means the measuring point).

	BS1			BS2
	2G	**UMTS**	**LTE**	
Frequency [MHz]	956	2127	1819	945
Base station antenna height [m]		43		32
ERP [dBm]	38.81	62.82		40.01
h_{UE} **[m]**		1.3		
h_b **[m]**		6 − 41		12 − 22
d_{BS} **[m]**		< 450		
$d_{BS\text{-}b}$ **[m]**	< 321	< 302		<382

Tab. 1. Experimental parameters.

were carried out in each point and the results were averaged. The experiment scheme is shown in Fig. 1. The $h_{BS\text{-}b}$ means the distance from the BS to the building far edge.

The various errors were calculated in order to evaluate the results accuracy according to (3): the relative error δ, the root-mean-square of the measurement result, the root-mean-square of the measurements' mathematical expectation, dispersion σ and skewness γ:

$$\begin{cases} \delta = \dfrac{\dfrac{1}{n}\left|RSL_i - \langle RSL\rangle\right|}{\langle RSL\rangle}\cdot 100\%, \\[2mm] \sigma = \dfrac{1}{n}\sum_{i=1}^{n}\left(RSL_i - \langle RSL\rangle\right)^2, \\[2mm] S_{RSL} = \sqrt{\dfrac{\sum_{i=1}^{n}\left(RSL - \langle RSL\rangle\right)^2}{n-1}}, \\[2mm] S_{\langle RSL\rangle} = \dfrac{S_{RSL}}{n}, \\[2mm] \gamma = \dfrac{\dfrac{1}{n}\sum_{i=1}^{n}\left(RSL_i - \langle RSL\rangle\right)^3}{\left(S_{RSL}\right)^3}. \end{cases} \quad (3)$$

where RSL_i is the signal strength of the separate measuring, dBm; $\langle RSL\rangle$ is the mathematical expectation, dBm; n is the number of measurements.

By summarizing all received errors it was found that the average relative error δ is about 4% for all measurements, the root-mean-square S_{RSL} is about 1.63 dBm, the root-mean-square of the measurement mathematical expectation is $S_{\langle RSL\rangle}$ is about 0.47 dBm, the dispersion σ is about 9.55.

4. Results

The path loss dependence on technology and the distance d_{BS} is shown in Fig. 2. Very clear dependence on technology can be seen there. The biggest path loss is observed on LTE technology and the lowest to GSM (EDGE). It is seen that the path losses react stronger to the increase of d_{BS} for GSM technology than for UMTS or LTE.

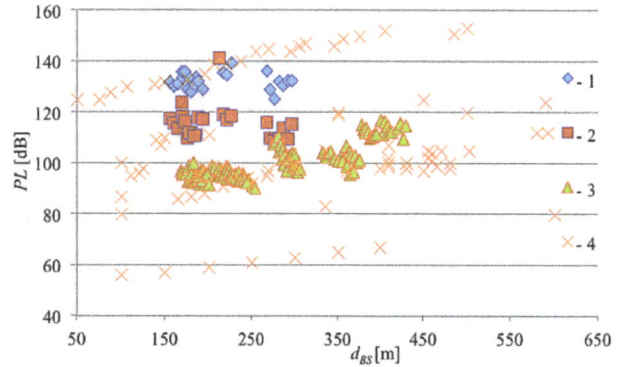

Fig. 2. Comparison of experimental results with the results of [5–7, 19–21]: LTE path loss – 1; UMTS path loss – 2; GSM path loss – 3; results of [5–7, 19–21] – 4.

It can be seen that the results in Fig. 2 are similar to "swarm", as in the results of the other authors. Such "swarm" of the results is due to the fact that results are shown without taking into account the measurement environment of a particular case: the influence of buildings, the measuring point location behind the buildings, the building height, whether the measured path loss is on the line of sight or non-line of sight with the base station, influence of trees, etc. All these environmental factors influence the path losses and, in order to make very accurate assessment of path losses, it is necessary to take into account these factors.

The cumulative distribution function CDF dependence on the path losses for different mobile technologies is shown in Fig. 3; it shows that using more sophisticated signal-forming technology shifts curve to the right. This fact statistically demonstrates that the higher-generation mobile technologies lead to higher path loss. The slope of the curves shows that the smallest errors are received for LTE technology, when the largest errors are for GSM technology. It seems that the highest technology gives the better accuracy.

The experimental path losses results are compared with the most popular 7 models results (Fig. 4); these models approximated the experimental results with relatively high errors. Table 2 summarizes the main propagation model expressions.

However, there is at least one model for the different technologies that allows to evaluate the experimental re-

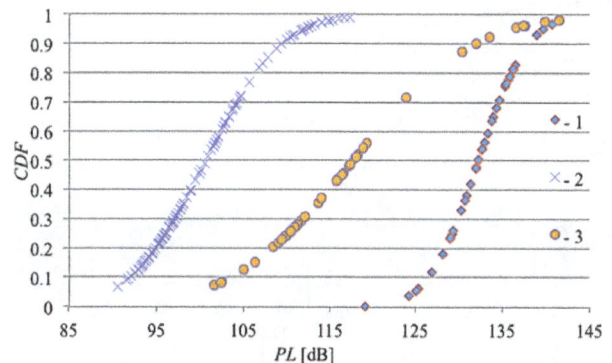

Fig. 3. Experimental path loss CDF for different mobile technologies: LTE – 1, UMTS – 2, GSM – 3.

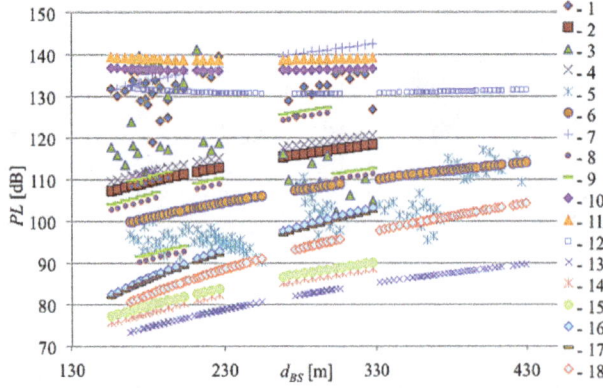

Fig. 4. Comparison of experimental results with some models prediction results: LTE experiment – 1, COST231-Hata (LTE) – 2, UMTS experiment – 3, COST231-Hata (UMTS) – 4, GSM experiment – 5, Okumura Hata (GSM) – 6, Multislope (LTE) – 7, Walfish-Ikegami (LTE) – 8, Walfish-Ikegami (UMTS) – 9, ECC 33 (LTE) – 10, ECC 33 (UMTS) – 11, ECC 33 (GSM) – 12, Clutter Factor (GSM) – 13, Clutter Factor (LTE) – 14, Clutter Factor (UMTS) – 15, Two slope (UMTS) – 16, Two slope (LTE) – 17, Two slope (GSM) – 18.

sults with reasonable errors. For the GSM, the best results (δ is about 18%) are achieved with the SUI and Okumura-Hata models; Hata model gives the best results (δ is about 30%) for UMTS; and Multi-slope model gives the best results (δ more than 10%) for LTE.

The other fact is also seen in Fig. 4. Although in general, the path losses increase with the d_{BS} increasing. The path losses behind the buildings relatively decrease with the distance from BS.

A few things can be seen after analyzing the path losses variation for the different building height and for the different mobile technologies. In particular, this path loss variation can be approximated by linear equation $PL = a \cdot d_{BS} + b$. In addition, when the height of the buildings h_b is the same, but d_{BS} is different, the PL variation can be approximated with the parallel straight lines i.e., the slope of the straight line which is defined by coefficient a does not depend on d_{BS}, but it depends on h_b. Moreover, this coefficient does not strongly depend on mobile technology. The typical examples of these experiments are shown in Fig. 5.

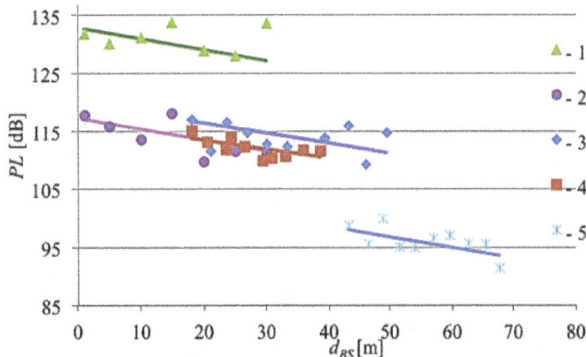

Fig. 5. Path losses variation vs distance behind the building for 18 m high buildings: LTE – 1, UMTS – 2, GSM – 3-5, approximation of results – solid lines.

Model	Description
Okumura Hata	$PL = 69.55 + 26.16 \cdot \lg(f) - 13.82 \cdot \lg(h_b) - a(h_m) + [44.9 - 6.55 \cdot \lg(h_b)] \cdot \lg(d)^b,$ $a(h_m) = 3.2 \cdot [\lg(11.75 \cdot h_m)]^2 - 4.97),$ $b = 1, d \le 20 \text{ km}$
COST231-Hata	$PL = 46.3 + 33.9 \cdot \lg(f) - 13.82 \cdot \lg(h_b) - a(h_m) + [44.9 - 6.55 \cdot \lg(h_b)] \cdot \lg(d) + C_m;$ $a(h_m) = 3.2 \cdot [\lg(11.75 \cdot h_m)]^2 - 4.97.$
Two slope	$L_{NEAR} = L_{OH}(d_{BP}) + slope_L [\lg(d) - \lg(d_{BP})],$ $slope_L = \dfrac{L_{BP}(d_{BP}) \cdot \lg(d_{BP}) - L_{FS}(d_{20}) \cdot \lg(d_{20})}{\lg(d_{BP}) - \lg(d_{20})}$
Multi-slope	$L_{3S} = 32.44 + 20 \cdot \lg(f) + 10 \cdot \lg\left[\dfrac{d^2 + (h_b - h_m)^2}{10^6}\right],$ $d < 0.04 \text{ km};$ $L_{3S} = L(d_{40}) + \dfrac{\lg(d) - \lg(d_{40})}{\lg(d_{100}) - \lg(d_{40})} \cdot [L(d_{100}) - L(d_{40})],$ $0.04 \text{ km} < d < 0.1 \text{ km}$
ECC 33	$PL = A_{fs} + A_{bm} - G_t - G_r;$ $A_{fs} = 92.4 + \lg(d) + \lg(f);$ $A_{bm} = 20.41 + 9.83 \cdot \lg(d) + 7.89 \cdot \lg(f) + 9.56 \cdot [\lg(f)]^2;$ $G_t = \lg\left(\dfrac{h_b}{200}\right) \cdot [13.958 + 5.8 \cdot (\lg(d))^2];$ $G_r = [42.5742 + 13.7 \cdot \lg(f)] \cdot [\lg(h_m) - 0.585].$
Walfish-Ikegami	$L_b = \begin{cases} L_o + L_{rts} + L_{msd}, & L_{rts} + L_{msd} > 0; \\ L_o, & L_{rts} + L_{msd} \le 0. \end{cases}$
Clutter Factor	$L = 40 \cdot \lg(d) + 20 \cdot \lg(f_c) - 20 \cdot \lg(h_b) + L_m;$ $h_b, h_m << d, L_m = 76.3 - 10 \cdot \lg(h_m).$

Tab. 2. Propagation model expressions.

The path losses variation for 18 m high buildings at different distances from the behind the buildings d_{BS} are shown there. Qualitatively similar results are observed in all other experiments.

The coefficient b determines the initial path loss level at the first measurement point behind the building and it should depend on the distance from the building and frequency. The variation of the straight coefficients a and b, depending on height of building and distance from the building, respectively is shown in Fig. 6 and Fig. 7.

Figure 6 shows that the coefficient a decreases with the increasing of the building height and the decrease is approximated according to line equation (4) with the mean relative error δ of about 7.96%.

$$a = -(0.0037 \cdot h_b + 0.1249). \qquad (4)$$

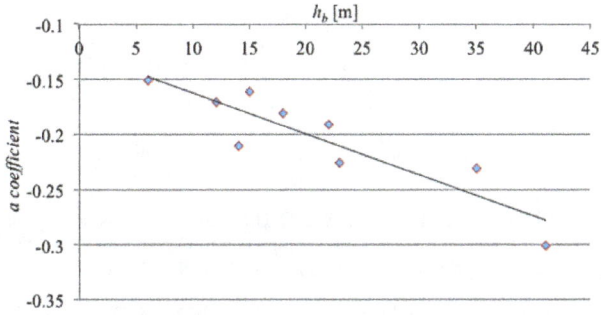

Fig. 6. *a* coefficient variation vs the building high h_b: experimental data – 1, approximation of results – solid line.

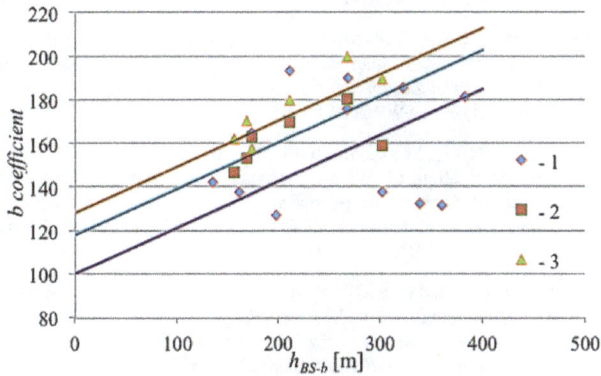

Fig. 7. *b* coefficient variation vs distance from BS $d_{BS\text{-}b}$: GSM – 1, UMTS – 2, LTE – 3, approximation of results – solid lines.

The variation of the coefficient b, as can be seen from Fig. 7, depends on the distance from the BS to the building $d_{BS\text{-}b}$, mobile technology and frequency f. The variation of the coefficient b could be approximated with the parallel line equations (5) for all experimental cases independently from the mobile technology:

$$b = \left(0.2125 \cdot d_{BS\text{-}b} + 20\lg f_{[\text{MHz}]} + \begin{cases} 40.4...\text{for GSM} \\ 51.5...\text{for UMTS} \\ 62.8...\text{for LTE} \end{cases}\right). \quad (5)$$

Comparing the results of the model with the experimental results it was noticed that in some cases the error exceeds 10%. It was observed that when $\tg\alpha$ (Fig. 8) is less than 0.05, then there is the need to add the correction factor CF. In this case equation (5) is:

$$b = \left(0.2125 \cdot d_{BS\text{-}b} + 20\lg f_{[\text{MHz}]} + \begin{cases} 40.4...\text{for GSM} \\ 51.5...\text{for UMTS} \\ 62.8...\text{for LTE} \end{cases}\right) \cdot$$
$$(-8.98 \cdot \tg\alpha + 1.436). \quad (6)$$

Equation (5) and (6) approximate experimentally determined values of coefficient b with the following mean relative error: for GSM δ is about 6.47%, for UMTS δ is about 4.75% and for LTE δ is about 3.85%. Thus, by evaluating (4), (5) and (6), the path losses behind the buildings in microcells can be approximated by the equation:

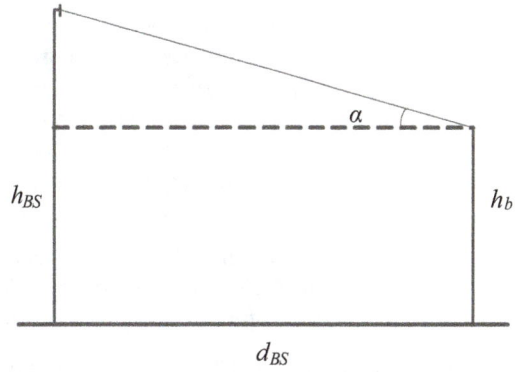

Fig. 8. Diagram explaining the angle α.

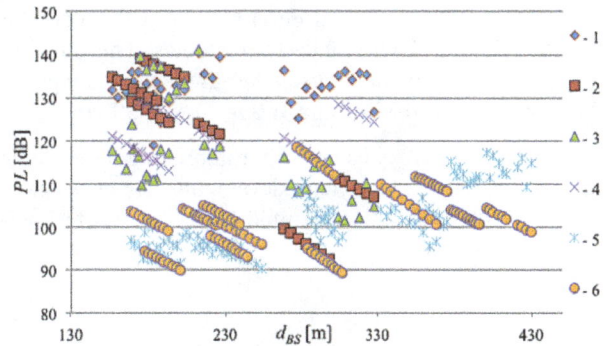

Fig. 9. Comparison of experimental results with the proposed model (7) results: LTE experimental results – 1, LTE model – 2, UMTS experimental – 3, UMTS model – 4, GSM experimental – 5, GSM model – 6.

$$PL = -\left(0.0037 \cdot h_b + 0.1249\right) \cdot d_{BS} +$$
$$\left(0.2125 \cdot d_{BS\text{-}b} + 20\lg f_{[\text{MHz}]} + \begin{cases} 40.4...\text{for GSM} \\ 51.5...\text{for UMTS} \\ 62.8...\text{for LTE} \end{cases}\right) \cdot CF \quad (7)$$

where CF is 1 if $\tg\alpha > 0.05$ and CF is $(-8.98 \cdot \tg\alpha + 1.436)$ if $\tg\alpha \le 0.05$.

The comparisons of the experimental results with the proposed model results are shown in Fig. 9.

The results coincide well enough and the average relative error does not exceed 6.5%. The calculated skewness γ shows that the deviation of the experimental results from the mean (mathematical expectation) is low, because $\gamma_{\text{LTE}} < 3.2 \times 10^{-7}$; $\gamma_{\text{UMTS}} < 7.2 \times 10^{-7}$; $\gamma_{\text{GSM}} < 1.1 \times 10^{-5}$, i.e. in all cases $\gamma \to 0$.

5. Conclusions

Path losses variation with the distance from BS behind the building with different heights for GSM, UMTS and LTE mobile technologies were compared with the results of seven models. It was found that no one of the models describes the experimental results with acceptable accuracy. In addition, these models do not explain certain experimental results scattering, which has the pattern character.

It was found that the path losses decrease behind the building, when the distance from base station increases. This decreasing may be affected of such electromagnetic wave propagation mechanisms as shadowing and diffraction. Such decreasing, in our view, would take place up to position behind the house, where the line-of-sight conditions would start.

It was found that in addition to the generally accepted parameters, such as: ERP, f, d_{BS}, h_{BS}, h_{UE}, h_b, path losses are affected by the mobile technology.

The new path losses evaluation model in microcells behind the buildings is proposed. The comparison of the experimental and modeling results gives the error of approximately 6%. This model describes the path losses for GSM, UMTS and LTE mobile technologies. The limits of the results validity are: $h_{BS} > h_b$, frequency range is 900 MHz $< f <$ 2200 MHz, and 100 m $< d_{BS} <$ 500 m.

These results may be useful for improving the existing propagation loss models and developing the new more accurate models, including the models for the new technologies (such as 5G).

References

[1] NOKIA NETWORKS. *Nokia Networks Deployment for Coverage (white paper)*. 20 pages. [Online] Cited 2016-03-09. Available at: http://networks.nokia.com/sites/default/files/document/nokia_depl oyment_for_coverage_white_paper_0.pdf

[2] CISCO SYSTEMS, INC. *Cisco Visual Networking Index: Global Mobile Data Traffic Forecast Update, 2015–2020 (white paper)*. 39 pages. [Online] Cited 2016-03-17. Available at: http://www.cisco.com/c/en/us/solutions/collateral/service-provider/visual-networking-index-vni/mobile-white-paper-c11-520862.html

[3] 4G AMERICAS. *5G Spectrum Recommendations (white paper)*. 28 pages. [Online] Cited 2016-03-18. Available at: http://www.4gamericas.org/files/6514/3930/9262/4G_Americas_5 G_Spectrum_Recommendations_White_Paper.pdf

[4] JAPERTAS, S., PILIPAVICIUS, K., JANARTHANAN, D. Signal propagation model for microcells at 900 MHz frequency range. *Elekronika ir Elektrotechnika*, 2015, vol. 21, no. 4, p. 65–68. DOI: 10.5755/j01.eee.21.4.12786

[5] OKOROGU, V. N., ONYISHI, D. U., NWALOZIE, G. C., et al. Empirical characterization of propagation path loss and performance evaluation for co-site urban environment. *International Journal of Computer Applications*, 2013, vol. 70, no. 10, p. 34–41. DOI: 10.5120/12001-7888

[6] HAMID, M., KOSTANIC, I. Path loss models for LTE and LTE-A relay stations. *Universal Journal of Communications and Network*, 2013, vol. 1, no. 4, p. 119–126. DOI: 10.13189/ujcn.2013.010401

[7] CHEBIL, J., LAWAS, A. K., RAFIQUL ISLAM, M. D. Comparison between measured and predicted path loss for mobile communications in Malaysia. *World Applied Sciences Journal*, 2013, no. 21, p. 123–128. DOI: 10.5829/idosi.wasj.2013.21.mae.99936

[8] RATURI, P., GUPTA, V., ERAM, S. Proposed propagation model for Dehradun region. *International Journal of Soft Computing and Engineering*, 2014, vol. 3, no. 6, p. 236–240. ISSN: 2231-230

[9] OSENI, F. O., POPOOLA, S. I., ABOLADE R. O., et al. Comparative analysis of received signal strength prediction models for radio network planning of GSM 900 MHz in Ilorin, Nigeria. *International Journal of Innovative Technology and Exploring Engineering*, 2014, vol. 4, no. 3, p. 45–50. ISSN: 2278-3075

[10] MAWJOUD, S. A. Path loss propagation model prediction for GSM network planning. *International Journal of Computer Applications*, 2013, vol. 84, no. 7, p. 30–33. DOI: 10.5120/14592-2830

[11] ISABONA, J., KONYEHA C. C. Urban area path loss propagation prediction and optimization using Hata model at 800 MHz. *IOSR Journal of Applied Physic*, 2013, vol. 3, no. 4, p. 08–18.

[12] TURKKA, J., RENFORS, M. Path loss measurements for a non-line-of-sight mobile-to mobile environment. In *Proceedings of 8th International Conference on ITS Telecommunications (ITST-2008)*. Hilton Phuket (Thailand), 2008, p. 274–278. DOI: 10.1109/ITST.2008.4740270

[13] ZHAO, X., RAUTIAINEN, T., KALLIOLA, K., et al. Path loss models for urban microcells at 5.3 GHz. *IEEE Antennas and Wireless Propagation Letters*, 2006, vol. 5, no. 1, p. 152–154. DOI: 10.1109/LAWP.2006.873950

[14] BARBIROLI, M., CARCIOFI, C., ESPOSTI, V. D., et al. Characterization of WIMAX propagation in microcellular and picocellular environments. In *Proceedings of the Fourth European Conference on Antennas and Propagation*. Barcelona (Spain), 2010, p. 1–5. ISBN: 978-1-4244-6431-9

[15] KLOZAR, L., PROKOPEC, J. Propagation path loss models for mobile communication. In *Proceedings of the 21st International Conference Radioelektronika 2011*. Brno (Czech Republic), 2011, p. 1–5. DOI: 10.1109/RADIOELEK.2011.5936478

[16] POLAK, L., KLOZAR, L., KALLER, O., et al. Study of coexistence between indoor LTE femtocell and outdoor-to-indoor DVB-T2-Lite reception in a shared frequency band. *EURASIP Journal on Wireless Communications and Networking*, 2015, vol. 114, p. 1–14. DOI: 10.1186/s13638-015-0338-x

[17] SAMIMI, M. K., RAPPAPORT, T. S., MACCARTNEY, G. R. Probabilistic omnidirectional path loss model for millimeter-wave outdoor communications. *IEEE Wireless Communications Letters*, 2015, vol. 4, no. 4, p. 357–360. DOI: 10.1109/LWC.2015.2417559

[18] MACCARTNEY, G. R., ZHANG, J., NIE, S., et al. Path loss models for 5G millimeter wave propagation channels in urban microcells. In *Proceedings of IEEE Global Communications Conference, Exhibition & Industry Forum (GLOBECOM)*. Atlanta (USA), 2013, p. 3948–3953. DOI: 10.1109/GLOCOM.2013.6831690

[19] NISSIRAT, L. A., ISMAIL, M., NISIRAT, M., et al. Lee's path loss model calibration and prediction for Jiza Town, South of Amman City, Jordan at 900 MHz. In *Proceedings of 4th IEEE International RF and Microwave Conference (RFM)*. Seremban (Malaysia), 2011, p. 412–415. DOI: 10.1109/RFM.2011.6168779

[20] MARDENI, R., PEY, L. Y. Path loss model development for urban outdoor coverage of Code Division Multiple Access (CDMA) system in Malaysia. In *Proceedings of International Conference on Microwave and Millimeter Wave Technology (ICMMT)*. Chengdu (China), 2010, p. 441–444. DOI: 10.1109/ICMMT.2010.5525001

[21] NISIRAT, M. A., ISMAIL, M., NISSIRAT, L., et al. A Hata based model utilizing terrain roughness correction formula. In *Proceedings of 6th International Conference on Telecommunication Systems, Services, and Applications (TSSA)*. Bali (Indonesia), 2011, p. 284–287. DOI: 10.1109/TSSA.2011.6095451

About the Authors...

Saulius JAPERTAS was born in 1959. He received his Ph.D. from the Lithuanian Energy Institute in 1991. He is head of Telecommunications Department of Kaunas University of Technology, Lithuania. His research interests include wireless networks, security and protection of electronics and telecommunications measures.

Vitas GRIMAILA was born in 1974. He received his M.Sc. and Ph.D. from the Kaunas University of Technology in 1999 and 2004, respectively. His research interests are in the area of mobile technologies, signal propagation, and spectral resource optimization.

4

Interference Analysis between Mobile Radio and Digital Terrestrial Television in the Digital Dividend Spectrum

Alberto TEKOVIC [1], Davor BONEFACIC [2], Gordan SISUL [2], Robert NAD [2]

[1] Access and Transport Network Engineering, VIPnet Ltd., Vrtni put 1, 10 000 Zagreb, Croatia

[2] Dept. of Radiocommunications, Faculty of Electrical Engineering and Computing, Unska 3, 10 000 Zagreb, Croatia

a.tekovic@vipnet.hr, {davor.bonefacic, gordan.sisul, robert.nadj}@fer.hr

Abstract. *This paper is concerned with the analysis of adjacent channel interference of the Long Term Evolution (LTE) mobile system operating in the Digital Dividend into Digital Video Broadcasting – Terrestrial (DVB–T) system. Field measurements in the real LTE network have been conducted in order to define the most significant scenarios and for each of these, Protection Ratios have been quantified. Variable load on the LTE base station has been taken into consideration. Therefore, Protection Ratios for the LTE base station in idle state, and fully dedicated mode have been calculated. Interference mitigation techniques have been reviewed, and an effective deployment method has been proposed.*

Keywords

Adjacent-channel interference, LTE FDD, DVB-T, Digital Dividend, Protection Ratio, Protection Distance, mitigation technique

1. Introduction

Due to greater spectrum efficiency, switching to digital terrestrial TV broadcasting frees part of the UHF spectrum from 790 MHz to 862 MHz called "digital dividend". The transition process is already completed in many countries. Ever increasing market interest for mobile broadband communications was a main driver for allocating the digital dividend to mobile services in several regions of the world, which occurred at the World Radiocommunication Conference in 2007 (resolution 749) [1]. Although this allocation began in 2015, some EU countries were allowed to utilize this allocation before 2015 [2], with necessary technical coordination with neighboring countries. This decision was very positive from the market point of view, but raised new co-existence issues that need to be carefully analyzed and evaluated:

- co-channel interference between neighboring countries or regions, one of them using the digital dividend band for mobile systems and the other for analog or digital terrestrial television [3], [4]

- adjacent channel interference within a given territory, where frequencies up to 790 MHz will be used for television and those immediately above this limit will be used for mobile radio communications

- co-channel interference within a given territory in the digital dividend band between mobile systems and DVB-C2 [5].

The theoretical analysis of co-channel interference between mobile system and digital terrestrial television has been presented in [6–8]. This is followed by experimental studies based on Monte Carlo simulations, aiming to define co-existence thresholds. Several scenarios between DVB-T and LTE FDD, such as co-channel and adjacent-channel interferences, variation of distance between DVB-T and LTE devices, and different directions of link have been studied in [9]. Co-existence in terms of DVB-T access coverage loss and outage probability has been studied in [10] and [11].

In [12] and [13], simulation analyses are carried out to estimate the adjacent channel interfering effects of LTE Base Station (eNodeB) and User Equipment (UE) on DVB-T receiver systems, through the computation of the correspondent Protection Distance (PD). In both works, the power density spectrum of LTE signals has been approximated using the spectrum Block Emission Masks (BEM) reported on the ETSI recommendations ETSI TS 136 104 V8.7.0 for LTE base stations. In recent work [14], the adjacent channel interference effects of the LTE FDD DL on DVB–T home receivers (only for channel 60) have been simulated, Protection Ratio (PR) as well as the Protection Distance between eNodeB and DVB–T receivers have been calculated. The LTE Downlink Link Level Simulator tool [15] has been used to generate the transmission from a base station to two registered users in a 2×2 MIMO mode. Coexistence between digital terrestrial television and LTE network in the new spectrum allocated from mobile communications 700 MHz band has been investigated in [16].

After theoretical analysis and simulations, in order to measure the impact of the interfering LTE system on the DVB-T system, appropriate experimental laboratory test beds have been used and results have been presented in [17–23]. Results from laboratory and conducted field tests cannot be directly compared, due to different DTT standard and LTE parametrization used in those tests. For example, high dependence of the digital TV signal power to interfering signal power ratio on the bandwidth of the interfering signal and applied FEC protection has been measured in [18]. However, significant PR value offset between channel 60, and channels 58-59 has been confirmed, both in laboratory [17–23] and conducted field test.

Taking into account the power density spectrum load dependency of an eNodeB signal, the present paper aims to provide an effective contribution in the comparison of the PR and PD results obtained from the field measurements with results obtained from the simulations [12–14], and results obtained from laboratory measurements [17–23]. Conclusions from [14] on the irrelevancy of adjacent channel interference cases LTE FDD UL interfering the DVB-T at the analyzed TV receiver, and DVB-T transmitter interfering the LTE FDD DL at the considered UE receiver have been confirmed in the conducted field measurements. Therefore, the adjacent channel interference case where LTE FDD DL interferes with DVB-T at the considered TV receiver is in the main focus of this paper.

The rest of this paper is structured as follows. Section 2 introduces terminology used throughout this paper and defines the main configuration parameters for DVB-T and LTE systems used in the test bed. Section 3 describes investigated interference scenarios and presents PR results. Section 4 explains proposed interference mitigation techniques. Finally, Section 5 presents our conclusions.

2. Measurement Setup

2.1 DVB-T QoS Metrics Definition

In order to evaluate the impact of interference, parameter Carrier-to-Noise ratio (C/N) defined as the ratio between the total received DVB-T power and total received LTE signal power has been used. The minimum Carrier-to-Noise ratio that assures a stable and sufficient service quality at the DVB-T receiver defines the Protection Ratio (PR) [24]. Quality of Service can be expressed on an objective or subjective basis. In the first case, the parameters Bit Error Rate (BER), the Packet Error Rate (PER), and the Modulation Error Rate (MER) can be used [20]. Unfortunately, BER, PER, and MER values usually are not available in consumer-grade DVB-T receivers. Therefore, the subjective method has been used in the interference analysis, and the most common metrics are the Picture Failure (PF), and the Subjective Failure Points (SFP) [21]. In order to be able to compare field measurement and simulation results, Protection Distance (PD) has

also been calculated in Sec. 4 The PD parameter is defined as the minimum distance between an LTE Base Station (BS) or User Equipment (UE) antenna and a DVB-T receiving antenna in order to make the interference effects at the DVB-T front-end acceptable in terms of quality of service.

2.2 DVB-T and LTE Configuration

DVB-T is the victim system in the analyzed scenarios. The conducted, and in this paper presented, measurement results are part of a study conducted for a mobile operator. The study goal was to clarify the magnitude of expected interference issues. Therefore, parameters for the wanted signal source (Tab. 1) have been selected in order to be compliant with operating DVB-T network in the Republic of Croatia. The desired DVB-T signal has been generated using the Rohde&Schwarz SFE 100 test transmitter for channels 50 to 60. Two typical DVB-T signal level (at DVB-T receiver) scenarios have been considered: −50 dBm and −70 dBm.

The DVB-T receiver side is characterized by several parameters including receiver Selectivity (S_x) and the minimum Carrier-to-Noise ratio (C/N_{min}) required [25] for a satisfying signal reception quality, presented in Tab. 2 for 64-QAM modulation.

Due to limited time available for carrying out the planned field test measurements, the number of DVB-T receivers used in the measurements has also been limited. Therefore, two products belonging to consumer-middle quality range (Nytro Box NB-4001T, Strong SRT 8100 HD), and one product belonging to consumer-lower quality range (NotOnlyTV- Scart DVB-T REC) have been used in field tests.

LTE is the interfering system in this analysis. As well as for DVB-T system, parameters for LTE system have been selected (Tab. 3) in accordance to radio network parameters planned for the rollout of the commercial LTE

Parameter	Value
Multiple Access	OFDM
FEC	3/4
FFT points	8K
Guard interval	1/4
Signal Level [dBm]	−50 / −70
Modulation	64 QAM

Tab. 1. DVB-T parameters used in field tests.

Modulation	Code Rate	Required C/N (dB) for BER=2×10⁻⁴ after Viterbi QEF after Reed- Solomon		
		Gaussian channel	Ricean channel (F1)	Rayleigh channel (P1)
64- QAM	1/2	13.8	14.3	16.4
64- QAM	2/3	16.7	17.3	20.3
64- QAM	3/4	18.2	18.9	23.0
64- QAM	5/6	19.4	20.4	26.2
64- QAM	7/8	20.2	21.3	28.6

Tab. 2. DVB-T Minimum C/N for 64-QAM [21].

Parameter	Value
Multiple Access	OFDMA
Duplex mode	FDD
Channel bandwidth [MHz]	10
Number of resource blocks	50
Number of OFDM sub-carriers	12
Sub- carrier bandwidth [kHz]	15
Channel modulation	64 QAM
Output power [dBm]	23 – 43 (1dB step)
MIMO	2 × 2
Number of users (active test)	1
eNodeB	Ericsson 6601 (main- remote)

Tab. 3. LTE parameters used in field tests.

network in the digital dividend spectrum. Ericsson's 6601 main-remote eNodeB solution has been used, and for the antenna system RFS antenna model APXV9R20B-C [26] has been selected. Tests have been carried out in 3 consecutive 10 MHz bands starting from 791 MHz (band A: 791÷801 MHz; band B: 801÷811 MHz; band C: 811÷821 MHz).

Having in mind that the data load on an eNodeB will vary during the day, two uttermost cases have been observed:

- Idle mode (Fig. 1) represents the eNodeB mode when there are no user data in the downlink direction to be carried out. Therefore, during this state, the following has been transmitted over the air interface: Control Format Indicator (CFI) mapped to the Physical Control Format Indicator Channel (PCFICH). CFI's sole purpose is the dynamic indication of a number of OFDMA symbols reserved for control information. Downlink Control Information (DCI) with different formats basically controls all the physical layer uplink and downlink resource allocation, and it is mapped on the Physical Downlink Control Channel (PDCCH). Last are Reference Signals (RS), with their purpose to deliver the reference point for the downlink power. The number of reference signals depends on the antenna system configuration deployed [27].

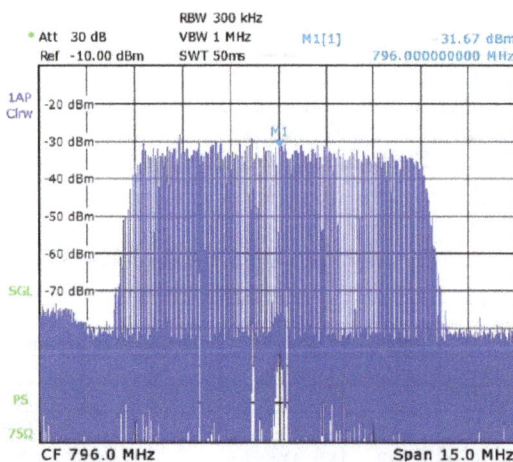

Fig. 1. LTE DL signal spectrum in idle mode (band A).

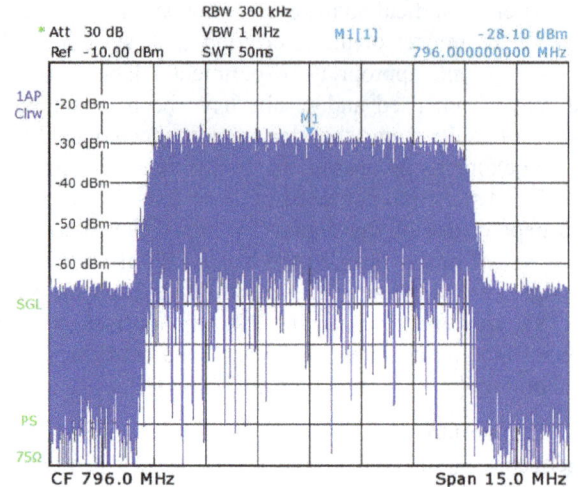

Fig. 2. LTE DL signal spectrum in dedicated mode (band A).

- Dedicated mode (Fig. 2) represents the eNodeB mode, where user data have been transmitted in the downlink direction on the Physical Downlink Shared Channel (PDSCH). Exclusive test terminal access to eNodeB has been assured by using the eNodeB service lock mode functionality (IMSI filtering). The use of the User Datagram Protocol (UDP) download service assured continuous allocation of all available Physical Resource Blocks (PRB) to the test client. For the UDP download service on the client side, a notebook with the USB data card Huawei E392U-12 (category 4) has been used. In order to assure continuous maximum allowed (set) eNodeB output power transmission, the UDP client has been placed at the LTE cell edge (RSPR ≤ –110 dBm).

The signal power has been measured using a spectrum analyzer with the following parameter setup: Span = 15 MHz, RBW = 100 kHz, VBW = 1 MHz. For the DVB-T signal power has been measured in the 8 MHz channel bandwidth, and for the LTE, in the 10 MHz channel bandwidth.

3. Scenarios and PR Results

The scope of these field measurements was to investigate all possible scenarios which could have a significant impact on DVB-T reception, to define most critical scenarios and to calculate PR and PD for those scenarios. Scenarios where LTE FDD DL is interfering the DVB-T at the considered TV receiver is considered the most critical. In order to deeply analyze this scenario, three most common DVB-T receiving antenna system configurations have been considered:

- DVB-T receiving antenna system uses outdoor mounted antenna without masthead amplifier,

- DVB-T receiving antenna system uses outdoor mounted antenna with masthead amplifier,

- DVB-T receiving antenna system uses indoor active antenna.

The last antenna system configuration which uses indoor active antenna is not considered an approved DVB-T receiving antenna system configuration by the Croatian authorities, and therefore does not have a legal right to protection, but it is expected to be used in significant number of households, as active indoor antennas are widely available on the market. In order to achieve results that will apply to real life conditions, the desired DVB-T and interfering LTE signal have been generated over the air interface. At first, measurements have been conducted in the clear optical view condition, which represents a typical Gaussian channel environment. Measurement in the Gaussian channel for the two most relevant scenarios have been presented in this paper.

In order to measure PR, the desired DVB-T signal has been fixed to –50 dBm and –70 dBm power respectively. The interfering LTE signal power has been set to minimum (23 dBm), and then increased in steps of 1 dB until the PF occurred. Then, the LTE signal power was decreased to the SFP, which has been defined as the minimum Carrier-to-Noise Ratio value that guarantees a 60-second period free from picture artifacts. Results presented in this paper correspond to the mean value of three consecutive independent measurements for every SFP point.

3.1 LTE DL Interfering DVB-T System using Outdoor Receiver Antenna without Masthead Amplifier

Figure 3 represents the scenario where, for DVB-T reception, an outdoor mounted antenna without a masthead amplifier has been used. The scenario is typical for private households in areas with a satisfying DVB-T signal level.

The PR for each DVB-T receiver has been presented with a different color in Fig. 4. PR for idle measurement has been displayed with a solid line, while PR for dedicated measurement with a dashed line. The mark on each graph line presents the case in which interference caused Picture Failure (PF). There is no mark on channels where PF has not occurred even for the highest LTE output power.

Despite the fact that the LTE DL channel signal strength for an eNodeB operating in idle mode is 9 dB lower than the dedicated mode, it can be concluded (Fig. 4)

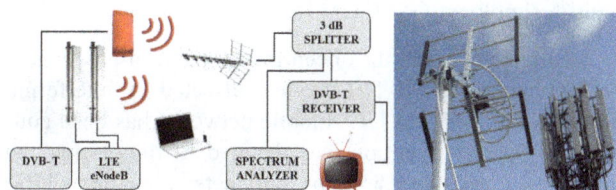

Fig. 4. Protection Ratio (PR) for DVB-T received signal –70 dBm, DVB-T antenna system without masthead amplifier, LTE idle (solid line); LTE dedicated (dashed line).

that the DVB-receivers are more sensitive to the idle LTE channel. A similar manifestation has been noticed in [17] for the LTE-UL scenario. Therefore, approximately 10 dB higher PR is needed for the idle state for eNodeB operating in band A, and approximately 5 dB for eNodeB operating in band B or C.

Visible variation in PR thresholds for various DVB-T receivers can be observed, with lower quality receivers requiring higher PR. As expected, the worst performance has been achieved while LTE was operating in band A, since this is the band closest to the DVB-T operating band. PR increases for channels 50, 51, and 52 due to the image frequency issue, which is typical for superheterodyne DVB-T receivers. Superheterodyne tuners are susceptible to image channel interference when the interferer appears nine channels above the wanted channel. Silicon tuners have a low, or zero, intermediate frequency and do not suffer from the same $N + 9$ image problems as superheterodyne receivers [28].

In PR measurement for the DVB-T received signal –50 dBm (Fig. 5), the interfering LTE signal was not strong enough to provoke PF for consumer-middle quality range receivers (NB 40001T, SRT 8100 HD). Therefore, PF only for consumer-low quality range receiver (NotOnlyTV-Scart DVB-T REC) has been recorded.

In order to increase the interfering signal level, measurements have been continued in the lab environment

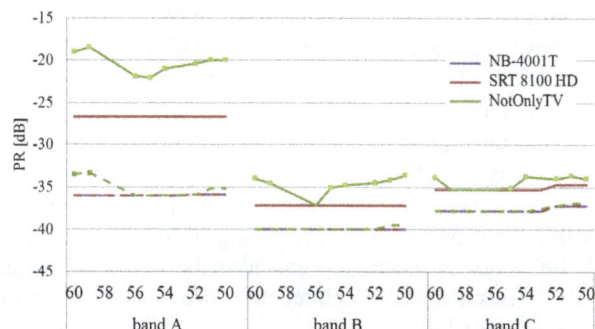

Fig. 3. DVB-T receiver system without masthead amplifier, schematic (left) and actual test setup (right).

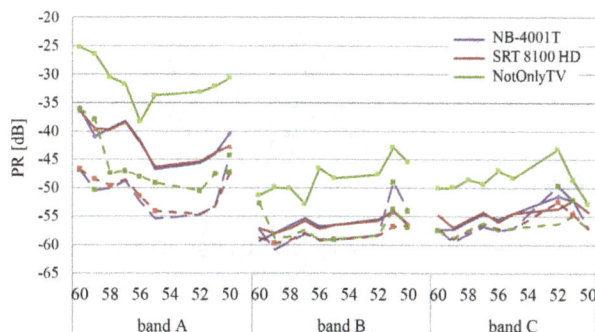

Fig. 5. Protection Ratio (PR) for DVB-T received signal –50 dBm, DVB-T antenna system without masthead amplifier, LTE idle (solid line); LTE dedicated (dashed line).

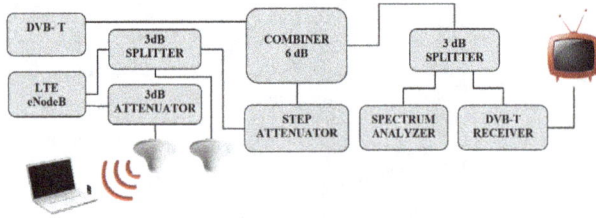

Fig. 6. Measurement setup for DVB-T received signal –50 dBm in lab environment.

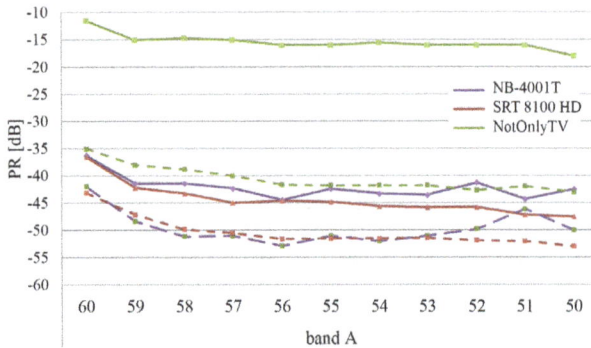

Fig. 8. DVB-T receiver system with masthead amplifier, schematic (left) and actual test setup (right).

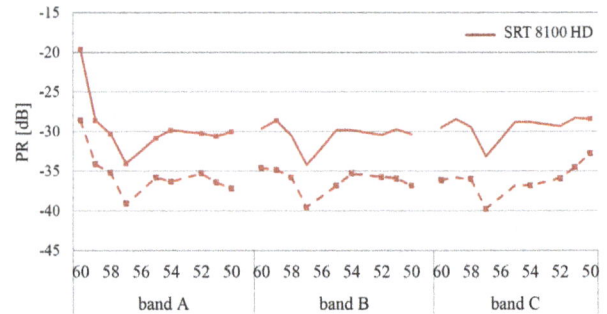

Fig. 7. Protection Ratio (PR) for DVB-T received signal –50 dBm, DVB-T antenna system without masthead amplifier, LTE idle (solid line); LTE dedicated (dashed line) - laboratory measurements.

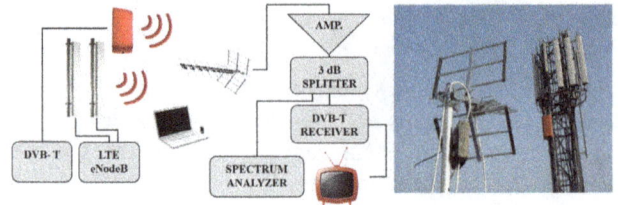

Fig. 9. Protection Ratio (PR) for DVB-T received signal –70 dBm, DVB-T antenna system with masthead amplifier, LTE idle (solid line); LTE dedicated (dashed line).

(Fig. 6), and results for PR have been presented in Fig. 7. Here only LTE DL in band A has been considered, as, according to previous measurements, in that band the largest PR is expected.

PR results (Figs. 4, 5, 7) for the dedicated measurement mode are compliant to results obtained in [9] and [24].

3.2 LTE DL Interfering DVB-T System using Outdoor Receiver Antenna with Masthead Amplifier

Figure 8 represents the scenario where the DVB-T receiving antenna system uses an outdoor mounted antenna with a masthead amplifier [29]. The scenario is typical for private households in areas where DVB-T signal level or quality is low, and on multi-tenant buildings regardless of the DVB-T signal quality condition. For this test scenario, wideband (channels 21÷69) masthead amplifier with 22 dB gain has been used. The wideband amplifier amplifies the desired DVB-T, but also the interfering LTE signal. Due to the increased signal level, the amplifier will operate in its nonlinear range, and it will generate intermodulation products. A mixture of amplified desired signal together with interference and intermodulation products will reach the DVB-T input port [21]. Increased signal level will reduce DVB-T receiver selectivity, and also the nonlinear range could be reached. All of the mentioned will cause a signal degradation and PF.

In this scenario (Fig. 9), only the SRT 8100 HD receiver has been tested, because in the previous measure-

ment both consumer-middle quality range receivers behaved similarly, and much worse behavior has been observed for the consumer-low quality range receiver. Comparing Fig. 4 and Fig. 9, significant PR increase can be observed for the scenario when the masthead amplifier has been used. For this scenario, it is also true that the PR is higher for channels 59 and 60.

Once again, the worst performance can be observed for the eNodeB operating in band A and an idle LTE DL interfering signal presents the worst case. PF occurred for idle and dedicated case while eNodeB was operating in band A. For eNodeB operating in bands B and C, PF occurred only in idle case.

3.3 LTE DL and GSM900 DL Interfering DVB-T System using Outdoor Receiver Antenna with Masthead Amplifier

During measurements, another important scenario has been identified. If the DVB-T antenna with a masthead amplifier receives simultaneous signals from interfering LTE FDD DL and GSM900 DL, a masthead amplifier overload occurs, causing PF. In general, due to the high rental and building expenses, mobile operator sites are multi-technology (GSM/ UMTS/ LTE).

Similar behavior has been investigated in [18] where performance of DVB-T/H services affected by interfering products of GSM and LTE mobile networks has been considered. Special attention was devoted to monitoring the interference with transmission parameter signaling (TPS) carriers, used in DVB-T/H system as reference information for the receiver [25].

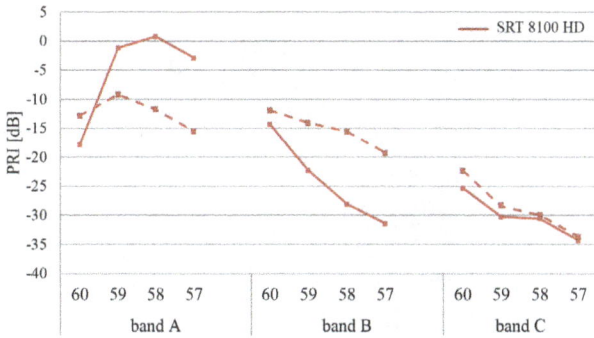

Fig. 10. Protection Ratio (PR) for DVB-T received signal –70 dBm, DVB-T antenna system with masthead amplifier, simultaneous operation of GMS 900 and LTE idle (solid line) and GSM900 and LTE dedicated (dashed line).

Protection ratios in this scenario (Fig. 10) are extremely high, especially for LTE transmitting in band A and B. Another fact which differentiates this case from all previous scenarios is that greater interference is caused by the LTE signal in dedicated mode. This behavior can be explained by an overload effect caused by simultaneous presence of LTE and GSM signals, which has a major impact on the masthead amplifier in this case.

4. Interference Mitigation Techniques

In the process of DVB-T planning in a certain geographical area, the operator needs to fulfill minimum electric field strength E, expressed in V/m (dBV/m), or more often μV/m (dBμV/m). Electric field strength in free space produces the flux density S. The available power at the receiver P_r is the product of the effective area of the receiving antenna A_{ef}, and the flux density S at the location of the antenna:

$$P_r = S\,A_{ef} = \left(\frac{E^2}{\eta}\right)\cdot\left(G_r\frac{\lambda^2}{4\pi}\right) = \frac{G_r c^2 E^2}{4\pi\eta f^2}. \qquad (1)$$

The second equation in (1) is obtained by substituting the flux density and effective area of the antenna according to common expressions [30], while in the third one the signal wavelength λ [m] is expressed by the signal frequency f [Hz] and the speed of light c [m/s]. Here η [Ω] is the free space intrinsic impedance (377 Ω) and G_r [dBi] is the receiving antenna gain.

In order to avoid degradation of the DVB-T signal reception and finally picture failure (PF), the following protection ratio (PR) criterion must be fulfilled:

$$\frac{P_{r,DVB-T}}{P_{r,interference}} \geq PR. \qquad (2)$$

In (2) $P_{r,DVB-T}$ is the received power of the desired television signal, while $P_{r,interference}$ is the total received power of all interfering signals (e.g. LTE, GSM, or any other). Both received powers P_r in (2) are expressed in linear power units (e.g. mW or μW) while PR is dimensionless quantity.

The alternative form of (2) in logarithmic units (e.g. dBm, dB) can be found in [18].

With the aim of making it easier to perceive measured PR thresholds, the results have been represented in terms of PD, which is defined as the minimum required distance between interfering LTE DL antenna and receiving DVB-T antenna. The worst case scenario with LTE and DVB-T antennas (APXV90R20B-C [26], DTM-27 [34]) with main lobes looking one to the other and with clear line-of-sight in between has been assumed. Therefore, the Friis transmission equation (3) can be used:

$$P_r = \frac{P_t G_t G_r \lambda^2}{(4\pi d)^2} \qquad (3)$$

where P_t is the transmit power, G_t [dBi] is transmitting LTE antenna gain, G_r [dBi] receiving antenna gain and d [m] is the distance between transmitting and receiving antennas. For calculation purposes, it has been assumed that the eNodeB transmits 2×20 W (2×2 MIMO), which together with G_t, gives 61.8 dBm of EIRP. Feeder loss has been ignored due to the main-remote eNodeB configuration, because remote radio unit has been installed on the top of the mast, close to the antenna. An additional 3 dB polarization loss has been taken into account.

An average PR (according to Figs 4, 5, 7) of –40 dB has been selected. DVB-T signal strength of 55 dBμV/m (Tab. 4), for the selected receiving DVB-T antenna is equal to –70 dBm signal power at the receiver. For the selected parameters and using (3), minimum distance d between interfering LTE DL and receiving DVB-T antenna is 2668 m. A DVB-T signal strength of 75 dBμV/m (–50 dBm) reduces the minimum distance to 270 m. However, in urban areas, a 270 m radius around the LTE site without at least a few DVB-T receiving antenna is almost unthinkable. Therefore, if criterion (2) cannot be fulfilled for the specific $P_{r,DVB-T}$, measures have to be applied in order to attenuate $P_{r,interference}$. Consequently, the required

DVB-T field strength					55 [dBμV/m]	
DVB-T signal strength at the receiver (G_r= 10 dB)					–70 [dBm]	
eNodeB output power					2×20 [W]	
LTE antenna gain (G_t)					15.8 [dBi]	
Polarization Loss					3 [dB]	
Additional interference attenuation (filtering) [dB]	0	10	20	30	40	50
PR [dB]	PD [m]					
–20	26678	8436	2668	844	267	84
–25	15002	4744	1500	474	150	47
–30	8436	2668	844	267	84	27
–35	4744	1500	474	150	47	15
–40	2668	844	267	84	27	8
–45	1500	474	150	47	15	5
–50	844	267	84	27	8	3
–55	474	150	47	15	5	2
–60	267	84	27	8	3	1

Tab. 4. Protection distance (PD).

additional interference attenuation (Tab. 4) has been calculated for the specific PR in order to satisfy the PD criteria. For example, required PR is –45 dB. The distance between LTE interfering antenna and DVB-T receiving antenna is 200 m, and cannot be optimized. Therefore, in order to satisfy minimum PD criteria, a filter with 20 dB attenuation should be applied in the DVB-T antenna system.

Several measures have been considered and proposed in CEPT reports 21, 22 [31], 23 and 30. Some of these relate to the way in which the LTE system already works (e.g. power control), and some require specific measures to be implemented at the LTE base station or DVB-T receiver side. Measures have been briefly discussed in [31]. However, applying antenna separation, coordination of antenna azimuths and tilts, and use of filtering have been identified as the most practical and cost effective measures to implement, and they have been analyzed in detail in studies [32] and [33].

4.1 Applying Antenna Separation and Coordination of Antenna Azimuth and Tilt

In this analysis, two typical commercial antenna products have been used: for the LTE system, antenna APXV90R20B-C [26], and for the DVB-T receiving system, antenna DTM-27 F [34].

The coordination of DVB-T and LTE antenna system can be considered in the horizontal and vertical plane. Theoretically, interference in the vertical plane can be reduced by positioning the DVB-T receiving antenna in the second null of the interfering LTE antenna.

From Fig. 11 a), interference suppression by positioning DVB-T antenna in the second null of interfering LTE antenna of 37 dB has been achieved, and an additional 3 dB interference suppression are coming from the DVB-T antenna vertical pattern. In Fig. 11 b), an additional 20 dB have been achieved by the DVB-T antenna vertical pattern due to the high back lobe suppression.

In the horizontal plane, the idea is to avoid placing the DVB-T receiving antenna in the LTE antenna horizontal main lobe. Since the LTE site has typically 3 sectors covering 360° (Fig. 12), sufficient interference suppression

Fig. 12. Interference suppression in horizontal plane, suppression of 30 dB achieved in a) and suppression of 15 dB achieved in b) and c).

cannot be accomplished in practice. In the best case (Fig. 12 a), 30 dB of interference suppression can be reached, and in the two other cases, only 15 dB of interference suppression should be expected (Fig. 12 b, Fig. 12 c).

The proposed theoretical solution does not apply in practice because of the fact that mobile operators, due to optimization needs, often change their antennas azimuth and tilt. Furthermore, on a multitenant building, there is often more than one DVB-T receiving antenna system, which makes the proposed solutions impossible to apply.

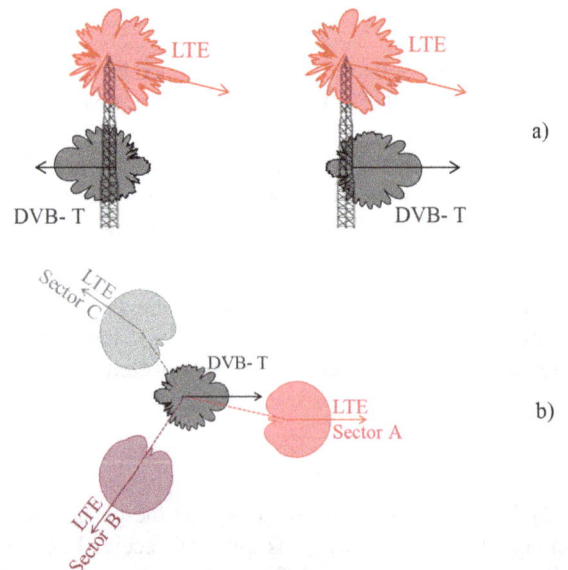

Fig. 13. Practical interference suppression implementation in vertical a) and horizontal b) plane.

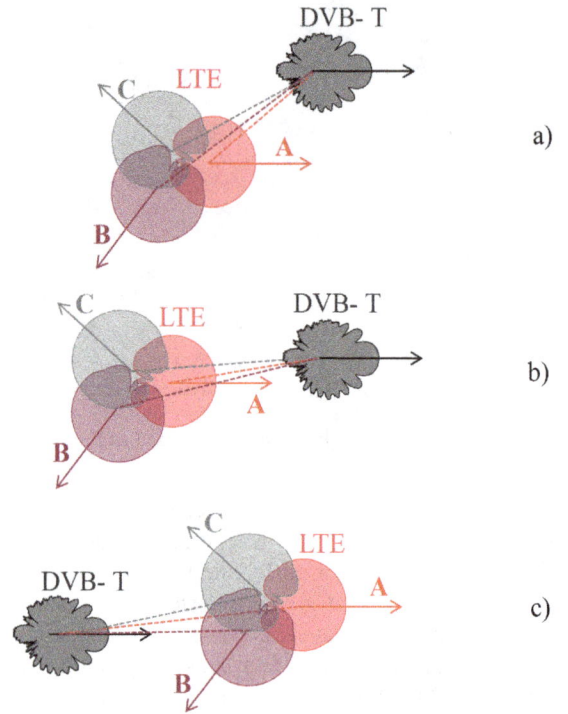

Fig. 11. Interference suppression in vertical plane, exploiting the second null of the interfering LTE antenna in a) and high DVB-T antenna back lobe suppression in b).

However, modification of this proposed solution does have a strong practical implementation value (Fig. 13). Firstly, by collocating LTE and DVB-T antenna on the same mast (Fig. 13 a), 35-40 dB suppression thanks to the LTE antenna vertical pattern, and in addition 30 dB thanks to the DVB-T antenna vertical pattern can be achieved [32].

On multitenant buildings, mobile operator masts are usually placed in the middle of the building, for example, on the elevator housing. Sufficient isolation in the horizontal plane can be achieved (Fig. 13 b) by moving the LTE antenna sectors toward the corner of those building roofs. The number and orientation of DVB-T antennas do not affect the result. Furthermore, the mobile operator will achieve better coverage in the area close to site. The proposed solutions are cost effective too.

4.2 Use of Filtering

Filtering for downlink or power amplifier linearization techniques can be used to reduce unwanted emissions from a base station. Similarly, filtering at the DVB-T receiver side can be applied in order to reduce in-band interference from the base station. The second method has been identified as the most practical solution. Therefore, a study on the deviation of the filter response characteristics due to (environmental) temperature variation has been conducted [33]. According to the results of that study, special attention needs to be paid in the filtering product selection.

5. Conclusion

In the present paper, field measurements have been conducted for the fixed DVB-T reception environment to determine the most relevant interference scenarios and to quantify the Protection Ratio for each of these.

It has to be emphasized that DVB-T reception is more sensitive to eNodeB operating in idle mode, which could be related to the way the Automatic Gain Control (AGC) of the DVB-T receiver works. Significant PR increase occurs when a masthead amplifier has been used, and for interference generated from eNodeB and GSM900 base station simultaneously. A large number of households is still using masthead amplifiers and most mobile operator sites will be multi-technology (GSM/ UMTS/ LTE) in the near future. Therefore, this effect has to be seriously taken into account, and further investigated in future research.

Applying antenna separation, coordination of antenna azimuths or tilts and use of filtering have been elaborated and identified as the most practical and cost effective measures, and an effective deployment method has been proposed.

References

[1] ITU. *Final Acts of the World Radiocommunication Conference (WRC-07)*. World Radiocommunication Conference, Geneva (Switzerland), 2007, p. 478–479. ISBN: 9261122019

[2] EUROPEAN COMMISSION - INFORMATION SOCIETY AND MEDIA DIRECTORATE-GENERAL, RADIO SPECTRUM COMMITTEE. *Opinion of the RSC Pursuant to Article 4.2 of Radio Spectrum Decision 676/2002/EC*. 6 pages. [Online] Cited 2015-06-04. Available at: http://ec.europa.eu/information_society/newsroom/image/3_april_2008_5769_draft_7481.pdf

[3] TEKOVIC, A., SIMAC, G., SAKIC, K. LTE downlink system performance measurement with intersystem interference caused by DVB-T signal. In *Proceedings of the 54th International Symposium ELMAR*. Zadar (Croatia), 2012, p. 255–258. ISSN 1334-2630

[4] SAKIC, K., GOSTA, M., GRGIC, S. Cross-border interference between broadcasting and mobile services. In *Proceedings of the 51st International Symposium ELMAR*. Zadar (Croatia), 2009, p. 229–232. ISBN 978-953-7044-10-7

[5] TEKOVIĆ, A. LTE in Digital Dividend deployment challenges-DVB-C2 case. In *Proceedings of the 54th International Symposium ELMAR*, Zadar (Croatia), 2012, p. 251–254.

[6] KANG, D. H., ZHIDKOV, S. V., CHOI, H. J. An adaptive detection and suppression of co-channel interference in DVB-T/H system. *IEEE Transactions on Consumer Electronics*, 2010, vol. 56, no. 3, p. 1320–1327. DOI: 10.1109/TCE.2010.5606265

[7] GUIDOTTI, A., GUIDUCCI, D., BARBIROLI, M., et al. Coexistence and mutual interference between mobile and broadcasting systems. In *Proceedings of the IEEE 73rd Vehicular Technology*, Budapest (Hungary), 2011, p. 1–5. DOI: 10.1109/VETECS.2011.5956540

[8] SAKIC, K., GRGIC, S. The influence of the LTE system on DVB-T reception. In *Proceedings of the 52nd International Symposium ELMAR*. Zadar (Croatia), 2010, p. 235–238.

[9] SETIAWAN, D., GUNAWAN, D., SIRAT, D. Interference analysis of guard band and geographical separation between DVB-T and E-UTRA in Digital Dividend UHF band. In *Proceedings of the International Conference on Instrumentation, Communications, Information Technology, and Biomedical Engineering (ICICI-BME)*. 2009, p. 1–6. DOI: 10.1109/ICICI-BME.2009.5417258

[10] CHEN, Y. X., XIAO, L., SUN, Y. Interference simulation from LTE to digital terrestrial television. In *Proceedings of the 7th Wireless Communications, Networking and Mobile Computing Conference (WiCOM)*. Wuhan (China), 2011, p. 1–4. DOI: 10.1109/wicom.2011.6040068

[11] DAE-HEE KIM, SEONG-JUN OH, JUNGSOO WOO. Coexistence analysis between IMT system and DTV system in the 700MHz band. In *Proceedings of the International Conference ICT Convergence (ICTC)*. Jeju (South Korea), 2012, p. 284–288. DOI: 10.1109/ICTC.2012.6386840

[12] ALOISI, A., CELIDONIO, M., PULCINI, L., et al. Experimental study on protection distances between LTE and DVB-T stations operating in adjacent UHF frequency bands. In *Proceedings of the Wireless Telecommunications Symposium (WTS)*. New York City (USA), 2011, p. 1–7. DOI: 10.1109/WTS.2011.5960859

[13] CELIDONIO, M., PULCINI, L., RUFINI, A. LTE and DVB-T coexistence: A simulation study in the UHF frequency band. *Journal of Communication and Computer*, 2012, vol. 9, no. 4, p. 444–455.

[14] BARUFFA, G., FEMMINELLA, M., MARIANI, F., et al. Protection ratio and antenna separation for DVB—T/LTE coexistence issues. *IEEE Communications Letters*, 2013, vol. 17, no. 8, p. 1588 to 1591. DOI: 10.1109/LCOMM.2013.070113.130887

[15] MEHLFUHRER, C., WRULICH, M., IKUNO, J. C., et al. Simulating the Long Term Evolution physical layer. In *Proceedings of the 17th European Signal Processing Conference* (EUSIPCO), 2009, p. 1471–1478.

[16] FUENTES, M., GARCIA-PARDO, C., GARRO, E., et al. Coexistence of digital terrestrial television and next generation cellular networks in the 700 MHz band. *IEEE Wireless Communications*, 2014, vol. 21, no. 6, p. 63–69. DOI: 10.1109/MWC.2014.7000973

[17] RIBADENEIRA-RAMIREZ, J., MARTINEZ, G., GOMEZ-BARQUERO, D., et al. Interference analysis between Digital Terrestrial Television (DTT) and 4G LTE mobile networks in the digital dividend bands. *IEEE Transactions on Broadcasting*, 2016, vol. 62, no. 1, p. 24–34. DOI: 10.1109/TBC.2015.2492465

[18] POLAK, L., KALLER, O., KLOZAR, L., et al. Mobile communication networks and digital television broadcasting systems in the same frequency bands: Advanced co-existence scenarios. *Radioengineering*, 2014, vol. 23, no. 1, p. 375–386.

[19] POLAK, L., KALLER, O., KLOZAR, L., et al. Influence of mobile network interfering products on DVB-T/H broadcasting services. In *Proceedings of the Wireless Days (IFIP)*. 2012, p. 1–5. DOI: 10.1109/WD.2012.6402860

[20] KRISTEL, J., POLAK, L., KRATOCHVIL, T. Co-channel coexistence between DVB-T/H and LTE standards in a shared frequency band. In *Proceedings of the 25th International Conference RADIOELEKTRONIKA*. Pardubice (Czech Rep.), 2015, p. 184–190. DOI: 10.1109/RADIOELEK.2015.7129004

[21] De VITA, A., MILANESIO, D., SACCO, B., et al. Assessment of interference to the DTT service generated by LTE signals on existing head amplifiers of collective distribution systems: A real case study. *IEEE Transactions on Broadcasting*, 2014, vol. 60, no. 2, p. 420–429. DOI: 10.1109/TBC.2014.2321677

[22] POLAK, L., KALLER, O., KLOZAR, L., et al. Coexistence between DVB-T/T2 and LTE standards in common frequency bands. *Wireless Personal Communication*, 2016, vol. 88, no. 3, p. 669–684. DOI: 10.1007/s11277-016-3191-2

[23] POLAK, L., KALLER, O., KLOZAR, L., et al. Study of coexistence between indoor LTE femtocell and outdoor-to-indoor DVB-T2-Lite reception in a shared frequency band. *EURASIP Journal of Wireless Communication and Networking*, 2015, no. 114, 14 p. DOI: 10.1186/s13638-015-0338-x

[24] ELECTRONIC COMMUNICATIONS COMMITTEE (ECC). *Measurements on the Performance of DVB–T Receivers in the Presence of Interference from the Mobile Service- Especially from LTE (ECC Report 148)*. 32 pages. [Online] Cited 2015-06-04. Available at: http://www.erodocdb.dk/Docs/doc98/official/pdf/ECCREP148.pdf

[25] EUROPEAN TELECOMMUNICATIONS STANDARDS INSTITUTE (ETSI). *Digital Video Broadcasting (DVB); Framing Structure, Channel Coding and Modulation for Digital Terrestrial Television (DVB-T) (Recommendation ETSI EN 300 744 V1.6.1)*. 66 pages. [Online] Cited 2015-06-04. Available at: http://www.etsi.org/deliver/etsi_en/300700_300799/300744/01.06.01_60/en_300744v010601p.pdf

[26] RADIO FREQUENCY SYSTEMS (RFS). *APXV9R20B-C Antenna Model (datasheet)*. 2 pages [Online] Cited 2015-06-04. Available at: http://www.rfsworld.com/websearch/Datasheets/pdf/?q=APXV9R20B-C

[27] HOLMA, H., TOSKALA, A. *LTE for UMTS OFDMA and SC-FDMA Based Radio Access*. 1st ed. John Wiley & Sons Ltd., 2009. ISBN: 9780470994016

[28] PARKER, I., MUNDY, S. *Assessment of LTE 800 MHz Base Station Interference into DTT Receivers (ERA Technology for OFCOM, Report 2011-0351)*. 41 pages. [Online] Cited 2015-06-04. Available at: http://stakeholders.ofcom.org.uk/binaries/consultations/dtt/annexes/Ite-800-mhz.pdf

[29] FTE MAXIMAL *LG 222 Reception Masthead Amplifier (datasheet)*. 2 pages [Online] Cited 2015-07-17. Available at: http://ftemaximal.com/images/files/soporte-servicios/Documentacion-tecnica/MALG227.pdf

[30] BALANIS, A. C., *Antenna Theory: Analysis and Design*. 3rd ed. New Jersey (USA): John Wiley & Sons Ltd., 2005. ISBN 978-0-471-66782-7

[31] EUROPEAN CONFERENCE OF POSTAL AND TELECOMMUNICATIONS ADMINISTRATIONS (CEPT). *Technical Considerations Regarding Harmonization Options for the Digital Dividend (CEPT Report 22)*. 52 pages. [Online] Cited 2015-06-04. Available at: http://www.erodocdb.dk/docs/doc98/official/pdf/CEPTRep022.pdf

[32] BONEFAČIĆ, D., ŠIŠUL, G. *Mobile Communications System (LTE) in the Frequency Range of the Digital Dividend with the Digital Television System (DVB-T) Interference Analysis (Report 2012-4500243244)*. Faculty of Electrical Engineering and Computing for VIPnet Ltd., 97 pages. (in Croatian)

[33] BONEFAČIĆ, D. *Testing of Filters for Interference Suppression from LTE System into DVB-T System. (Report 2013- 4500267607)*. Faculty of Electrical Engineering and Computing for VIPnet Ltd., 44 pages. (in Croatian)

[34] ISKRA. *DTM-27 F Antenna Model* (datasheet). 1 page. [Online] Cited 2015-06-04. Available at: http: http://www.iskra.eu/

About the Authors...

Alberto TEKOVIĆ received his Mag. Ing. El. and M. Sc. in Electrical Engineering from the Faculty of Electrical Engineering, University of Zagreb, Croatia in 2000 and 2005, respectively. He is currently employed as a Radio Network Optimization Expert in the mobile operator VIPnet Ltd. He is also engaged as a senior lecturer at the University College for Applied Computer Engineering and Polytechnic of Zagreb, and teaching activities include several subjects on mobile networks: Mobile Radiocommunications and Wireless Computer Networks. He published one book and more than 10 scientific papers in journals and conference proceedings. From 2012 he organizes a special session related to modern mobile networks, as a part of the International Symposium ELMAR. In 2015 he served as a committee member on the Microwave and Radio Electronics Week – MAREW. His research interests are in the field of wireless communications, mobile network performance optimization and radio propagation.

Davor BONEFAČIĆ received his Dipl. Ing., Mr. Sc. and Dr. Sc. degrees in Electrical Engineering from the University of Zagreb, Faculty of Electrical Engineering and Computing (FER), Zagreb, Croatia, in 1993, 1996 and 2000, respectively. In 1993 he joined the Department of Wireless Communications at the Faculty of Electrical Engineering and Computing in Zagreb. Today he is Full Professor at the same Department. In 1996 he was a visiting researcher at the Third University of Rome, Rome, Italy. His research interests are in the field of small and integrated antennas, textile antennas and waveguiding structures, and on-body communications. He is currently leading a research project on textile antennas financed by the Croatian Science Foundation. He published one book chapter and more than 90 scientific papers in journals and conference proceedings. In the 2003-2014 period, he served as a member of the editorial board of Radioengineering journal. He was also the editor of eleven international conference proceedings in printed and electronic form. He is co-author of one university textbook. His teaching activity includes several subjects on microwave engineering, RF systems, antennas and radars at graduate and doctoral studies. His professional activities are in the field of EMC, EMI estimation and suppression, EM field measurement and estimation of human exposure to EM fields. He authored and co-authored more than 35 professional and technical papers for government entities and companies. He is a member of the Croatian Academy of Engineering (HATZ) and he was elected Secretary of the Dept of Communication Systems of the Academy for the 2013-2017 term. He is a senior member of IEEE, chair of the MTT Chapter of the IEEE Croatia Section for the term 2011-2015 and vice-chair of the joint AES/GRS Chapter for the term 2015-2017. He received the silver plaque "J. Lončar" from FER for outstanding master thesis in 1996. In 2009 he received the "Rikard Podhorsky" award from the Croatian Academy of Engineering (HATZ) for excellence in scientific and professional achievements.

Gordan ŠIŠUL received his B.Sc., M.Sc. and Ph.D. in Electrical Engineering from the Faculty of Electrical Engineering, University of Zagreb, Croatia in 1996, 2000 and 2004, respectively. He is currently employed as an Associate Professor at the same Faculty. His academic interests include wireless communications, signal processing applications in communications, modulation techniques and radio propagation.

Robert NAĐ received his Dipl.Ing. Mr.Sc. and Dr.Sc. degrees in Electrical Engineering from the University of Zagreb, Faculty of Electrical Engineering and Computing (FER), Zagreb, Croatia. He is currently employed as a Full Professor at the same Faculty. He is co-author of two university textbooks. His teaching activity includes several subjects on modern mobile networks at graduate and doctoral studies: Mobile Communications, Mobile Systems Planning, Signal Equalization in Wireless Transmission and Spread Spectrum Systems. His academic interests are related to the analysis of physical phenomena in radio channel and its impact on modulation and radio propagation and wireless communications in general.

5

Investigation of Unequal Planar Wireless Electricity Device for Efficient Wireless Power Transfer

Mohd Hidir MOHD SALLEH[1], Norhudah SEMAN[1], Dyg Norkhairunnisa ABANG ZAIDEL[2], Akaa Agbaeze ETENG[1]

[1]Wireless Communication Centre (WCC), Universiti Teknologi Malaysia, 81310, Johor, Malaysia
[2]Dept. of Electrical and Electronics Engineering, Faculty of Engineering, Universiti Malaysia Sarawak, 94300 Kota Samarahan, Sarawak, Malaysia

eday_89@yahoo.com.my, huda@fke.utm.my, azdnorkhairunnisa@unimas.my, akaaet@yahoo.com

Abstract. *This article focuses on the design and investigation of a pair of unequally sized wireless electricity (Witricity) devices that are equipped with integrated planar coil strips. The proposed pair of devices consists of two different square-shaped resonator sizes of 120 mm × 120 mm and 80 mm × 80 mm, acting as a transmitter and receiver, respectively. The devices are designed, simulated and optimized using the CST Microwave Studio software prior to being fabricated and verified using a vector network analyzer (VNA). The surface current results of the coupled devices indicate a good current density at 10 mm to 30 mm distance range. This good current density demonstrates that the coupled devices' surface has more electric current per unit area, which leads to a good performance up to 30 mm range. Hence, the results also reveal good coupling efficiency between the coupled devices, which is approximately 54.5 % at up to a 30 mm distance, with both devices axially aligned. In addition, a coupling efficiency of 50 % is achieved when a maximum lateral misalignment (LM) of 10 mm, and a varied angular misalignment (AM) from 0° to 40° are implemented to the proposed device.*

Keywords

Coupling efficiency, magnetic resonance coupling, misalignment, wireless power transfer, Witricity

1. Introduction

Wireless power transfer (WPT) has been a subject of interest among researchers since the feasibility of transmitting power wirelessly was reported by Nikolai Tesla in 1900 [1], [2]. This research interest is further stimulated by the evolution of wireless technologies from fourth generation (4G) to fifth generation (5G) technologies, expected to become mainstream in the year 2020. Compared to 4G, 5G technologies are more focused on the device-to-device communication (D2D) and wearable devices [3]. To facilitate wearable devices, various designs of WPT systems have been introduced through the use of such methods as strong magnetic resonance coupling, also known as Wireless Electricity (Witricity) [4], [5], conventional inductive and capacitive coupling [6], [7] and rectifying antennas (rectennas) [8]. The strong magnetic resonance technique uses two coupled magnetized objects within a non-radiative near-field region at megahertz (MHz) frequencies. As reported in [4], the most efficient way to realize this technique is to couple two identical designs on a similarly resonant frequency. The majority of commercial modern gadgets, including smart phones, tablets, notebooks, and pacemakers, possess the capacity to utilize Witricity. These devices are usually designed to be compact in size. While Witricity transmitter sizes are typically not critically constrained, it is crucial for the Witricity receiver to be as compact as possible in order to be placed in the intended device to be powered. However, there is a drawback when designing a compact Witricity device. The range of transferred power firstly depends on Witricity size, which corresponds to the wavelength. Thus, smaller Witricity device tends to transfer less efficient power compared to larger Witricity device. Therefore, the feasibility of coupling unequally sized transmitter and receiver Witricity devices is shown in this article. Nevertheless, the level of transferred power is not limited by the size of the device, but also influenced by the transfer distance and alignment of the transmitter and receiver as presented investigation.

In a bid to achieve compact designs, planar-type Witricity devices have been proposed in [9–12]. These designs use flat single or multilayer coils, with different conductor levies, to create a homogenous magnetic field between transmitter and receiver [9]. In [10], the authors have proposed a Witricity design with equal transmitter and receiver (162×162 mm^2) sizes by using a plastic lamella board, and investigate the performance of the Witricity devices when angular and lateral misalignments occur. Copper tape is used as the inductor coil and capacitor plates. For the single coil, copper wire has been attached to the capacitor plates. This single coil provides high inductance due to its loop's large diameter and thick cross-sectional area. Another Witricity design has been proposed in [11], which is a quarter of the size (80×80 mm^2) of the design reported in [10]. Double-sided copper coated FR-4 substrate has been used in this more compact Witricity device design, thereby eliminating the need for thick copper wire. Similar to the designs proposed in [10] and [11], an equal-sized transmitter and receiver design has been proposed in [12]. The design, which is stated to be more compact in size (40 × 40 mm²), however, replaces the cheaper FR-4 substrate with a high performance and pricy Rogers RO3010 substrate.

Thus, to overcome the problem of the high cost of the Rogers RO3010 substrate, this work proposes the coupling of two with unequal-sized Witricity devices. The proposed arrangement uses a smaller receiver device, and provides a comparable performance to a larger design proposed in [10], yet utilizing low cost Flame Retardant 4 (FR-4) as the substrate. Furthermore, given the small form factor requirements of modern gadgets, the proposed design of the Witricity device has an integrated planar coil in the resonant unequal-sized transmitter and receiver. This article also investigates the impact of misalignments between the coupled unequal-sized Witricity devices.

2. Design of Witricity Devices

In this article, a larger Witricity device is proposed to function as a transmitter, while a smaller one functions as a receiver. A larger-sized transmitter is needed to maximize the transferable energy distance, as the distance is always proportional to its coil size. The reduction of transmitter size will lead to the reduction of maximum transferable energy distance. The receiver is purposely designed to have smaller dimension compared to the transmitter in order to suit today's modern devices, which typically compact in size. Figure 1 depicts a transmission scheme from the first port to the second port, while Figure 2 shows the arrangement of the proposed Witricity device in three-dimensional (3D) simulation tool of CST Microwave Design Studio, as well as its fabricated prototype.

In Fig. 1 and 2, each transmitter and receiver that consists of two conducting layers: top and bottom are separated by the air gap. The top layer consists of a rectangular spiral coil inductor (Tx and Rx coil), while the bottom layer has four capacitor plates that are diagonally structured and attached to a single turn coil (Tx and Rx loop). The chosen Flame Retardant 4 (FR-4) substrate, with a relative permittivity of 4.41, loss tangent of 0.025, 1.6 mm thickness, and 35 μm thickness of copper coating, is sandwiched between these two conducting layers.

Both transmitter and receiver need to be operated on a similar operating frequency to allow the strongly coupled magnetic resonance technique to function properly. Based on [11], the parameters such as dimension of capacitor plate and the number of turns of spiral coil can be tuned for the device to operate at a specific frequency. A larger design, which commonly tends to have a lower frequency, can operate at a higher resonance frequency by tuning the above-mentioned parameters. The operating resonance fre-

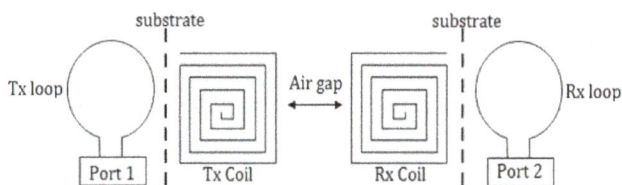

Fig. 1. Witricity device transmission scheme.

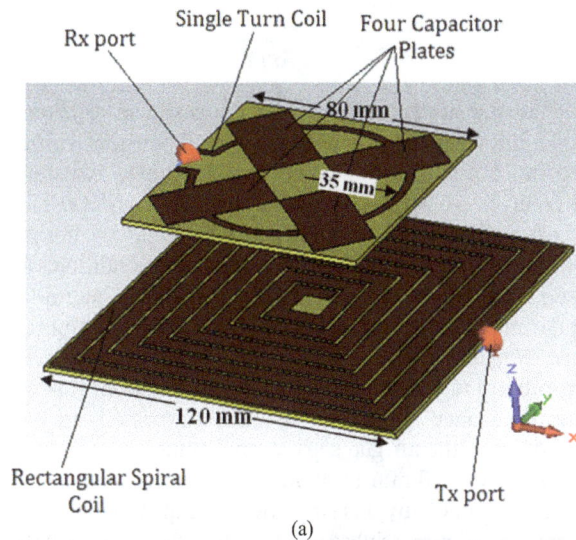

(a)

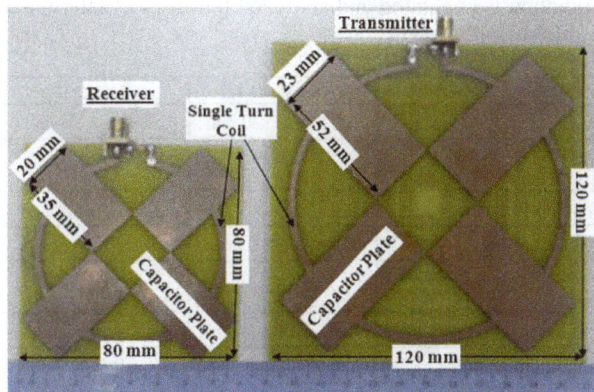

(b)

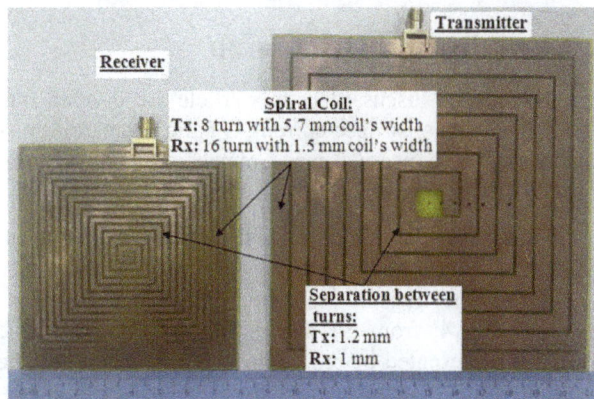

(c)

Fig. 2. The unequally sized Witricity device design: (a) the generated 3D layout in CST Microwave Design Studio, (b) the bottom view, and (c) the top view of the fabricated prototype.

quency denoted by f_r [11], strongly depends on the capacitance C and the inductance L value, which is described in (1). C depends on the dimensions of the capacitor plates and the thickness of the substrate, and L relies on the spiral coils, which consequently influence the device's induction strength. In addition, the capacitor plates store energy from the single turn coil prior to the transferal of energy to the spiral coils located at the top layer of the device.

$$f_r = \frac{1}{2\pi\sqrt{LC}} . \tag{1}$$

The transmitter and receiver are designed to resonate at a similar frequency according to the dimensions summarized in Tab. 1. The individual and decoupled transmitter and receiver are set to resonate around 18–20 MHz. These two unequally sized resonators are then aligned horizontally or vertically to perform at optimal condition. The optimal condition can be achieved without lateral and angular misalignment. It is expected that the coupling resonance frequency will shift to a higher band when the transmitter and receiver are coupled. In addition, this coupling resonance frequency is also expected to shift depending on the air gap separation distance, which in this design is from 10 mm to 40 mm. The performance of the device is studied by varying the air gap distances from 10 mm to 40 mm. Subsequently, the device's coupling efficiency is evaluated and analyzed for cases of lateral and angular misalignment.

Parameter	Transmitter	Receiver
Size	120 mm × 120 mm	80 mm × 80 mm
Number of turns of spiral coil	8	16
Width of spiral coil	5.7 mm	1.5 mm
Capacitor plate	52 mm × 23 mm	35 mm × 20 mm
Inner radius of single turn coil	43 mm	35 mm
Width of single turn coil	2.5 mm	2.5 mm

Tab. 1. The summarized dimensions of the proposed Witricity device.

3. Results and Discussion

The results discussed in this article are divided into three sub-sections: surface current, S-parameters, and misalignment. The concerned measurement setup of the proposed Witricity device is shown in Fig. 3.

The surface current indicates the current density level, and the distribution of the current on the conductor's surface during the transfer process. The color sequence as shown in Fig. 4, from the highest current density to the lowest, is represented by red, yellow, green, and blue. The insertion loss, S_{21} parameter is used to analyze the efficiency of the Witricity as in (2). Equation (2) presents the efficiency of the output power at the receiver referenced to input power at the transmitter [13], [14]:

$$\eta_{21} = |S_{21}|^2 \times 100\% . \tag{2}$$

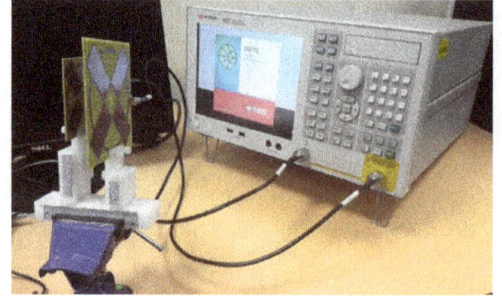

Fig. 3. The measurement setup of the proposed Witricity device.

The S_{11} and S_{22} describe the reflection coefficients at the transmitter and receiver, respectively. Lateral and angular misalignments are introduced to the device to evaluate its performance.

3.1 Surface Current

Figure 4 shows the current distribution on the conductive surface of the Witricity device over a 10 mm to 40 mm air gap distance. The surface current density, $\tilde{\mathbf{J}}_s$ can be expressed by (3):

$$\tilde{\mathbf{J}}_s = \hat{\mathbf{n}} \times \left(\tilde{\mathbf{H}}_2 - \tilde{\mathbf{H}}_1 \right) \tag{3}$$

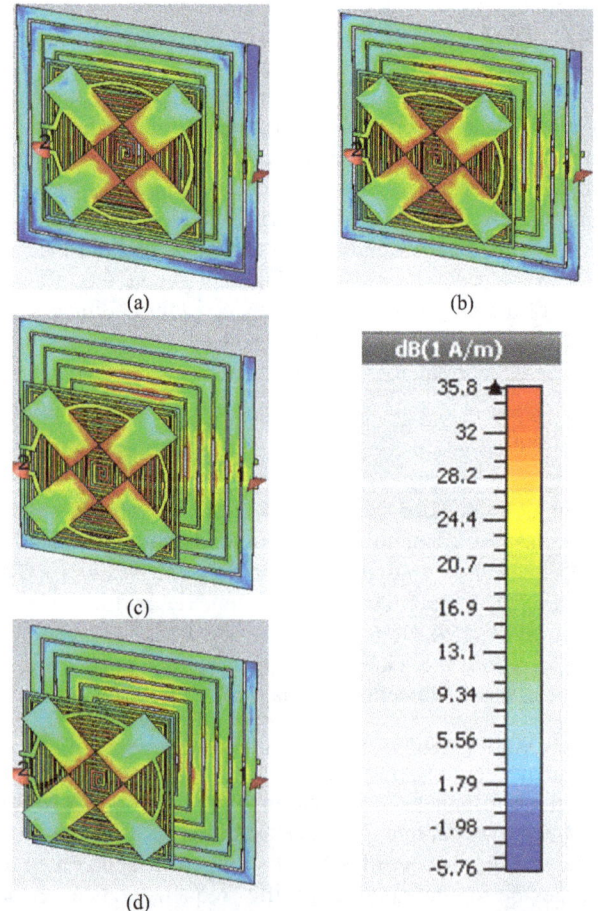

Fig. 4. The Witricity current distribution with different air gap separation distances: (a) 10 mm, (b) 20 mm, (c) 30 mm, and (d) 40 mm.

where $\hat{\mathbf{n}}$ and $\tilde{\mathbf{H}}_1$ and $\tilde{\mathbf{H}}_2$ are a normal unit vector pointing out of the medium at the interface and magnetic-field intensity in two mediums, accordingly.

The highest density of current distribution observed, which reflects the highest received power, is concentrated in the middle area of the device. Otherwise, the majority of the transmitter and receiver current densities fall within a mid-range of approximately 13 Am^{-1} to 17 Am^{-1}. Observations reveal that the receiver has a better current density resulting from higher electric current at its smaller area of cross section due to its small size, which allows for full-coverage areas of coupling from the transmitter. This better current density allows for better coupling and matching at the receiver side. A similar pattern can be noted up to 30 mm, which speaks to a good coupling efficiency. A lower current density is observed for an air gap distance of 40 mm, implying that the receiver (the smaller size resonator) tends to experience more power at distances shorter than 40 mm. The transmitted power is limited to certain transferable distances that become weaker as it travels further. It can be inferred that increasing air gap distance has reduced the effectiveness of power coupled to the receiver and the non-transferrable power will be reabsorbed by the transmitter.

3.2 Scattering Parameters

A comparison of the simulated and measured coupling effect between the transmitter and the receiver is shown in Fig. 5. The –3 dB line in Fig. 5 indicates the benchmark of 50 % coupling efficiency. The simulation result in Fig. 5 shows that the maximum transferable distance is 30 mm, with 52.2 % coupling efficiency, and an effective bandwidth between 19.5 to 20.4 MHz. This result is verified by measurement at 30 mm, with 54.5 % efficiency in the 19.5 to 20.1 MHz band. This has proved that the proposed Witricity device transfers power wirelessly at a maximum of 30 mm separation distance with more than 50 % coupling efficiency, and agrees with the surface current findings in Sec. 3.1. The operating frequency at the 40 mm air gap distance is not considered, as the coupling efficiency results are less than 50 % (approximately 31.6 %) for both the simulation and measurement. The summary of the coupling efficiency, derived from $|S_{21}|^2$ as in (2) is tabulated in Tab. 2.

The transmitter and receiver reflection coefficients when coupled together are shown in Fig. 6 and Fig. 7, respectively, with the corresponding return losses tabulated in Tab. 2. The larger positive of return loss indicates the small amount of reflected power relative to its incident power. Both simulation and measurement results of the return loss show good agreement, which complies with the 10 dB reference when the separation distance is varied up to 10 mm distance for Port 1 at the transmitter and 30 mm for Port 2 at the receiver. Figure 6 shows the magnitude of the reflection coefficient at Port 1, which is at the transmitter when both transmitter and receiver are coupled. It can be seen that S_{11} is less than –10 dB at up to a 10 mm

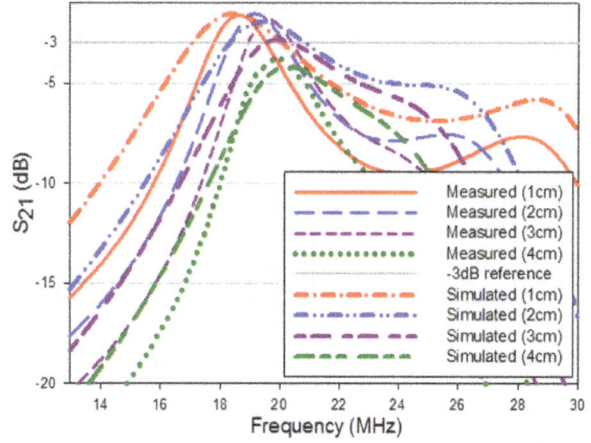

Fig. 5. Simulated and measured results of S_{21} of the unequal size of Witricity device from 10 mm to 40 mm separation distance.

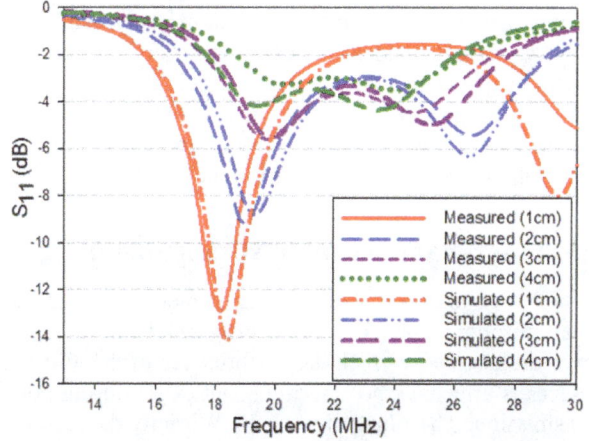

Fig. 6. Simulated and measured results of S_{11} of the unequal size of Witricity device from 10 mm to 40 mm separation distance.

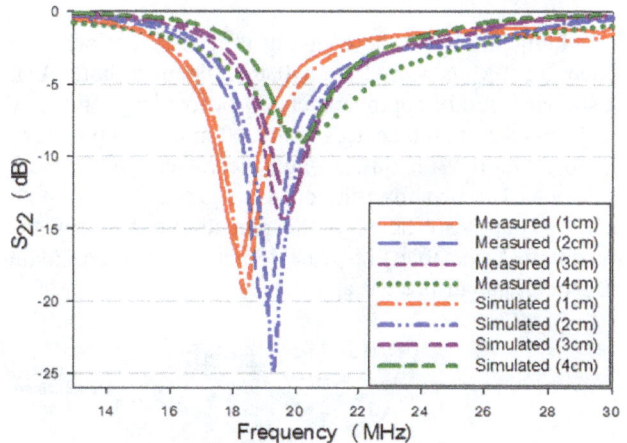

Fig. 7. Simulated and measured results of S_{22} of the unequal size of Witricity device from 10 mm to 40 mm separation distance.

distance in both the simulation and measurement results. Due to the different sizes of the transmitter and receiver, the signals are easily distorted and reflected back to the transmitter when the distance is more than 10 mm, which will affect the matching at Port 1 (as indicated in Fig. 6 and Tab. 2).

Distance (mm)	Operating Frequency (MHz)		Coupling Efficiency (%)		Return Loss at Port 1 (dB)		Return Loss at Port 2 (dB)	
	S	M	S	M	S	M	S	M
10	17.1-20.0	18.1-19.6	70.3	64.1	≥ 10	≥ 10	≥ 10	≥ 10
20	18.3-20.8	19.1-20.4	64.2	62.2	≥ 8	≥ 8	≥ 10	≥ 10
30	19.5-20.4	19.5-20.1	52.2	54.5	≥ 5	≥ 5	≥ 10	≥ 10
40	-	-	31.6	31.6	≥ 3.7	≥ 3.5	≥ 8	≥ 8

*S = Simulation, M = Measurement

Tab. 2. Tabulated results of coupling efficiency and return loss at Port 1 and 2.

The return loss of the receiver side (Port 2) is shown in Tab. 2 based on its reflection coefficient magnitude (plotted in Fig. 7). It can be clearly seen that the resonance frequency increases along with the increment of separation distance, with an acceptable return loss of 10 dB up to 30 mm distance. At 40 mm, the return loss is still quite close to 10 dB, which indicates the good impedance-matching of the receiver at this distance. Both simulation and measurement results show good performance of return loss at Port 2 up to 30 mm. Hence, good coupling efficiency up to a maximum 30 mm distance is expected.

3.3 Lateral and Angular Misalignment

The condition of lateral misalignment (LM) and angular misalignment (AM) is investigated in order to observe the degree to which the performance of the Witricity devices is affected when misalignment exists during power transmission. The illustration of the Witricity devices with AM and LM is shown in Fig. 8. The investigation focuses on coupling efficiency when the devices experience LM from 5 mm to 30 mm distance and an angular misalignment ranging from 0° to 40°.

Figure 9 shows the coupling efficiency performance when the LM is varied from 0 to 30 mm without AM. A similar result of approximately 70 % coupling efficiency is observed from the devices with a 10 mm air gap for LM up to 20 mm. With this, a slight decrement of coupling efficiency is noted from the other air gap distances. With a 50 % coupling efficiency, the plotted results show the maximum acceptable LM to be 15 mm when a maximum 30 mm air gap distance is applied.

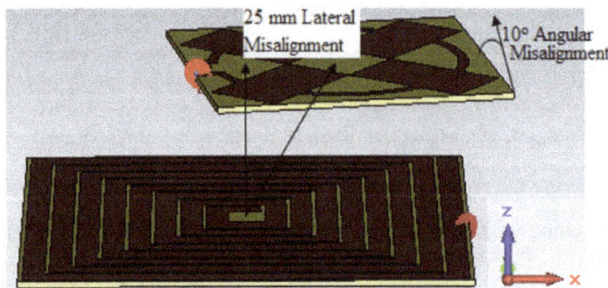

Fig. 8. Configuration of Witricity devices with 10° AM and 25 mm LM.

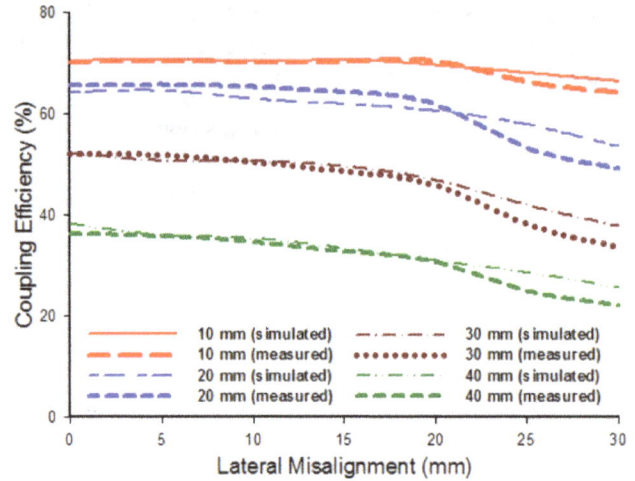

Fig. 9. The coupling efficiency performance for a 10 mm to 40 mm air gap separation distance with a 0 to 30 mm lateral misalignment (LM).

The effect of AM variation from 0° to 40° is depicted in Fig. 10. Aside from varying the AM, the figure also shows the effect of LM varying from 5 mm to 20 mm. In this investigation, the distance between transmitter and receiver is set to be at 30 mm. Shown in the figure, 50 % coupling efficiency is achievable when 0° to 40° AM variation is introduced and a maximum LM of 10 mm exists. With constant air gap separation between transmitter and receiver, as well as similar LM and 0° to 40° AM, the Witricity devices demonstrate a 1.5 % ripple size of the oscillated coupling efficiency. However, the measured results show nearly constant and less oscillating results compared to the simulation. As expected, when the transmitter and receiver do not experience any LM but only AM, the facing area between transmitter and receiver is nearly similar, which results in almost constant coupling efficiency.

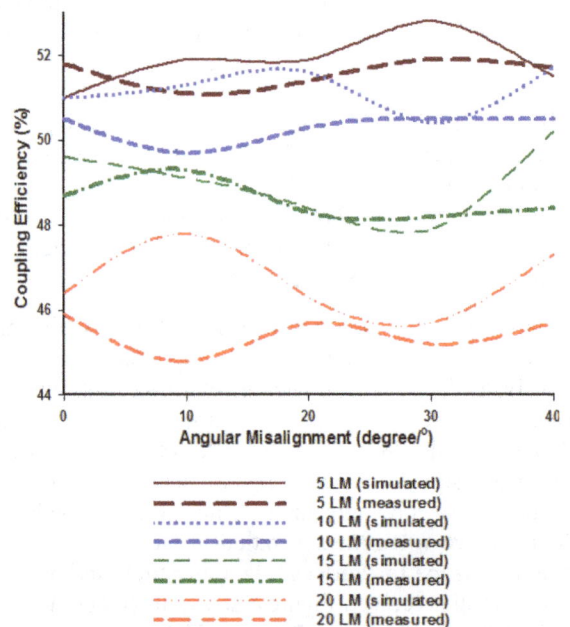

Fig. 10. The coupling efficiency performance for 10° to 40° of AM and 5 to 20 mm of LM.

Work	Transmitter Size (mm^2)	Receiver Size (mm^2)	Simulation (Distance-efficiency)	Measurement (Distance-efficiency)
[10]	162 × 162	162 × 162	Not specified	50 mm-52.3%
[11]	80 × 80	80 × 80	30 mm-52.6%	Not specified
[12]	40 × 40	40 × 40	30 mm-30.0%	30 mm-43.0%
Proposed work	120 × 120	80 × 80	30 mm-52.2%	30 mm-54.5%

Tab. 3. The size and performance comparison between the proposed and related works in [10], [11] and [12].

Table 3 summarizes a comparison of the transmitter and receiver size, transferable distance and efficiency between the proposed design and other works reported in [10], [11] and [12]. In [10], the measurement shows a good result for up to 50 mm distance. However, the size is too large to be fitted to modern devices. Therefore, more compact design has been proposed in [11] and its simulation results present a good performance for up to 30 mm distance, which is an adequate amount in wireless power transfer application. Nonetheless, no measurement has been performed in [11]. In [12], the authors proposed a smaller Witricity design compared to [11], however, the design is good for only up to 20 mm distance. Even though [12] shows a possibility of achieving a more compact design, nevertheless, the material that has been used is expensive and also the results show a decrement in the maximum transferable distance. Hence, this work proposed a fairly design size and lower cost material, which is similar to the material proposed in [11] with better and reliable measurement performance for up to 30 mm distance, 10 mm lateral misalignment and 40° angular misalignment. This work also shows that, to achieve better performance, the transmitter size can be increased as this work performs better compared to [11], which is 54.5 % measurement results compared to 52.6 % simulation results performed by [11].

4. Conclusion

In this article, the design of a pair of unequally sized Witricity devices integrated with a thin coil strip has been presented. The simulation and measurement results show that the device is coupled at a maximum air gap of 30 mm with 54.5 % efficiency. With regard to the performance of the Witricity device given lateral and angular misalignment, a maximum LM of 10 mm can be achieved with the variation of AM from 0° to 40°. The results indicate a potential for efficient wireless power transfer to devices that require more stringent receiver size restrictions.

Acknowledgments

This work is carried out with the financial support from the Malaysian Ministry of Higher Education (MOHE) via MyPhD program and Universiti Teknologi Malaysia (UTM) via Research University Grant (RUG), Flagship Grant and HiCoE Grant with Vote Numbers of 05H43, 03G41 and 4J212.

References

[1] TESLA, N. *System of Transmission of Electrical Energy. US Patent* 645576, 1900.

[2] POON, A. S. Y. A general solution to wireless power transfer between two circular loops. *Progress in Electromagnetics Research*, 2014, vol. 148, p. 171–182. ISSN: 1070-4698. DOI: 10.2528/PIER14071201

[3] BANGERTER, B., TALWAR, S., AREFI, R., STEWART, K. Networks and devices for the 5G era. *IEEE Communications Magazine*, 2014, vol. 52, no. 2, p. 90–96. ISSN: 0163-6804. DOI: 10.1109/MCOM.2014.6736748

[4] JANG, B.-J., LEE, S., YOON, H. HF-band wireless power transfer system: concept, issues, and design. *Progress in Electromagnetics Research*, 2012, vol. 124, p. 211–231. ISSN: 1070-4698. DOI: 10.2528/PIER11120511

[5] KURS, A., KARALIS, A., MOFFATT, R., et al. Wireless power transfer via strongly coupled magnetic resonances. *Science Journal*, 2007, vol. 31, no. 5834, p. 83–86. ISSN: 1095-9203. DOI: 10.1126/science.1143254

[6] LIU, C., HU, A. P., NAIR, N. K. C. Modelling and analysis of a capacitively coupled contactless power transfer system. *IET Power Electronics*, 2011, vol. 4, no. 7, p. 808–815. ISSN: 1755-4535. DOI: 10.1049/iet-pel.2010.0243

[7] SON, H. W., PYO, C. S. Design of RFID tag antennas using an inductively coupled feed. *Electronics Letters*, 2005, vol. 41, no. 18, p. 994–996. ISSN: 0013-5194. DOI: 10.1049/el:20051536

[8] CHA, H. K., PARK, W. T., JE, M. A CMOS rectifier with a cross-coupled latched comparator for wireless power transfer in biomedical applications. *IEEE Transactions on Circuits and Systems II: Express Briefs*, 2012, vol. 59, no. 7, p. 409–413. ISSN: 1549-7747. DOI: 10.1109/TCSII.2012.2198977

[9] KRACEK, J., MAZANEK, M. Possibilities of wireless power supply. *International Journal of Microwave and Wireless Technologies*, 2010, vol. 2, no. 2, p. 153–157. ISSN: 1759-0787. DOI: 10.1017/S1759078710000255

[10] WANG, J., HO, S. L., FU, W. N., SUN, M. Analytical design study of a novel Witricity charger with lateral and angular misalignments for efficient wireless energy transmission. *IEEE Transactions on Magnetics*, 2011, vol. 47, no. 10, p. 2616–2619. ISSN: 0018-9464. DOI: 10.1109/TMAG.2011.2151253

[11] MOHD SALLEH, M. H., SEMAN, N., DEWAN, R. Reduced-size Witricity charger design and its parametric study. In *Proceedings of the 2013 IEEE International RF and Microwave Conference*. Penang (Malaysia), 2013, p. 387–390. DOI: 10.1109/RFM.2013.6757290

[12] MOHD SALLEH, M. H., SEMAN, N., ABANG ZAIDEL, D. N. Design of a compact planar Witricity device with good efficiency for wireless applications. In *Proceedings of the 2014 Asia-Pacific Microwave Conference (APMC)*. Sendai (Japan), 2014, p. 1369 to 1371. ISBN: 978-4-9023-3931-4

[13] KIM, J. W., SON, H. C., KIM, K. H., PARK, Y. J. Efficiency analysis of magnetic resonance wireless power transfer with intermediate resonant coil. *IEEE Antennas and Wireless Propagation Letters*, 2011, vol. 10, p. 389–392. ISSN: 1536-1225. DOI: 10.1109/LAWP.2011.2150192

[14] MORIWAKI, Y., IMURA, T., HORI, Y. Basic study on reduction of reflected power using DC/DC converters in wireless power transfer system via magnetic resonant coupling. In *IEEE 33rd*

International Telecommunications Energy Conference (INTELEC). Amsterdam (Netherlands), 2011, p. 1-5. ISSN: 2158-5210. DOI: 10.1109/INTLEC.2011.6099737

About the Authors ...

Mohd Hidir MOHD SALLEH obtained his first degree from the Universiti Teknologi Malaysia (UTM) in Electrical Engineering (Telecommunication) in 2012 with first class honor. He currently pursues his Ph.D. in Electrical Engineering in UTM since 2012 under fastrack programme.

Norhudah SEMAN received the B.Eng. in Electrical Engineering (Telecommunications) degree from the Universiti Teknologi Malaysia, Johor, Malaysia, in 2003 and M.Eng. degree in RF/Microwave Communications from The University of Queensland, Brisbane, St. Lucia, Qld., Australia, in 2005. In September 2009, she completed her Ph.D. degree at The University of Queensland. In 2003, she was an Engineer with Motorola Technology, Penang, Malaysia, where she was involved with the RF and microwave components design and testing. Currently, she is a Senior Lecturer in Wireless Communication Centre (WCC), Universiti Teknologi Malaysia. Her research interests concern the design of microwave circuits for biomedical and industrial applications, UWB technologies, and mobile communications.

Dyg Norkhairunnisa ABANG ZAIDEL received her B.Eng. in Electrical Engineering (Telecommunications) degree from the Universiti Teknologi Malaysia, Johor, Malaysia, in 2010. She completed her Ph.D. degree in the same university in 2014. She is now currently a Senior Lecturer in the Universiti Malaysia Sarawak. Her research interests include microwave devices and smart antenna beam forming system.

Akaa Agbaeze ETENG received his B.Eng. degree from the Federal University of Technology Owerri, Nigeria, in 2002, and his M.Eng. degree in Telecommunications and Electronics from the University of Port Harcourt, Nigeria, in 2008. He is currently pursuing his Ph.D. degree at the Wireless Communication Center, Universiti Teknologi Malaysia in the area of near-field wireless energy transfer.

6

Power Parameters and Efficiency of Class B Audio Amplifiers in Real-World Scenario

Hristo ZHIVOMIROV [1], Nikolay KOSTOV [2]

[1] Dept. of Theory of Electrical Engg. and Measurements, Technical Univ. of Varna, Sudentska Street 1, Varna, Bulgaria
[2] Dept. of Telecommunications, Technical University of Varna, Sudentska Street 1, Varna, Bulgaria

hristo_car@abv.bg, n_kostov@mail.bg

Abstract. *Consumer audio amplifiers are intended to operate with various loudspeaker loads, i.e. the load impedance profile of the audio amplifier is a priori unknown. We propose the power parameters analysis of the class B audio amplifiers to be carried out in the realistic worst-case (RWC) scenario of operation with the minimal value of the impedance and a RWC type of signal, instead of the nominal impedance of the loudspeaker and a sine-wave signal. Experimental validation, carried out for different types of signals and loudspeaker loads, demonstrate the advantages of the proposed RWC-based power parameters estimation. Furthermore, we provide a way of assessing the safe-operating area (SOA) boundaries, based on the output I-V loci of the amplifier and by means of an equivalent load line (ELL).*

Keywords

Audio, class B amplifier, power parameters, estimation, realistic worst-case scenario

1. Introduction

The consumer audio amplifiers are intended to operate with music and speech signals and various loudspeaker loads (incl. an individual transducer plus an enclosure), instead of sine-waves and resistive loads. Therefore, it makes sense to estimate the amplifier's parameters in real-world or close to real-world conditions. Here we focus on the power parameters estimation of the widely used class B audio amplifiers as these influence the choice of active components, heat sinks, and the overall power supply unit (PSU) design. The critical quantity for the active components selection is the instantaneous power dissipation while for the heat sink design this is the average power dissipation and for the PSU design – the instantaneous and the average current through the load [1].

Since both music and speech signals are stochastic and non-stationary, and the loudspeakers complex impedance is frequency-depended, exact analytical expressions for the average power parameters are hardly possible.

In [2] Raab presents expressions for the average power parameters, using the signal probability density function, in case of operation with music signals and resistive load. Zee [3] and Self [1] elaborated the idea, but still assuming that the amplifier load is purely resistive. In [4] Benjamin presents and discusses experimental results of operation of a class B amplifier in real-world conditions, but mathematical apparatus is not derived.

One way to address the above mentioned problem is by using a mathematical model and further statistical processing as shown in a previous related work [5]. A key condition for successful use of such a model is the availability of *prior* knowledge of the amplifier's load impedance response $\dot{Z}_L(\omega)$. For instance, this is the case when the audio amplifier is being manufactured along with the loudspeaker(s) (e. g. subwoofer systems, radio receivers, etc.). However, in the following we consider that the load impedance profile is unknown in advance which is a common case, so this approach is inconvenient.

Here, we propose a method for analysis of the power parameters of class B audio amplifier, based on relations derived in a previous related work [6]. These relations are used for addressing the case of real-world signals and initially unknown loudspeaker loads. Since there are virtually infinity combinations of signals and loads, we suggest a realistic worst-case (RWC) scenario of amplifier operation. In addition, we propose a way for charting the safe-operating area (SOA) boundaries of the active components in the amplifier's output stage, which allows an estimation of the instantaneous power dissipation and the peak collector current to be made, in the light of SOA boundaries.

2. Proposed Method

We propose the power parameters analysis to be carried out for the RWC scenario of operation with the minimal value $Z_{Lmin} = \min(Z_{Lnom})$ of the loudspeaker impedance, instead of the nominal impedance Z_{Lnom}, as well as with RWC type of music signals, instead of sine-wave. Typical minimal values of the loudspeaker impedance are $Z_{Lmin} \in \{2.75, 5.5\}$ Ω for $Z_{Lnom} \in \{4, 8\}$ Ω. Our considerations are: the operation with sine-wave signals is not representative for the amplifier exploitation conditions, and since the sine-wave parameters are worse (as shown below) in comparison with these of music signals, the amplifier power parameters would be overestimated. Furthermore, the loudspeaker impedance modulus in a wide frequency range is below Z_{Lnom}, so the analysis should consider the worst case scenario of operation with Z_{Lmin}. Then, the average values of the power parameters of a class B audio amplifier could be calculated as for a sine-wave operation [7], using two correction coefficients ξ_{eq} and χ defined in [6]. The equivalent coefficient of the power supply voltage ξ_{eq} is defined as

$$\xi_{eq} = \frac{CF_{sin}}{CF_{rand}} \cdot \frac{\max\left(\left|u_{out}(t)\right|\right)}{U_{cc}} = \frac{CF_{sin}}{CF_{rand}}\xi \quad (1)$$

where CF_{sin} and CF_{rand} are the crest-factors for sine-wave and random signals , $u_{out}(t)$ is the amplifier output voltage, U_{cc} is the power supply voltage, and ξ is the coefficient of the power supply use for sine-wave signals.

The coefficient of average rectified values compliance χ is given as

$$\chi = \frac{FF_{sin}}{FF_{rand}} \quad (2)$$

where FF_{sin} and FF_{rand} are the signal form-factors for sine-wave signals, and for signals with random distribution, respectively. The values of CF and χ for certain signals of interest are: (i) $CF = 3.01$ dB, $\chi = 1$ for a sine-wave signal; (ii) $CF = 10 \div 20$ dB, $\chi = 0.5\sqrt{\pi}$ (assuming Gaussian amplitude distribution) for music signals [1–3], and (iii) $CF = 15 \div 25$ dB, $\chi = 0.25\pi$ (assuming Laplace amplitude distribution) for speech signals [3], [8]. Ultimately we consider music with $CF = 6$ dB as the RWC type of signal, since there is no evidence smaller CF to be found in practice.

The definitions for the average power parameters and efficiency of a class B audio amplifier for resistive loads and any type of signals [6] are as follows:

$$P_{DC} = \frac{2U_{cc}^2}{\pi R_L}\xi_{eq}\chi , \quad (3)$$

$$P_L = \frac{U_{cc}^2}{2R_L}\xi_{eq}^2 , \quad (4)$$

$$P_D = P_{DC} - P_L = \frac{U_{cc}^2}{R_L}\left(\frac{2\xi_{eq}\chi}{\pi} - \frac{\xi_{eq}^2}{2}\right), \quad (5)$$

$$\eta = \frac{P_L}{P_{DC}} = \frac{\pi}{4}\frac{\xi_{eq}}{\chi}. \quad (6)$$

In (3)-(6), P_{DC} is the power drawn from the PSU, P_L is the power delivered to the load, P_D accounts for the losses in the amplifier, and η is the amplifier efficiency.

In order to assess the maximum value of the instantaneous collector current $i_C(t)$ and the SOA boundaries we use assessment based on the output I-V loci of the amplifier (collector current i_C vs. collector-emitter voltage u_{CE} plots) and an equivalent load line (ELL) [1]. However, here we consider the RWC-signal and not sine-wave signals, and Z_{Lmin} instead of Z_{Lnom}. Therefore, we propose the ELL to be drawn on the amplifier output I-V locus by defining two points with Cartesian coordinates $A (0, I_{Cmax} = U_{cc}/Z_{Lmin})$, $B (U_{CEmax} = 2U_{cc}, 0)$.

One should be aware that the RWC output I-V loci are not below the ELL by default and the ELL could be exceeded, although this rather rarely happens [1, 4].

3. Analytical, Numerical and Measurement Results

Equations (3)-(6) were evaluated for two representative cases: (i) the classic design with a sine-wave signal ($R_L = Z_{Lnom} = 8 \Omega$, $CF = 3.01$ dB, $\chi = 1$) and (ii) the RWC-based design ($R_L = Z_{Lmin} = 5.5 \Omega$, $CF = 6$ dB, $\chi = 0.5\sqrt{\pi}$). In both cases $U_{cc} = 12$ V and $\xi = 0.85$.

We carried out numerical experiments using the SiPoLo-model [5, 9] (cf. Fig. 1) in order to assess the average power parameters, the efficiency and the output I-V locus of a given class B audio amplifier for different types of signals and loudspeaker loads.

Two test signals are used: (i) a test signal defined by CEA-426-B (formerly EIA-426-B) standard [10] (which is considered to be a good approximation of the music signals [3, 11]) with duration 30 min and $CF = 6$ dB, and (ii) music (hours of different genres from commercial CDs). The amplifier loads were various types of loudspeakers, incl. subwoofers, mid-ranges and tweeters with impedance response modeling as is shown in the Appendix.

The numerical experiments were repeated as real-word measurements under the same conditions using audio power amplifier based on IC LM3886 [12] in a follower configuration. The output voltage and the current through the load were measured using data acquisition system NI USB-6211 and the amplifier parameters of interest were evaluated.

Some of the results of the analytical, numerical and real-world estimation procedures are shown in Tab. 1, Fig. 1 and Fig. 2. We present the results and the output I-V loci for the CEA-426-B test signal and for 90 min music signal from the DVD "Stars of the 90's Volume 01" (DVD

SiPoLo Model

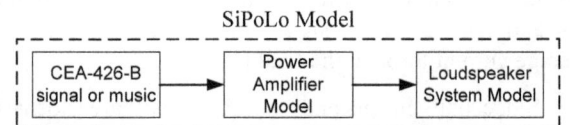

Fig. 1. Block diagram of the SiPoLo-model used for numerical estimation of the amplifier power parameters for different types of signals and loads.

Test conditions (TC)		P_{DC} [W]	P_D [W]	P_L [W]	η [%]
1.	Classic estimation	9.74	3.24	6.50	66.8
2.	RWC-based estimation	8.90	4.15	4.75	53.4
3.	18W/8424G00 in 4th order bandpass double chamber box, CEA-426-B signal (simulation)	5.98	3.72	2.26	37.7
4.	18W/8424G00 in 4th order bandpass double chamber box, music (simulation)	2.22	1.83	0.38	17.3
5.	Unknown single loudspeaker system, CEA-426-B signal (measurement)	5.73	3.36	2.37	41.3
6.	Unknown single loudspeaker system, music (measurement)	3.13	2.34	0.79	25.2

Tab. 1. Results from analytical, numerical and real-world estimations of power parameters of a class B audio amplifier.

10501200501) played on loudspeaker 18W/8424G00 [13] mounted in 4th order bandpass double chamber box and a single loudspeaker system with unknown loudspeaker and enclosure parameters.

4. Discussion

The analytical study and the experimental results led us to the following:

1. The power parameters values of the amplifier operating with CEA-426-B test signal are always greater than these of music signals (cf. Tab. 1 TC 3-4 and TC 5-6). This holds true for subwoofers, mid-ranges and tweeters, regardless of the different mechanical resonant frequencies. Hence, the CEA-426-B test signal could be successfully used as a RWC-signal.

2. The power parameters in all tests are well below the values calculated via RWC-based estimation, especially for music signals (cf. Tab. 1 TC 2-4 and TC 2-6). Therefore, for the class B audio amplifiers the RWC-approach could be considered as a good estimator of the power parameters in a real-world scenario.

3. The values of P_{DC} and P_L obtained via RWC-based estimation are always lower than these obtained by the classic approach (cf. Tab. 1 TC 1-2). This allows lowering the design requirements for both the amplifier PSU and the loudspeaker.

4. The actual value of P_D could be greater than the one obtained through the classic estimation, which has to be taken into account in the design of the amplifier's heat sink(s) (cf. Tab. 1 TC 1-3 and TC 1-5).

Consequently, the presented RWC-based estimation of the power parameters and efficiency of the class B audio amplifiers is more relevant than the classic approach, because it approximates more accurately the real-world parameters of interest.

Furthermore, we can make the following conclusions based on observations of the amplifier loci:

1. There is no evidence that the maximum value of the instantaneous collector current $i_C(t)$ would exceed the value I_{Cmax} calculated by (7) (cf. Fig. 1 and Fig. 2), which facilitates the active components selection and the PSU design.

2. Virtually all the I-V loci of the amplifier are below the ELL (cf. Fig. 1 and Fig. 2) with minor exceptions, generally when operating with the CEA-426-B signal.

Hence, the ELL could be considered as a good estimator for SOA limitations of the amplifier in real-world conditions. Our conclusions are in good agreement with the empirical results shown in [4], performed with numerous loudspeakers and hundred hours of music.

There is possibility to decrease the dissipated power P_D and to increase the efficiency η of the class B power amplifiers via loudspeaker system's impedance compensation using an appropriate Zobel network [14, 15]. Thus the impedance profile of the amplifier's load will have a (close to) resistive nature (in the frequency band of operation) and this is proved to be the optimum case of operation in the light of amplifier efficiency [16]. Also, this approach is

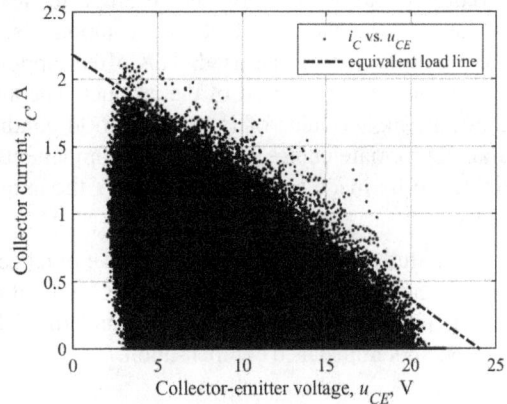

Fig. 2. Output I-V locus of the analyzed amplifier playing CEA-426-B test signal on 18W/8424G00 mounted in 4th order bandpass double chamber box.

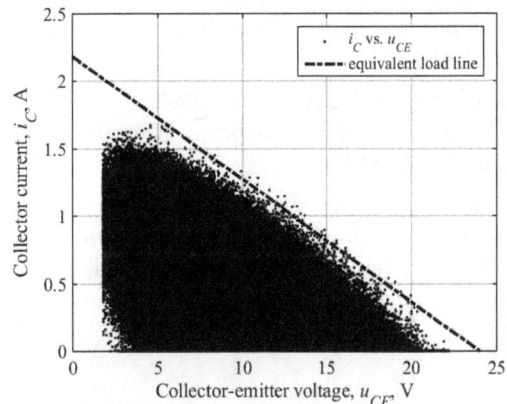

Fig. 3. Output I-V locus of the analyzed amplifier playing music on 18W/8424G00 mounted in 4th order bandpass double chamber box.

extremely effective in reducing of the peak instantaneous power dissipation of the amplifier [4]. One must be aware that this is possible only if there is a *prior* knowledge of the amplifier's load impedance response $\dot{Z}_L(\omega)$.

5.   Conclusion

A new approach for estimation of the power parameters of the class B amplifiers is proposed, based on RWC scenario of operation with the minimal value of the loudspeaker impedance and a RWC type of signal, instead of the nominal impedance of the loudspeaker and a sine-wave signal, respectively. It is shown that the considered RWC-estimation approximates more accurately the real-world power parameters of the audio amplifier. This is of particular importance when the load impedance profile of the audio amplifier is *a priori* unknown, which is the common case in the practice. Also, assessing of the SOA boundaries based on the output I-V loci of the amplifier and by means of ELL is presented.

The analytical study and the experimental results show that the values of P_{DC} and P_L obtained via the RWC-estimation are always lower than these obtained via the classic approach, but P_D could be greater. Lowering the design requirements to the PSU of the amplifier and the loudspeaker allows cheaper electronic components, heat sinks, and loudspeakers to be used. This offers opportunities for reducing the prime cost of the product. The awareness about the higher value of P_D in real-world conditions allows an appropriate choice of the active components and heat sink(s) to be made and hence improves the amplifier robustness.

Further study should be carried to derive practical design of class B amplifiers, based on the RWC-estimation of the power parameters presented here, including loudspeaker system's impedance compensation.

Appendix

In this Appendix, we briefly illustrate the loudspeaker system impedance response modeling technique used in the SiPoLo-model.

The simplified electrical equivalent circuit of a 4th order bandpass double chamber loudspeaker system is shown in Fig. A1 [17].

The electrical impedance of the system includes the sum of the voice coil electrical impedance, the impedance introduced by the mechanical system and one by the acoustic volume [17]

$$\dot{Z}_L(j\omega) = R_e + j\omega L_e +$$

$$\frac{(B.l)^2}{R_{ms} + \dfrac{1}{j\omega C_{ms}} + \dfrac{S_d^2}{j\omega C_{ab1}} + j\omega M_{ms} + \dfrac{S_d^2}{j\omega C_{ab2} + \dfrac{1}{\dfrac{1}{R_{aL}} + \dfrac{1}{j\omega M_{ap2}}}}} . \quad (A1)$$

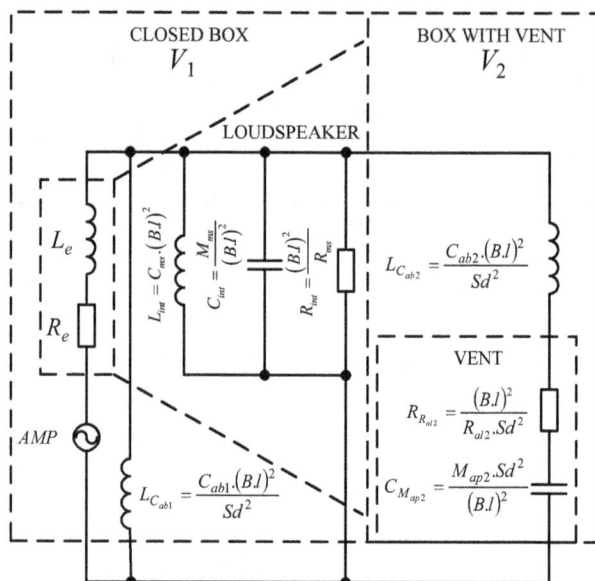

Fig. A1. Simplified electrical equivalent circuit of the 4th order bandpass double chamber loudspeaker system.

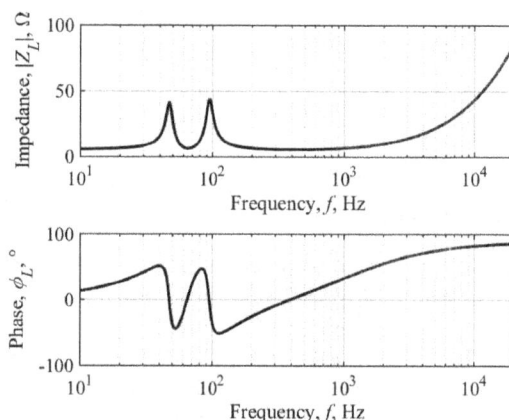

Fig. A2. Impedance response (magnitude and phase) of the 18W/8424G00 loudspeaker mounted in 4th order bandpass double chamber box.

The amplitude and phase of the impedance response of the 18W/8424G00 loudspeaker mounted in 4th order bandpass double chamber box used in the experiments, is shown in Fig. A2. Additional information on loudspeaker system impedance modeling could be found in [18–21].

Acknowledgments

The authors would like to express their deep gratitude to Assoc. Prof. Ph.D. Todor Ganchev for the appropriate recommendations and advices.

References

[1] SELF, D. *Audio Power Amplifier Design*. Abingdon (UK): Focal Press, 2013. (Chapter 16). ISBN: 978-0-240-52614-0

[2] RAAB, F. Average efficiency of class-G power amplifiers. *IEEE Transactions on Consumer Electronics*, 1986, vol. CE-32, no. 2, p. 145–150. DOI: 10.1109/TCE.1986.290146

[3] ZEE, R. *High Efficiency Audio Power Amplifiers; Design and Practical Use*. Enschede (Netherlands), PhD Thesis in Dept. Elect. Eng. in University of Twente, 1999. ISBN: 90-36512875

[4] BENJAMIN, E. Audio power amplifiers for loudspeaker loads. *Journal of Audio Engineering Society*, 1994, vol. 42, no. 9, p. 670 to 683. ISSN: 0004-7554

[5] ZHIVOMIROV, H., VASILEV, R. Power parameters and efficiency of class B amplifier operating with complex load and random signal. In *Proceedings of the UNITECH'2014*. Gabrovo (Bulgaria), Nov. 2014, vol. 2, p. 53–58. ISSN: 1313-230X

[6] ZHIVOMIROV, H. Power parameters and efficiency of class B amplifier operating with resistive load and random signal. *TEM Journal*, 2015, vol. 4, no. 1, p. 16–21. ISSN: 2217-8309

[7] SEDRA, A., SMITH, K. *Microelectronic Circuits*. Oxford (UK): Oxford University Press, 2015. ISBN: 978-0-19-933913-6

[8] HANSLER, E., SCHMIDT, G. *Acoustic Echo and Noise Control: A Practical Approach*. Hoboken (NJ, USA): John Wiley & Sons Inc., 2004. ISBN: 0-471-45346-3

[9] ZHIVOMIROV, H. *Power Analysis of Class B Power Amplifier with Matlab Implementation, version 1.1*. [Online] Cited 2016-06-30. Available at: http://www.mathworks.com /matlabcentral/fileexchange/47438-power-analysis-of-class-b-power-amplifier-with-matlab-implementation

[10] KEELE, D., Jr. Development of test signals for the EIA-426-B loudspeaker power rating compact disk. In *111th Audio Engineering Society Convention*. New York, 21–24 September, 2001

[11] BOYCE, T. *Introduction to Live Sound Reinforcement: The Science, the Art, and the Practice*. Victoria (BC, Canada): FriesenPress, 2014. ISBN: 978-1-4602-3890-5

[12] TEXAS INSTRUMENTS. *LM3886 High-Performance 68 W Audio Power Amplifier (datasheet)*. 31 pages. [Online] Cited 2016-06-30. Available at: http://www.ti.com/product/LM3886

[13] SCAN-SPEAK. *18W/8424G00 Midwoofer (datasheet)*. 2 pages. [Online] Cited 2016-06-30. Available at: http://www.scan-speak.dk/datasheet/pdf/18w-8424g00.pdf

[14] LEACH, W. M., Jr. Impedance compensation networks for the lossy voice-coil inductance of loudspeaker drivers. *Journal of Audio Engineering Society*, April 2004, vol. 52, no. 4, p. 358–365. ISSN: 0004-7554

[15] ZOBEL, O. Theory and design of uniform and composite electric wave filters. *Bell System Technical Journal*, January 1923, vol. 2, p. 1–46. DOI: 10.1002/j.1538-7305.1923.tb00001.x

[16] HÁJEK, K. Efficiency of the class B-CE and class B-CC high-voltage wideband amplifiers with a capacitive load. In *Proceedings of 21st International Conference Radioelektronika 2011*. Brno (Czech Rep.) 2011, p. 137–140. DOI: 10.1109/RADIOELEK.2011.5936430

[17] SIRAKOV, E. Band-pass loudspeaker systems with single vent. In *Proceedings of Papers of International Scientific Conference on Information, Communication and Energy Systems and Technologies ICEST 2009*. Sofia (Bulgaria), 2009, vol. 1, p. 235–238.

[18] BERKHOFF, A. Impedance analysis of subwoofer systems. *Journal of Audio Engineering Society*, 1994 Jan./Feb., vol. 42, no. 1/2, p. 4–14. ISSN: 0004-7554

[19] THIELE, A. Loudspeakers in vented boxes. Parts I and II. *Journal of Audio Engineering Society*, 1971 May/Jun., vol. 19, no. 5, p. 382–392, p. 471–483. ISSN: 0004-7554

[20] SMALL, R. Closed-box loudspeaker systems. Part I: Analysis. *Journal of Audio Engineering Society*, 1972 Dec., vol. 20, no. 10, p. 798–808. ISSN: 0004-7554

[21] SMALL, R. Vented-box loudspeaker systems. Part I: Small-signal analysis. *Journal of Audio Engineering Society,* 1973 Jun., vol. 21, no. 5, p. 363–372. ISSN: 0004-7554

About the Authors ...

Hristo ZHIVOMIROV was born in Varna, Bulgaria, in 1987. He received his B.Sc. and M.Sc. degrees in Communication Equipment and Technologies from the Technical University of Varna in 2010 and 2012, respectively. In 2016 he received his Ph.D. degree in Theory of Communication. Mr. Zhivomirov is currently an Assistant Professor at the Department of Theory of Electrical Engineering and Measurements, Technical University of Varna. Mr. Zhivomirov is a member of IEEE, USB Bulgaria and FSTU Bulgaria. His research interests include the field of signal processing, circuits and systems, electrical measurements and Matlab programming (signal processing, data acquisition, data visualization, etc.).

Nikolay KOSTOV was born in 1969 in Varna, Bulgaria. He received Dipl. Eng. and Dr. Eng. degrees from the National Military University of Veliko Tarnovo, Bulgaria, in 1993 and 1999, respectively. He is an Associated Professor and a Head of the Department of Telecommunications, Technical University of Varna, Bulgaria. His research interests include analog and digital signal processing, Matlab programming and simulations, source/ channel coding, modulation techniques and communication over fading channels.

7

Compact Dual-mode Microstrip Bandpass Filter Based on Greek-cross Fractal Resonator

Hongshu LU, Weiwei WU, Jingjian HUANG, Xiaofa ZHANG, Naichang YUAN

College of Electronic Science and Engineering, National University of Defense Technology,
Changsha, Hunan, 410073, China

Luhongshu0321@163.com

Abstract. *A geometrically symmetrical fractal structure is presented in this paper to provide an alternative approach for the miniaturization design of microstrip bandpass filters (BPFs). The generation process of the geometric geometry is described in detail, and a new fractal resonator called Greek-cross fractal resonator (GCFR) is produced by etching the proposed fractal configuration on the surface of the conventional dual-mode meandered loop resonator. Four microstrip BPFs based on the first four iterations GCFR are modeled and simulated. The simulation results show that with the increase of the number of iterations, the central frequency of the BPF is gradually moving towards the low frequency, which indicates that the proposed fractal resonator has the characteristic of miniaturization. In addition, the parameter optimization and surface current density distribution are also analyzed in order to better understand the performance of the BPF. Finally, a compact dual-mode BPF based on the third iteration GCFR is designed, fabricated and measured. The measurement results are in good agreement with the simulation ones.*

Keywords

Dual-mode, miniaturization, bandpass filter (BPF), fractal geometry, Greek-cross

1. Introduction

Microstrip bandpass filter (BPF) is an indispensable part in the RF front end of various communication systems. It plays a vital role in filtrating the desired signals in the specified frequency band and suppressing the out of band clutter and interference signals. Consequently, the design of microstrip BPFs is always a hot research topic in wireless communication systems. With the progressive development of wireless communication, the demand for microstrip bandpass filter with excellent performance and compact structure becomes more and more urgent [1].

Miniaturization of BPF is one of fundamental requirements in communication systems. Since the dual-mode resonator was firstly introduced by Wolff [2], the BPFs with dual-mode resonators have been extremely attractive. This is because each dual-mode resonator can be used as a doubly tuned resonant circuit, so the number of resonators required for a given degree of filter is reduced to half, resulting in a compact filter structure [3]. In addition, dual-mode resonators also have advantageous features such as wide passband, low radiation loss, and easy-to-design layout because transmission-line theory and design tools can easily be exploited [4]. In these literatures [5–9], the filters based on the dual-mode resonator are all built by using the coupling between the two degenerate modes, and the two degenerate modes can be excited by introducing a slot along the orthogonal plane or attaching a capacitive patch to the resonator. A pair of crossed slots with unequal widths acted as the perturbation elements had been embedded in a patch resonator [5]. By adjusting the width of the slots, two degenerate modes were excited, and a pair of transmission zeros (TZs) was also generated, which improved the selectivity of the BPF performance. R.Q. Zhang [6] had applied the arc- and radial-oriented slots to bring down the TM_{01}-like mode and to split the TM_{11}-like mode and its degenerate mode, respectively. By using a pair of square etched areas (SEAs) acted as perturbation elements, the dual mode filtering performance had been excited and two TZs located on either side of the passband were clearly observed [7]. The two TZs could effectively improve the stopband rejection, thus avoiding the interference from other communication systems. A novel capacitance loaded square loop resonator (CLSLR) with spurious response suppression and size reduction had been proposed in [8], and the conception of using the inner patch perturbation elements to split the degenerate modes to form two passbands was firstly put forward. In literature [9], the elliptical and linear phase filtering characteristics could be obtained in the first passband by using the perturbation elements in the form of patch or corner cut, respectively.

The term fractal was firstly proposed by Mandelbrot to represent a class of seemingly irregular geometries [10], since then, many fractal geometries have been widely studied and applied in various microwave devices, such as antennas [11–14], frequency selective surfaces (FSSs) [15], and microstrip BPFs [16]. Unlike Euclidean geometry, the fractal geometry has two essential properties: space-filling and self-similarity. The self-similarity can be used to achieve multi-band characteristics and the space-filling property can be employed for miniaturization design [17]. Yordanov et al. had made a pioneering work in the application of fractal geometry in the design of filters [18]. Based on the investigation of the Cantor fractal geometry, they predicted the fractal could provide a new and promising breakthrough for the design of filters and reflectors. Prior to this, most of the research efforts had been devoted to the application of fractal structure in antennas. Since then, intensive research efforts have been focused on the application of fractal structures in filters. A compact dual-mode T-shape fractal microstrip resonator was proposed in literature [19] and a shorting pin was added to the resonator to excite the degenerate modes in the lower band. By adjusting the size and position of the shorting pin, the proposed resonator could well meet the design requirements. A Minkowski-island-based (MIB) fractal patch resonator was used to design a dual-mode BPF for the applications of wireless local area network (WLAN) [20]. By the perturbation and T-couple / inner-digital coupling, the wide-band and dual-band responses were obtained, respectively. In order to realize the miniaturization of filter and improve the capability of harmonic suppression, the Koch fractal shaped structure was applied to design a compact microstrip BPF [21]. What's more, the Koch fractal electromagnetic bandgap (KFEBG) structures were applied to design improved low-pass filter (LPF) [22]. Juan de Dios Ruiz et al. had designed high-performance BPFs based on the substrate integrated waveguide (SIW) and half mode SIW (HMSIW) with KFEBG patterns etched on the waveguide surface [23]. Moore curve fractal-shaped spiral resonator could provide 49% size reduction as compared to conventional split ring resonators [24], and the BPFs based on the second and third iterations Moore space-filling curve [25] had narrow band frequency responses, high selectivity and blocked harmonics in out-of-band regions. Hilbert-fork resonator [26] was applied to design tri-band bandstop and bandpass filters with excellent performance, small overall sizes, as well as the possibility of independent control of the passbands. Several dual-mode BPFs based on Sierpinski carpet fractal geometry were presented in [27]. With the increase of the number of iterations, both the insertion loss (IL) and fractional bandwidth (FBW) of the BPFs are reduced, and the return loss and the frequency selectivity are improved.

In this paper, a geometrically symmetrical fractal structure is introduced. A new resonator called Greek-cross fractal resonator in this paper is formed by embedding the fractal configuration in the surface of the conventional meandered loop resonator. A dual-mode BPF based on the third iteration GCFR is designed, fabricated and measured. Measurement results show a very close agreement with the simulated ones.

This paper is organized as follows: The generation process and characteristics of the proposed fractal structure are illustrated in Sec. 2 and four dual-mode BPFs based on

the first four iterations GCFR are modeled and studied in Sec. 3. Section 4 provides the simulation and measurement results. The parameters optimization, mode-splitting characteristic and the surface current distribution are also discussed in this section. Finally, a conclusion is drawn in Sec. 5.

2. Generation Process of Greek-cross Fractal Geometry

In the aspects of the applications of fractals in microwave circuits, the most interesting iteration is the one based on the generator. In that case, two components are needed – the initiator and the generator. The initiator is a set of linear segments that comprise a starting shape of a fractal, whilst the generator is an arranged collection of scaled copies of the initiator [28].

The generation process of the Greek-cross fractal geometry is illustrated in Fig. 1. It can be seen that this fractal structure can be generated by using the following steps:

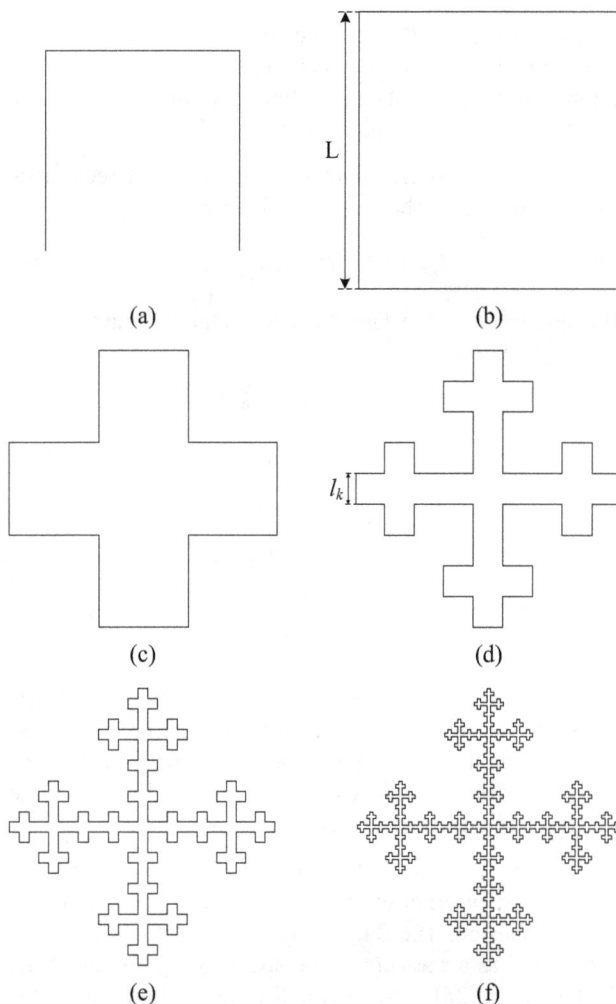

Fig. 1. Generation process of Greek-cross fractal geometry: (a) Generator, (b) Initiator, (c) First iteration, (d) Second iteration, (e) Third iteration, and (f) Fourth iteration.

- The iterative process is started with the initiator, i.e. the zeroth iteration which is a square with the side length of L as shown in Fig. 1(b).

- In the first iteration, each segment of the initiator is replaced by a copy of the generator which is depicted in Fig. 1(a). The dimensions and positions of the copy of the generator are changed so that its end points coincide with the end points of the segment which is to be replaced. Figure 1(c) shows the first iteration of fractal structure.

- In the second iteration, the procedure of copying of the generator is repeated for every segment of the curve obtained after the first iteration. In addition, some line segments which are proportional to the ones of the initiator are added to form a closed curve. Figure 1(d) depicts the second iteration of fractal structure.

- Continuing the above process can obtain the following iterative fractal structures, and the fractal structure of the third and fourth iterations are shown in Fig. 1(e) and Fig. 1(f), respectively.

Theoretically, the iterative process can continue infinitely. However, it has been concluded, in practice, that the number of iterations should be limited to only a few, since otherwise additional complexities arise [29, 30].

The enclosed area donated by A_k of the Greek-cross fractal structure can be derived and given by:

$$A_k = (5/9)^k L^2 \ [\text{mm}^2], k \geq 0 \tag{1}$$

The perimeter of the Greek-cross fractal structure can be derived and given by:

$$P_k = l_k \times n_k \ [\text{mm}], k \geq 0 \tag{2}$$

$$l_k = 3^{-k} \times L \ [\text{mm}], k \geq 0 \tag{3}$$

$$\begin{cases} n_0 = 4 \times 1 = 4, k = 0 \\ n_1 = 4 \times 3 = 12, k = 1 \\ n_k = 4 \times \left(3 \times n_{k-1} + 4 \times \sum_{i=0}^{k-2} \left(3^i \times n_{k-2-i} \right) \right), k \geq 2 \end{cases} \tag{4}$$

where the integer k is the number of iterations, L the side length of the initiator, P_k the perimeter of the fractal geometry after the k-th iteration, n_k the number of the line segments after the k-th iteration, l_k the length of the line segments after the k-th iteration.

From the aspect of miniaturization of circuits, the most important criterion in the selection of a fractal curve is its dimension. The dimension of a fractal curve can be understood as a measure of the space-filling ability of the fractal curve [28]. The higher the fractal dimension, the better the fractal curve fills the given area, therefore achieving higher compactness. The dimension D can be determined as:

$$D = \left[\log(N) \right] / \left[\log(r) \right] \tag{5}$$

where N is the number of self-similar segments obtained from one segment after each iteration and r is the number of segments obtained from one segment in each iteration. According to Fig. 1, we can obtain $N = 5$ and $r = 3$, therefore, the dimension of the proposed Greek-cross fractal structure can be calculated as $D = 1.465$.

3. Design of Fractal BPF

3.1 Greek-cross Fractal Resonator (GCFR)

A conventional meandered loop resonator as shown in Fig. 2(a) is considered to construct the resonator based on Greek-cross fractal geometry. The reason for this is that the meandered loop resonator can be seen as the first iteration GCFR, which is obtained by etching the first iteration Greek-cross fractal geometry on the surface of a cross-shaped patch. The corresponding iterative fractal resonator can be obtained by etching different iterative fractal geometries on the identical cross-shaped patch. The geometrical structures of fractal resonator based on the first four iterations are depicted in Fig. 2(a) to (d), respectively.

It's obvious that with the increase of the iteration, the surface current path length (the perimeter of the fractal geometry) increases and the width of slot line is reduced. This conclusion can also be verified by (2). Therefore, in

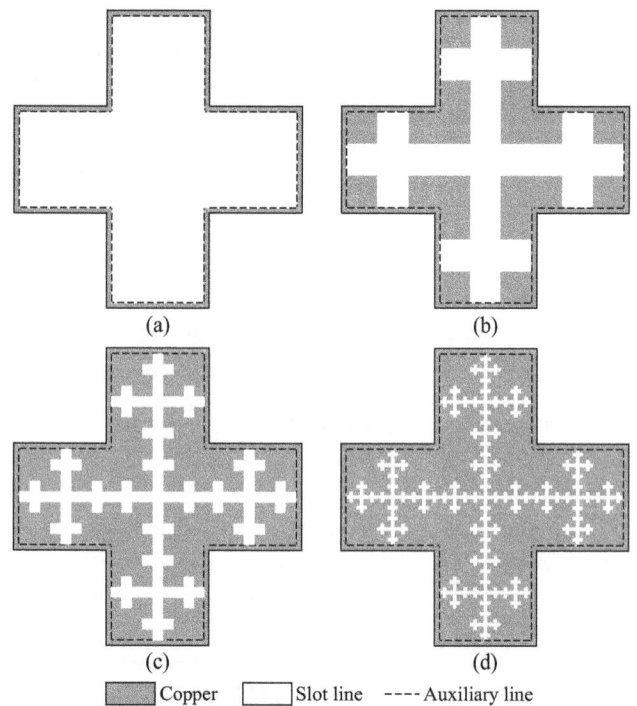

(a)   (b)

(c)   (d)

▬ Copper   ▢ Slot line   ---- Auxiliary line

Fig. 2. Geometrical structure of the proposed GCFR based on the first four iterations: (a) First iteration: meandered loop resonator, (b) Second iteration, (c) Third iteration, and (d) Fourth iteration.

the design of the proposed fractal resonator, the main factor that restricts the iterative process is the width of the slot line. More accurately, the main factor that limits the iterative process is the machining precision.

3.2 Characteristics of GCFR

In order to design a BPF based on the proposed fractal resonator with high performance, the fractal resonator should be analyzed firstly. The quality factor Q is an important parameter for the resonant circuit. It can be seen as a measure of the loss of the resonant circuit. In other words, the lower loss means higher Q. In addition, higher Q can also lead to a narrower bandwidth and a more steep response curve. The unloaded quality factor Q_u for the proposed fractal resonator can be obtained from the measurement using the circuit shown in Fig. 3. The Q_u can be calculated by [31]:

$$Q_u = \frac{Q_l}{1 - |S_{21}|}, \tag{6}$$

$$Q_l = \frac{f_{res}}{\Delta f_{3\text{-}dB}} \tag{7}$$

where Q_l is the loaded quality factor, S_{21} is the IL at the resonance frequency, f_{res} is the resonance frequency, and $\Delta f_{3\text{-}dB}$ is the 3-dB bandwidth of the f_{res}.

In addition, in the process of measuring the Q_u, it can be observed that with the increase of the number of iterations, the resonance frequency of the resonator is gradually shifted to a lower frequency. Figure 4 shows the effect of the number of iterations on the unloaded quality factor Q_u and normalized resonance frequency ratio R. The R can be found using the following formula:

$$R = \frac{f_{res-k}}{f_{res-1}} \tag{8}$$

where f_{res-k} is the resonance frequency of the k-th iteration resonator and f_{res-1} is the resonance frequency of the first

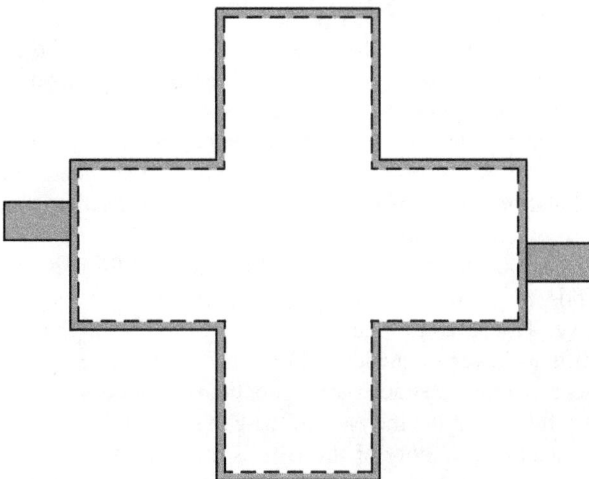

Fig. 4. Normalized resonance frequency ratio R and unloaded quality Q_u against the number k of iterations.

iteration resonator. It can be seen that the change trend gradually becomes more and more slowly with the increase of the number of iteration. So it can be predicted that when k increases to a certain value, R will obtain the minimum value. What's more, the Q_u reaches the maximum value when $k = 3$.

3.3 Filter Design

The geometrical structure of the proposed dual-mode BPF based on the third iteration GCFR is shown in Fig. 5. The other three BPFs are similar in configuration to this one. The only difference is the selected resonator, which is respectively corresponding to the first iteration GCFR, the second iteration GCFR and the fourth iteration GCFR. It has been supposed that these BPFs have been designed on

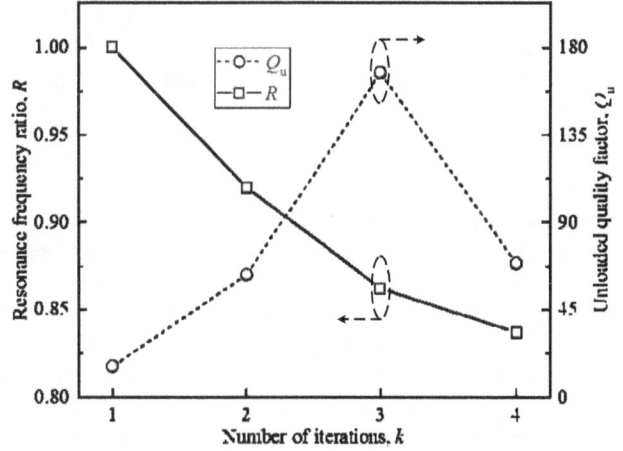

Fig. 3. Sketch map of Greek-cross fractal resonator for the unloaded quality factor Q_u measurement.

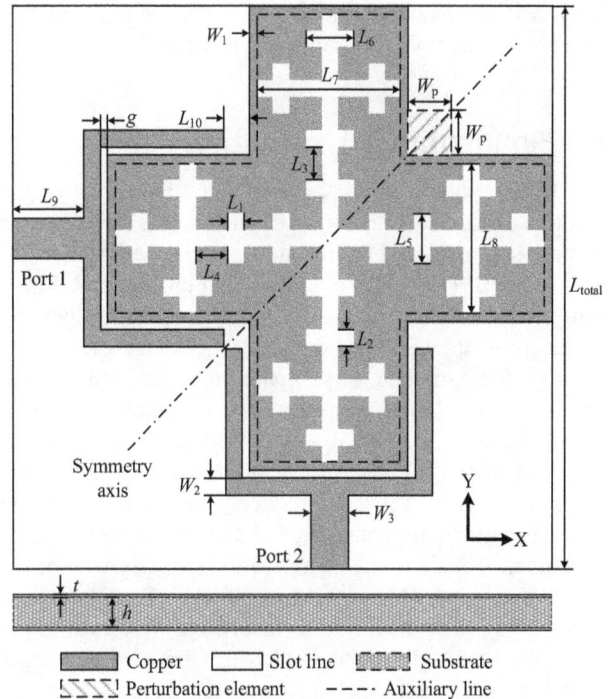

Fig. 5. Geometrical structure of the proposed dual-mode BPF based on the third iteration of GCFR.

the RO3210 substrate with a relative permittivity ε_r of 10.8, a thickness h of 1.27 mm and a loss tangent tanδ of 0.0027. The thickness t of the copper on the substrate is 35 μm.

The width W_3 of the feed line is chosen to be 0.98 mm, which corresponds to the characteristic impedance of 50 Ω. As depicted in Fig. 5, a pair of orthogonal feed lines as the input and output (I/O) ports are connected to two identical U-shaped coupling arms. The resonator is fed by the U-shaped coupling arms by gap coupling. According to the analysis in [32], the length of the coupling arms and the gap size between the coupling arms and the resonator have a great influence on the performance of the filter. By adjusting the length and the gap size adequately, the coupling strength and the frequency response can be optimized.

It's known that the nature of the coupling between the degenerate modes is an especially significant parameter for dual-mode filter design since it determines the filter characteristics [33]. What's more, the strength and nature of the coupling between the degenerate modes of the dual-mode resonator is mainly determined by the perturbation's size and shape [34]. Therefore, as depicted in Fig. 5, a perturbation element in the form of a small square patch is attached to a corner along the symmetry axis of the GCFR. Of course, the perturbation element can also be other shapes and locations to excite the degenerate modes. However, the manner in this paper is one of the most convenient manners to meet the design requirements. Consequently, the identical perturbation elements are added to the same position of other iterative resonators.

For comparison, the occupied area of the four BPFs is exactly the same, in other words, the length of each BPF donated by L_{total} remains the same. However, in order to obtain better filter performance, the size of some dimensions will be made a fine-tuning, such as L_{10} and W_p, whose variation will not change the value of L_{total}.

4. Performance Evaluation

In this paper, we have investigated the effect of the length of U-shaped coupling arms and the gap size between the coupling arms and the resonator on the performance of the fractal BPF as shown in Fig. 5. Figure 6 and Fig. 7 show the simulation results for five cases from changing the length of the U-shaped coupling arms with a fixed gap size ($g = 0.2$ mm) and varying the gap size with a fixed length ($L_{10} = 0.9$ mm), respectively. It can be seen from Fig. 6(a) that only single mode is excited when L_{10} is less than 0.85 mm. With the increase of L_{10}, the length of the coupling arms is gradually reduced, the two degenerate modes are gradually being excited and moving away from each other. What's more, the reflection characteristic of the BPF is degraded. However, the variation of the length of the coupling arms has little influence on the transmission characteristics of the BPF, which can be evidently concluded from Fig. 6(b). In other words, no matter how the length of the coupling arms changes, the insertion loss in band remains unchanged.

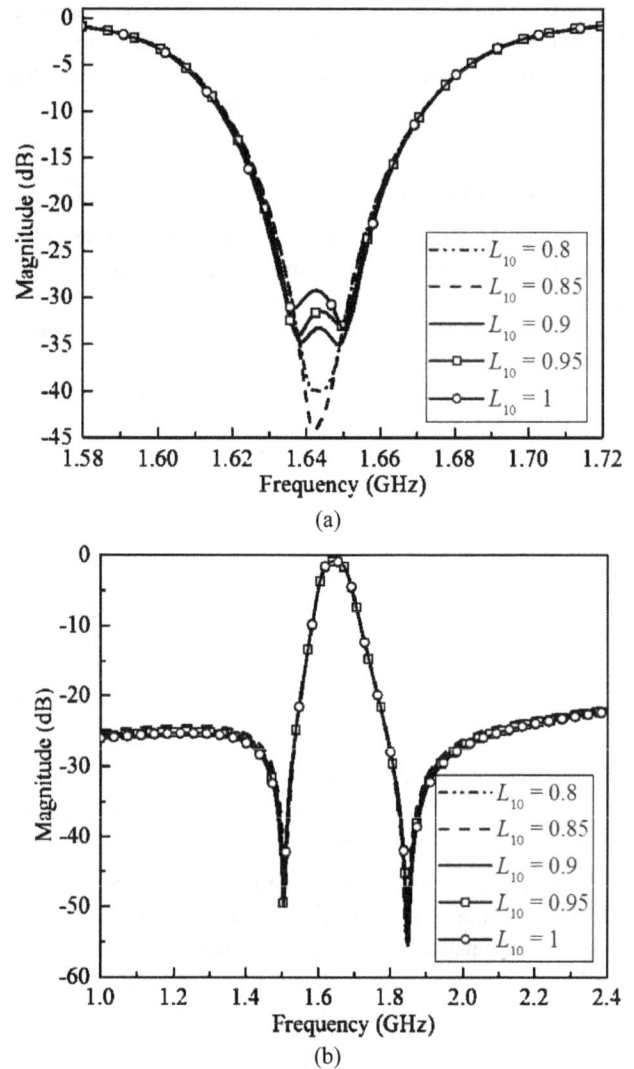

Fig. 6. Simulation results for the performance of the BPF by adjusting the length of the U-shape coupling arms with a fixed gap size ($g = 0.2$ mm). ($W_p = 2.15$ mm) (a) Reflection characteristics (S11). (b) Transmission characteristics (S21).

It's clearly observed from Fig. 7(a) that only single mode is excited when $g = 0.2$ mm, and mode splitting occurs when the value of g is larger than 0.2 mm. With the increase of gap size, the resonance frequencies of the two degenerate modes move away from each other. As shown in Fig. 7(a), higher mode frequency shifts to much higher frequency, while lower mode frequency remains basically constant. As a result, the bandwidth is expanded. The simulation results of the transmission characteristics of the BPF by changing the gap size are shown in Fig. 7(b). Two TZs are located on each side of the passband for these five kinds of conditions. And as the gap size increases, the lower TZs shifts to higher frequency, while the higher TZs shifts to lower frequencies. Combined with the changes in the reflection characteristics mentioned above, it can be concluded that the increase of the gap size can improve the frequency selectivity of the BPF, so that the BPF can obtain sharp passband skirts. Additionally, it's evident from the partial enlarged detail in Fig. 7(b) that with the increase of the gap size g, the single passband gradually splits into

(a)

(b)

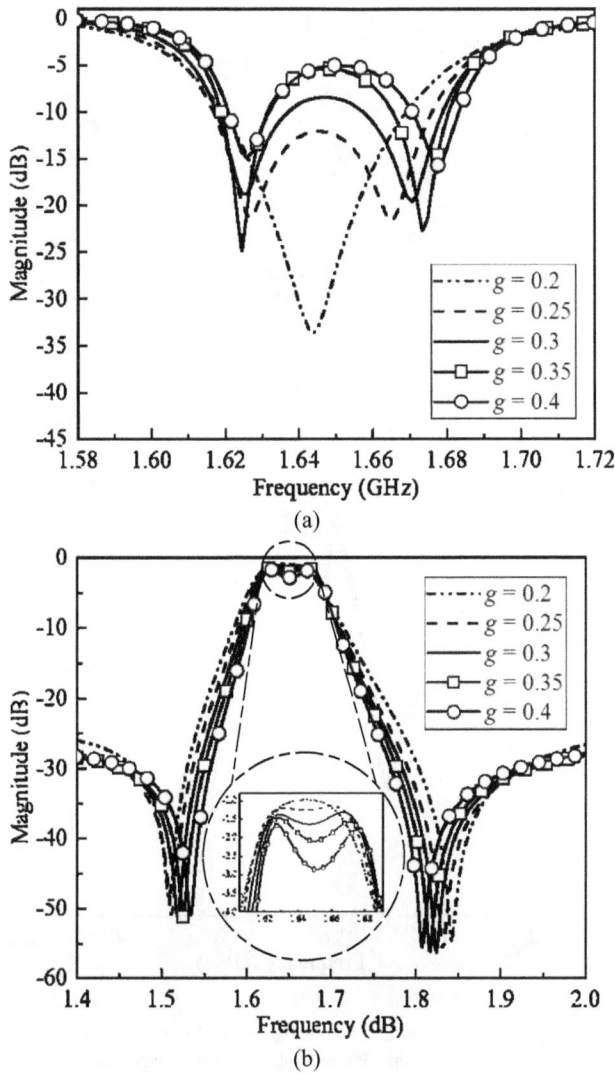

Fig. 7. Simulation results for the performance of the BPF by adjusting the gap size with a fixed length of the U-shape coupling arms ($l_3 = 0.9$ mm), ($W_p = 2.15$ mm): (a) S11, (b) S21.

two frequency bands. The reason for this is that the coupling strength gradually weakens with the increase of the gap size.

In summary, both the length of the coupling arms and the gap size have a certain effect on the characteristics of the BPF. In the design of the BPF, we should focus on the optimization of the two parameters in order to obtain high performance BPF.

Figure 8 shows the simulation results of the change in mode frequencies and coupling coefficient between the degenerate modes with respect to the perturbation size W_p. The coupling coefficient K can be calculated by [35]:

$$K = \frac{f_1^2 - f_2^2}{f_1^2 + f_2^2} \qquad (9)$$

where f_1 and f_2 are the resonance frequencies of the mode-I and mode-II, respectively. As it can be seen from Fig. 8, $f_1 = f_2 = 1.644$ GHz and $K = 0$ when $W_p \leq 2.1$ mm. In other words, only single mode is excited and the corresponding

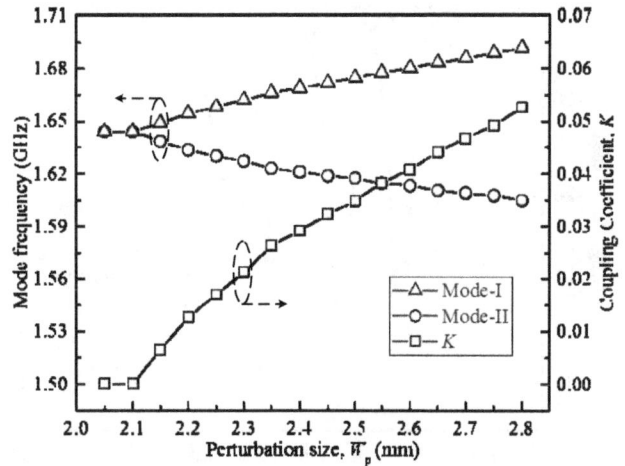

Fig. 8. Simulation results of mode frequency and coupling coefficient between the degenerate modes with respect to the perturbation size W_p ($g = 0.2$ mm, $L_{10} = 0.9$ mm).

coupling coefficient is zero when the perturbation size is less than 2.1 mm. With the increasing of the perturbation size, the two degenerate modes are gradually excited and moving away from each other. It is interesting that the degree of change of the resonance frequencies of the mode-I and mode-II are almost the same. This is because the square perturbation element is symmetrically distributed along the diagonal, and the variation of the perturbation size has the same effect on the electromagnetic field distribution for the mode-I and mode-II. In addition, the coupling coefficient also increases with the increase of the perturbation size.

Based on the above analysis, the physical dimensions of each BPF after optimization are listed in Tab. 1. The simulation results of the frequency responses of these four BPFs are depicted in Fig. 9.

It is obvious from Fig. 9 that all the four fractal BPFs have elliptical frequency responses. In other words, there are two TZs for the passband in the real frequencies. With the increase of the number of iterations, the central frequency of the BPF is shifted to the lower frequency, which is changed from 1.907 GHz in the first iteration to 1.596 GHz in the fourth iteration. Additionally, the corresponding 3-dB bandwidth gradually decreases with the increase of the number of iterations. That is to say, a higher

Dimensions [mm]	First iteration	Second iteration	Third iteration	Fourth iteration
$L_1 = L_2$				0.6
$L_3 = L_4$				1.2
$L_5 = L_6$			1.8	
$L_7 = L_8$			5.4	
L_9			2	
L_{10}	0.9	0.85	0.9	0.8
W_1			0.2	
W_2			0.4	
W_3			0.98	
W_p	2.2	2.2	2.2	2.1
g			0.2	
L_{total}			19.2	

Tab. 1. Comparison table for different parameters.

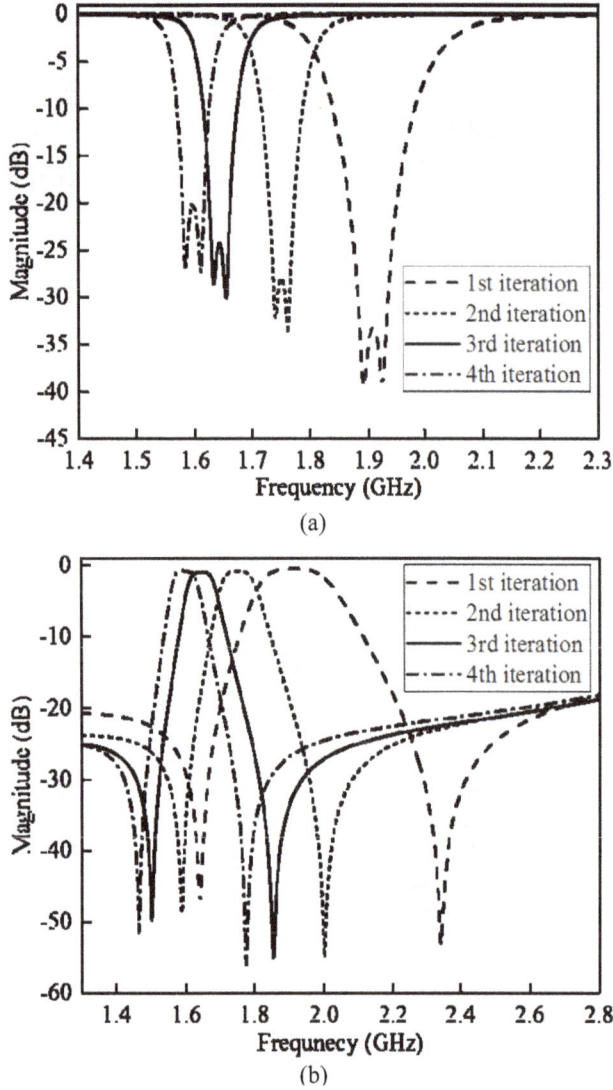

(a)

(b)

Fig. 9. Simulation results for the frequency responses of the BPF based on GCFRs with different iterations: (a) S11, (b) S21.

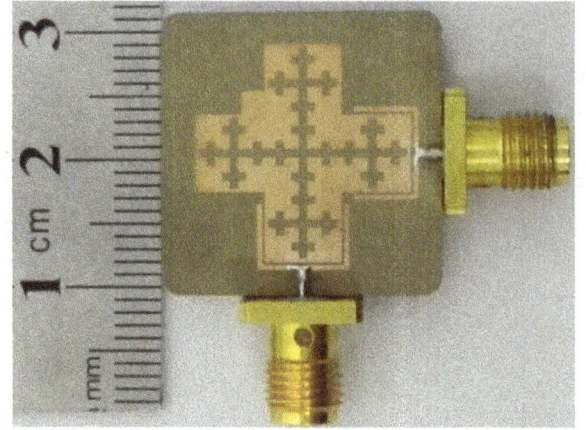

(a)

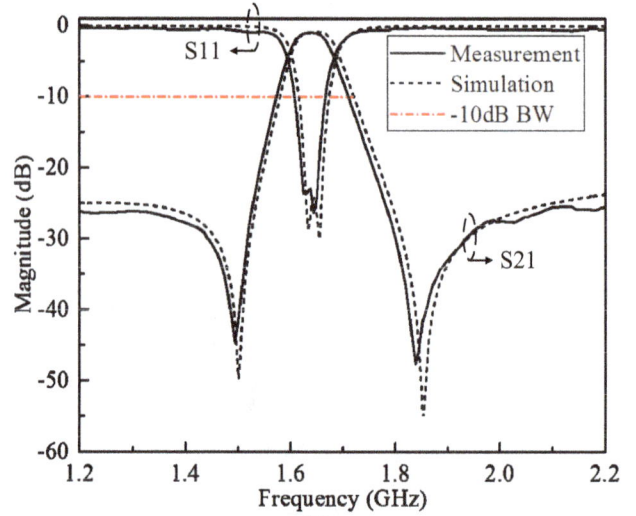

(b)

Fig. 10. Final dual-mode BPF based on the third iteration of GCFR. (a) Photograph of the fabricated BPF. (b) Simulation and measurement results for the proposed BPF.

iterative BPF can achieve a more steep response curve than the lower one. It can be obtained that the 3-dB bandwidth of these four fractal BPFs is 227 MHz, 111 MHz, 87 MHz and 83 MHz, respectively. Therefore, the corresponding FBW is 11.9%, 6.3%, 5.3% and 5.2%, respectively.

According to the above analysis, a dual-mode BPF based on the third iteration GCFR has been fabricated and measured. The measurement results are achieved by using the Agilent network analyzer PNA-L. The photograph of the fabricated BPF and the simulation and measurement results are illustrated in Fig. 10(a) and (b), respectively. The side length of the BPF is about 20 mm, which is approximate $0.267\lambda_g$, where λ_g is the guided wavelength at the design frequency f_0. And λ_g can be calculated by the equation [35]:

$$\lambda_g = \frac{c}{f_0\sqrt{\varepsilon_{eff}}} \tag{10}$$

where c is the speed of light and ε_{eff} is the effective dielectric constant, given by [35] as:

$$\varepsilon_{eff} = \frac{\varepsilon_r + 1}{2} \tag{11}$$

where ε_r is dielectric constant of the substrate.

As can be seen from Fig. 10(b), the dual-mode BPF operates at the central frequency of 1.65 GHz with a FBW of 5.1%. Two TZs with respective frequency location at 1.507 GHz and 1.851 GHz can be clearly observed. At the lower TZ, the measured S_{21} is about –44.7 dB; while for the higher TZ, the measured S_{21} is about –49.2 dB, and the proposed BPF can effectively suppress harmonic response with a better than 25 dB suppression. Additionally, there are two poles in the passband, and the corresponding frequency and IL are 1.637 GHz with –23.9 dB and 1.656 GHz with –26.5 dB.

In order to get insight into the nature of current density distributions at the surface of the fabricated BPF, the simulation results for the surface current density at four different frequencies of operation are depicted in Fig. 11. In these figures, the maximum current density magnitude indicates the strongest coupling effects while the minimum magnitude indicates the weakest ones. As shown in Figure

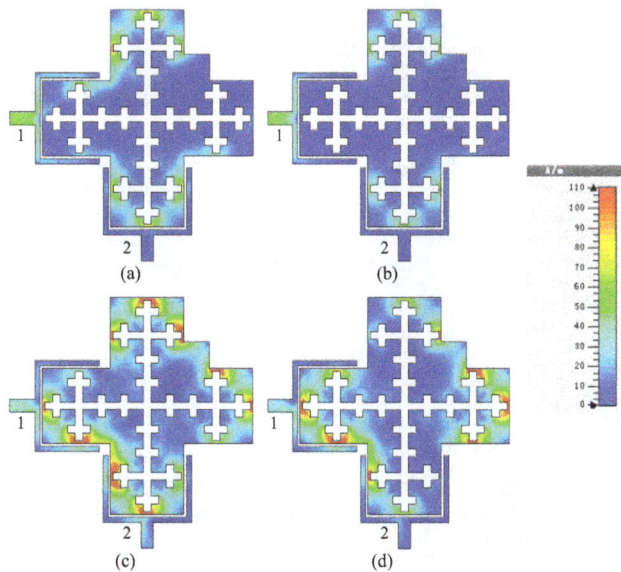

Fig. 11. Simulation results for the current density distribution at the surface on the filter for Port 1: (a) at 1.507 GHz; (b) at 1.851 GHz; (c) at 1.637 GHz; (d) at 1.656 GHz.

11(a) and Fig. 11(b), they are the surface current density distribution of the two TZs at 1.507 GHz and 1.851 GHz, respectively. It's obvious that the energy is mainly distributed on the coupling arm of Port 1 and the upper and lower edges of the resonator, and almost no energy is distributed in the Port 2. This is because the energy is almost reflected back to the Port 1 at the TZs. Figure 11(c) and (d) are the surface current density distribution of the two poles in the passband at 1.637 GHz and 1.656 GHz, respectively. It can be seen that the energy is coupled from Port 1 to Port 2. It's worth noting that the surface current density distributions at 1.637 GHz and 1.656 GHz are different. In addition to the surface current density distribution neat the two ports, the energy for 1.637 GHz is mainly distributed on the upper edge of the resonator, while the energy for 1.656 GHz is mainly distributed on the right edge of the resonator. This is because resonance frequencies of two degenerate modes in a resonator have a 90° phase offset, which is the same as their current distributions.

5. Conclusions

In this paper, a compact dual-mode BPF based on the third iteration GCFR is presented. The new fractal resonator has the property of miniaturization in the design of microstrip BPFs. In addition, the generation process of the fractal structure is described in detail. After parameters optimization, a dual-mode BPF operating at 1.65 GHz with a FBW of 5.1% has been designed, fabricated and measured. The measurement results are in a close agreement with the simulation ones. In addition to the machining accuracy, the deviation between the dielectric constant of the substrate and the theoretical value, which is not considered in the simulation, leads to the small deviation between simulation and measurements. Therefore, the proposed fractal resonator can be used in the miniaturized design of

microwave components in modern satellite and wireless communication systems.

Acknowledgments

This work is supported by the National Science Foundation of China under Grant No.61302017. The authors would like to thank the editor, the associate editor, and the anonymous reviewers for their valuable comments and suggestions that greatly improved this paper.

References

[1] NAGHAR, A., AGHZOUT, O., ALEJOS, A.V., et al. Design of compact multiband bandpass filter with suppression of second harmonic spurious by coupling gap reduction. *Journal of Electromagnetic Waves and Applications*, 2015, vol. 29, no. 14, p. 1813–1828. DOI: 10.1080/09205071.2015.1043029

[2] WOLFF, I. Microstrip bandpass filter using degenerate modes of a microstrip ring resonator. *Electronic Letters*, 1972, vol. 8, no. 12, p. 302–303. DOI: 10.1049/el:19720223

[3] WANG, J.-P., WANG, L., GUO, Y.-X., et al. Miniaturized dual mode bandpass filter with controllable harmonic response for dual band applications. *Journal of Electromagnetic Waves and Applications*, 2009, vol. 23, no. p. 1525–1533. DOI: 10.1163/156939309789476482

[4] ZHU, L., WECOWSKI, P.-M., WU, K. New planar dual-mode filter using cross-slotted patch resonator for simultaneous size and loss reduction. *IEEE Transactions on Microwave Theory and Techniques*, 1999, vol. 47, no. 5, p. 650–654. DOI: 10.1109/22.763171

[5] WU, S., WENG, M.-H., JHONG, S.-B., et al. A novel crossed slotted patch dual-mode bandpass filter with two transmission zeros. *Microwave and Optical Technology Letters*, 2008, vol. 50, no. 3, p. 741–744. DOI: 10.1002/mop.23219

[6] ZHANG, R., ZHU, L., LUO, S. Dual-mode dual-band bandpass filter using a single slotted circular patch resonator. *IEEE Microwave and Wireless Components Letters*, 2012, vol. 22, no. 5, p. 233–235. DOI: 10.1109/lmwc.2012.2192419

[7] SU, Y.-K., CHEN, J.-R., WENG, M.-H., et al. Design of a miniature and harmonic control patch dual-mode bandpass filter with transmission zeros. *Microwave and Optical Technology Letters*, 2008, vol. 50, no. 8, p. 2161–2163. DOI: 10.1002/mop.23604

[8] FU, S., WU, B., CHEN, J., et al. Novel second-order dual-mode dual-band filters using capacitance loaded square loop resonator. *IEEE Transactions on Microwave Theory and Techniques*, 2012, vol. 60, no. 3, p. 477–483. DOI: 10.1109/tmtt.2011.2181859

[9] KARPUZ, C., GORUR, A.K., SAHIN, E. Dual-mode dual-band microstrip bandpass filter with controllable center frequency. *Microwave and Optical Technology Letters*, 2015, vol. 57, no. 3, p. 639–642. DOI: 10.1002/mop.28914

[10] MANDELBROT, B. B. *The Fractal Geometry of Nature.* 1st ed., New York (USA): W. H. Freeman and Company, 1983. ISBN: 0716711869

[11] SONG, C.T.P., HALL, P.S., GHAFOURI-SHIRAZ, H., WAKE, D. Sierpinski monopole antenna with controlled band spacing and input impedance. *Electronic Letters*, 1999, vol. 36, no. 13, p. 1036 to 1037. DOI: 10.1049/el:19990748

[12] PUENTE, C., ANGUERA, J., BORJA, C., et al. Fractal-shaped antennas and their application to GSM 900/1800. *The Journal of the Institution of British Telecommunication Engineers*, 2001, vol. 2, no. 3, p. 92–95.

[13] ANGUERA, J., PUENTE, C., BORJA, C., et al. Fractal-shaped antennas: A review. *Encyclopedia of RF and Microwave Engineering*, 2005, vol. 2, p. 1620–1635. DOI: 10.1002/0471654507.eme128

[14] ANGUERA, J., DANIEL, J. P., BORJA, C., et al. Metallized foams for antenna design: application to fractal-shaped Sierpinski-carpet monopole. *Progress in Electromagnetics Research*, 2010, vol. 104, p. 239–251. DOI: 10.2528/pier10032003

[15] WERNER, D. H., LEE, D. Design of dual-polarized multiband frequency selective surfaces using fractal elements. *Electronic Letters*, 2000, vol. 36, no. 6, p. 487–488. DOI: 10.1049/el:20000457

[16] NEETHU, S., SANTHOSH KUMAR, S. Microstrip bandpass filter using fractal based hexagonal loop resonator. In *2014 Fourth International Conference on Advances in Computing and Communications (ICACC 2014)*. Kochi (India), 2014, p. 319–322. DOI: 10.1109/ICACC.2014.81

[17] ORIZI, H. SOLEIMANI, H. Miniaturisation of the triangular patch antenna by the novel dual-reverse-arrow fractal. *IET Microwaves, Antennas and Propagation*, 2015, vol. 9, no. 7, p. 627–633. DOI: 10.1049/iet-map.2014.0462

[18] YORDANOV, O. I., ANGELOV, I., KONOTOP, V. V., et al. Prospects of fractal filters and reflectors. In *1991 Seventh International Conference on (IEE) Antennas and Propagation (ICAP 91)*. New York (USA), 1991, p. 698–700.

[19] AHMED, E. S. Dual-mode dual-band microstrip bandpass filter based on fourth iteration T-square fractal and shorting pin. *Radioengineering*, 2012, vol. 21, no. 2, p. 617–623.

[20] LIU, J.-C., CHANG, C. C., KUEI, C.-P., et al. Dual-mode wide-band and dual-band resonators with Minkowski-island-based fractal patch for WLAN systems. In *Cross Strait Quad-Regional Radio Science and Wireless Technology Conference (CSQRWC 2011)*. Harbin (Heilongjiang, China), 2011, p. 583–585. DOI: 10.1109/CSQRWC.2011.6037017

[21] ESA, M., THAYAPARAN, D., ABDULLAH, M. S., et al. Miniaturized microwave modufied Koch fractal hairpin filter with harmonic suppression. In *2010 IEEE Asia-Pacific Conference on Applied Electromagnetics (APACE 2010)*. Port Dickson (Negeri Sembilan, Malaysia), Malaysia, 2010, p. 1-4. DOI: 10.1109/APACE.2010.5719750

[22] de DIOS-RUIZ, J., MARTÍNEZ-VIVIENTE, F. L., HINOJOSA, J. Optimisation of chirped and tapered microstrip Koch fractal electromagnetic bandgap structures for improved low-pass filter design. *IET Microwaves, Antennas and Propagation*, 2015, vol. 9, no. 9, p. 889–897. DOI: 10.1049/iet-map.2014.0453

[23] de DIOS-RUIZ, J., MARTINEZ-VIVIENTE, F. L., ALVAREZ-MELCON, A., et al. Substrate integrated waveguide (SIW) with Koch fractal electromagnetic bandgap structures (KFEBG) for bandpass filter design. *IEEE Microwave and Wireless Components Letters*, 2016, vol. 25, no. 3, p. 160–162. DOI: 10.1109/LMWC.2015.2390537

[24] GHATAK, R., PAL, M., GOSWAMI, C., et al. Moore curve fractal-shaped miniaturized complementary spiral resonator. *Microwave and Optical Technology Letters*, 2013, vol. 55, no. 8, p. 1950–1954. DOI: 10.1002/mop.27682

[25] MEZAAL, Y. S., ALI, J. K., EYYUBOGLU, H. T. Miniaturised microstrip bandpass filters based on Moore fractal geometry. *International Journal of Electronics*, 2015, vol. 102, no. 8, p. 1306–1319. DOI: 10.1080/00207217.2014.971351

[26] JANKOVIC, N., GESCHKE, R., CRNOJEVIC-BENGIN, V. Compact tri-band bandpass and bandstop filters based on Hilbert-Fork resonators. *IEEE Microwave and Wireless Components Letters*, 2013, vol. 23, no. 6, p. 282–284. DOI: 10.1109/LMWC.2013.2258005

[27] SONI, V., KUMAR, M. New kinds of fractal iterated and miniaturized narrowband bandpass filters for wireless applications. In *2014 International Conference on Advances in Computing Communications and Informatics (ICACCI 2014)*. Delhi (India), 2014, p. 2786–2792. DOI: 10.1109/ICACCI.2014.6968632

[28] CRNOJEVIC-BENGIN, V. *Advances in Multi-Band Microstrip Filters*. 1st ed. United Kingdom: Cambridge University Press, 2015. ISBN: 978-1-107-08197-0

[29] FALCONER, K. *Fractal Geometry: Mathematical Foundations and Applications*. 1st ed. Chichester (UK): John Wiley and Sons Ltd., ISBN: 2003. 0-470-84861-8

[30] PEITGEN, H.-O., JURGENS, H., SAUPE, D. *Chaos and Fractals*. 1st ed. New York (USA): Springer-Verlag, 2004. ISBN: 978-1-4684-9396-2

[31] YE, C.S., SU, Y.K., WENG, M.H., et al. Resonant properties of the Sierpinski-based fractal resonator and its application on low-loss miniaturized dual-mode bandpass filter. *Microwave and Optical Technology Letters*, 2009, vol. 51, no. 5, p. 1358–1361. DOI: 10.1002/mop.24321

[32] HSIEH, L.-H., CHANG, K. Dual-mode quasi-elliptic-function bandpass filters using ring resonators with enhanced-coupling tuning stubs. *IEEE Transaction on Microwave Theory and Techniques*, 2002, vol. 50, no. 5, p. 1340–1345. DOI: 10.1109/22.999148

[33] GORUR, A. Description of coupling between degenerate modes of a dual-mode microstrip loop resonator using a novel perturbation arrangement and its dual-mode bandpass filter applications. *IEEE Transactions on Microwave Theory and Techniques*, 2004, vol. 52, no. 2, p. 671–677. DOI: 10.1109/TMTT.2003.822033

[34] MANSOUR, R. R. Design of superconductive multiplexers using single-mode and dual-mode filters. *IEEE Transactions on Microwave Theory and Techniques*, 1994, vol. 42, no. 7, p. 1411 to 1418. DOI: 10.1109/22.299738

[35] HONG, J.-S., LANCASTER, M. J. *Microstrip Filters for RF/Microwave Applications*. 1st ed. New York (USA): John Wiley and Sons Ltd., 2001. ISBN: 0-471-38877-7

About the Authors ...

Hongshu LU was born in 1988. He received his M.S. degree in Electronic Science and Technology from the National University of Defense Technology in 2013. Currently he is working towards the Ph.D. degree in the College of Electronic Science and Engineering, National University of Defense Technology, Changsha, Hunan, China. His research interests include passive microwave circuits design and wireless communication.

Weiwei WU was born in 1981. She received her M.S. and Ph.D. degree in Electronic Science and Technology from the National University of Defense Technology in 2008 and 2011, respectively. Currently she is a teacher in the College of Electronic Science and Engineering, National University of Defense Technology, Changsha, Hunan,

China. Her research interests include antennas design and wave propagation.

Jingjian HUANG was born in 1983. He received his Ph.D. degree in Electronic Science and Technology from the National University of Defense Technology in 2014. Currently he is a teacher in the College of Electronic Science and Engineering, National University of Defense Technology, Changsha, Hunan, China. His research interests include antennas design and wave propagation.

Xiaofa ZHANG was born in 1978. He received his Ph.D. degree in Electronic Science and Technology from the National University of Defense Technology in 2007. Currently he is a teacher in the College of Electronic Science and Engineering, National University of Defense Technology, Changsha, Hunan, China. His research interests include microwave circuits and antennas design.

Naichang YUAN was born in 1965. He received his M.S. and Ph.D. degree in Electronic Science and Technology from the University Science and Technology of China in 1991 and 1994, respectively. He is currently a professor with the College of Electronic Science and Engineering, National University of Defense Technology, Changsha, Hunan, China. His research interests include microwave circuits design, wireless communication, antennas and wave propagation.

8

2-D DOA Estimation of LFM Signals Based on Dechirping Algorithm and Uniform Circle Array

Kaibo CUI, Weiwei WU, Xi CHEN, Jingjian HUANG, Naichang YUAN

College of Electronic Science and Engineering, National University of Defense Technology,
Changsha, Hunan, 410073, China

764608294@qq.com

Abstract. *Based on Dechirping algorithm and uniform circle array (UCA), a new 2-D direction of arrival (DOA) estimation algorithm of linear frequency modulation (LFM) signals is proposed in this paper. The algorithm uses the thought of Dechirping and regards the signal to be estimated which is received by the reference sensor as the reference signal and proceeds the difference frequency treatment with the signal received by each sensor. So the signal to be estimated becomes a single-frequency signal in each sensor. Then we transform the single-frequency signal to an isolated impulse through Fourier transform (FFT) and construct a new array data model based on the prominent parts of the impulse. Finally, we respectively use multiple signal classification (MUSIC) algorithm and rotational invariance technique (ESPRIT) algorithm to realize 2-D DOA estimation of LFM signals. The simulation results verify the effectiveness of the algorithm proposed.*

Keywords

2-D DOA estimation, Dechirping algorithm, LFM signal, FFT, MUSIC algorithm, mode-space, ESPRIT algorithm

1. Introduction

Because of the advantages of UCA such as fewer array elements, the ability of 2-D DOA estimation, non-oriented fuzzy (uniform circular array is non-oriented fuzzy when the elements is an odd number greater than 5 or an even number greater than 8 [1]) etc., it has been widely used [2–7]. LFM signals have been widely used in sonar, radar and other detection equipments [8–12]. Especially for the development of imaging technology, LFM signals have become the main choice of the radar. So the DOA estimation of LFM signals based on UCA becomes an important issue. However, as the LFM signal belongs to a typical non-stationary signal, the traditional subspace algorithms which are based on the stationary signals cannot be applied to such condition. With the development of signal processing technology, people have developed a series of DOA estimation algorithms suitable for wideband signals [13–19]. In [13], the author uses a triangular array to realize DOA estimation of broadband LFM signals. There is no need for such algorithm to estimate the signal parameters and solve the signal spectrum. In [14], the author uses a space-time extended MUSIC estimation algorithm to realize DOA estimation of wideband signals. The algorithm shows a good ability to estimate a number of sources that exceed the number of sensors in the array. In [15–19], people use the sparse matrix theory to realize DOA estimation of wideband signals. Since Amin introduces the time-frequency analysis tool to DOA estimation field in 1999 [20], [21], people also have developed a series of DOA estimation algorithms based on the time-frequency analysis tool. In [22–24], the authors proceed DOA estimation by using Wigner-Ville distribution (WVD). But the calculation of WVD is very complex and the cross-terms seriously affect the estimation accuracy in the case of multiple signals. In [25], [26], the authors study the DOA estimation algorithm based on short-time Fourier transform (STFT). These kinds of algorithms avoid cross terms interference of WVD, but it is difficult to select an appropriate time-frequency point. Especially, the irreconcilable conflicts between the STFT window length (the calculation complexity of the algorithm) and the estimate accuracy limit further development of such algorithm. In [27–32], the authors study DOA estimation of LFM signals based on fractional Fourier transform (FRFT). They construct a novel array data mode in the fractional Fourier transform domain (FRFD) based on a fact that LFM signals have the energy concentrated performance in the FRFD and then use MUSIC algorithm to obtain the 2-D DOA estimation of multiple LFM sources. But in their papers, they just discuss the linear and rectangular array model. It is necessary to increase the number of array elements to achieve high accuracy, which is difficult to realize. Also, FRFT is difficult to understand and project implementation. So these kinds of algorithms have not been widely used.

This paper studies 2-D DOA estimation algorithm of LFM signals based on Dechirping algorithm and UCA. Firstly, we use the Dechirping algorithm to process LFM signals received by each array element and transform signals from time domain to frequency domain using Fourier transform (FFT). Then, we construct a new array data model through extracting the prominent parts of the impulse in the frequency domain. Finally, we realize 2-D DOA estimation of LFM signals using MUSIC algorithm and ESPRIT algorithm respectively. For the LFM signals, the time-frequency analysis tools are very suitable to process them [33]. So the main DOA estimation algorithms of LFM signals are based on the time-frequency analysis tools. Dechirping algorithm can reduce the quantity of data [34], so we choose it to estimate DOA of LFM signals in this paper. Also in the method proposed in this paper the cross-term interference does not exist which exists in the DOA algorithms based on WVD. It also does not have the puzzle of selecting the appropriate window function compared to the DOA algorithms based on STFT. The algorithm also does not require selecting the correct time frequency points compared to the DOA algorithms based on WVD and STFT. Unlike the existing methods which are based on more computationally expensive approach, the proposed one is significantly more efficient in terms of computational complexity. The algorithm also uses the relatively simple FFT operator and has the capability of estimating DOA of multiple LFM signals simultaneously with a high precision which are superior to the DOA algorithm based on FRFT. So the proposed algorithm is prone to be easily realized in engineering. The simulation results showed the good performance of the algorithm proposed by this paper.

The paper is organized as follows. In Sec. 2, we introduce the new array data model of LFM signals based on Dechirping algorithm and we proceed detailed derivation in this part. In Sec. 3, we realize 2-D DOA estimation of LFM signals using MUSIC algorithm and ESPRIT algorithm respectively. For ESPRIT algorithm, we also study the mode-space algorithm. In Sec. 4, we proceed the numerical simulation on the algorithm proposed by this paper. Finally, Section 5 concludes the paper.

2. The Novel UCA Data Model of LFM Signals Based on Dechirping Algorithm

The UCA model with N sensors is shown in Fig. 1. The UCA radius is r and $A_1, A_2,...,A_N$ are elements of the UCA separately. Without loss of generality, we can set the angle between A_1 and X axis to be w_0 and the angle between the array elements is w, so $w = 2\pi/N$. The angle between the far-field incident wave and Z axis is β, which is called the elevation angle. The angle between the projection of the incident wave in XOY plane and X axis is α, which is called the azimuth angle.

If there are M LFM signals from far-field, the output of the n th element can be written as:

$$x_n(t) = \sum_{m=1}^{M} s_m(t - \tau_{nm}) + n_n(t), n = 1, 2,...,N , \quad (1)$$

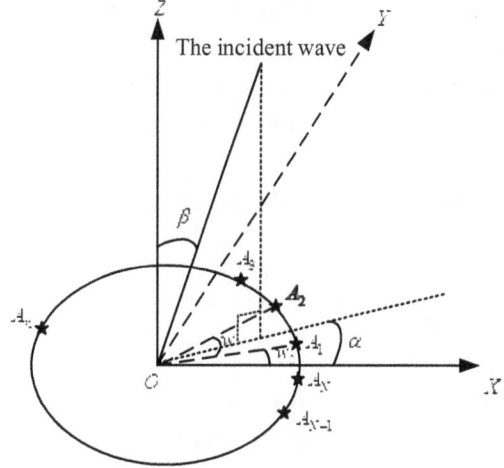

Fig. 1. The uniform circular array model.

$$s_m(t) = \exp\left[j\pi(2f_m t + \gamma_m t)\right] . \quad (2)$$

In (1) and (2), $n_n(t)$ is the noise, $s_m(t)$ is the mth LFM signal, f_m is the central frequency of the mth LFM signal, γ_m is the modulated frequency of the mth LFM signal, τ_{nm} is the time delay of the mth LFM signal on the nth element respect to the reference element. If we regard O as a virtual reference element and the speed of light is c, the time delay caused by the wave path difference for each element respect to the reference element is:

$$\tau_{nm}(t) = \frac{r}{c}\sin(\theta_{em})\cos(\theta_{am} - \frac{2\pi(n-1)}{N}) . \quad (3)$$

In (3), θ_{am} is the azimuth angle of the mth LFM signal and θ_{em} is the elevation angle of the mth LFM signal.

In this paper, we assume that the modulated frequencies of LFM signals are not identical, namely:

$$\gamma_i \neq \gamma_j, i \neq j, i, j \in 1, 2,...,M . \quad (4)$$

Dechirping algorithm is a kind of time-domain transform technique used in high-resolution radar, which can greatly reduce the amount of data, so it has been widely used in the imaging field. Firstly, we construct a LFM signal whose central frequency and modulated frequency are the same as the original LFM signal. Then, we process difference frequency treatment between the original signal and the reference signal. Finally, we can get a single-frequency signal. For the LFM signals received by each array element, we use the ith LFM signal received by the reference element to carry through Dechirping process. So we can get:

$$y_n(t) = x_n(t) * s_i^*(t) = \sum_{m=1}^{M} z_{nm}(t) . \quad (5)$$

$$z_{nm}(t) = \quad (6)$$
$$\exp\left\{j\pi\left[(\gamma_m - \gamma_i)t^2 + 2(f_m - f_i - \gamma_m \tau_{nm})t + \gamma_m \tau_{nm}^2 - 2f_m \tau_{nm}\right]\right\}.$$

Due to the different modulated frequency of all LFM signals, $z_{nm}(t)$ is a single-frequency signal just when $m = i$, otherwise $z_{nm}(t)$ is a LFM signal. So the received signals of each element are the superposition between $(M-1)$ LFM signals and a single-frequency signal after Dechirping process.

For (5) and (6), we can execute FFT on both ends of the equal sign to get the frequency-domain expression.

$$Y_n(\omega) = 2\pi \exp\left[j\pi(\gamma_i \tau_{ni}{}^2 - 2 f_i \tau_{ni}) \right] \delta(\omega + 2\pi\gamma_i\tau_{ni}) +$$

$$\sum_{m=1,m\neq i}^{M} \frac{1}{\sqrt{|Q_m|}} \exp\left[j\pi\left(C_{nm} + \frac{1}{4} \right) \right] \exp\left[-j\pi \frac{\left(\frac{\omega}{2\pi} - B_{nm} \right)^2}{Q_m} \right]. \quad (7)$$

In (7), δ is the impulse function, $Q_m = \gamma_m - \gamma_i$, $B_{nm} = f_m - f_i - \gamma_m \tau_{nm}$, $C_{nm} = \gamma_m \tau_{nm}{}^2 - 2 f_m \tau_{nm}$.

Since each LFM signal owns a certain bandwidth, the result of (7) is an isolated impulse point when $\omega = -2\pi\gamma_i\tau_{ni}$. So we can regard this point as the output of the array, namely:

$$X_n = Y_n(\omega)\big|_{\omega = -2\pi\gamma_i\tau_{ni}}$$

$$= 2\pi\exp\left[j\pi(\gamma_i\tau_{ni}{}^2 - 2 f_i\tau_{ni}) \right] + 2\pi\sum_{m=1,m\neq i}^{M} D_{nm}, \quad (8)$$

$$D_{nm} = \frac{1}{2\pi\sqrt{|Q_m|}} \exp\left[j\pi\left(C_{nm} + \gamma_i\tau_{ni} + B_{nm} + \frac{1}{4} \right) \right]. \quad (9)$$

For (8), we only take the prominent parts of the isolated impulse for each LFM signals.

$$X_{ni} = 2\pi A_{ni}, \quad (10)$$

$$A_{ni} = \exp\left[j\pi(\gamma_i\tau_{ni}{}^2 - 2 f_i\tau_{ni}) \right]. \quad (11)$$

The second term of A_{ni} is very small, so we generally ignore it in practice, namely:

$$A_{ni} \approx \exp\left[-j2\pi f_i\tau_{ni} \right]. \quad (12)$$

We can let i traverse from 1 to M and get X_{ni} through the above method. In this way, we can get a new array data model.

$$\begin{cases} \mathbf{X} = \mathbf{A}\mathbf{S} + \mathbf{N} \\ \mathbf{X} = [\mathbf{X}_1, \ldots, \mathbf{X}_N]^{\mathrm{T}} \\ \mathbf{X}_n = [X_{n1}, \ldots, X_{nM}] \\ \mathbf{A} = [\mathbf{A}_1, \ldots, \mathbf{A}_M] \\ \mathbf{A}_m = [A_{1m}, \ldots A_{Nm}]^{\mathrm{T}} \\ \mathbf{S} = \mathrm{diag}\{\overbrace{2\pi, \ldots, 2\pi}^{M}\} \end{cases} \quad (13)$$

3. 2-D DOA Estimation Algorithm Based on Dechirping Algorithm

3.1 Dechirping-MUSIC Algorithm

From (13), we can get the correlation matrix of the array output [35].

$$\mathbf{R}_{\mathbf{X}} = E[\mathbf{X}\mathbf{X}^{\mathrm{H}}] = \mathbf{A}\mathbf{R}_{\mathbf{S}}\mathbf{A}^{\mathrm{H}} + \sigma^2\mathbf{I}. \quad (14)$$

In (14), $\mathbf{R}_{\mathbf{S}}$ is the correlation matrix of the LFM signals, σ^2 is the power of Gaussian white noise, $\mathbf{I}$ is the unit matrix. Since the signal and noise are independent, $\mathbf{R}_{\mathbf{X}}$ can be decomposed into two parts: signal and noise. So, we can proceed features decomposition and get the following expression.

$$\mathbf{R}_{\mathbf{X}} = \mathbf{U}_{\mathbf{S}}\sum\nolimits_{\mathbf{S}}\mathbf{U}_{\mathbf{S}}{}^{\mathrm{H}} + \mathbf{U}_{\mathbf{N}}\sum\nolimits_{\mathbf{N}}\mathbf{U}_{\mathbf{N}}{}^{\mathrm{H}}. \quad (15)$$

In (15), $\mathbf{U}_{\mathbf{S}}$ and $\mathbf{U}_{\mathbf{N}}$ respectively are signal subspace and noise subspace. So we can construct the MUSIC spatial spectrum [31] as follows.

$$P(\theta_{am}, \theta_{em}) = \frac{1}{\mathbf{A}_m{}^{\mathrm{H}}(\theta_{am}, \theta_{em})\mathbf{U}_{\mathbf{N}}\mathbf{U}_{\mathbf{N}}{}^{\mathrm{H}}\mathbf{A}_m(\theta_{am}, \theta_{em})}. \quad (16)$$

We can proceed two-dimensional search based on (16) and find the angle $(\theta_{am}, \theta_{em})$ when its value is maximum. $(\theta_{am}, \theta_{em})$ is the DOA of the mth LFM signal.

From (16), we can also see that $\mathbf{A}_m(\theta_{am}, \theta_{em})$ is different for different LFM signals, so in the case of multiple signals, the Dechirping-MUSIC algorithm needs to proceed the two-dimensional search repeatedly, which causes huge computation. So it is necessary to research the algorithm based on Dechirping algorithm and ESPRIT algorithm.

3.2 Dechirping-ESPRIT Algorithm

ESPRIT algorithm needs rotational invariance of the array [36], [37]. But the circular array does not have this structural characteristic, so we use mode-space algorithm firstly in this paper. Mode-space algorithm is a kind of spectral estimation algorithm for the UCA [38], [39] and its central idea is to equalize a UCA to a virtual uniform linear array (ULA) through matrix transformation so that the DOA estimation algorithm based on the ULA can be applied to a UCA. We ignore the details of the mode-space algorithm in the main body and show the results as (17). The details of UCA equalization to ULA are included in Appendix: The derivation of (17).

$$
\begin{cases}
\mathbf{Y} = \mathbf{TX} = \mathbf{A}_c\mathbf{S} + \mathbf{N} \\[4pt]
\mathbf{A}_c = \begin{bmatrix} J_{-K}(-\beta_1)\exp(-jK\theta_{e1}) & \cdots & J_{-K}(-\beta_M)\exp[-jK\theta_{eM}] \\ \cdots & \ddots & \vdots \\ J_K(-\beta_1)\exp[jK\theta_{e1}] & \cdots & J_K(-\beta_M)\exp[jK\theta_{eM}] \end{bmatrix} \\[6pt]
\mathbf{T} = \dfrac{1}{N}\mathbf{J}^{-1}\mathbf{F}^{H} \\[6pt]
\mathbf{F} = \begin{bmatrix} \mathbf{w}_{-K} & \mathbf{w}_{-K+1} & \cdots & \mathbf{w}_K \end{bmatrix} \\[6pt]
\mathbf{w}_q = \left[1 \quad \exp\left[-j\dfrac{2\pi}{N}q\right] \quad \cdots \quad \exp\left[-j\dfrac{2\pi(N-1)}{N}q\right] \right]^{H} \\[6pt]
\mathbf{J} = \mathrm{diag}\{j^{-K},\ldots,j^{K}\}
\end{cases} \tag{17}
$$

In (17), $\mathbf{Y}$ is the output of the virtual ULA, $\mathbf{T}$ is the transformation matrix, $\mathbf{A}_c$ is the manifold matrix of the virtual linear array, $\mathbf{F}$ is the spatial discrete Fourier transform (DFT) transformation matrix, $J_m(x)$ is the order m of Bessel function, β_m is a factor related with the incoming wave parameters. K is the maximum phase mode excited by the UCA, so the number of phase mode excited is $2K+1$.

$$
\beta_m = \frac{2\pi r f_m}{c}\sin\theta_{em}. \tag{18}
$$

$$
K = \left\lfloor \min\left[\frac{2\pi r f_m}{c}\right] \right\rfloor. \tag{19}
$$

In (19), $\min\lfloor\ \rfloor$ expresses the minimum, $\lfloor\ \rfloor$ is the rounded down symbol.

In this way, the UCA is converted to be a ULA whose element number is $2K+1$.

As the parameters of LFM signals are different, so:

$$
\beta_i \neq \beta_j, i \neq j, i,j = 1,2,\ldots,M. \tag{20}
$$

Based on (20), we can see that $\mathbf{A}_c$ is not a general manifold matrix of the ULA. So we cannot simply divide the array into several sub-arrays and use ESPRIT algorithm. Bessel functions have the recursive nature of the order, namely:

$$
J_{m-1}(x) + J_{m+1}(x) = \frac{2m}{x}J_m(x). \tag{21}
$$

We can get the following expressions from (17) and (21).

$$
\begin{cases}
\begin{bmatrix} \mathbf{C}_1 & \mathbf{C}_3 \end{bmatrix}\begin{bmatrix} \boldsymbol{\xi} & \boldsymbol{\xi}^* \end{bmatrix}^{T} = \mathbf{L}\mathbf{C}_2 \\[4pt]
\mathbf{L} = \mathrm{diag}\{2(-K+1) \quad \cdots \quad 2(K-1)\} \\[4pt]
\boldsymbol{\xi} = \mathrm{diag}\{\beta_1\exp(j\theta_{a1}) \quad \cdots \quad \beta_M\exp(j\theta_{aM})\}
\end{cases} \tag{22}
$$

In (22), $\mathbf{C}_1$ is a $(2K-1)\times M$ sub-array cut out at the beginning of the first line of $\mathbf{A}_c$, $\mathbf{C}_2$ is a $(2K-1)\times M$ sub-array cut out at the beginning of the second line of $\mathbf{A}_c$, and $\mathbf{C}_3$ is a $(2K-1)\times M$ sub-array cut out at the beginning of the third line of $\mathbf{A}_c$.

Because $\mathbf{A}_c$ is the manifold matrix of the virtual ULA, so $\mathbf{C}_1$, $\mathbf{C}_2$ and $\mathbf{C}_3$ separately are the manifold matrix of the

three sub-array, which correspond to their own signal subspace, so (22) can also be written as:

$$
\begin{cases}
\begin{bmatrix} \mathbf{U}_{S_1} & \mathbf{U}_{S_3} \end{bmatrix}\boldsymbol{\psi} = \mathbf{L}\mathbf{U}_{S_2} \\[4pt]
\boldsymbol{\psi} = \begin{bmatrix} \boldsymbol{\xi} & \boldsymbol{\xi}^* \end{bmatrix}^{T}
\end{cases} \tag{23}
$$

In (23), $\mathbf{U}_{S_1}$, $\mathbf{U}_{S_2}$ and $\mathbf{U}_{S_3}$ separately are the signal subspaces of the three sub-arrays.

We can use the least square (LS) method to solve the above equation and the result is:

$$
\boldsymbol{\psi}_{LS} = \begin{bmatrix} \boldsymbol{\psi}_{LS1} & \boldsymbol{\psi}_{LS2} \end{bmatrix}^{T} = \begin{bmatrix} \mathbf{U}_{S_1} & \mathbf{U}_{S_3} \end{bmatrix}^{+}\mathbf{L}\mathbf{U}_{S_2}. \tag{24}
$$

In (24), $^{+}$ represents the generalized inverse matrix. $\boldsymbol{\psi}_{LS1}$ is the first M rows data of $\boldsymbol{\psi}_{LS}$ and $\boldsymbol{\psi}_{LS2}$ is the second M rows data of $\boldsymbol{\psi}_{LS}$. We can separately proceed features decomposition for $\boldsymbol{\psi}_{LS1}$ and $\boldsymbol{\psi}_{LS2}$. The eigenvalues of them respectively are λ_{1m} ($m = 1,\ldots,M$) and λ_{2m} ($m = 1,\ldots,M$), so the DOA of the mth LFM signal is:

$$
\begin{cases}
\theta_{am} = \arg(\lambda_{1m}/\lambda_{2m})/2 \\[4pt]
\theta_{em} = \arcsin\left(\dfrac{\mathrm{real}(\lambda_{1m})c}{2\pi r f_m \cos\theta_{am}}\right).
\end{cases} \tag{25}
$$

In (25), $\arg()$ represents a phase angle function of a plural and $\mathrm{real}()$ means to take the real part of a plural.

We can see from (25) that the Dechirping-ESPRIT algorithm does not need to proceed eigen decomposition repeatedly in the case of multiple signals.

For the readers' convenience, the step by step procedure of the proposed algorithm is given in Tab. 1.

If there are M LFM signals arriving at a UCA with N sensors and the sampling numbers are L for each signal. We can analyze the computational complexity of the algorithm based on the procedure of the algorithm in Tab. 1. For the DOA estimation algorithms based on time-frequency analysis tools, the differences of the computational complexity are mainly concentrated on the process of

1)	Select the ith LFM signal as the reference signal to proceed Dechirping process.
2)	Execute FFT on the Dechirping results and get the prominent parts of the isolated impulse.
3)	Let i traverse from 1 to M and repeat (1) – (2) to get the novel array data model. For Dechirping-MUSIC algorithm, proceed (4). For Dechirping-ESPRIT algorithm, proceed (5)-(7).
4)	Proceed the two-dimensional search according to (16) and get the DOA estimation.
5)	Proceed the mode-space transformation based on (17) and get the new manifold matrix.
6)	Construct the rotational invariant equation according to (23) and get $\boldsymbol{\psi}_{LS}$ according to (24).
7)	Proceed eigen decomposition for $\boldsymbol{\psi}_{LS}$ and obtain the DOA estimation based on (25).

Tab. 1. The complete procedure of the algorithm proposed.

establishing the new array model. So the algorithm computational complexity in this paper is the computational complexity of establishing the array model. Based on the procedure of the algorithm in Tab. 1, M LFM signals need to be processed by Dechirping algorithm in each sensor firstly and the computation is MNL. Then, the Dechirping results are transformed from time domain to frequency domain using FFT whose computation is $MNL\log_2 L$. Finally, we can construct a new array data model through extracting the prominent parts of the impulse in the frequency domain and the computation is MN. So the computational complexity of the proposed algorithm is $O(MNL\log_2 L)$. For the DOA estimation algorithms based on WVD [20–24], the computational complexity is $O(MN^2 L^2\log_2 L)$. For the DOA estimation algorithms based on STFT [25], [26], the computational complexity is $O(MNL^2\log_2 L)$. For the DOA estimation algorithms based on FRFT [27–32], the computational complexity is $O(NL + MN^2)$. So the computational complexity of the proposed algorithm achieves a low level contrast to the other DOA estimation algorithms based on time-frequency analysis tools, which is prone to realization in engineering.

4. Numerical Simulation

We proceed the numerical simulation of the proposed algorithm in this paper and use a UCA with sixteen sensors

Tab. 2. The parameters of three LFM signals.

Signal number	Frequency [Hz]	Modulated frequency [Hz/s]	Sampling rate [Hz]
1	10×10^8	6×10^{12}	10^8
2	8×10^8	-2×10^{12}	10^8
3	9×10^8	3×10^{12}	10^8
Signal number	Snapshot numbers	SNR [dB]	DOA [°]
1	300	20	(−25,−35)
2	300	20	(25,−15)
3	300	20	(35,15)

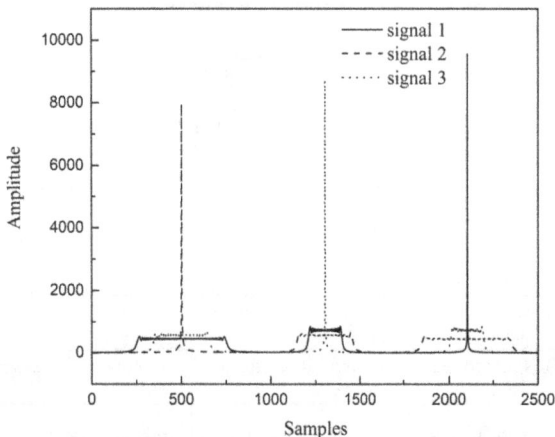

Fig. 2. The results of the reference element after Dechirping process and FFT.

whose radius is 0.4 m. There are three far-field LFM signals whose parameters are shown in Tab. 2.

Firstly, the LFM signals received by each element are processed by the Dechirping algorithm and the frequency domain result of the reference element is shown in Fig. 2.

As can be seen from the figure, the three LFM signals correspond to the three isolated impulse points in the frequency domain after Dechirping process. We can take the prominent parts of the isolated impulse for each array element to get the new array output matrix.

4.1 Dechirping-MUSIC Algorithm Simulation

We use the Dechirping-MUSIC algorithm to estimate DOA of the three LFM signals and the MUSIC spectrum is shown in Fig. 3. In order to facilitate the observation, the MUSIC spectrum of these three signals is displayed on a map, which means that the value of the azimuth angle and elevation angle is changed every three points.

We proceed the two-dimensional search on the MUSIC spectrum and the search range are all from −90° to 90° in the azimuth and elevation direction. The search step is 0.01°. We can get the estimates of DOA by searching the peak point which respectively are: (−25.02°,−35.01°), (25.02°,−14.99°), and (35.02°, 15.01°). The estimates match with the DOA set in Tab. 2.

In order to further examine the performance of the algorithm, we also analyze the relationship between the root mean square error (RMSE) of DOA estimation and signal to noise ratio (SNR) of LFM signals. In addition to an incremental SNR of LFM signals, the other parameters remain unchanged. In order to reduce the effect of random errors, we proceed a total of 1000 times Monte Carlo simulation and the results are shown in Fig. 4.

In Fig. 4(a) there is the RMSE of azimuth angle along with SNR and in Fig. 4 (b) there is the RMSE of elevation angle along with SNR. For comparison purpose, the Cramer-Rao lower bound (CRLB) [40–44] is also presented in Fig. 4. As can be seen from the figure, the RMSE

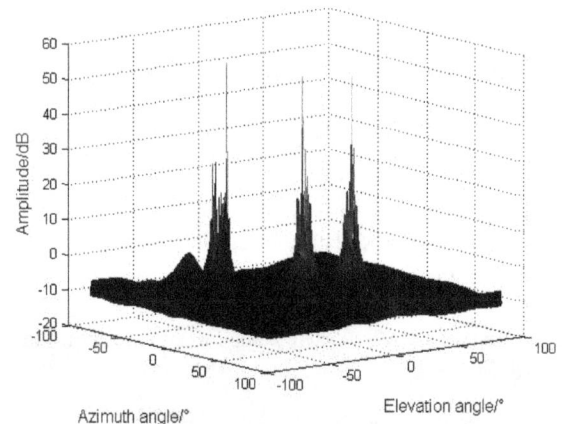

Fig. 3. The Dechirping-MUSIC spectrum of these three LFM signals.

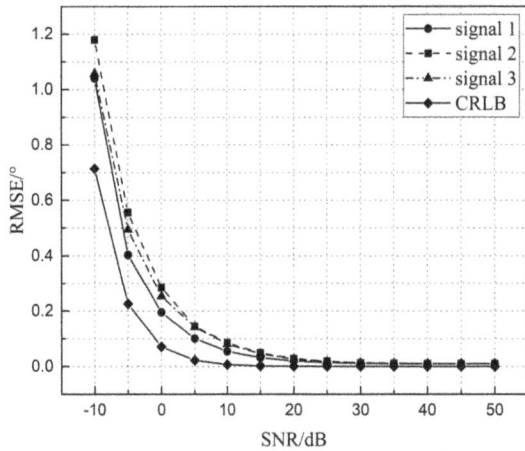

(a) Azimuth angle

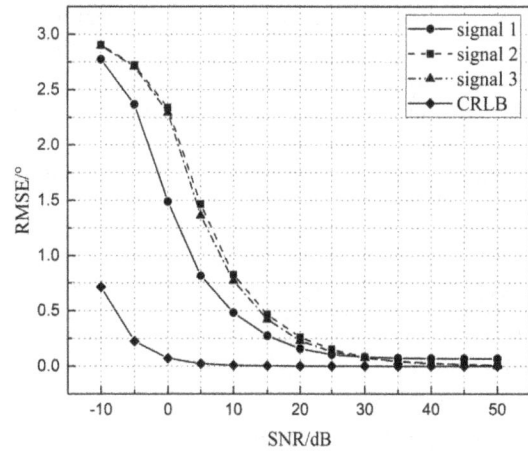

(b) Elevation angle

Fig. 4. RMSE of DOA estimation along with SNR using Dechirping-MUSIC algorithm.

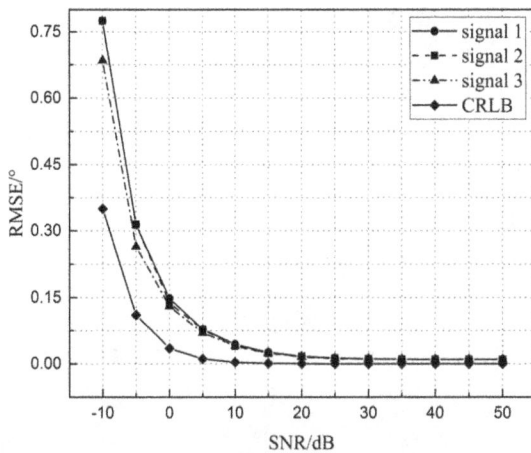

(a) Azimuth angle

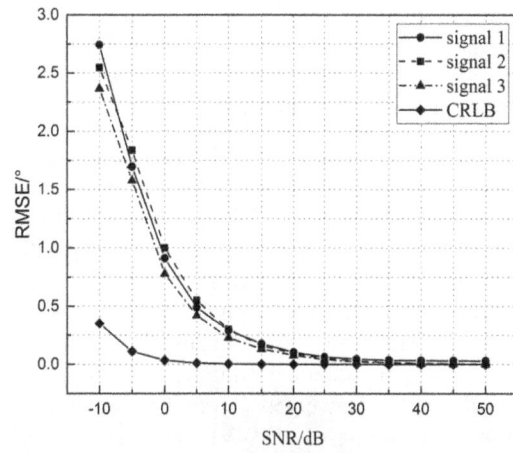

(b) Elevation angle

Fig. 5. RMSE of DOA estimation along with SNR using Dechirping-ESPRIT algorithm.

of DOA estimations decrease rapidly as the SNR of signals increase and achieve the convergence condition (The RMSE is infinitely close to zero) ultimately. In the case of low SNR (less than 0 dB), the algorithm still has a good estimation performance. For a UCA, it has isotropy for the waves of different direction in theory, however we use the approximation in (12) when constructing the new array data model, so the DOA estimation performance for the waves of different direction are not the same. The simulation results are in line with expectations.

4.2 Dechirping-ESPRIT Algorithm Simulation

We use the Dechirping-ESPRIT algorithm to estimate DOA of the three LFM signals. According to (19), we can get that the value of K is 6, which means the number of phase mode excited is 13, so we can equalize this UCA to a ULA with 13 elements based on (17). Then following the steps 6) and 7), we can obtain the estimates of DOA for these three LFM signals, which are: $(-25.28°, -34.96°)$, $(25.19°, -14.94°)$, and $(35.26°, 14.93°)$. The estimates match with the angle set in Tab. 2.

This paper also simulates the relationship between RMSE of DOA estimation of Dechirping-ESPRIT algorithm and SNR of these three LFM signals. In addition to an incremental SNR, the other signal parameters remain unchanged. We proceed a total of 1000 times Monte Carlo simulation and the results are shown in Fig. 5.

In Fig. 5(a) there is the RMSE of azimuth angle along with SNR and in Fig. 5(b) there is the RMSE of elevation angle along with SNR. As can be seen from the figure, the RMSE of DOA estimations decrease rapidly as the SNR of signals increase and achieve the convergence condition ultimately which are similar to the Dechirping-MUSIC algorithm. Compared with the Dechirping-MUSIC algorithm, the RMSE of DOA estimation of Dechirping-ESPRIT algorithm is larger, which means the estimation precision is lower. This is due to the inherent high precision of MUSIC algorithm [44]. But the advantage of Dechirping-ESPRIT algorithm is that the calculation amount is less. Especially in the case of multiple signals, the Dechirping-ESPRIT algorithm does not need to proceed eigen decomposition repeatedly, which reduces the computational complexity greatly. When using Dechirping-ESPRIT algorithm, we use the approximation not only when constructing the new array data model but also when proceeding mode-space conversion in (17), so the DOA estimation performance for the waves of different direction are not the same and the differences are more obvious compared with Dechirping-MUSIC algorithm according to Fig. 4 and Fig. 5. The simulation results are in line with expectations.

4.3 Algorithm Comparison Simulation

In order to further examine the performance of the algorithm, we proceed the comparison simulation between WVD-MUSIC algorithm in [21], WVD-ESPRIT algorithm in [24], STFT-ESPRIT algorithm in [25], STFT-MUSIC algorithm in [26], FRFT-ESPRIT algorithm in [27], FRFT-MUSIC algorithm in [29] and the algorithm proposed by this paper. We select signal 1 in Tab. 2 as the incoming wave signal and the parameters of uniform circle array are unchanged. We proceed a total of 1000 times Monte Carlo simulation and the results are shown in Fig. 6.

In Fig. 6(a) there is the RMSE of azimuth angle along with SNR based on MUSIC algorithm, (b) is the RMSE of elevation angle along with SNR based on MUSIC algorithm, (c) is the RMSE of azimuth angle along with SNR based on ESPRIT algorithm, (d) is the RMSE of elevation angle along with SNR based on ESPRIT algorithm. We can see from Fig. 6 that the estimation precision of four kinds of DOA estimation algorithm is every high and they all can achieve the convergence condition finally. By contrast, the algorithm based on WVD owns the highest estimation precision, followed by the algorithm based on STFT, the algorithm proposed in this paper and the algorithm based on FRFT. Meanwhile, the estimation precision differences of these four kinds of algorithm are not big. Especially in high SNR situation, the estimation precision is almost the same. But based on the above analysis, the algorithm proposed in this paper owns the lowest computational complexity and uses the relatively simple FFT operator. In addition, when constructing the new array data model of LFM signals, the algorithm proposed in this paper extracts only the prominent parts of the impulse.

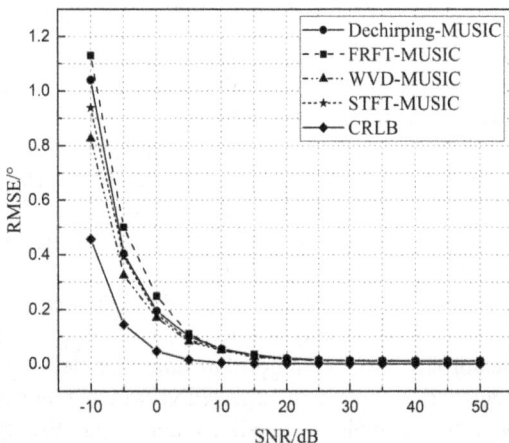

(b) Elevation angle based on MUSIC algorithm

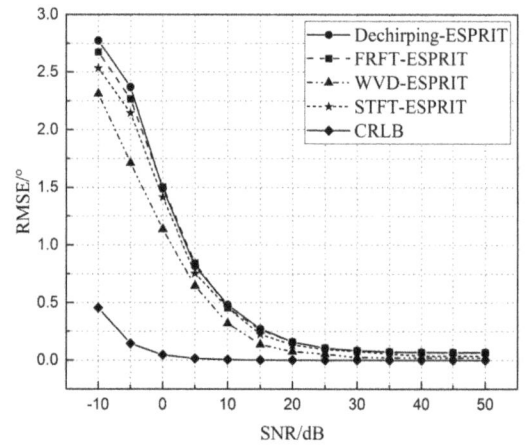

(c) Azimuth angle based on ESPRIT algorithm

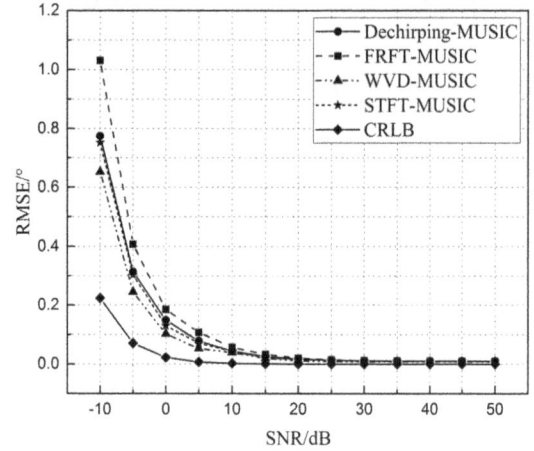

(d) Elevation angle based on ESPRIT algorithm

Fig. 6. RMSE of DOA estimation along with SNR using different algorithms.

So the proposed algorithm is prone to be easily realized in engineering.

In conclusion, we can contrast the main DOA estimation algorithm in many ways, such as the estimation precision, the computational complexity, the sampling rate, the cross-term interference and so on. The comparison results are shown in Tab. 3.

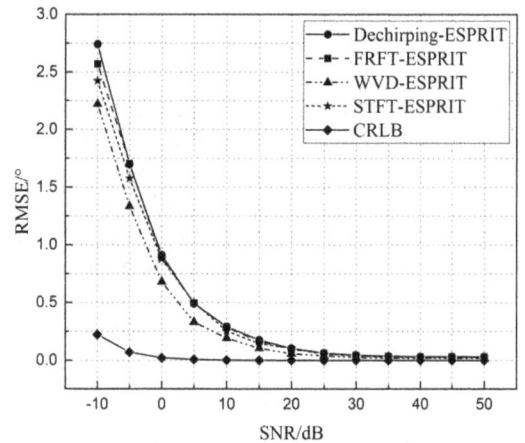

(a) Azimuth angle based on MUSIC algorithm

Algorithm	Estimation precision	Computational complexity	Cross-term interference
Dechirping	3rd	$O(MNL\log_2 L)$	N
WVD	1st	$O(MN^2L^2\log_2 L)$	Y
STFT	2nd	$O(MNL^2\log_2 L)$	N
FRFT	4th	$O(NL+MN^2)$	N

Algorithm	Sampling rate	Time-frequency point	Window function
Dechirping	Normal	N	N
WVD	High	Y	N
STFT	Normal	Y	Y
FRFT	Normal	N	N

Tab. 3. The comparison results of four algorithms.

In Tab. 3, "1st", "2nd", "3rd" and "4th" stand for the order of the estimation precision which means the algorithm based on WVD owns the highest estimation precision, followed by the algorithm based on STFT, our proposed algorithm and the algorithm based on FRFT. "N" stands for no and "Y" stands for yes. From Tab. 3, we can see that the proposed method owns high estimation precision and low computational complexity. Also it does not exist the cross-term interference and the puzzle of selecting the appropriate time-frequency point. The requirement for the system sampling is not high and it does not use the window function. The proposed algorithm also uses the relatively simple FFT operator. So it is easy to be realized in engineering.

5. Conclusion

In this paper, we proposed a new DOA estimation algorithm of multiple LFM signals based on the thoughts of Dechirping algorithm. Firstly, we proceed the difference frequency treatment between LFM signals received by each element and a specific reference signal. The reference signal is the signal to be estimated. It is received by the reference element. We can get a single-frequency pulse related to the signal to be estimated after Dechirping process. Then, we transform the Dechirping results from time domain to frequency domain by FFT to obtain an impulse pulse. We construct a new array data model through extracting the prominent parts of the impulse to obtain the time-invariant steering vector matrix. Finally, we realize 2-D DOA estimation of LFM signals by using MUSIC algorithm and ESPRIT algorithm respectively. The method proposed by this paper does not exist the cross-term interference and the puzzle of selecting the appropriate time-frequency point and window function. The computational complexity of the proposed algorithm achieves a low level contrast to the other DOA estimation algorithms based on time-frequency analysis tools. The algorithm also uses the relatively simple FFT operator and has the capability of estimating DOA of multiple LFM signals simultaneously with a high precision. The simulation results verify the effectiveness of the algorithm and it can be applied to DOA estimation of LFM signals.

Appendix. The Derivation of (17)

If there are M LFM signals from far-field, according to (13), the output of the nth element can be written as:

$$x_n = \left[\exp(-j2\pi f_1\tau_{n1})s_1 \quad \cdots \quad \exp(-j2\pi f_M\tau_{nM})s_M\right]. \quad (26)$$

We can carry out N point discrete Fourier transform (DFT) on the output of the array element and get (27).

$$
\begin{aligned}
v_q &= \sum_{n=0}^{N-1} x_n \exp\left(-j\frac{2\pi}{N}nq\right) \\
&= \left[\begin{array}{c}\sum_{n=0}^{N-1}\exp(-j2\pi f_1\tau_{n1})\exp\left(-j\frac{2\pi}{N}nq\right)s_1,\ldots,\\ \sum_{n=0}^{N-1}\exp(-j2\pi f_M\tau_{nM})\exp\left(-j\frac{2\pi}{N}nq\right)s_M\end{array}\right] \\
&\approx \left[Nj^{-q}J_{-q}(-\beta_1)\exp(-jq\theta_1)s_1,\ldots,Nj^{-q}J_{-q}(-\beta_M)\exp(-jq\theta_M)s_M\right].
\end{aligned}
\quad (27)
$$

In (27), $J_m(\cdot)$ stands for the order m of Bessel function, β_m is a factor related to the incoming wave parameters. $\beta_m = 2\pi\sin(\theta_{em})rf_m/c$, $\beta = \max[2\pi rf_m/c]$, $q \in [-\lfloor\beta\rfloor, \lfloor\beta\rfloor]\cap Z$ and $\lfloor\ \rfloor$ is the rounded down symbol. Z is the mathematical set of whole numbers. If $K = \lfloor\beta\rfloor$, the phase modes excited by this UCA [44] are: $-K, -K+1, \ldots, K$. So the number is $2K+1$, which means the UCA can be equivalent to a ULA with $2K+1$ elements.

If $u_q = v_{-q}$, (27) can be expressed as matrix form:

$$\mathbf{U} = NJ\mathbf{A}_c\mathbf{S}, \quad (28)$$

$$\mathbf{U} = \left[u_{-K},\ldots,u_K\right]^T, \quad (29)$$

$$\mathbf{J} = \mathrm{diag}\{j^{-K},\ldots,j^K\}, \quad (30)$$

$$\mathbf{A}_c = \left[\begin{array}{ccc} J_{-K}(-\beta_1)\exp(-jK\theta_{e1}) & \cdots & J_{-K}(-\beta_M)\exp(-jK\theta_{eM}) \\ \cdots & \ddots & \vdots \\ J_K(-\beta_1)\exp(jK\theta_{e1}) & \cdots & J_K(-\beta_M)\exp(jK\theta_{eM}) \end{array}\right]. \quad (31)$$

We can also express the definition of DFT as matrix form:

$$\mathbf{U} = \mathbf{F}^H\mathbf{X}, \quad (32)$$

$$\mathbf{F} = \left[\mathbf{w}_{-K} \quad \mathbf{w}_{-K+1} \quad \cdots \quad \mathbf{w}_K\right], \quad (33)$$

$$\mathbf{w}_q = \left[1 \quad \exp\left(-j\frac{2\pi}{N}q\right) \quad \cdots \quad \exp\left(-j\frac{2\pi(N-1)}{N}q\right)\right]^H. \quad (34)$$

Based on (28) and (32), we can get:

$$\mathbf{Y} = \mathbf{TX} = \mathbf{A}_c\mathbf{S}. \quad (35)$$

In (35), $\mathbf{T} = \mathbf{J}^{-1}\mathbf{F}^H / N$, which is called transformation matrix. In this way, we complete the mode-space transformation of UCA and the UCA with N elements is equivalent to a ULA with $2K+1$ elements. Then we can get (17).

References

[1] XIAO, W., XIAO, X. C., TAI, H. M. Rank-1 ambiguity DOA estimation of circular array with fewer sensors. In *Proceedings of the 45th IEEE Midwest Symposium on Circuits and Systems (MWSCAS-2002)*. Tulsa (USA), 2002, p. III-29 – III-32. DOI: 10.1109/MWSCAS.2002.1186962. ISBN: 0-7803-7523-8

[2] DU, W., SU, D. L., XIE, S. G., et al. A fast calculation method for the receiving mutual impedances of uniform circular arrays. *IEEE Antennas and Wireless Propagation Letters,* 2012, vol. 11, p. 893–896. DOI: 10.1109/LAWP.2012.2211329

[3] WANG, P., LI, Y. H., VUCETIC, B. Millimeter wave communications with symmetric uniform circular antenna arrays. *IEEE Communications Letters,* 2014, vol. 18, no. 8, p. 1307–1310. DOI: 10.1109/LCOMM.2014.2332334

[4] DORSEY, W. M., COLEMAN, J. O., PICKLES, W. R. Uniform circular array pattern synthesis using second-order cone programming. *IET Microwaves, Antennas & Propagation*, 2015, vol. 9, no. 8, p. 723–727. DOI: 10.1049/iet-map.2014.0418

[5] JACKSON, B. R., RAJAN, S., LIAO, B. J., et al. Direction of arrival estimation using directive antennas in uniform circular arrays. *IEEE Transactions on Antennas and Propagation*, 2015, vol. 63, no. 2, p. 736–747. DOI: 10.1109/TAP.2014.2384044

[6] WANG, M., MA, X. C., YAN, S. F., et al. An auto calibration algorithm for uniform circular array with unknown mutual coupling. *IEEE Antennas and Wireless Propagation Letters*, 2016, vol. 15, p. 12–15. DOI: 10.1109/LAWP.2015.2425423

[7] PAN, Y. J., ZHANG, X. F., XIE, S. Y., et al. An ultra-fast DOA estimator with circular array interferometer using lookup table method. *Radioengineering*, 2015, vol. 24, no. 8, p. 850–856. DOI: 10.13164/re.2015.0850

[8] JAIN, V., BLAIR, W. D. Filter design for steady-state tracking of maneuvering targets with LFM waveforms. *IEEE Transactions on Aerospace and Electronic Systems*, 2009, vol. 45, no. 2, p. 765–773. DOI: 10.1109/TAES.2009.5089558

[9] WANG, P., LI, H. B., DJUROVIC, P., et al. Integrated cubic phase function for linear FM signal analysis. *IEEE Transactions on Aerospace and Electronic Systems*, 2010, vol. 46, no. 3, p. 963–977. DOI: 10.1109/TAES.2010.5545167

[10] TAO, R., ZHANG, N., WANG, Y. Analyzing and compensating the effects of range and Doppler frequency migrations in linear frequency modulation pulse compression radar. *IET Radar, Sonar and Navigation*, 2011, vol. 5, no. 1, p. 12–22. DOI: 10.1049/iet-rsn.2009.0265

[11] NGUYEN, V. K., TURLEY, M. D. E. Bandwidth extrapolation of LFM signals for narrowband radar systems. In *International Conference on Radar*. Adelaide (SA), 2013, vol. 51 no. 1, p. 702–712. DOI: 10.1109/RADAR.2013.6651975

[12] SU, J., TAO, H. H., RAO, X., et al. Coherently integrated cubic phase function for multiple LFM signals analysis. *Electronics Letters*, 2015, vol. 51 no. 5, p. 411–413. DOI: 10.1049/el.2014.4164

[13] YUAN, X. Direction-finding wideband linear FM sources with triangular arrays. *IEEE Transactions on Aerospace and Electronic Systems*, 2012, vol. 48, no. 3, p. 2416–2425. DOI: 10.1109/TAES.2012.6237600

[14] LAGHMARDI, N., HARABI, F., MEKNESSI, H., et al. A space-time extended music estimation algorithm for wide band signals. *Arabian Journal for Science and Engineering*, 2013, vol. 38, no. 3, p. 661–667. DOI: 10.1007/s13369-012-0328-9

[15] SHA, Z. C., LIU, Z. M., HUANG, Z. T., et al. Covariance-based direction-of-arrival estimation of wideband coherent chirp signals via sparse representation. *Sensors*, 2013, vol. 13, no. 9, p. 11490–11497. DOI: 10.3390/s130911490

[16] HE, Z. Q., SHI, Z. P., HUANG, L., et al. Underdetermined DOA estimation for wideband signals using robust sparse covariance fitting. *IEEE Signal Processing Letters*, 2015, vol. 22, no. 4, p. 435–439. DOI: 10.1109/LSP.2014.2358084

[17] PAN, Y. J., TAI, N., YUAN, N. C. Wideband DOA estimation via sparse Bayesian learning over a Khatri-Rao dictionary. *Radioengineering*, 2015, vol. 24, no. 2, p. 552–557. DOI: 10.13164/re.2015.0552

[18] CHEN, H., WAN, Q., FAN, R., et al. Direction-of-arrival estimation based on sparse recovery with second-order statistics. *Radioengineering*, 2015, vol. 24, no. 1, p. 208–213. DOI: 10.13164/re.2015.0208

[19] WANG, L., ZHAO, L. F., BI, G. A., et al. Novel wideband DOA estimation based on sparse Bayesian learning with Dirichlet process priors. *IEEE Transactions on Signal Processing*, 2016, vol. 64, no. 2, p. 275–289. DOI: 10.1109/TSP.2015.2481790

[20] AMIN, M. G. Spatial time frequency distributions for direction finding and blind source separation. In *The International Society for Optical Engineering (Proc Spie)*. Orlando (USA), 1999, vol. 3723, p. 62–70. DOI: 10.1117/12.342958

[21] BELOUCHRANI, A., AMIN, M. G. Time frequency MUSIC. *IEEE Signal Processing Letters*, 1999, vol. 6, no. 5, p. 109–110. DOI: 10.1109/97.755429

[22] ZHANG, Y. M. D., AMIN, M. G., HIMED, B. Joint DOD/DOA estimation in MIMO radar exploiting time-frequency signal representations. *EURASIP Journal on Advances in Signal Processing*, 2012, vol. 2012, no. 1, p. 1–10. DOI: 10.1186/1687-6180-2012-102

[23] KHODJA, M., BELOUCHRANI, A., ABED-MERAIM, K. Performance analysis for time-frequency MUSIC algorithm in presence of both additive noise and array calibration errors. *EURASIP Journal on Advances in Signal Processing*, 2012, vol. 2012, no. 1, p. 1–11. DOI: 10.1186/1687-6180-2012-94

[24] LIN, J. C., MA, X. C., YAN, S. F., et al. Time-frequency multi-invariance esprit for DOA estimation. *IEEE Antennas and Wireless Propagation Letters*, 2015, vol. 15, p. 770–773. DOI: 10.1109/LAWP.2015.2473664

[25] LI, L. P., HUANG, K. J., CHEN, T. Q. 2-D DOA estimation of coherent wideband FM signals based on STFT. *Journal of Electronics and Information Technology*, 2005, vol. 27, no. 11, p. 1760–1764. (in Chinese)

[26] ZHANG, H. J., BI, G. A., CAI, Y. L., et al. DOA estimation of closely-spaced and spectrally-overlapped sources using a STFT-based MUSIC algorithm. *Digital Signal Processing*, 2016, vol. 52, p. 25–34. DOI: 10.1016/j.dsp.2016.01.015

[27] YANG, X. M., TAO, R. 2D DOA estimation of LFM signals based on fractional Fourier transform and ESPRIT algorithm. *Acta Armamentarii*, 2007, vol. 28, no. 12, p. 1438–1442. (in Chinese)

[28] YANG, X. M., TAO, R. 2-D DOA estimation of LFM signals based on fractional Fourier transform. *Acta Electronica Sinica*, 2008, vol. 36, no. 9, p. 1737–1740. (in Chinese)

[29] YANG, W., SHI, Y. W. FRFT based method to estimate DOA for wideband signal. *Advanced Materials Research*, 2013, vol. 712-715, p. 2716–2720.

[30] CUI, Y., WANG, J., F. Wideband LFM interference suppression based on fractional Fourier transform and projection techniques. *Circuits Systems and Signal Process*, 2014, vol. 33, no. 2, p. 613–627. DOI: 10.1007/s00034-013-9642-z

[31] YU, J. X., ZHANG, L., LIU, K. H., et al. Separation and localization of multiple distributed wideband chirps using the

fractional Fourier transform. *EURASIP Journal on Wireless Communications and Networking*, 2015, vol. 266, p. 1–8. DOI: 10.1186/s13638-015-0497-9

[32] YU, J. X., ZHANG, L., LIU, K. H. Coherently distributed wideband LFM source localization. *IEEE Signal Processing Letters*, 2015, vol. 22, no. 4, p. 504–508. DOI: 10.1109/LSP.2014.2363843

[33] TANG, X. H., LI, Q. L. *Time Frequency Analysis and Wavelet Transform*. Beijing (China): Science Press, 2016. ISBN: 978-7-03-047542-8. (in Chinese)

[34] BAO, Z., XING, M. D., WANG, T. *Radar Imaging Technology*. Beijing (China): Publishing of Electronics Industry, 2014. ISBN: 978-7-121-01072-9. (in Chinese)

[35] SCHMIDT, R. O. Multiple emitter location and signal parameter estimation. *IEEE Transactions on Antennas and Propagation*, 1986, vol. 34, no. 3, p. 276–280. DOI: 10.1109/TAP.1986.1143830

[36] ROY, R., PAULRAJ, A., KAILATH, T. ESPRIT-a subspace rotation approach to estimation of parameters of cissoids in noise. *IEEE Transactions on Acoustics, Speech, and Signal Processing*, 1986, vol. 34, no. 5, p. 1340–1342. DOI: 10.1109/TASSP.1986.1164935

[37] ROY, R., KAILATH, T. ESPRIT-estimation of signal parameters via rotational invariance techniques. *IEEE Transactions on Acoustics, Speech, and Signal Processing*, 1989, vol. 37, no. 7, p. 984–995. DOI: 10.1109/29.32276

[38] GRIFFITHS, H. D., EIGES, R. Sectoral phase modes from circular antenna arrays. *Electronics Letters*, 1992, vol. 28, no. 17, p. 1581–1582. DOI: 10.1049/el:19921006

[39] MATHEWS, C. P., ZOLTOWSKI, M. D. Eigenstructrure techniques for 2-D angle estimation with uniform circular arrays. *IEEE Transactions on Signal Processing*, 1994, vol. 42, no. 9, p. 2395–2407. DOI: 10.1109/78.317861

[40] TOMIC, S., BEKO, M., DINIS, R. RSS-based localization in wireless sensor networks using convex relaxation: noncooperative and cooperative schemes. *IEEE Transactions on Vehicular Technology*, 2015, vol. 64, no. 5, p. 2037–2050. DOI: 10.1109/TVT.2014.2334397

[41] TOMIC, S., BEKO, M., DINIS, R. Distributed RSS-AoA based localization with unknown transmit powers. *IEEE Wireless Communications Letters*, 2016, vol. 5, no. 4, p. 392–395. DOI: 10.1109/LWC.2016.2567394

[42] TOMIC, S., BEKO, M., DINIS, R., et al. A closed-form solution for RSS/AoA target localization by spherical coordinates conversion. *IEEE Wireless Communications Letters*, 2016, vol. 5, no. 6, p. 680–683. DOI: 10.1109/LWC.2016.2615614

[43] TOMIC, S., BEKO, M., DINIS, R., et al. Distributed algorithm for target localizion in wireless sensor networks using RSS and AoA measurements. *Pervasive and Mobile Computing*, 2016. DOI: 10.1016/j.pmcj.2016.09.013

[44] WANG, Y. L., CHEN, H., PENG, Y. N., et al. *Spatial Spectrum Estimation*. Beijing (China): Tsinghua University Press, 2004. ISBN: 7-302-09209-5. (in Chinese)

About the Authors ...

Kaibo CUI was born in 1990. He received his M.S. degree in Electronic Science and Technology from the National University of Defense Technology in 2013. Currently he is working towards the Ph.D. degree in the College of Electronic Science and Engineering, National University of Defense Technology, Changsha, Hunan, China. His research interests include signal processing and spatial spectrum estimation.

Weiwei WU was born in 1981. She received her M.S. and Ph.D. degree in Electronic Science and Technology from the National University of Defense Technology in 2008 and 2011, respectively. Currently she is a teacher in the College of Electronic Science and Engineering, National University of Defense Technology, Changsha, Hunan, China. Her research interests include array signal processing and antenna design.

Xi CHEN was born in 1983. He received his Ph.D. degree in Electronic Science and Technology from the National University of Defense Technology in 2013. Currently he is a teacher in the College of Electronic Science and Engineering, National University of Defense Technology, Changsha, Hunan, China. His research interests include signal processing.

Jingjian HUANG was born in 1983. He received his Ph.D. degree in Electronic Science and Technology from the National University of Defense Technology in 2014. Currently he is a teacher in the College of Electronic Science and Engineering, National University of Defense Technology, Changsha, Hunan, China. His research interests include antenna design.

Naichang YUAN was born in 1965. He received his M.S. and Ph.D. degree in Electronic Science and Technology from the University of Science and Technology of China in 1991 and 1994, respectively. He is currently a professor with the College of Electronic Science and Engineering, National University of Defense Technology, Changsha, Hunan, China. His research interests include array signal processing, signal processing in radar.

9

Soft – Partial Frequency Reuse Method for LTE-A

Slawomir GAJEWSKI

Dept. of Radio Communication Systems and Networks, Faculty of Electronics, Telecommunications and Informatics,
Gdansk University of Technology, Narutowicza 11/12, 80233 Gdansk, Poland

slagaj@eti.pg.gda.pl

Abstract. *In the paper a novel SPFR frequency reuse method is proposed which can be used for improvement of physical resources utilization efficiency in LTE-A. The proposed method combines both SFR and PFR giving the possibility of more flexible use of frequency band in different regions of a cell. First, a short study on the problem of frequency reuse in cells is discussed including bibliography overview. In next section the principle of the proposed SPFR method is described. Then the simulation model is discussed and simulation parameters are expected. In the last part, results of simulation of SPFR efficiency in comparison to known frequency reuse methods are presented. Presented results include both capacity and throughput for single connection. The proposed method eliminates main disadvantages of both SFR and PFR methods and gives significantly greater capacity of radio interface in boundary region of cells.*

Keywords

Frequency reuse, LTE, SPFR, FFR, resource management, SFR, PFR, ICI reduction

1. Introduction

The problem of Multiple Access Interference (MAI) is one of the most important issues in mobile networks of new generation. In modern communication systems single frequency networks are typically implemented where the same frequency channel is used in each cell. But using the same channels in neighbor cells causes inter-cell interferences (ICI). Interferences decrease transmission performance, reduce overall cell capacity and destroy spectral efficiency. It is strongly important in downlink transmission from base station (BS) to mobile stations (MS), especially, when MSs are located at cell edge sometimes called boundary area of cell (BAC). In general, it's well known that in BACs high power of ICI is received from neighbor BSs. Furthermore, both the cell capacity and transmission rate are strongly reduced [1–3].

In LTE and LTE-A the Fractional Frequency Reuse (FFR) methods can be used for improving signal performance, capacity and throughput in cells. The FFR is used to intelligent signal spectrum allocation that reduces the effect of ICI on signal performance. In general, the FFR is based on the allocation of small part of available frequency band for cell-edge connections. Two main methods are known, and their modifications, called the Soft and Partial Frequency Reuse (SFR and PFR). Sometimes, the PFR is called the Strict FFR. Each of these methods has some advantages and disadvantages what determines their usefulness in practical applications. A comprehensive analysis of known frequency reuse methods is presented in survey [4] but there are many more publications about FFR.

In [5] a contribution to analytical quantification of interference cancellation and SINR (Signal to Noise and Interference Ratio) estimation in LTE is presented and a cell coverage improvement is identified when using FFR. A comparison of SFR and strict FFR was done in [6] where analytical evaluation of both methods is presented. Moreover, in [7] some modifications of PFR in which the band reserved for BAC is doubled are proposed, as a cost of some additional interference. Authors of [8] propose an interference avoidance scheme for SFR used in LTE downlink. Additionally, in [9] a mechanism that selects efficient FFR scheme based on the user throughput and user satisfaction is proposed while in [10] efficient algorithms for bandwidth and power allocation depending on user's selection are analyzed.

Interesting concepts are studied in [11] where modification of SFR to increase resource usage efficiency called Enhanced Fractional Frequency Reuse (EFFR) is proposed. It is partially based on Incremental Frequency Reuse (IFR) presented before in [12]. The IFR scheme reuses effectively frequency bands through systematic segment allocation over a cluster of adjoining cells. Some improvement of SFR is done in [13] where a number of power density levels, achieving better interference pattern and increasing data rate at cell edge is studied.

In [14] the proposal of SFR enhancement for interference management in femto-cells located inside macro-cells is discussed while in [15] resource allocation scheme that gradually varies frequency resource share with distance from BS for both macro-cells and femto-cells is analyzed.

Authors of [16] give original contribution to performance analysis of frequency planning for FFR methods due to subcarrier collisions. Moreover, in [17] some resource adaptation concepts study for the case of uplink transmission using adaptive SFR is discussed. In [18] a proposal for frequency bands planning that simplifies the problem of subcarrier allocation with frequency reuse in OFDMA networks is analyzed.

Authors of [19] propose a self-organizing algorithm for sub-carrier and power allocation that achieves ICI avoidance while in [20] a new scheme for efficient location for Relay Nodes in LTE is discussed.

It's obvious that there are many modifications of FFR for improving signal performance and network capacity. In this paper basic concepts were taken into account because they are used to compare with the modified method called the Soft-Partial Frequency Reuse (SPFR).

In the case of PFR, the overall frequency band (a given radio channel) is divided into two parts. The first part is used in central area of cell (CAC) only (in each cell) and the second part is available in BAC. Additionally, the part of band reserved for BAC is divided into three parts which are allocated to different groups of cells. This minimizes the effect of ICI on signal performance and increases the transmission rate achievable in each cell of a network. But, the use of PFR increases the frequency reuse factor and decreases the capacity of a network (a cell) what is clearly presented in e.g. [2], [6]. A major advantage of PFR is approximately almost total interference isolation for cell-edge users. The principle of PFR is clearly shown in [2].

In the case of SFR, full frequency band can be used to all connections in CAC but only a small part of this band can be used to BAC connections. Additionally, this small part of frequency band is different for adjacent cells [2, 6].

In general, the SFR gives better capacity but the PFR gives better results of throughput for single connection what was presented in [2]. In PFR, the result of reservation of the part of available frequency band to its use in BAC only (excluding CAC) is large capacity reduction both in CAC and BAC. On the other hand, the use of SFR guarantees better values of the capacity in BAC but decreases transmission rate achieved in this area.

2.   SPFR Concept

The main goal of this paper is to present SPFR method for FFR improvement. The main advantage of SPFR is overall cell capacity enhancement, especially, increasing the capacity of BAC. The SPFR is the method in which some features of SFR and PFR were applied. The principle of SPFR is to make decisions about the resource allocation for various MSs on the basis of users' location information in different cell regions and on the basis of the load measurement in a cell. Thus, for the realization of SPFR, MSs monitoring is necessary and dynamic creation of the location map. But, we know that MSs location information is continuously reported to BS because the same

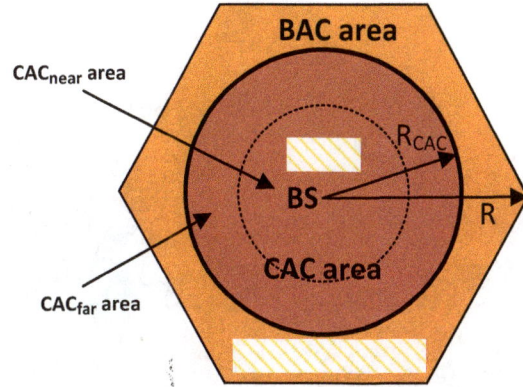

Fig. 1.   A cell division into CAC_{near}, CAC_{far} and BAC for SPFR; R_{CAC} is the radius of CAC, and R is the total cell radius.

problem is with users location when both SFR and PFR standardized methods are used.

For the creation of users location map a cell division into different regions is necessary as shown in Fig. 1.

In practice, the problem of SPFR parameters configuration consists in distance measurement for CAC_{near}/CAC_{far}/BAC configuration and it is basically the same problem as in the case of both SFR and PFR. Typically, configuration is made using radio path distance information. Of course, in this situation allocation of users to different regions not fully corresponds to their actual geographical location but this is the simplest way for the use of FFR methods in cells. Additionally, it is contemplated the use of location service. Then the problem of time alignment is partially resolved. The location service is available in LTE and still optimized.

In the case of SPFR users can be mapped into three different regions of a cell:
- BAC region for cell-edge users, as in PFR,
- CAC_{near} region for users allocated close to the BS,
- CAC_{far} region for users between CAC_{near} and BAC.

A cell radius is denoted as R, and it is the result of a network design. The CAC radius is denoted as R_{CAC} and determines the geometrical relationship between CAC and BAC. The R_{near} radius of CAC_{near} is designed in such a way that the CAC_{near} area is the same as the CAC_{far} area. Thus, the relationship between R_{near} and R_{CAC} is defined as

$$R_{near} = \frac{1}{\sqrt{2}} R_{CAC}. \qquad (1)$$

However, it is obvious that it can be designed in another way and can be optimized due to the actual network operating conditions and geometrical aspects.

2.1   Frequency Allocation

In the case of SPFR, frequency band allocation is made in different way in comparison to PFR. We use three types of frequency bands depending on the place of their

a)

b)

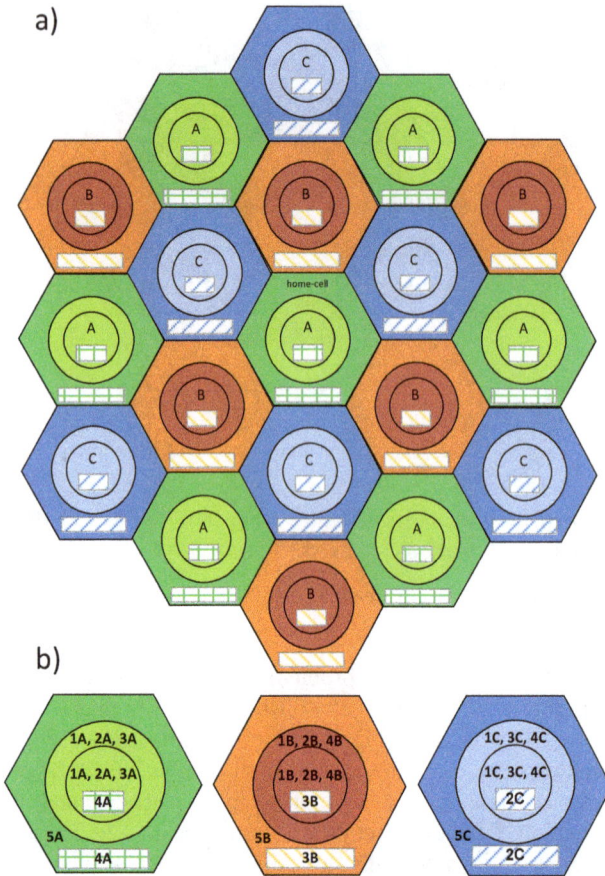

Fig. 2. General concept of frequency band allocation in SPFR:
a) division into groups of cells, b) band allocation for
cells from different groups.

use in CAC or BAC. Additionally, the way of band allocation depends on groups of cells (or sectors) as shown in
Fig. 2. These groups are denoted as A, B and C.

Both in PFR and SPFR, the 10 % of the total channel
band is allocated to BAC region only. It can be used in
a single group of cells and this band is different for each
group. Note that signals in these bands are transmitted at
increased power (e.g. 3 dB over the power of signals
transmitted at the rest of a band, excluding pilot signals).
The 20 % of the total band is an unused band and it can be
used in other groups of cells only.

The novelty is that for BAC we can use not only the
small part of total band mentioned before (as in PFR) but,
additionally, it is possible to use some soft-band for both
BAC and CAC. This soft-band equals to 20 % of the total
band and it is different for various groups of cells. This is
similar to the SFR but the difference is that the soft-band
can be used for users allocated close to the serving BS in
the CAC_{near} region only (not in CAC_{far}), and for users in
the BAC.

It is presented in Fig. 3 where the soft-band is denoted as 4A, 3B or 2C depending on the group of cells.

The remaining part of band which is 50% of the total
band is allocated for CAC area only and it can be used both
in CAC_{near} and CAC_{far} regions. The first 10 % part of this

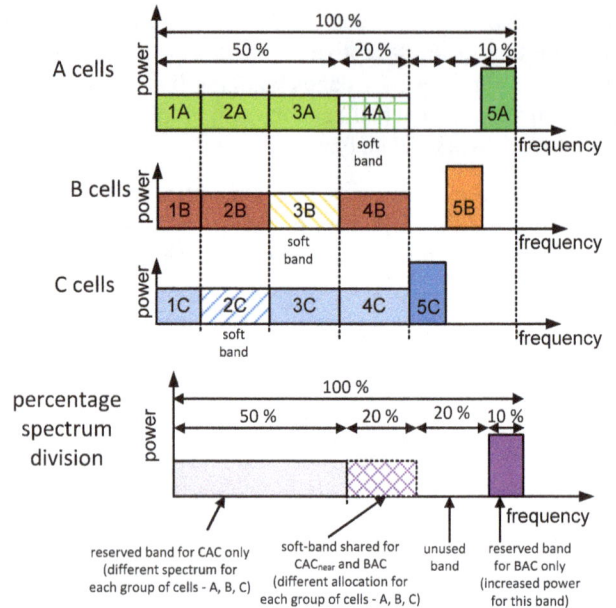

Fig. 3. Detailed SPFR spectrum allocation for different groups
of cells.

band is the same for each group of cells and this is denoted
in Fig. 3 as 1A, 1B, and 1C. Additionally, we can use 2
parts of 20 % of the total band in the CAC region. For the
A group of cells, these parts are 2A and 3A bands, and for
the B group these are 2B and 4B, and for the C group these
are 3C and 4C bands. So, in each CAC region we can use
the part of band reserved for the use in a given group of
cells and this band is 50% of the total band. And, additionally, we can use the soft-band which is 20% of the total
band but it can be used in CAC_{near} part of CAC only and
this band is shared between CAC_{near} and BAC. Furthermore, for BAC we can use both the reserved-band (10% of
the total band) and the soft-band.

The use of soft-band in the CAC causes additional ICI
in a network compared to PFR. But, due to its use in
CAC_{near} zone only, additional ICI are not critical. While it
gives large gain in capacity achievable in BAC region
compared to strict PFR. The decision to the use of these
parts is taken dynamically by a system and it is dependent
on a cell load. The load is measured for both the bands
allocated to CAC and BAC regions and results from the
actual frequency band occupation.

2.2 The Principle of SPFR

At the beginning, the soft-band is dedicated to users
allocated in the CAC_{near} region only. If the load in the frequency band allocated to BAC-users grows then the soft-
band is progressively allocated to these connections, if
necessary.

The power of signals transmitted in soft-band at BAC
is not-increased compared to the power of signals transmitted in bands reserved for BAC only. Thus, the best
situation is when soft-band is allocated to MSs located
close the CAC_{far} region (the boundary between CAC_{far} and

BAC). While users located at cell edge should better use the bands reserved for BAC which are denoted as 5A, 5B or 5C. Note that it is not critical and depends on load and signal quality measurements. In simulations it was seen that the SINR measured for signals in soft-band at BAC and reserved-band at BAC do not differ significantly in most situations, and sometimes the SINR is better for soft-band, sometimes for reserved-band.

On the other hand, the soft-band can be used in CAC_{near} and BAC only (not in CAC_{far}). It takes place because the use of soft-band in CAC_{far} causes additional ICI interferences to other groups of cells. So, the isolation between BAC and CAC_{near} regions is recommended in each cell. For instance, the use of the 4A soft-band in the A group of cells causes ICI both to 4B and 4C what we can see in Fig. 2 and Fig. 3. If the soft-band was used in CAC_{far} then the distance from BS to MSs located in CAC_{far} regions is less compared to the situation when 4A, 4B, and 4C bands are used in CAC_{near} regions only. The result of this is additional ICI received in CAC_{far} regions in each group of cells, received signals quality deterioration and throughput reduction. Thus, the use of soft-band for CAC_{near} and for BAC is recommended but the power of signals in soft-band should not be increased when allocated to BAC.

The use of soft-band in BAC regions depends on the load measured in 5A, 5B and 5C bands. The expanded soft-band is used if the load increases over 70%. Additionally, the soft-band is allocated in different way for each group of cells as we can see in Fig. 2 and Fig. 3.

In conclusion, we can use in a cell the bands denoted in Fig. 3 as:

- 1A, 1B and 1C which are the same for each group of cells (A, B and C), and these bands are reserved for both CAC_{near} and CAC_{far},

- 4A, 3B or 2C (depending on the group of cells) which can be used as soft-bands both in CAC_{near} and BAC,

- 2A, 3A, 2B, 4B, 3C, 4C which can be used in CAC only,

- 5A, 5B, and 5C which are reserved for BAC only.

One can see that a different way of soft-band placing in each group of cells gives some reduction of ICI received in various cells. It's clear that ICI are not completely eliminated. But we see that in SPFR the band allocated to MSs placed in BAC can be soft-expanded. The expansion can be made using the part of band allocated for both CAC and BAC called the soft-band. And it is a major difference between SPFR and strict PFR.

3. Simulation Results

The simulation model allows research of throughput-coverage characteristics of the OFDMA-based LTE network (see [2]). Simulation process is based on Monte Carlo

method. The aim of simulation is to prove that the use of SPFR method allows increasing the efficiency of known frequency reuse methods.

3.1 Simulation Model

The network model was implemented as a set of cells including the central cell (denoted as the home-cell in Fig. 2) and two tiers of surrounding cells which introduce ICI. Each cell is divided into CAC_{near}, CAC_{far} and BAC areas. The R_{CAC} radius sets the geometrical relationship between CAC and BAC in each cell. Thus, R_{CAC} can be changed during simulation from 0.1 R to 0.9 R. Each MS is located using geometrical relations between BSs and MSs and it defines its placing to different cells and regions of the home-cell.

The network simulation is based on random process of mobile stations (MS) placing in cells, estimation of ICI received by MSs at different locations and on the estimation of signal quality given by SINR. On the basis of actual SINR, we can estimate throughput-coverage characteristics for connections established in a cell. If a time of simulation is sufficiently long, then we can estimate average value of throughput for connections and evaluate real network capacity. The home-cell connections are observed because this cell allows us to estimate the total power of ICI. So, the home-cell is a reference cell in overall simulation process. The distribution of users in cells is uniform. Erceg-Greenstein propagation loss model [21] for suburban environment was implemented. Slow fadings were simulated using the log-normal distribution.

Received signals are analyzed in the spatial grid where we have a large number of points represented by small squares having sides equal to 2 m. In Fig. 4 we can see example results of SINR visualization for SFR.

The basic set of simulation parameters is presented in Tab. 1.

As we can see in the next sections, for better interpretation of results, we took into account the distance d/R from BS to MS, normalized by R (maximum cell radius)

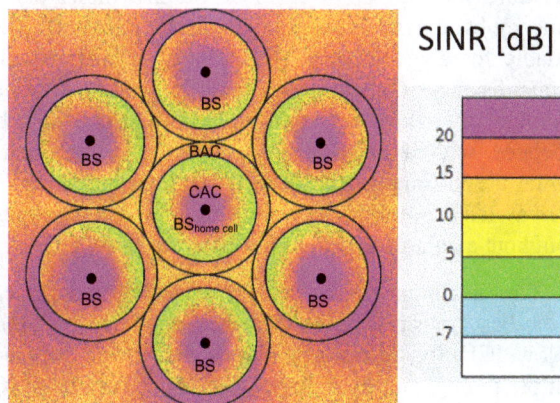

Fig. 4. Example of SINR distribution in modeled network; the home-cell and single tier of adjacent cells are shown.

Parameter	Setting
Reuse methods	PFR, SFR, SPFR
Frequency band	10 MHz
Central frequency	2.6 GHz
Maximum BS power	30 dBm
Transmission type	Single antenna connection
Designed cell radius R	400 m
Propagation model	Erceg-Greenstein
Antenna type	Omni-directional
Slow fluctuations	Log-normal distribution (8 dB signal deviation)
Scheduling	Proportional Fair
Users distribution	Uniform

Tab. 1. Set of basic simulation parameters.

and the throughput for single connection C_{MS}/C_{MSmax} normalized by maximum throughput achievable in a network for this connection. As well as the overall cell capacity, C/C_{max} is normalized by the maximum capacity available in a cell. So, both the C_{MS}/C_{MSmax} and C/C_{max} are observed at various distances d/R.

3.2 Simulation Results of Throughput for Single Connection

This simulation scenario corresponds to the case of low-loaded network (in home-cell) when there is a possibility of the transmission of signals for single connection with the transmission rate as high as possible in a radio interface. It means that other home-cell connections do not affect transmission rate for analyzed, single connection.

The throughput for single connection is observed for downlink transmission from BS to the reference MS. Three methods were evaluated: SFR, PFR in classic form and the proposed SPFR. MS is observed in different locations and at various distances from BS in home-cell. So, we receive signals with different SINR and various power of ICI from adjacent cells.

The maximum throughput C_{MSmax}, used to normalizations, is calculated for single connection at different distances from home BS. This it is the maximum throughput achievable for best CQI (Channel Quality Indicator) in frequency band available at a given distance from BS (i.e. available for a given MS). Thus, C_{MSmax} is different for various distances (MS locations) and, in general, it is not maximum throughput available in a cell and it cannot be associated with a cell capacity. Thus, presented charts provide the information how the throughput is reduced at different distances from BS compared to the maximum throughput available at this distance for the best CQI.

The maximum transmitted power of BS is sufficient for the best transmission performance estimated by SINR for low ICI. So, during simulations the power does not limit the throughput, excluding situations of very large ICI received (i.e. in BAC). In modeled case, the network is not capacity limited what means that theoretical capacity is available, and radio interface is not fully-loaded by users.

These simulation conditions include the type of connection, possible coding-modulation scheme (dependent on CQI), maximum BS power, signal loss and fadings, etc.

The situation is possible when ICI growth causes the situation in which BS power cannot be sufficient for interference compensation. Thus, SINR is reduced and throughput is decreased due to different CQI as in a real LTE-A network [2]. Moreover, the analyzed throughput for single connection can be reduced if capacity is too small in the observed area (i.e. ICI cannot be compensated by the transmitted BS power). It's obvious that the throughput is greater when the MS is located close to the BS in the home-cell and decreases when it moves away due to ICI growth. However, when MS moves to BAC depending on the simulated R_{CAC}, the throughput strongly grows because the power of ICI is relatively small in this region.

It was assumed that the MS is located in CAC when its geometrical distance to the BS is between $0 < d/R \leq R_{CAC}/R$ ($0 < d \leq R_{CAC}$). At greater distance, from $R_{CAC}/R < d/R \leq 1$ ($R_{CAC} < d \leq R$), the MS is located in BAC. Additionally, in the CAC area MSs are mapped to CAC_{near} or CAC_{far} areas, depending on their actual location, taking into account the R_{near} radius.

Simulation results are presented in Fig. 5–9, as the throughput C_{MS}/C_{MSmax} available for single connection normalized by its maximum value, appointed for different R_{CAC}/R configurations.

As we can see, the achieved throughput for single connection is the greatest for PFR. It is possible because the simulated network is not fully-loaded in the presented experiment. As previously mentioned, it takes place because throughput for single connection is the strength point of the PFR method, especially, compared to SFR (see [2]).

On the other hand, we know that for the SFR the throughput in BAC is small. This is the main SFR disadvantage. One can see that the use of PFR gives better stability of throughput for single connection compared to other methods – especially SFR. The reservation of ICI-resistant frequency band guarantees higher values of SINR especially in the BAC region.

As we see, SPFR gives a little worse results of throughput for single connection achieved in BAC, compared to PFR. However, deterioration in BAC is from a dozen % to approximately 20%. This is the price we must pay for the modification of the strict PFR. At the same time, it should be noted that the throughput is much greater in comparison to SFR. So, we can notice that the SPFR strongly reduces the main disadvantage of SFR in the BAC region while deterioration of throughput for PFR is not too large.

In the CAC region, moreover, we have approximately the same results of throughput for both PFR and SPFR. So, mentioned earlier deterioration concerns the BAC region only. Note, that in this case both PFR and SPFR methods give better results than SFR.

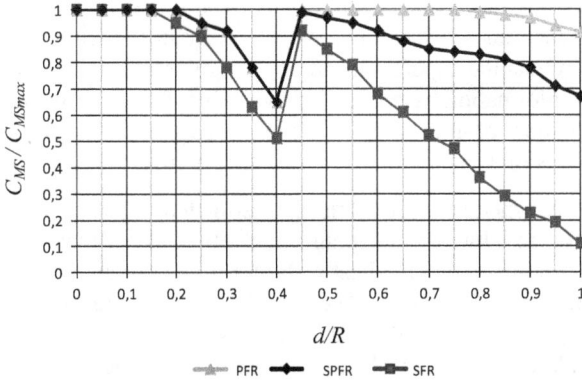

Fig. 5. Maximum throughput for single connection for PFR, SPFR and SFR; $R_{CAC} = 0.4R$.

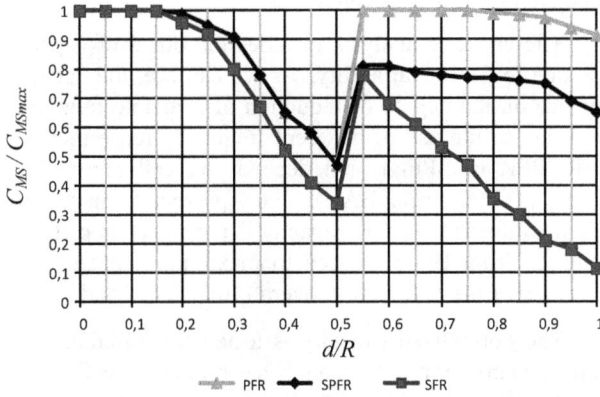

Fig. 6. Maximum throughput for single connection for PFR, SPFR, and SFR; $R_{CAC} = 0.5R$.

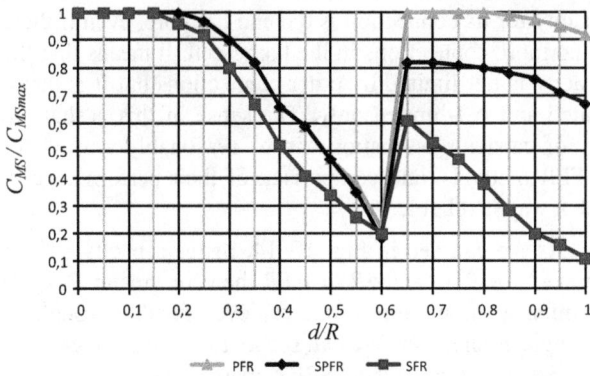

Fig. 7. Maximum throughput for single connection for PFR, SPFR, and SFR; $R_{CAC} = 0.6R$.

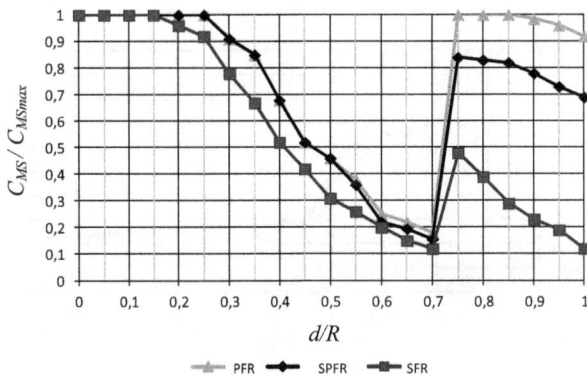

Fig. 8. Maximum throughput for single connection for PFR, SPFR, and SFR; $R_{CAC} = 0.7R$.

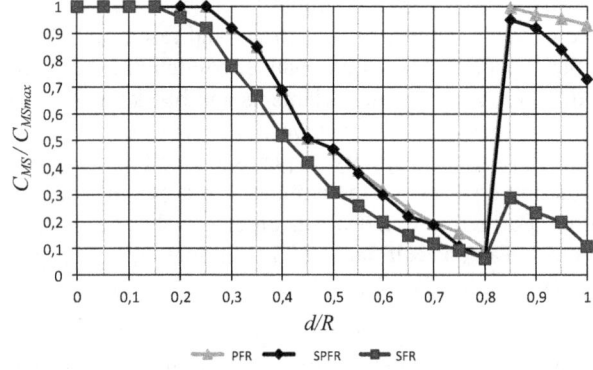

Fig. 9. Maximum throughput for single connection for PFR, SPFR, and SFR; $R_{CAC} = 0.8R$.

As we can see in presented figures, the R_{CAC} radius plays a great role in the design process and achievable transmission efficiency. The use of large R_{CAC} is not recommended because throughput for single connection is rather better for $R_{CAC} = 0.5R$ or $R_{CAC} = 0.6R$ than for larger values. On the other hand, we can see that BAC should not be too small in comparison to CAC. It is strongly important not only for uniform users' distribution in a cell. We know that in most cases the number of users in BAC is greater than their number in CAC. So, rather R_{CAC} radiuses less than $0.7R$ are preferred. Additionally, we should not use too small R_{CAC} due to high capacity reduction as we can see in the next section.

3.3 Cell Capacity Analysis

The Shannon capacity bound in AWGN channel, applied to practical LTE system implementation was clearly considered in [1], [3]. So, the spectral efficiency can be estimated using the formula

$$S_{\text{eff,max}} = B_{\text{eff}} \sum_{k=1}^{\min(l_{Tx},l_{Rx})} \log_2\left(1+\frac{SINR_k}{SINR_{\text{eff},k}}\right)\left[\text{bits/s/Hz}\right] \quad (2)$$

where:

- l_{Tx} and l_{Rx} are the numbers of transmit and receive antennas, respectively (when MIMO is used),
- $SINR_k$ is signal-to-interference and noise ratio resulting from the k-th spatial subchannel,
- $SINR_{\text{eff},k}$ is the factor of SINR implementation efficiency for the k-th subchannel [3],
- B_{eff} is bandwidth efficiency factor of LTE [3].

Typical value of B_{eff} is 0.9 as was explained in [3]. It is obvious that full SINR efficiency is not possible in LTE due to the limited code block length and it may be determined by the simulation research. Additionally, the bandwidth efficiency is reduced by the overhead of the use of cyclic prefix and the pilot assisted channel estimation as well as other overheads.

In practice, the capacity of LTE air interface can be understood as the maximum throughput available for all users in a given frequency channel, at a given moment of time. In OFDMA based radio interface the channel is composed of some number of subcarriers and resource blocks. So, the capacity depends on channel bandwidth, the number of subcarriers, resource elements and blocks as well as on the number of OFDM symbols allocated in a time slot (frame). Also, the capacity depends on coding rate and modulation type, i.e. on the number of bits per modulation symbol. Furthermore, it depends on the number of transmit and receive antennas if the MIMO (Multiple Input Multiple Output) technique is used.

A major capacity estimation problem results from the determination of the number of available physical resource blocks for a given user from the plan of the structure of data transmitted in a single time slot as well as from the number of OFDM symbols available for this user with the exception of control data overhead. When the number of available resource blocks has been planned, and data structure has been designed, we may to estimate the total capacity C_p [bps] as maximum throughput available for all users. Thus, the capacity can be found using the formula

$$C_p = n_b \; \eta_{cod} \; k_{MIMO} \; l_{SC\,RB} \; v_{loss} \; {l_{symb}}\Big/{T_{sub}} \tag{3}$$

where:

- n_b [b/symbol] is the number of bits dependent on modulation type (QPSK: 2 bits per symbol, 16QAM: 4 bits per symbol, 64QAM: 6 bits per symbol),

- η_{cod} is a channel coding rate,

- k_{MIMO} is the factor of growth of spectral efficiency when MIMO is used (corresponds to the number of transmit and receive antennas), $k_{MIMO} = 1$,

- $l_{SC\,RB}$ is the number of subcarriers assigned for a given service in an available frequency band,

- l_{symb} is the number of OFDM symbols assigned for a given service in a single subframe,

- $T_{sub} = 1$ ms is the subframe time duration,

- v_{loss} – the loss factor resulting from the use of certain OFDM symbols by the control data, i.e. reference symbols et al. [3] (minimum 2 symbols per subframe; thus $v_{loss} = 0.86$).

In presented simulations, the capacity is calculated as the maximum throughput available for all connections in home-cell at a given distance from BS to MS in overall frequency band. Thus, the simulation supposes that the capacity at a given distance is calculated for all terminals located at this distance. The capacity is different for users located in CAC and BAC areas at various places because it depends on the received SINR and available frequency band. Moreover, the SINR determines achievable CQI which in turn sets available coding-modulation scheme and throughput. Additionally, it is different for various frequency reuse methods.

On the other hand, the overall cell capacity can be calculated as the sum of throughputs of all users which depends on many random variables. This method of capacity estimation gives the information about overall cell performance and it does not give the clear information how ICI affects the capacity in different regions of a cell. Thus, this information is much less valuable for performance evaluation of frequency reuse methods compared to the analysis of capacity at different distances from BS.

As mentioned before, the estimated capacity C is normalized by the maximum capacity C_{max} available in a cell achieved for the best CQI at a given distance d/R in overall frequency band. The C and C_{max} capacities as well as C_{MS} and C_{MSmax} throughputs for single connection are calculated using (3). Thus, the method of calculation is similar but the difference is that both the maximum capacity and maximum throughput for single connection are calculated for different conditions.

The capacity analysis includes the effect of capacity loss due to SINR variability. As we can see, the capacity depends on the degree of reduction of available frequency band due to the use of different frequency reuse methods (SFR, PFR or SPFR), and depends on cell region type (CAC$_{near}$, CAC$_{far}$ and BAC). Thus, presented charts provide the information how capacity C is reduced at different distances from BS compared to maximum capacity C_{max} available for the best CQI in overall frequency band.

The normalized capacity is tested for different locations of many terminals in analyzed regions of cells but it is estimated for different distances. For instance, if $C/C_{max} = 0.6$ for $d/R = 0.35$ then the information is that if a terminal is at this distance then the maximum achievable capacity for its connection is 0.6 and no more, even if there are no more connections in the home-cell. It means that the capacity is not limited by other connections but it is band-limited and ICI-limited only. The sense of this is that the way of capacity estimation can be reasonably unified for all FFR methods, and comparison of their performance in such a way is reliable.

As we can see in Fig. 10–14, better capacity can be achieved for $R_{CAC} = 0.5R$ or $0.6R$ than for greater R_{CAC}. It is similar to simulation results achieved for the throughput for single connection. We can see that the capacity gain for BAC is very large when SPFR method is implemented. The use of SPFR gives, in general, more than 2 times greater capacity in BAC in comparison to both PFR and SFR.

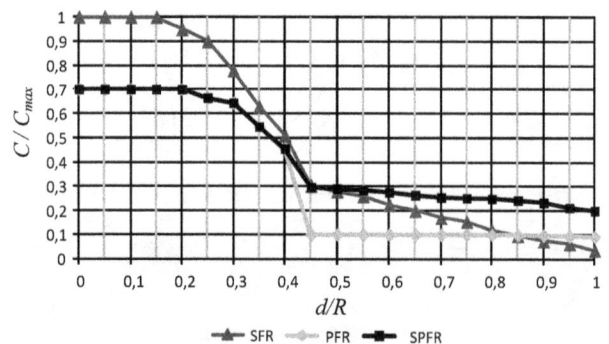

Fig. 10. Normalized capacity C/C_{max} for PFR, SPFR, and SFR; $R_{CAC} = 0.4\,R$.

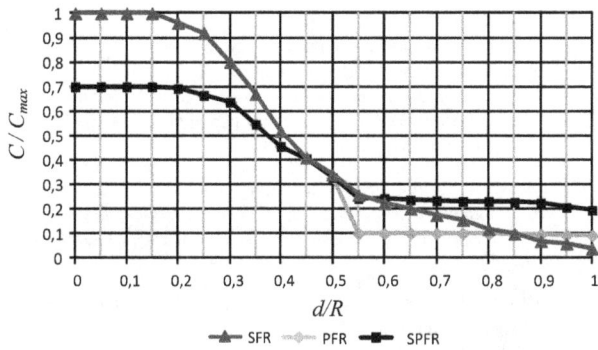

Fig. 11. Normalized capacity C/C_{max} for PFR, SPFR, and SFR; $R_{CAC} = 0.5\,R$.

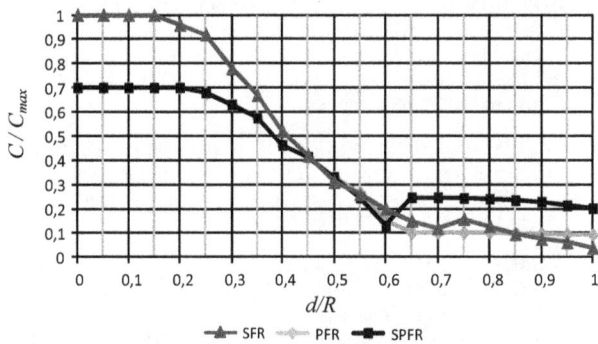

Fig. 12. Normalized capacity C/C_{max} for PFR, SPFR, and SFR; for $R_{CAC} = 0.6\,R$.

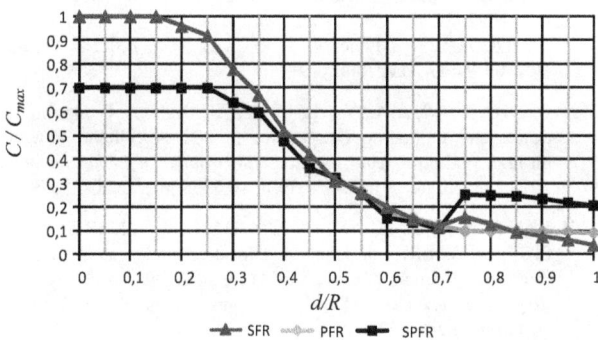

Fig. 13. Normalized capacity C/C_{max} for PFR, SPFR, and SFR; $R_{CAC} = 0.7\,R$.

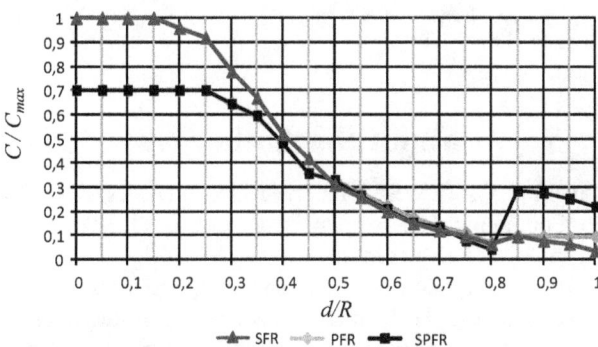

Fig. 14. Normalized capacity C/C_{max} for the PFR, SPFR and SFR; $R_{CAC} = 0.8\,R$.

Both PFR and SPFR give capacity reduction in CAC compared to SFR due to band limitation. But it is obvious that capacity growth in BAC is much more important than

its reduction in CAC. For SPFR the largest capacity in BAC is observed when $R_{CAC} = 0.5R$. Additionally, the use of SPFR does not degrade the capacity in CAC more than PFR.

Of course, the presented results strongly depend on the designed R_{CAC} and it should be taken into account during the network design process and its optimization.

Note, that it is possible to reach the maximum value of $C/C_{max} = 1$ because the capacity depends on the reported CQI which, additionally, depends on SINR. If the CQI is of the maximum value then the system uses the best performance coding-modulation scheme. Thus, the capacity and throughput are maximal. In simulations, in this case SINR fluctuations were observed but SINR was greater than its value guaranteeing the transmission with the best coding-modulation scheme what we can see in Fig. 4.

In general, the overall cell capacity for PFR is less than the capacity for SPFR and it may be even more than two times greater for the BAC region. We see that SPFR significantly eliminates a major disadvantage of PFR. However, the price for this is some deterioration of throughput for single connection.

4. Conclusions

From the presented considerations we can see that the SPFR method significantly improves the BAC capacity compared to both strict PFR and SFR. It is very important because low capacity in BAC is the main disadvantage of these methods. The cost of this is some reduction of the available throughput for single connection compared to strict PFR and some reduction of the CAC capacity compared to SFR. But, the CAC capacity is approximately the same as achieved for PFR, and the throughput is much greater than the throughput in BAC achieved for SFR.

So, we can see that SPFR allows eliminating the capacity degradation in BAC regions of cells. In general, the results show more than twofold increase of capacity in BAC at the cost of a relatively small reduction of throughput for a single connection compared to PFR.

In conclusion, therefore, we can say that the main features of SPFR are:

- Not less than twofold increasing the total BAC capacity compared to both PFR and SFR,
- Decreasing of throughput for single connections in BAC compared to PFR which does not exceed 20%,
- Greater throughput for single connections in BAC in comparison to the use of SFR,
- Little growth of algorithm implementation and frequency planning complexity,
- The need for monitoring of MS location in different cell regions although this is performed in BSs, and some complications in algorithm configuration due to problems with BS-MS distance measurement.

Thus, a major conclusion is that SPFR is a good candidate for frequency reuse method implemented in OFDMA-based networks of 4G and 5G. It can be used to improve the radio access network performance as a good compromise between SFR and PFR.

References

[1] SCHOENEN, R., ZIRWAS, W., WALKE, B. W. Capacity and coverage analysis of 3GPP-LTE multihop deployment scenario. In *Proceedings of IEEE International Conference on Communications, ICC Workshops*. Beijing (China), 2008, p. 31–36. DOI: 10.1109/ICCW.2008.11

[2] GAJEWSKI, S. Throughput-coverage characteristics for soft and partial frequency reuse in the LTE downlink. In *Proceedings of 36th International Conference on Telecommunications and Signal Processing (TSP 2013)*. Rome (Italy), 2013, p. 199–203. DOI: 10.1109/TSP.2013.6613919

[3] MOGENSEN, P., WEI, N., KOVACS, I., et al. LTE capacity compared to the Shannon bound. In *Proceedings of IEEE 65th Vehicular Technology Conference (VTC 2007-Spring)*. Dublin (Ireland), 2007, p. 1234–1238. DOI: 10.1109/VETECS.2007.260

[4] PORJAZOSKI, M., POPOVSKI, B. Analysis of intercell interference coordination by fractional frequency reuse in LTE. In *Proceedings of International Conference on Software, Telecommunications and Computer Networks (SoftCOM)*. Dubrovnik (Croatia), 2010, p. 160–164.

[5] MAO, X., MAAREF, A., TEO, K. H. Adaptive soft frequency reuse for intercell interference coordination in SC-FDMA based 3GPP LTE uplinks. In *Proceedings of IEEE Global Telecommunications Conference (GLOBECOM)*. New Orleans (USA), 2008, p. 1–6. DOI: 10.1109/GLOCOM.2008.ECP.916

[6] GHAFFAR, R., KNOPP, R. Fractional frequency reuse and interference suppression for OFDMA networks. In *Proceedings of 8th International Symposium on Modeling and Optimization in Mobile, Ad Hoc and Wireless Networks (WiOpt)*. Avignon (France), 2010, p. 273–277.

[7] RAHMAN, M., YANIKOMEROGLU, H., WONG, W. Interference avoidance with dynamic inter-cell coordination for downlink LTE system. In *Proceedings of IEEE Wireless Communications and Networking Conference (WCNC)*. Budapest (Hungary), 2009, p. 1–6. DOI: 10.1109/WCNC.2009.4917761

[8] BILIOS, D., BOURAS, C., KOKKINOS, V., et al. Optimization of fractional frequency reuse in long term evolution networks. In *Proceedings of IEEE Wireless Communications and Networking Conference (WCNC)*. Paris (France), 2012, p. 1853–1857. DOI: 10.1109/WCNC.2012.6214087

[9] XIE, Z., WALKE, B. Frequency reuse techniques for attaining both coverage and high spectral efficiency in OFDMA cellular systems. In *Proceedings IEEE Wireless Communications and Networking Conference (WCNC)*. Sydney (Australia), 2010, p. 1–6. DOI: 10.1109/WCNC.2010.5506110

[10] NOVLAN, T. D., GANTI, R. K., GHOSH, A., ANDREWS, J. G. Analytical evaluation of fractional frequency reuse for OFDMA cellular networks. *IEEE Transactions on Wireless Communications*, 2011, vol. 10, no. 12, p. 4294–4395. DOI: 10.1109/TWC.2011.100611.110181

[11] HINDIA, M. N., KHANAM, S., REZA, A., W., et al. Frequency reuse for 4G technologies: A survey. In *Proceedings of the 2nd International Conference on Mathematical Sciences & Computer Engineering (ICMSCE 2015)*. Langkawi (Malaysia), 2015.

[12] KRASNIQI, B., MACLENBRAUKER, C. F. Efficiency of partial frequency reuse in power used depending on user's selection for cellular networks. In *Proceedings of IEEE 22nd International Symposium on Personal Indoor and Mobile Radio Communications (PIMRC)*. Toronto (Canada), 2011, p. 268–272. DOI: 10.1109/PIMRC.2011.6139963

[13] YANG, X. A multilevel soft frequency reuse technique for wireless communication systems. *IEEE Communications Letters*, 2014, vol. 18, no. 11, p. 1983–1986. DOI: 10.1109/LCOMM.2014.2361533

[14] SELIM, M. M., EL-KHAMY, M., EL-SHARKAWY, M. Enhanced frequency reuse schemes for interference management in LTE femtocell networks. In *Proceedings of International Symposium on Wireless Communications Systems (ISWCS)*. Paris (France) 2012, p. 326–330. DOI: 10.1109/ISWCS.2012.6328383

[15] ELAYOUBI, S. E., BEN HADDADA, O., FOURESTIE, B. Performance evaluation of frequency planning schemes in OFDMA-based networks. *IEEE Transactions on Wireless Communications*, 2008, vol. 7, no. 5, p. 1623–1633. DOI: 10.1109/TWC.2008.060458

[16] ALI, S. H., LEUNG, V. C. M. Dynamic frequency allocation in fractional frequency reused OFDMA networks. *IEEE Transactions on Wireless Communications*, 2009, vol. 8, no. 8, p. 4286–4295. DOI: 10.1109/TWC.2009.081146

[17] KIM, K. T., OH, S. K. An incremental frequency reuse scheme for an OFDMA cellular system and its performance. In *Proceedings of IEEE Vehicular Technology Conference (VTC 2008-Spring)*. Marina Bay (Singapore), 2008, p. 1504–1508. DOI: 10.1109/VETECS.2008.352

[18] STOLYAR, A. L., VISWANATHAN, H. Self-organizing dynamic fractional frequency reuse in OFDMA systems. In *Proceedings of the IEEE 27th Conference on Computer Communications (INFOCOM)*. Phoenix (USA), 2008. DOI: 10.1109/INFOCOM.2008.119

[19] ALDHAIBANI, J. A., YAHYA1, A., AHMAD, A. B. Optimizing power and mitigating interference in LTE-A cellular networks through optimum relay location. *Elektronika ir Elektrotechnika*, 2014, vol. 20, no. 7, p. 73–79. DOI: 10.5755/j01.eee.20.7.3379

[20] KAWSER, M. T., ISLAM, M. R., AHMED, K. I., KARIM, M. R., SAIF, J. B. Efficient resource allocation and sectorization for fractional frequency reuse (FFR) in LTE femtocell systems. *Radioengineering*, 2015, vol. 24, no. 4, p. 940–947. DOI: 10.13164/re.2015.0940

[21] ERCEG, V., GREENSTEIN, L. J., TJANDRA, S. Y., et al. An empirically based path loss model for wireless channels in suburban environments. *IEEE Journal on Selected Areas in Communications*, 1999, vol. 17, no. 7, p. 1205–1211. DOI: 10.1109/49.778178

About the Author ...

Slawomir GAJEWSKI received his Ph.D. degree in Radio Communication Systems from Gdansk University of Technology (Poland) in 2004. Now, he works in the Dept. of Radio Communication Systems and Networks, Faculty of Electronics, Telecommunications and Informatics in Gdansk. He is an author of more than 140 publications in radio communications published in Poland and in many countries. He is a member of scientific committees and reviewer of many international conferences and journals (IEEE, IARIA, and Mosharaka). He is the TCP member of IEEE Vehicular Technology Conference. His research interests include radio resource management in modern radio communication systems, scheduling techniques and subcarriers allocation, interference management and cancellation in OFDMA and CDMA systems, physical implementation of OFDMA-based radio communications systems, design of cellular networks, MIMO techniques as well as signal processing, security in wireless systems, radio communication systems design for public safety and military applications, wireless monitoring and transport systems telematics.

10

Non-Invasive Microwave Sensors for Biomedical Applications: New Design Perspectives

Sandra COSTANZO

DIMES, University of Calabria, Via P. Bucci cubo 42C, 87036 Rende (CS), Italy

costanzo@dimes.unical.it

Abstract. *The basic operation principles of non-invasive microwave sensors are summarized in this work, with specific emphasis on health-care systems applications. Design criteria to achieve reliable results in terms of biological parameters detection are specifically highlighted. In particular, the importance to adopt accurate frequency models for the complex permittivity (in terms of both dielectric constant as well as loss tangent) in the synthesis procedure of the microwave sensor is clearly motivated. Finally, an application example of the outlined new perspectives in the framework of glucose monitoring to face diabete disease is deeply discussed.*

Keywords

Microwaves, sensors, biomedical applications, dielectric characterization

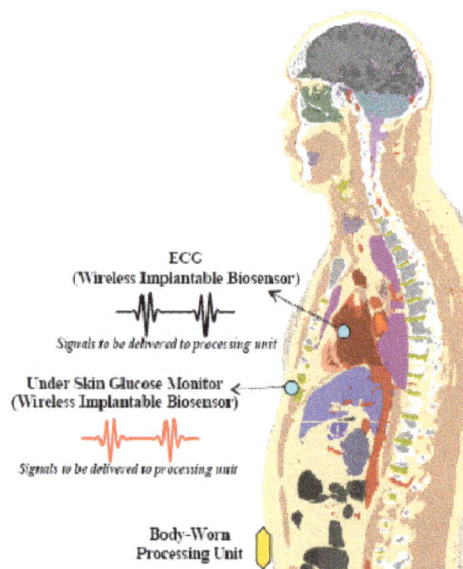

Fig. 1. A description of e-health monitoring (courtesy of http://antennas.eecs.qmul.ac.uk/research/body-centric-wireless-commnication-and-networks/)

1. Introduction

The increase in chronic pathologies related to actual lifestyle demands for urgent changes in the evolution of health-care systems. A primary challenge to satisfy this need is the achievement of high-care at reduced costs, by focusing on the promotion of prevention and effective disease management, rather than on specialized care systems for the treatment of late-stage pathologies. Health-care providers and users are increasingly interested in the adoption of new communication technologies (e.g. smartphones), thus moving towards a new paradigm of 'e-health' monitoring (Fig. 1), which however is at an early stage and is still in development. In order to provide accurate and early diagnoses, e-health devices should guarantee a continuous monitoring without interference in daily life, and they should also avoid the need of external people for the biomedical control. To this end, microwave biosensors can give a valid alternative to standard chemical devices, typically having short lifetime, and based on the use of fluid probe which limits the application to discontinuous monitoring.

The basic principle of microwave biosensors relies on the specific property of electromagnetic fields to interact with matter, in a different way depending on its molecular structure, thus leading to investigate the tissues compositions by examining the variation of their dielectric properties in response to the applied excitation field.

Specific advantages can be identified in the adoption of microwaves and millimeter waves. First of all, the ability of electromagnetic fields to penetrate into biological media can be exploited to implement noninvasive measurements. Furthermore, microwaves represent nonionizing fields, thus avoiding dangers related to the adoption of other types of radiation, such as X-rays. When exposed to microwave fields, biological molecules become polarized similarly to water, and the excitation wave travels more slowly with respect to a free-space environment. Microwave biosensors are so designed to convert these changes in the wave propagation speed through the biological medium into a quantifiable signal giving the variation of a specific bio-parameter.

In its practical configuration, a microwave biosensor is composed by two specific parts, namely the microwave sensor element and the probe circuit. Various architectures can be adopted to implement this latter component, which is essentially demanded at replacing in a compact minia-turized form the functionality of a vector network analyzer (VNA). Practically, it should be able to characterize the propagation features of the microwave sensor in the presence of the biological medium, in order to extract its dielectric properties. Since the early 1970s, various architectures have been introduced in literature, such as those based on heterodyne architecture [1] or the six-port technique [2].

All existing approaches are well suitable to be soon applied in a large class of biomedical applications [3–6], by also exploiting emerging integration techniques to realize devices in a very compact form. However, further advancements are still required for a more selective correlation of the different stages of a specific pathology to the changes in the dielectric features of the involved biological medium. These challenging advancements directly involve a more accurate design of the microwave sensor element, through the development of more reliable dielectric models for biomaterials, to be successively translated into more reliable correlations to the different stages of the monitored disease.

2. Basic Principle and Design Criteria

The ability of microwave sensors to monitoring biological parameters relies on the fact that all biological molecules have a different dielectric behavior as compared to air or water, thus changes in the wave propagation speed (or, equivalently, in the dielectric permittivity) can be in principle correlated to the human tissues compositions. However, to achieve reliable results in terms of biological parameter changes, accurate models are strongly required to define the variation of dielectric permittivity versus frequency for different biological materials. Actually, only approximated models exist, essentially derived from empirical evaluations.

At microwave and millimeter-wave frequencies, the interaction between electromagnetic field and matter caused two specific phenomena, namely:

- a reorientation motion of molecular dipoles, which is a polarization effect modeled by the real part of permittivity;

- a translational motion of free electric charges, which is modeled by an equivalent conductivity, giving rise to an imaginary part in the permittivity.

For polar molecules such as water, the frequency behavior of permittivity is described in literature by the Debye equation as [7]:

$$\varepsilon_{\mathrm{r}} = \varepsilon_\infty + \frac{\varepsilon_{\mathrm{s}} - \varepsilon_\infty}{1 + \mathrm{j}\omega\tau} + \frac{\sigma}{\mathrm{j}\omega\varepsilon_0} \qquad (1)$$

where ε_{s} is the low-frequency permittivity; ε_∞ is the high-frequency permittivity; τ gives the relaxation time; σ represents the ionic conductivity; ε_0 is the free-space permittivity.

Even if biological materials contain large quantities of water (typically 70-80%), their dielectric properties at microwave and millimeter-wave frequencies exhibit a behavior which differs from the simple model given by (1). So, to take into account for the different relaxation processes, the Cole-Cole model can be adopted for an efficient dielectric representation of biological tissues over frequency, as given by the following expression [8]:

$$\varepsilon_{\mathrm{r}} = \varepsilon_\infty + \frac{\varepsilon_{\mathrm{s}} - \varepsilon_\infty}{1 + (\mathrm{j}\omega\tau)^{1-\alpha}} + \frac{\sigma}{\mathrm{j}\omega\varepsilon_0} \qquad (2)$$

where the new parameter α models the broadening of the dispersion lines.

For better approximations, a description in terms of multiple Cole-Cole dispersion can be adopted, as given by [8]:

$$\varepsilon_{\mathrm{r}} = \varepsilon_\infty + \sum_n \frac{\varepsilon_{sn} - \varepsilon_\infty}{1 + (\mathrm{j}\omega\tau_n)^{1-\alpha}} + \frac{\sigma}{\mathrm{j}\omega\varepsilon_0}. \qquad (3)$$

Most of existing works on microwave biosensors do not assume the complete form of the Cole-Cole model, but the ionic conductivity σ, giving rise to the loss tangent, is typically neglected. This causes an important approximation error in the retrieval of biological parameter.

The adoption of accurate dielectric models for the tissue structure to be monitored has a fundamental importance to retrieve reliable information and diagnose pathologies at an early stage. As a matter of fact, the microwave sensor should be designed by properly taking into account the lossy behavior of the specific biological material to be monitored, which acts as radiation medium. This will lead to establish a reliable relationship between the response of the microwave sensor and the biological parameter to be detected for health-care purposes. However, some specific points still remain open to further advancements. First of all, most of existing microwave biosensors in literature are designed on the basis of full-wave simulations performed through cad software, with the optimization achieved numerically and typically assuming only the frequency variation of the real part of permittivity for different tissues structures. To improve accuracy, the variation of the imaginary part of permittivity should be also included in the simulation model, and the design process of the microwave sensor should be enhanced by adopting analytical models which assume the biological material as lossy radiation medium. On the other side, further developments should be also performed to investigate dielectric models for inhomogeneous biomaterials, thus improving the accuracy of future implementation of microwave biosensors.

3. Application Example: Glucose Monitoring

Diabete is a chronic disease resulting in the inability of affected patients to control the glucose level in the blood stream. The increasing aging of population and the rise of obesity negatively act to a continuous increment in the number of diabete patients, and a proper monitoring of blood glucose levels is strongly required for managing the disease and avoiding complications.

To achieve a continuous monitoring of glucose levels, implantable biosensors have been introduced [6–9] which rely on the interstitial fluid within the dermis to perform the measurement. However, many degradation factors, including also tissue fibrosis and inflammation, lead to a very limited implantation time (max 10 days), thus turning the attention of research to the investigation of alternative non-invasive procedures. To this end, various configurations of microwave sensors have been proposed in literature, all based on the principle that blood permittivity is affected by its glucose concentration, so glucose levels can be retrieved from the detection of frequency shift in the reflection response of the resonant microwave sensor. Many papers can be found which prove the feasibility of non-invasive blood glucose monitoring through the adoption of a resonant antenna [5], [10–12], but further work is still required to achieve true reliable implementations.

A primary approximation of existing approaches relies on the fact that the adopted design procedure only assumes the frequency variation of the real part of permittivity, while neglecting the effect of glucose concentration on the loss tangent of blood. This important aspect has been recently faced by the author in a preliminary work [13] and it is detailed here as illustration example.

To demonstrate the concept, a very simple configuration is assumed for the resonant microwave sensor, namely a standard inset-fed patch antenna working in the Industrial, Scientific, Medical (ISM) band around a frequency $f_0 = 2.4$ GHz. A high permittivity dielectric ($\varepsilon_r = 10$) is chosen as antenna substrate, in order to reduce the effect of environmental properties as much as possible. The optimization of antenna dimensions is performed on Ansys software to achieve the resonant condition at the design frequency f_0, but considering the proper frequency variation of the complex permittivity (dielectric constant as well as loss tangent) relative to the biological material, which acts as radiation medium.

As the objective is the demonstration of enhanced design principle, a water-glucose solution is assumed as radiation environment for this preliminary validation work. First, an experimental characterization of complex permittivity is performed on water solutions with different glucose concentrations (GC). Dielectric measurements are executed in the Microwave Laboratory at University of Calabria, by adopting the Anritsu VectorStar VNA and the open-ended coaxial probe Speag DAK, with an uncertainty

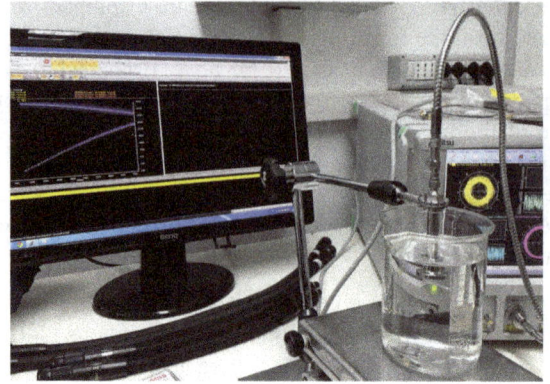

Fig. 2. Test setup for dielectric measurements (Microwave Laboratory at University of Calabria).

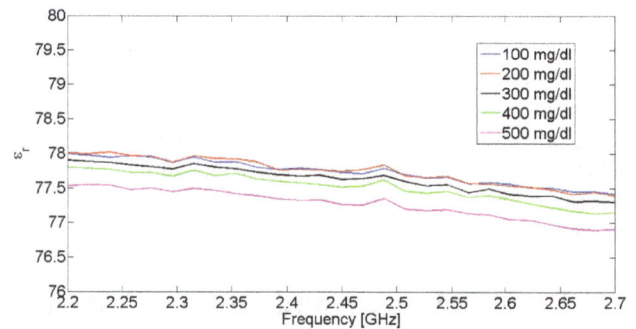

Fig. 3. Measured dielectric constant vs. frequency for water solutions with different glucose concentrations.

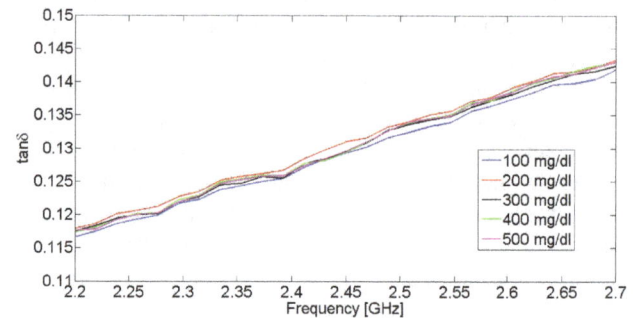

Fig. 4. Measured loss tangent vs. frequency for water solutions with different glucose concentrations.

error less than 1%. A photograph of the adopted test setup is illustrated in Fig. 2, while the measured frequency data of both dielectric constant as well as loss tangent are reported in Figs. 3 and 4, respectively.

From data represented in Figs. 3 and 4, it is straightforward to deduce that both real and imaginary parts of permittivity for water-glucose solutions are influenced by the glucose concentration, thus both information should be considered as input data in the synthesis process of the microwave resonant sensor. Furthermore, the dependency of both dielectric constant and loss tangent on the glucose level can be fruitfully considered to elaborate reliable correlations from which the glucose level can be accurately retrieved. On the basis of the above considerations, measured complex permittivity of water-glucose solutions, retrieved from the preliminary dielectric characterization stage, is used as input data in the synthesis procedure of the microwave sensor, acting to obtain the antenna dimensions which guarantee the optimum matching condition at the design frequency $f_0 = 2.4$ GHz.

Reflection measurements are then performed on the water solutions with different GC (the same as those adopted for complex permittivity characterization), and the experimental return loss curves are reported in Fig. 5, where a photograph of the realized microwave sensor is also illustrated. To avoid inaccuracies due to the positioning of the microwave sensor, a specific plexiglass box including a slit is adopted to contain the measured water-glucose solutions, as illustrated in Fig. 6.

Two significant observations can be done with reference to measured data of Fig. 5, namely:

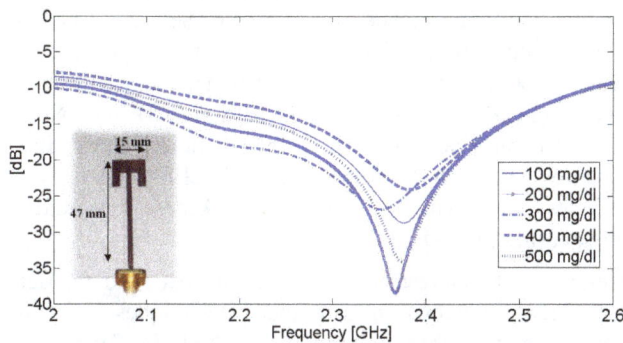

Fig. 5. Measured return loss of microwave sensor for various GC (realized prototype is visible on the lower left corner).

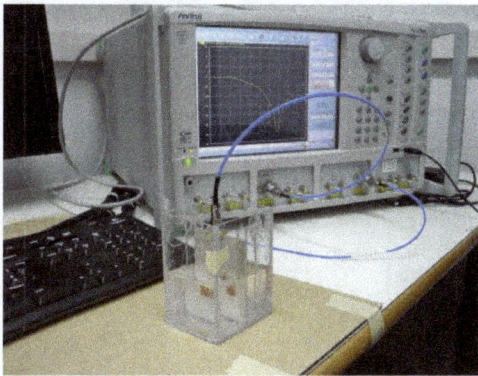

Fig. 6. Test setup for return loss measurements.

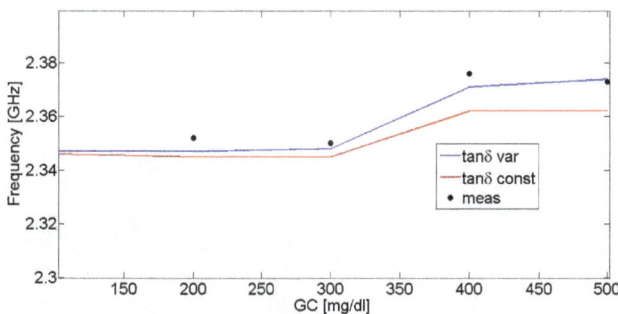

Fig. 7. Frequency variation vs. GC in the reflection response of microwave sensor.

- a frequency shift is clearly visible when varying the GC; in the existing works [5], [10–12], this is the only effect considered till now for the retrieval of glucose level;

- an amplitude variation in the curves is obtained for the different GC; this effect, related to the variation of loss tangent, is not actually taken into account in literature, and it could be strongly useful to elaborate future enhanced correlation models between the reflection response of resonant microwave sensor and the relative GC.

Finally, in Fig. 7, the frequency variation versus GC is reported, by comparing measured data of Fig. 5 with the simulated results (Ansys software) obtained in two distinct cases, namely with and without assuming the loss tangent variation. The comparison clearly demonstrates a more accurate result when the loss tangent variation is considered in the synthesis process of the microwave sensor.

4. Conclusion

The importance and physical principles related to the adoption of microwave and millimeter waves for non-invasive monitoring of biological parameters have been discussed in the present work. Existing limitations in the design criteria of microwave biosensors have been clearly outlined, together with new useful perspectives for improving the accuracy of future implementations. The importance of assuming the loss tangent frequency variation of the biological radiation medium in the design of microwave sensor is particularly highlighted, and validation results in the framework of glucose level monitoring have been discussed. The concepts outlined in the present work are intended to provide useful starting guidelines in the implementation of reliable non-invasive monitoring systems based on microwaves or millimeter waves as excitation source for biomedical applications.

References

[1] NEHRING, J., NASR, I., BORUTTA, K., WEIGEL, R., KISSINGER, D. A silicon integrated microwave vector network analyzer for biomedical sensor read-out applications. In *Proceedings of IEEE MTT-S International Microwave Symposium.* Tampa (FL), 2014. DOI: 10.1109/MWSYM.2014.6848254

[2] KOELPIN, A., VINCI, G., LAEMMLE, B., KISSINGER, D., WEIGEL, R. The six port in modern society. *IEEE Microwave Magazine*, 2010, vol. 11, no. 7, p. 35–43. DOI: 10.1109/MMM.2010.938584

[3] ROSEN, A., STUCHLY, M. A., VANDER VORST, A. Applications of RF/microwaves in medicine. *IEEE Transactions on Microwave Theory and Techniques*, 2002, vol. 50, no. 3, p. 963–974. DOI: 10.1109/22.989979

[4] GRENIER, K., DUBUC, D., POLENI, P.-E., KUMEMURA, M., TOSHIYOSHI, H., FUJII, T., FUJITA, H. Integrated broadband microwave and microfluidic sensor dedicated to bioengineering.

IEEE Transactions on Microwave Theory and Techniques, 2009, vol. 57, no. 12, p. 3246–3253. DOI: 10.1109/TMTT.2009.2034226

[5] JEAN, B. R., GREEN, E. C., MCCLUNG, M. J. A microwave frequency sensor for non-invasive blood-glucose measurement. In *Proceedings of IEEE Sensors Applications Symposium*. Atlanta (GA), 2008. DOI: 10.1109/SAS.2008.4472932

[6] AHMADI, M. M., JULLIEN, G. A. A wireless-implantable microsystem for continuous blood glucose monitoring. *IEEE Transactions on Biomedical Circuits and Systems*, 2009, vol. 3, no. 3, p. 169–180. DOI: 10.1109/TBCAS.2009.2016844

[7] COLE, K. S., COLE, R. H. Dispersion and absorption in dielectrics I. Alternating current characteristics. *Journal of Chemical Physics*, 1941, vol. 9, p. 341–351. DOI: 10.1063/1.1750906

[8] GABRIEL, S., LAU, R., W., GABRIEL, C. The dielectric properties of biological tissues: III. Parametric models for the dielectric spectrum of tissues. *Physics in Medicine and Biology*, 1996, vol. 41, no. 11, p. 2271–2293.

[9] KARACOLAK, T., HOOD, A., TOPSAKAL, E. Design of a dual-band implantable antenna and development of skin mimicking gels for continuous glucose monitoring. *IEEE Transactions on Microwave Theory and Techniques*, 2008, vol. 56, no. 4, p. 1001–1008. DOI: 10.1109/TMTT.2008.919373

[10] VENKATARAMAN, J., FREER, B. Feasibility of non-invasive blood glucose monitoring. In *Proceedings of IEEE International Symposium on Antennas and Propagation AP-S/URSI*. Spokane (WA), 2011, p. 603–606. DOI: 10.1109/APS.2011.5996782

[11] BAGHBANI, R., RAD, M. A., POURZIAD, A. Microwave sensor for non-invasive glucose measurements design and implementation of a novel linear. *IET Wireless Sensor Systems*, 2015, vol. 5, no. 2, p. 51–57. DOI: 10.1049/iet-wss.2013.0099

[12] CHOI, H., NAYLON, J., LUZIO, S., BEUTLER, J., BIRCHALL, J., MARTIN, C., PORCH, A. Design and in vitro interference test of microwave noninvasive blood glucose monitoring sensor. *IEEE Transactions on Microwave Theory and Techniques*, 2015, vol. 63, no. 10, p. 3016–3025. DOI: 10.1109/TMTT.2015.2472019

[13] COSTANZO, S. Loss tangent effect on the accurate design of microwave sensors for blood glucose monitoring. In *Proceedings of the 11th European Conference on Antennas and Propagation EuCAP*. Paris (France), March 2017.

About the Author ...

Sandra COSTANZO received the Laurea degree (summa cum laude) in Computer Engineering from the University of Calabria in 1996, and the Ph.D. degree in Electronic Engineering from the University of Reggio Calabria in 2000. Currently, she is an Associate Professor at the University of Calabria, Italy, where she teaches the courses of electromagnetic waves propagation, remote sensing and radar systems and electromagnetic diagnostics. At the same University, she is the Coordinator of Master Degree Course in Telecommunication Engineering.

Since 1996, she has been involved in many research projects funded by ESA (European Space Agency), ASI (Agenzia Spaziale Italiana), MIUR (Ministero dell'Istruzione, dell'Università e della Ricerca) and private companies. She is a Senior Member of IEEE, member of IEEE South Italy Geoscience and Remote Sensing Chapter, CNIT (Consorzio Nazionale Interuniversitario per le Telecomunicazioni) and SIEm (Società Italiana di Elettromagnetismo), and a Board Member of IEEEAP/ED/MTT North Italy Chapter, and of IEEE Information Theory Italy Chapter. She is an Associate Editor of IEEE Antennas and Wireless Propagation Letters, IEEE Access and Radioengineering journals.

Her research interests are focused on near-field far-field techniques, antenna measurement techniques, antenna analysis and synthesis, numerical methods in electromagnetics, millimeter-wave antennas, reflectarrays, synthesis methods for microwave structures, electromagnetic characterization of materials, innovative antennas and technologies for radar applications.

She has been Editor of two books and Lead Editor of three special issues on international journals. She has (co) authored more than 160 contributions in international journals, books and conferences.

11

Distributed Aggregate Function Estimation by Biphasically Configured Metropolis-Hasting Weight Model

Martin KENYERES[1], Jozef KENYERES[2], Vladislav SKORPIL[1], Radim BURGET[1]

[1] Dept. of Telecommunications, Brno University of Technology, Technická 12, 612 00 Brno, Czech Republic
[2] Sipwise GmbH, Europaring F15, 2345 Brunn am Gebirge, Austria

kenyeres@phd.feec.vutbr.cz, jkenyeres@sipwise.com, skorpil@feec.vutbr.cz, burgetrm@feec.vutbr.cz

Abstract. *An energy-efficient estimation of an aggregate function can significantly optimize a global event detection or monitoring in wireless sensor networks. This is probably the main reason why an optimization of the complementary consensus algorithms is one of the key challenges of the lifetime extension of the wireless sensor networks on which the attention of many scientists is paid. In this paper, we introduce an optimized weight model for the average consensus algorithm. It is called the Biphasically configured Metropolis-Hasting weight model and is based on a modification of the Metropolis-Hasting weight model by rephrasing the initial configuration into two parts. The first one is the default configuration of the Metropolis-Hasting weight model, while, the other one is based on a recalculation of the weights allocated to the adjacent nodes' incoming values at the cost of decreasing the value of the weights of the inner states. The whole initial configuration is executed in a fully-distributed manner. In the experimental section, it is proven that our optimized weight model significantly optimizes the Metropolis-Hasting weight model in several aspects and achieves better results compared with other concurrent weight models.*

Keywords

Distributed computing, aggregate function, average consensus algorithm, Metropolis-Hasting weight model, wireless sensor networks

1. Introduction

1.1 Wireless Sensor Networks

Wireless sensor networks (WSNs) are systems intended to perform a real-time detection of a stochastic event or to monitor physical quantities [1–2]. They are formed by battery-constrained nodes deployed in a geographical area where a phenomenon of interest is observed. These nodes are equipped with hardware components such as a wireless transceiver, a sensor to sense physical quantities, a central processor unit, a source of energy etc. [3–4] Thus, the nodes are able to obtain necessary information about the observed phenomenon, process it, mutually exchange data and make a meaningful decision on the examined physical quantity [4]. Due to their character, the WSNs find the application in various areas such as military surveillance, a natural disaster detection and its elimination, habitat monitoring, inventory tracking, an acoustic detection, pollution monitoring, medical systems, target tracking, a robotic exploration, a health care (especially, they find the usage in the scenarios considering monitoring elderly patients in a remote area), environment monitoring, a micro surgery, agriculture etc. [5–6]. In many applications, WSNs may be formed by hundreds of nodes potentially situated in inaccessible locations and therefore, a battery recharge or replacement may be complicated [7]. An exhausted battery results in a node death, which can decrease the quality of the final decisions or even prevent a whole system from fulfilling its functionality. As the results, the attention of many scientists has been focused on an optimization of the energy consumption aspect in the last years [8–11]. It is because an effective optimization can significantly increase the network lifetime of a WSN application [12].

In [13], the authors divide the architectures of a global event detection into three categories. The second and the third architecture require a complementary consensus algorithm to estimate aggregate functions in order to ensure a higher credibility of the measured outputs. These architectures do not assume the presence of a fusion center in a network. The implementation of this supplementary algorithm ensures a higher precision of the final decision on the observed phenomenon in many applications [14]. A decision made according to data obtained by independently-measuring nodes secures a more credible output than a decision made in terms of a single measurement and minimizes a change of an incorrect classification [14]. The importance of the consensus algorithm implementation for high-quality monitoring in WSNs is discussed in [15].

1.2 Average Consensus Algorithm

Due to the character of the WSNs, the modern applications are often based on the implementation of distributed mechanisms. The algorithms of distributed computing substitute the older frequently-implemented centralized manner of the computation [14]. Despite its reliability and high precision, the centralized algorithms do not pose the optimal solution for the implementation into the systems formed by battery-constrained devices.

One of the most appropriate distributed algorithms for WSN applications is average consensus, which is a fully distributed iterative algorithm primarily for estimation the average from the values of all the nodes present in a network [16]. This algorithm does not require the presence of any fusion center. The nodes are able to estimate the average by a mutual exchange of their inner states with the nodes situated in the adjacent area. The average consensus algorithm is characterized by a high flexibility because its execution is modifiable by the chosen weight model [17]. The weight models differ from each other in several aspects, for example, we can list the convergence rate of the algorithm, the process of the initial configuration, the information that is necessary for its proper functionality, the robustness etc. [17], [18]. In this paper, we focus on the Metropolis-Hasting model, which requires only the locally-available information for the initial configuration and therefore, it is one of the most preferred solutions for a real implementation into battery-constrained systems [19].

As mentioned earlier, the average consensus algorithm is primarily proposed for the estimation of the average value. However, tiny modifications can ensure that the algorithm is able to estimate other aggregate functions. One of the other frequently-used applications is the estimation of the network size. The information about the number of the nodes in a network is crucial for a proper functionality of many distributed systems [20]. In this case, the execution of the average consensus algorithm is modified in such a way that one of the nodes has the initial value set to the value equaling 1 (it is called the leader) [21]. The other nodes are set to 0. Subsequently, the nodes converge to the value equaling the reciprocal of the size of a network [21]. However, this modification causes several problems. One of the most significant ones is how to appoint the most suitable node as the leader. It often requires the implementation of other complementary mechanisms to determine this, which is not the optimal solution for battery-constrained devices [21]. As shown later in this paper, a bad choice of the leader can significantly decrease the convergence rate of the algorithm. Thus, an improvement of the leader selection can significantly optimize WSN applications by removing the necessity for other complementary algorithms.

1.3 Motivation

All the previously-discussed problems motive us to propose an optimization mechanism of the Metropolis-

Hasting weight model that improves the convergence rates, the number of the necessary messages and minimize the negative effect caused by an inappropriate choice of the leader. An optimization of these mentioned aspects can significantly accelerate and simplify the computation process and therefore optimizes the real-life applications of WSNs.

The choice of this weight model is affected by an effort to improve a weight model of the average consensus algorithm that finds the wide usage thanks to its specific character. The Metropolis-Hasting weight model fulfills these criteria because it does not require any global information about the network for its proper initial configuration and so it works in a fully distributed manner. Thus, it is an appropriate solution for an implementation into the WSNs. Thus, this was the main reason that motivated us to focus our research on this weight model.

In this paper, we introduce an optimized weight model derived from the Metropolis-Hasting weight model that improves the discussed aspects of this model. The optimized model modifies the weight matrix of the Metropolis-Hasting weight model by an additional step during the initial configuration. Thus, the weight matrix is initially configured twice. The first configuration is the default one defined within the Metropolis-Hasting weight model and the other one poses the novelty proposed by us. The other phase is based on recalculating the weights allocated to the adjacent nodes' incoming states at cost of decreasing the value of the weight of the inner states in a distributed way.

1.4 Paper Organization

In Sec. 2, we turn our attention to the latest papers related to the average consensus optimization. In the next section, we provide the mathematical tool used to model the average consensus algorithm executed in the WSNs, present the main features of this algorithm and adduce important theorems defined within the spectral graph theory. We also introduce concurrent weight models which the optimized weight model is compared with. In Sec. 4, we introduce our optimized weight model, provide mathematical tools to model it and derive the convergence proof. In Sec. 5, we examine the optimization of chosen aspects ensured by the optimized weight model. We focus on an optimization of the average estimation, the network size estimation and the range of the convergence rates caused by the choice of the leader. We compare this optimization in three types of the networks – weakly, averagely and strongly connected. All of them consist of ten randomly generated networks. In Appendix section, we adduce the complete results obtained within our numerical experiments.

2. Related Work

This section is devoted to an insight into an optimization mechanism proposed for the average consensus algorithm. We introduce the latest papers dealing with optimization mechanisms.

In [19], [22–25], the authors' attention focuses on the Metropolis-Hasting weight model (also in various applications as a complementary mechanism), which our optimized weight model is derived from. It was developed by Metropolis, Rosenbluth and Teller in 1953 and generalized by Hastings in 1970. It was originally defined within Markov chain Monte Carlo methods and proposed to simulate complex, non-standard multivariate distributions [26]. Its modification for the consensus achievement problem finds the usage in many applications due to its character. For its proper initial configuration, only locally-available information is required, i.e. the number of the neighbors of a particular node and the number of neighbors of the node from its adjacent area. This significantly simplifies the initial configuration phase. Thus, this weight model finds the usage in many applications. Additionally, it also poses a robust solution against a quantization noise [18].

In [27], an average consensus optimization based on the usage of the opportunistic inter-agent communication to achieve the consensus is presented. Each node is endowed by a local criterion determining when to broadcast the inner state to the nodes situated in the adjacent area. In paper [28], a novel consensus protocol is presented that achieves the average state consensus for multi-agent systems in finite time. The protocol contains a linear and a non-linear term. The state consensus is achieved by the non-linear term, while the performance optimization is ensured by the linear term to some degree. The authors of paper [29] present an optimization mechanism based on a division of the computation process into two phases. The first one is the phase of reaching the local consensuses and the second one is the phase of reaching the global consensus. Within the first phase, a network is reorganized into geographically close areas so-called packs. Here, each node converges to the value equaling the average of all the nodes present in a pack. Subsequently, each pack appoints one of the nodes as the head, which communicates with the other heads and converges to the average value. The authors of [30] present a novel continuous-time dynamic average consensus algorithm for networks with the interaction that can be described by weight-balanced directed graphs of a strong connectivity. The nodes are able to track the average of the dynamic inputs with some non-zero steady-error. Its size is controlled by exploiting a design parameter. In [31], a distributed algorithm for average consensus that solves the discrete-time average consensus problem on strongly connected weighted digraphs is presented. Its principle lays in the computation of the average value using the estimation of the left eigenvector associated with the zero eigenvalue of the Laplacian matrix. The authors of [32] built their optimization mechanism on the exploiting of the second-order neighbors. They focus their attention on both the continuous-time case, where the edges are chosen by solving a convex optimization problem formed by utilizing the convex relaxation method, and the discrete-time case, when the edges are chosen using the

brute force method. In [33], an optimization mechanism is presented that exploits the prediction of the future value of the inner states. This technique is based on the estimation of the states for the next iterations in terms of the values of the inner states from the previous iterations. The authors of [34] introduce IACA, which is a two-layer improved consensus algorithm of a multi-agent system. The authors propose a new distributed cost optimization method for loading shedding of an islanded microgrid considering cost. The technique solves distributed cost optimization of load shedding by exploiting the synchronization processing of IACA in the layer 2. The authors of [35] present an optimization mechanism minimizing the negative effects caused by a random packet loss. It is based on keeping track of the changes in the state variable, which the neighbors influence causes.

The papers [36–38] focus on the Maximum Degree weight model, which is a modification of the Constant weight model, and its applications. The Constant weight model is characterized by the parameter ε, which affects the convergence rate as well as the interval of the convergence. The higher value it takes, the faster the algorithm is. However, a too high value can cause the divergence of the algorithm. The divergence is a type of a failure when the convergence is not reached. Instead of it, the nodes diverge to infinite large values [39]. This error poses a serious problem that stunts a whole network [39]. The Maximum Degree weight model is based on the setting of the parameter ε to the value equaled to the reciprocal of the number of the neighbors of the best-connected node in a network. The initial configuration requires the knowledge about this value and therefore, it is necessary to implement a supplementary mechanism to determine it [19].

In [18], [40], the Best Constant weight model is discussed. Its optimized version is based on the utilization of the knowledge about the second smallest and the largest eigenvalue of the Laplacian matrix [18]. To compose it, it is necessary to know the information about the complete network topology. Thus, this weight model requires a particular centralization for its optimization.

3. Modeling of Average Consensus Algorithm in WSNs

In Sec. 3.1, we introduce the used mathematical model of the WSNs executing the average consensus algorithm and the main features of this algorithm. In Sec. 3.2, we discuss and mathematically describe the concurrent weight models of average consensus with which our optimized weight model is compared.

3.1 Used Mathematical Model

In order to model the WSNs, a mathematical tool defined within the spectral graph theory is used [41], [42]. A WSN is considered to be an indirect finite graph defined as $G = (\mathbf{V}, \mathbf{E})$. The set $\mathbf{V}$ is formed by all the vertexes, which are representatives of the particular nodes. Each node is labeled by the unique identity number v_i. We assume that the nodes are labeled with the numbers $1, 2,...,N$, where N is the size of a network and therefore $|\mathbf{V}| = N$. The mutual connectivity between the nodes is indicated by the existence of an edge. The set $\mathbf{E} \subset \mathbf{V} \times \mathbf{V}$ consists of all the edges present in a graph. The edge is labeled as (v_i, v_j) or e_{ij}. We assume the range homogeneity of the nodes and therefore the following statement holds:

$$e_{ij} \in \mathbf{E} \Leftrightarrow e_{ji} \in \mathbf{E} . \tag{1}$$

There are several tools to describe a network topology within the spectral graph theory. One of them is the Laplacian matrix for a description of the mutual connectivity among the nodes. It is a square symmetric matrix defined for all the indirect graphs as follows [43]:

$$[L]_{ij} = \begin{cases} -1, & \text{if } e_{ij} \in \mathbf{E} \\ d_i, & \text{if } i = j \\ 0, & \text{otherwise,} \end{cases} \tag{2}$$

Here, d_i is the degree of a vertex v_i and so, the number of the corresponding node's neighbors. Except for the mutual connectivity, the Laplacian matrix provides other useful information about the topology. Let us focus on the following sentence [44]:

Lemma 1: Let $G = (\mathbf{V}, \mathbf{E})$ be a graph and let $0 = \mu_1(\mathbf{L}) \le \mu_2(\mathbf{L}) \le \mu_3(\mathbf{L}) \le ...... \le \mu_N(\mathbf{L})$ be the ascendingly-ordered eigenvalues of the Laplacian matrix of this graph. Then, G is not connected if $\mu_2(\mathbf{L}) = 0$.

According to Lemma 1, only the networks with the Laplacian matrix whose second smallest eigenvalue is not equaled to 0 are connected. In the case when $\mu_2(\mathbf{L})$ equals 0, the average consensus algorithm does not estimate the average from the values of all the nodes but estimates the set of the local averages in each connected subpart of a graph. Therefore, we assume only topologies whose second smallest eigenvalue of the Laplacian matrix is not equaled to 0. The knowledge about the exact value of $\mu_2(\mathbf{L})$ and $\mu_N(\mathbf{L})$ is necessary for the optimized initial configuration of the Best Constant weight model [26]. A configuration of this model with a smaller positive value ensures the convergence but the execution of the algorithm is slower [18].

As mentioned above, the average consensus algorithm is an iterative distributed algorithm based on a mutual exchange of the current states among the nodes. The algorithm is modeled by the difference equation defined as follows [45]:

$$\mathbf{x}(k+1) = \mathbf{W} \times \mathbf{x}(k) . \tag{3}$$

Here, $\mathbf{W} \subset \mathbf{L}$ is a weight matrix of the algorithm and the time-variant vector $\mathbf{x}(k) \in \mathbb{R}^{N \times 1}$ gathers all the inner states at kth iteration. We assume that the initial states are labeled as $k = 1$. The elements of $\mathbf{W}$ depend on the used weight model. This matrix also provides useful information

about the network topology. The following lemma says about the connectivity of the topology [44]:

Lemma 2: Let $G = (\mathbf{V}, \mathbf{E})$ be a graph and let $1 = \lambda_1(\mathbf{W}) \ge \lambda_2(\mathbf{W}) \ge \lambda_3(\mathbf{W}) \ge ...... \ge \lambda_N(\mathbf{W}) \ge -1$ be the descendingly-ordered eigenvalues of the weight matrix. Then, G is not connected if $max\{\lambda_2(\mathbf{W}), -\lambda_N(\mathbf{W})\} = 1$.

Within the spectral theory, it is defined that the value $max\{\lambda_2(\mathbf{W}), -\lambda_N(\mathbf{W})\}$ equals the spectral radius ρ of the matrix determined as the difference between the matrix $\mathbf{W}$ and matrix defined as $1/N \cdot \mathbf{1} \times \mathbf{1}^T$ [46]. Thus, in terms of the previous statement, we can write as follows:

$$\rho(\mathbf{W} - \frac{1}{N} \cdot \mathbf{1} \times \mathbf{1}^T) = max\{\lambda_2(\mathbf{W}), -\lambda_N(\mathbf{W})\} . \tag{4}$$

Let us focus on the features of the weight matrix $\mathbf{W}$. According to [47], the weight matrix $\mathbf{W}$ is required to hold the following conditions:

$$\mathbf{W} \times \mathbf{1} = \mathbf{1}, \tag{5}$$

$$\mathbf{1}^T \times \mathbf{W} = \mathbf{1}^T, \tag{6}$$

$$\rho(\mathbf{W} - \frac{1}{N} \cdot \mathbf{1} \times \mathbf{1}^T) < 1 . \tag{7}$$

Here, the vector $\mathbf{1}$ is a column vector whose all elements are equaled to 1 (its size is implicitly assessable from the previous context). Fulfilling the formula (7) ensures the convergence of the average consensus algorithm, meanwhile, the formulae (5) and (6) imply that the weight matrix is doubly stochastic (sometimes, labeled as bistochastic) and determines the convergence point [47]. These two formulas also implicate the following statement:

$$\mathbf{W} = \mathbf{W}^T . \tag{8}$$

Within our analysis, we use also another descriptive tool defined within the spectral graph theory. It is called the adjacency matrix $\mathbf{A} \in \{0, 1\}^{N \times N}$ and contains the information about the mutual connectivity between the pairs of the nodes. It is a diagonally symmetric matrix of a square shape for all the indirect graphs. The direct connection (i.e. the existence of an edge) is indicated by the presence of 1 in the corresponding position. Thus, the presence of 0 is an indicator that two nodes are not directly connected to one another. Mathematically, the adjacency matrix is defined as follows [48]:

$$[A]_{ij} = \begin{cases} 1, & \text{if } (v_i, v_j) \in \mathbf{E} \\ 0, & \text{otherwise} \end{cases} \tag{9}$$

Another useful tool is the identity matrix $\mathbf{I} \in \{0,1\}^{N \times N}$ defined as (10) [49]. In literature, also the other notation $\mathbf{I} = diag(1,1,....1)$ can be found.

$$[I]_{ij} = \begin{cases} 1, & \text{if } i = j \\ 0, & \text{otherwise} \end{cases} \tag{10}$$

Usually, a lower index is allocated to indicate its size (the label $\mathbf{I}_N$ indicates that the underlying matrix has the size equaled to the number of the size a network).

As mentioned earlier, the average consensus algorithm is an iterative algorithm (regardless of the used weight model) executed in such a way that the nodes converge to the value determined as the average calculated from all the initial values [50]. Therefore, this behavior can be described as follows [45]:

$$\lim_{k \to \infty} \mathbf{x}(k) = \lim_{k \to \infty} \mathbf{W}^{k-1} \times \mathbf{x}(1) = \frac{\mathbf{1} \times \mathbf{1}^{\mathrm{T}}}{N} \times \mathbf{x}(1). \quad (11)$$

Only the existence of this limit ensures the convergence of the average consensus algorithm. As mentioned, it is achieved by using such a weight matrix that holds the conditions (5-7) [46].

As the algorithm convergences to the value in the infinite [51], it is necessary to implement a mechanism indicating the consensus. We use the mechanism defined as follows:

$$\left| \max\{\mathbf{x}(k)\} - \min\{\mathbf{x}(k)\} \right| < \delta. \quad (12)$$

Here, the parameter δ determines the precision. Its higher values ensure a higher precision at the cost of a slower convergence rate. In our experiments, we assume that its value is equaled to 0.0001.

3.2 Concurrent Weight Models

As mentioned, we compare our optimized weight model with other three concurrent ones. In order to distinguish these models from each other, the weight matrix of a particular model has an upper index with the abbreviated name of the model. We use the following abbreviations:

- Metropolis-Hasting weight model – MH
- Maximum Degree weight model – MD
- Best Constant weight model – BC
- Biphasically configured Metropolis-Hasting weight model – BMH

The first examined model is the Metropolis-Hasting weight model whose weight matrix is defined as follows [19], [22-25]:

$$[W^{\mathrm{MH}}]_{ij} = \begin{cases} (1 + \max\{d_i, d_j\})^{-1}, & \text{if } (v_i, v_j) \in \mathbf{E} \\ 1 - \sum_{k=1, k \neq i}^{N} [W^{\mathrm{MH}}]_{ik}, & \text{if } i = j \\ 0, & \text{otherwise} \end{cases} \quad (13)$$

As mentioned, our contribution optimizes this model.

The second model is the Maximum Degree weight model. It is derived from the Constant weight model in such a way that the weighting parameter ε is set to the value equaling the reciprocal of the degree of the best-connected node. Therefore, it is defined as follows [36-38]:

$$[W^{\mathrm{MD}}]_{ij} = \begin{cases} 1 / d_{\max}, & \text{if } (v_i, v_j) \in \mathbf{E} \\ 1 - d_i / d_{\max}, & \text{if } i = j \\ 0, & \text{otherwise} \end{cases} \quad (14)$$

Within the initial configuration phase, this model requires each node in a network to be aware of the number of the neighbors of the best-connected node. In order to get this information in a distributed manner, it is necessary to implement a complementary algorithm [19].

The last examined model is called the Best Constant weight model. We assume its optimized variant even though it requires the information about the second smallest and the largest eigenvalue of the Laplacian matrix. It is defined as follows [40]:

$$[W^{\mathrm{BC}}]_{ij} = \begin{cases} 2 / (\mu_2(\mathbf{L}) + \mu_N(\mathbf{L})), & \text{if } (v_i, v_j) \in \mathbf{E} \\ 1 - 2 \cdot d_i / (\mu_2(\mathbf{L}) + \mu_N(\mathbf{L})), & \text{if } i = j \\ 0, & \text{otherwise} \end{cases} \quad (15)$$

4. Biphasically Configured Metropolis-Hasting Weight Model

In this section, we introduce our Biphasically configured Metropolis-Hasting weight model. In Sec. 4.1, we explain the model mechanics and provide the mathematical description. In Sec. 4.2, the convergence conditions are presented.

4.1 Principle of Biphasically Configured Metropolis-Hasting Weight Model

This subsection focuses on an introduction of the main features of our optimized weight model. As mentioned above, it is called the Biphasically configured Metropolis-Hasting weight model and is derived from (as its name implies) the Metropolis-Hasting weight model. As already discussed, the Metropolis-Hasting model is appropriate for the implementation into real-life applications thanks to its simplified demands for the initial configuration. In order to correctly fulfill its functionality, each node has to be aware of the number of its neighbors as well as the number of the neighbors of the adjacent nodes. Thus, only locally available information is necessary for the correct initial configuration. There are several approaches to obtain this information (centralized one, distributed one assuming a phase when this information is distributed in the adjacent area, a manual configuration etc.). The most appropriate solution depends on a particular application.

Our optimized weight model is based on rephrasing the initial configuration process of the Metropolis-Hasting weight model into two phases. The first phase is identical to the default configuration of the Metropolis-Hasting weight model. The other phase consists of a recalculation of the weights allocated to the incoming values from the adjacent nodes. Within this phase, each node determines when to do this recalculation according to its unique identity. Thus, each node has to be additionally aware of the diagonal value of the weight matrix $\mathbf{W}^{\mathrm{BMH}}$ corresponding to all its neighbors – this value is locally available. The recalculation of $\mathbf{W}^{\text{BMH}}$ has to be executed sequentially, i.e. the node with the identity number equaled to 1 initiates the whole process, updates the matrix $\mathbf{W}^{\text{BMH}}$ (the active node updates its inner updating rules as well as informs its neighbors about new weights) and only then the node 2 can start the recalculation. Thus, we assume the variability of the weight matrix $\mathbf{W}^{\text{BMH}}(a)$ during the second phase of the configuration process. The parameter a takes the values from 1, 2,...,N and labels the active node (i.e. the one that is allowed to make the recalculation) as well as the round of the recalculation process (we assume that $\mathbf{W}^{\text{BMH}}(0) = \mathbf{W}^{\text{MH}}$ and each label of the round corresponds to the unique number of a node). Thus, we label the node currently allowed to make the update as v_a. The length of the recalculation process is determined by the size of a network (i.e. the process lasts N rounds).

Let us define the set $\mathbf{N}_a$ gathering all the nodes from the adjacent area of v_a. So, we can write as follows:

$$\mathbf{N}_a = \left\{ v_i \mid [A]_{ai} = 1 \right\}. \quad (16)$$

At the round when a node v_a is allowed to update $\mathbf{W}^{\text{BMH}}$, it has to be aware of the current value in the diagonal corresponding to it and all its neighbors. These values represent the weights of the current inner state. Subsequently, v_a calculates the growth coefficient defined as follows:

$$\chi(v_a) = \frac{\min\{[W^{\text{BMH}}(a-1)]_{aa}, \sum_{j=1}^{N}[W^{\text{BMH}}(a-1)]_{jj} \cdot [A]_{aj}\}}{\sum_{j=1}^{N}[W^{\text{BMH}}(a-1)]_{jj} \cdot [A]_{aj}}. \quad (17)$$

The choice of the minimal value from the weight of the inner state of the active node and the sum of weights of the inner states of its neighbors ensures that the growth coefficient is never greater than 1. Thus, the convergence of the weight model regardless of the underlying topology (see Sec. 4.2) is guaranteed.

Subsequently, the node currently making the recalculation decreases the weight of its inner state to the minimal possible value (the ideal scenario is when this weight is equaled to 0 after finishing this procedure) and distributes this value among the neighbors in terms of the weights of the inner states of these nodes. From the central view, this procedure is described according to the following rules:

$$[W^{\text{BMH}}(a)]_{ia} = [W^{\text{BMH}}(a-1)]_{ia} + \chi(v_a) \cdot [W^{\text{BMH}}(a-1)]_{ii} \quad (18)$$
$$\text{for } \forall v_i \in \mathbf{N}_a.$$

This formula describes an increase of the weights of the incoming values of the adjacent nodes. Their values are increased with the value equaled to the diagonal value of the adjacent nodes (i.e. the weight of their inner states) weighted by the growth coefficient. The Metropolis-Hasting weight model assumes a doubly stochastic matrix (i.e. an edge e_{ij} is allocated only one weight – the incoming state of v_a and v_i is weighted with the same value) and therefore, it is necessary to preserve this condition:

$$[W^{\text{BMH}}(a)]_{ai} = [W^{\text{BMH}}(a-1)]_{ai} + \chi(v_a) \cdot [W^{\text{BMH}}(a-1)]_{ii} \quad (19)$$
$$\text{for } \forall v_i \in \mathbf{N}_a.$$

After all the neighbors are allocated a new weight, the nodes with an increased weight have to decrease their diagonal value by the increase of their incoming value weight. Otherwise, the convergence conditions may not be preserved. From the central view, it is possible to describe the previous procedure using tools defined within the spectral graph theory as follows:

$$[W^{\text{BMH}}(a)]_{ii} = 1 - \sum_{j=1, j\neq i}^{N}[W^{\text{BMH}}(a)]_{ij}, \quad \text{for } v_i \in \mathbf{N}_a, v_a. \quad (20)$$

After all the nodes execute the recalculation described above (this procedure is repeated for all N nodes), the weight matrix for Biphasically configured Metropolis-Hasting weight model is completed and the average consensus algorithm can be executed according to (3).

4.2 Convergence Proof

In the following subsection, we provide the sufficient conditions for the convergence of average consensus algorithm. The average consensus algorithm whose weights are symmetric can be described using a weighted graph [52]. The non-zero elements of its adjacency matrix $\mathbf{A}^{\text{WG}}$ are allocated a strictly positive weight $[A^{\text{WG}}]_{ij} = w_{ij}$. Subsequently, it is possible to derive the weighted Laplacian matrix as [52]:

$$\mathbf{L}^{\text{WG}} = \text{diag}\{\mathbf{d}^{\text{WG}}\} - \mathbf{A}^{\text{WG}}. \quad (21)$$

Here, the $\mathbf{d}^{\text{WG}}$ is a weighted degree vector formed by the value of the degrees of the nodes. The weighted degree vector is defined as follows [52]:

$$\mathbf{d}^{\text{WG}} = \mathbf{A}^{\text{WG}} \times \mathbf{1}. \quad (22)$$

Firstly, we show that the weights of the Metropolis-Hasting weight model (13) ensure the convergence of the algorithm regardless of the underlying topology. Within the spectral graph theory, it is defined [52] that $\mathbf{W} = \mathbf{I} - \mathbf{L}^{\text{WG}}$ and therefore, $\mathbf{W}$ is doubly stochastic with an eigenvalue with the magnitude equaled to 1 and associated to the eigenvector with the values $N^{-1/2}$. Furthermore, the matrix $\mathbf{L}^{\text{WG}}$ is semidefinite and fulfills the following statement [52]:

$$\lambda_k(\mathbf{W}) = 1 - \lambda_k(\mathbf{L}^{\text{WG}}) \quad \text{for } \forall k. \quad (23)$$

Thus, the convergence condition (7) can be reformulated as follows [52]:

- $\rho(\mathbf{L}^{\text{WG}}) < 2$,

- Multiplicity of the zero eigenvalue of $\mathbf{L}^{\text{WG}}$ has multiplicity one.

The second statement is satisfied for all the connected graphs. It can be confirmed using the quadratic form defined as follows [52]:

$$\mathbf{v}^{\text{T}} \times \mathbf{L}^{\text{WG}} \times \mathbf{v} = \sum_{e_{ij} \in \mathbf{E}} w_{ij} \cdot (v_i - v_j)^2. \quad (24)$$

The vector $\mathbf{v}$ is an eigenvector associated to the weighted Laplacian matrix $\mathbf{L}^{WG}$. As seen, the quadratic form is equaled to the zero value if and only if $v_i = v_j$. This statement is valid for the weights of the positive values. This requirement is met by the unique normalized vector $\mathbf{v} = N^{-1/2}$. Let us focus on the first constraint. As (21) and (22) hold, the eigenvalues λ^{WG} satisfy the following condition according to Gershgorin circle theorem, which is defined as follows [53]:

$$\left| \lambda(\mathbf{L}^{WG}) - d_i^{WG} \right| \leq d_i^{WG}. \tag{25}$$

The parameter d_i^{WG} is the weighted degree of the node i defined as:

$$d_i^{WG} = \sum_{j=1}^{N} w_{ij}. \tag{26}$$

The value also presents the ith row of the vector $\mathbf{d}_i^{WG}$. In particular, the following statement is valid [52]:

$$\lambda(\mathbf{L}^{WG}) \leq 2 \cdot d_i^{WG}. \tag{27}$$

Thus, $d_i^{WG} \leq 1$ is a sufficient condition for the average consensus algorithm to convergence for all i [52] (when a graph is neither bipartite nor regular). Since it is improbable that the graph describing a WSN is bipartite regular [52], we do not deal with these critical graph topologies. From (13), it is clear that this condition holds for each topology. Thanks to the expression in the numerator in (17), our recalculation always ensures $d_i^{WG} \leq 1$. It secures that the sum in both the rows and the columns does not change despite the recalculation. This guarantees the convergence conditions for our mechanism.

Let us analyze the functionality of our optimized weight model. The minimal value from two parameters in the numerator of (17) is chosen to preserve the convergence conditions. The value of $[W^{BMH}(a-1)]_{aa}$ poses the maximal possible value with which v_a can decrease the weight of its inner state and distribute it among its adjacent nodes. A decrease with a value greater than $[W^{BMH}(a-1)]_{aa}$ always causes $d_i^{WG} \leq 1$ not to hold. Thus, in such a scenario, the convergence conditions are not fulfilled. Regarding all the positive values smaller than $[W^{BMH}(a-1)]_{aa}$: the convergence conditions are preserved but the optimization is less significant. Now, let us focus on $\Sigma_j[W^{BMH}(a-1)]_{jj}\cdot[A]_{aj}$. There can be a scenario when v_a can increase the weights of the incoming values of its adjacent nodes with a value that causes some of its neighbors to achieve a negative value in the diagonal after (20). In this scenario, the growth coefficient χ is greater than 1, which results in $d_i^{WG} > 1$. Therefore, parameter $\Sigma_j[W^{BMH}(a-1)]_{jj}\cdot[A]_{aj}$ ensures that the growth coefficient never exceed the value 1 and so, the convergence conditions are always preserved. The procedure in (20) must be executed in order to keep the weight matrix $\mathbf{W}$ doubly-stochastic. This step secures that (5) and (6) always hold.

5. Numerical Experiments and Discussion

In this section, we present the results of the numerical experiments executed in Matlab R2015a. All the used software was designed by the authors of this paper. In our experiments, three sets of networks are assumed with randomly generated topologies. We assume weakly, averagely and strongly connected networks. Each set consists of 10 unique topologies with the size of 200 nodes. Due to the limited range of the paper, only one representative of these sets is shown in Fig. 1, Fig. 2 and Fig. 3 respectively. The networks were generated as follows: each free position within the working area of a square shape was allocated the probability equaled to the reciprocal of the number of free positions. Thus, the choice of the position where a node was placed had a uniform distribution. Subsequently, the nodes situated in the transmission range of each node were labeled as its neighbors (i.e. there is an edge between them). In order to ensure a various average connectivity, the transmission range was changed.

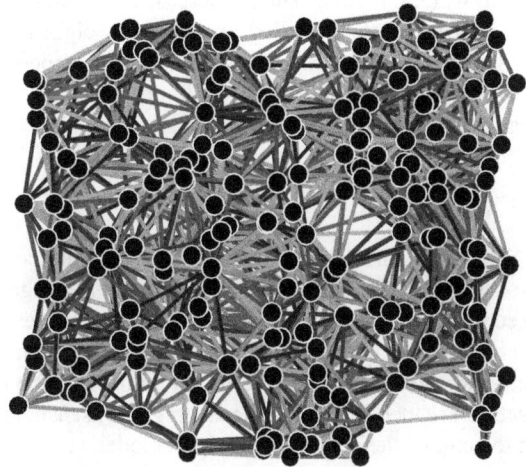

Fig. 1. Example representative of strongly connected topologies

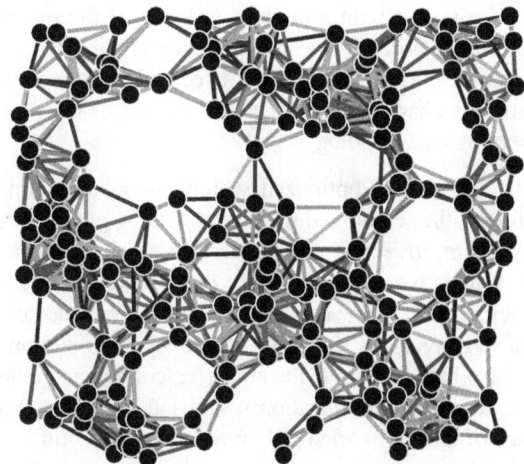

Fig. 2. Example representative of averagely connected topologies

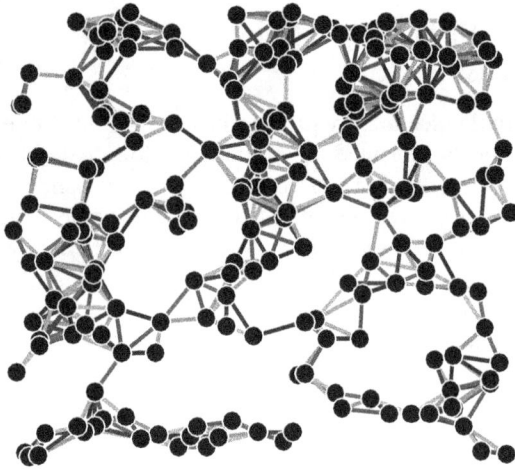

Fig. 3. Example representative of weakly connected topologies.

In order to evaluate the achieved optimization, we compare our mechanism with other discussed weight models. We compare it with the Metropolis-Hasting weight model, the Maximum-degree weight model, and the Best Constant weight model. These models were chosen because all of them are classified as constant weight models [25], frequently-used and discussed in other papers and optimize average consensus with a similar principle as our mechanism.

5.1 Estimation of Average Value

Within the first experiment, we examine the convergence rates and the number of the required messages. In this section, we draw our attention to the average estimation, compare our mechanism with the concurrent weight models and show the difference in the number of the messages that are necessary for the average consensus algorithm to achieve the consensus for each model.

As the initial configuration of our optimized weight model depends on the placement of the identity numbers, we use *randperm*, which is a built-in function in Matlab to generate a vector with a random position of the numbers. Therefore, in order to ensure a higher credibility of our conclusions, our mechanism was executed 100 times (each execution is characterized by a shuffled set of the identity numbers) in each topology.

In Tab. 1, our optimized weight model is compared with three other examined weight models. A positive value means an improvement ensured by our mechanism, meanwhile, a negative value indicates that a concurrent model achieves a faster convergence rate. We label these scenarios as positive, respectively, negative optimization. In Tab. 1, we have shown the average (calculated as the average value of the optimization of all ten networks within one set of the networks), the maximal (the optimization achieved in the network where the algorithm is optimized most significantly) and the minimal optimization (the optimization achieved in the network where the algorithm is

optimized worse or even negatively) as well as the range of the optimization (all are expressed in %). Within this comparison, the average calculated from 100 convergence rates obtained within these executions is chosen as a representative of 100 executions of the Biphasically configured Metropolis-Hasting weight model. The complete results are shown in Appendix A. The column labeled as CR contains the convergence rates expressed as the number of the iterations. The column OPT [%] is formed by the relative optimizations [%] of our mechanism compared with the concurrent weight models.

We can see from the results that our optimized weight model achieves a faster average convergence rate compared with all the concurrent models in all the sets of the networks. However, the Best Constant weight model achieves a faster rate in one of ten strongly connected topologies. The optimization is the most significant compared with the Maximum Degree weight model. Its average value ranges from 40.08% to 54.60%. The optimization of the Metropolis-Hasting model is from the range 8.19% to 15.45%. In this case, the most important fact is that our optimized weight model achieves a faster convergence rate in all the networks. The Best Constant weight model is optimized in the range 14.24% to 23.25%. Furthermore, we can see that (for all the weight models) the less connected the networks are, the higher average optimization our mechanism ensures. Let us focus on the range of the optimization (calculated as the difference between the maximal and the minimal optimization within one set). We can observe that this parameter achieves the highest value for the Best Constant weight model in all the sets. In Appendix A, we have also shown the number of the messages necessary for the average consensus algorithm to be completed. Since this parameter is closely related to the convergence rate, we do not provide a separate analysis.

Consequently, we analyze the effect of a random shuffle of the identity numbers on the convergence rates.

Convergence rate optimization of average estimation [%] in Weakly connected networks			
	MD	MH	BC
Average	54.60 %	15.45 %	23.25 %
Maximum	64.60 %	21.28 %	35.95 %
Minimum	47.38 %	12.65 %	8.77 %
Range	17.22 %	8.63 %	27.18 %
Convergence rate optimization of average estimation [%] in Averagely connected networks			
	MD	MH	BC
Average	46.84 %	11.15 %	15.36 %
Maximum	60.04 %	15.64 %	36.21 %
Minimum	39.71 %	7.80 %	1.42 %
Range	20.33 %	7.84 %	34.79 %
Convergence rate optimization of average estimation [%] in Strongly connected networks			
	MD	MH	BC
Average	40.08 %	8.19 %	14.24 %
Maximum	69.30 %	16.93 %	49.29 %
Minimum	31.47 %	4.29 %	-3.42 %
Range	37.83 %	12.64 %	52.62 %

Tab. 1. Comparison of BMH with others – average est.

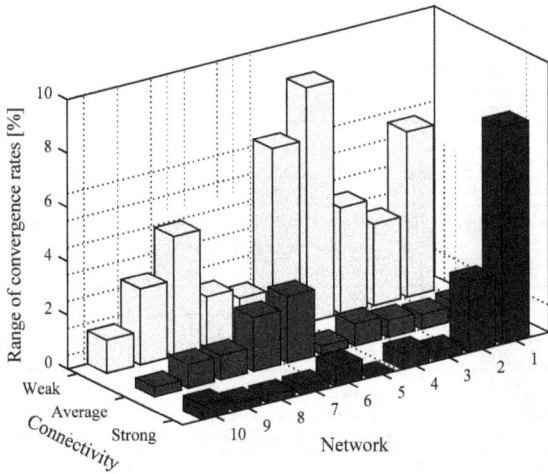

Fig. 4. Percentage range of convergence rates caused by shuffle of identity numbers (average est.).

The position of the identity numbers can affect only our optimized weight model – the other ones do not assume this value during the configuration process. In Fig. 4, we have depicted the range of the convergence rates calculated from the convergence rates obtained within 100 reparations (we depict the value of the ratio: the range/the average convergence rate expressed in % because the convergence rates differ from each other in different topologies).

We can see that the range does not exceed 9 % of the average value in any case. This primarily affects the convergence rate in the weakly connected topologies. In the averagely and the strongly connected networks, its impact is negligible except for one topology of a strong connectivity.

In the following part, we examine whether this shuffle can cause that there is an execution whose convergence rate is slower than the convergence rate of one of the concurrent models. Thus, we depict the slowest scenario within the Biphasically configured Metropolis-Hasting weight model with the fastest concurrent model in order to show that a shuffle does not cause a negative optimization when the average optimization is positive. The mentioned comparison for each set of the networks is depicted in Fig. 5 (weakly connected networks), Fig. 6 (averagely connected networks) and Fig. 7 (strongly connected networks).

The white column represents the slowest convergence rate within the Biphasically configured Metropolis-Hasting weight model, while, the black one represents the fastest concurrent weight model.

We can see that a positive optimization is preserved in all the cases when a positive average optimization is. Thus, a random allocation of the identity numbers has only a minimal impact on the convergence rate.

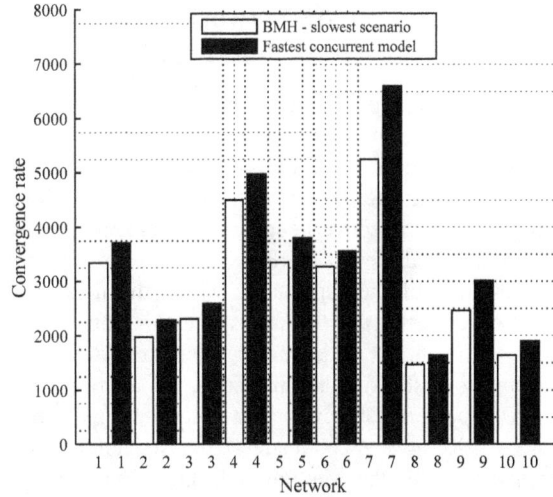

Fig. 5. Comparison of slowest scenario of BMH with fastest concurrent model – weak connectivity – average est.

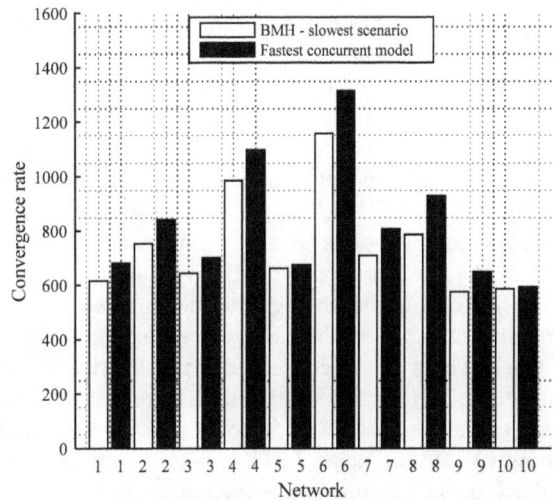

Fig. 6. Comparison of slowest scenario of BMH with fastest concurrent model – average connectivity – average est.

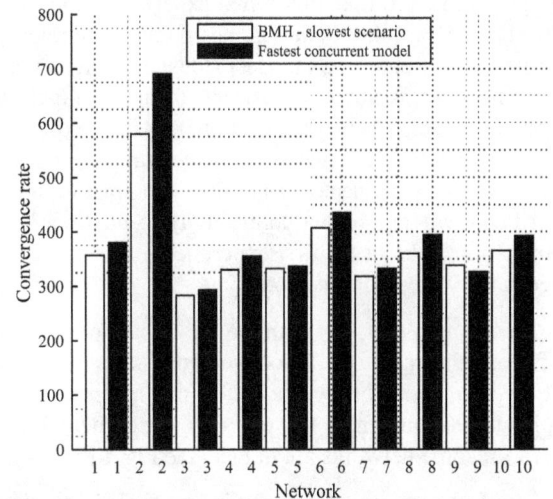

Fig. 7. Comparison of slowest scenario of BMH with fastest concurrent model – strong connectivity – average est.

5.2 Estimation of the Network Size: Convergence Rate Optimization

The second experiment is focused on the network size estimation. We examine the optimization ensured by our optimized weight model in the same topologies used in the previous experiment. In the first part of this section, an examination of the convergence rate optimization achieved by our optimized weight model is presented. As discussed in Sec. 1, one of the aspects by which the network size estimation differs from the average estimation is the necessity of the choice of the leader. Thus, in order to ensure a generality of the simulation results, the average consensus was repeated 200 times (i.e. executions use different leaders – 200 times = 200 leaders). Consequently, the convergence rates of the Maximum Degree, the Metropolis-Hasting and the Best Constant were examined. Within our first analysis, we choose the average of these values as a representative of the convergence rates. When the Biphasically configured Metropolis-Hasting was examined, the experiment was repeated 200 times for each shuffle. Therefore, within one topology, we made $200 \cdot 100 = 20\,000$ executions. Here, the average was again chosen as a representative. In Tab. 2, we have shown the optimization of our optimized weight model compared with the concurrent ones. In Appendix B, the complete results are depicted. As in the previous experiment, also the number of the messages is depicted in Appendix B. We can see from the results that our mechanism again achieves a positive average optimization in all the cases. However, the Best Constant achieves a faster convergence rate in two strongly connected networks. The Maximum Degree (the averages are in the range 41.80% – 55.05%) and the Metropolis-Hasting (within the range 7.87% – 16.62%) are optimized similarly as in the previous experiment (in the strongly connected networks, the MD achieves a small positive deviation) and with the same character compared with the average estimation. Like in the first experiment, the Metropolis-Hasting weight model is positively optimized for each network. Regarding the Best Constant weight model, the optimization is not as significant for the averagely and strongly connected networks as in the previous experiment (8.19% – 23.50%). The worst average results are obtained for the averagely connected networks in contrast to the first experiment, where a higher connectivity results in a higher optimization. Like in the first experiment, the optimization range is the widest for this model.

In the next part, we examine the impact of a random shuffle on the range of the convergence rates like in the first experiment. The character of this phenomenon is similar to the one from the previous experiment but the values for some networks are higher (see Fig. 8).

Furthermore, in order to show that there is no slower rate (when the average positive optimization is achieved) than the fastest concurrent model, Fig. 9 (weak connectivity), Fig. 10 (average connectivity) and Fig. 11 (strong connectivity) are shown.

Convergence rate optimization of network size estimation [%] in Weakly connected networks			
	MD	MH	BC
Average	55.05	16.62	23.50
Maximum	64.46	20.86	36.47
Minimum	47.75	13.95	13.94
Range	16.71	6.91	22.53
Convergence rate optimization of network size estimation [%] in Averagely connected networks			
	MD	MH	BC
Average	45.99	10.56	8.19
Maximum	60.00	13.21	32.14
Minimum	36.90	8.29	0.78
Range	23.10	4.92	31.36
Convergence rate optimization of network size estimation [%] in Strongly connected networks			
	MD	MH	BC
Average	41.80	7.87	9.23
Maximum	68.16	15.21	47.41
Minimum	30.76	5.25	-11.10
Range	37.40	9.96	58.51

Tab. 2. Comparison of BMH with others – network size est.

Fig. 8. Percentage range of convergence rates caused by shuffle of identity numbers (network size est.).

Fig. 9. Comparison of slowest scenario of BMH with fastest concurrent model – weak connectivity – size est.

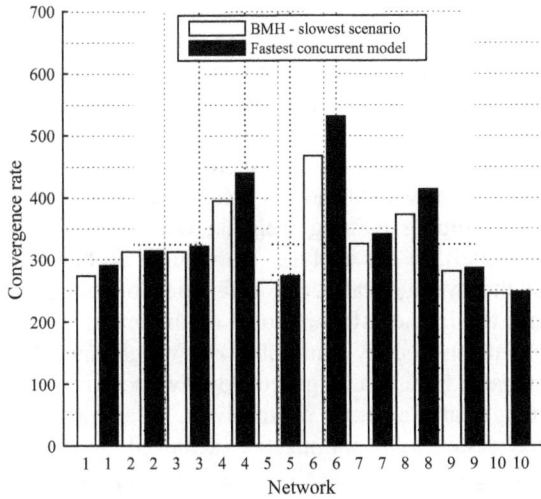

Fig. 10. Comparison of slowest scenario of BMH with fastest concurrent model – average connectivity – size est.

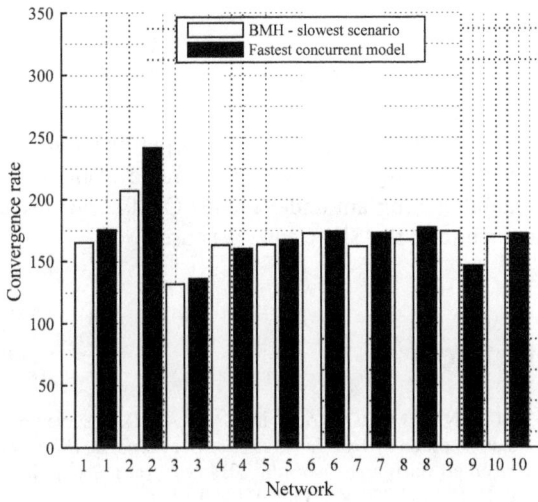

Fig. 11. Comparison of slowest scenario of BMH with fastest concurrent model – strong connectivity – size est.

Convergence rate optimization of network size estimation [%] in Weakly connected networks			
	MD	MH	BC
Average	53.08 %	15.41 %	52.02 %
Maximum	65.26 %	28.55 %	63.80 %
Minimum	38.19 %	4.84 %	37.66 %
Range	27.07 %	23.71 %	26.14 %
Convergence rate optimization of network size estimation [%] in Averagely connected networks			
	MD	MH	BC
Average	45.61 %	8.51 %	46.80 %
Maximum	68.08 %	16.28 %	66.57 %
Minimum	25.21 %	4.00 %	15.87 %
Range	42.87 %	12.28 %	50.70 %
Convergence rate optimization of network size estimation [%] in Strongly connected networks			
	MD	MH	BC
Average	52.55 %	11.92 %	51.36 %
Maximum	83.19 %	30.79 %	66.74 %
Minimum	22.72 %	1.64 %	19.47 %
Range	60.47 %	29.15 %	47.27 %

Tab. 3. Comparison of BMH with others – choice of the leader.

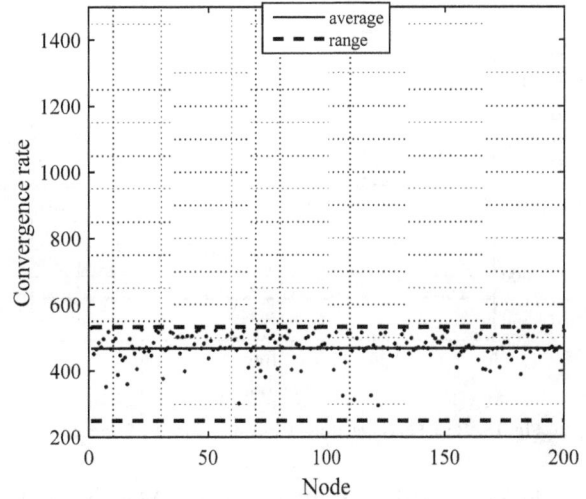

Fig. 12. Biphasically configured Metropolis-Hasting weight model – range of convergence rates.

5.3 Estimation of the Network Size: Impact of the Leader Choice

In the last experiment, it is examined how the choice of the leader affects the convergence rates of the algorithm. We choose the range of the obtained convergence rates as a quality indicator of this aspect. The ideal scenario would be the range equaled to 0. However, this state is unreachable for complicated structures and therefore, we compare the value of the range of our optimized weight model with the concurrent weight models. Its smaller value means that the choice of the leader has a smaller impact on the convergence rate – which is the desired outcome. The experiment is again executed on our optimized weight model, the Maximum Degree weight model, the Metropolis-Hasting weight model, and the Best Constant weight model.

In Tab. 3 and Appendix C, we show the optimization of our mechanism for all the topologies. We can see that our optimized weight model achieves a positive optimization for all 30 networks.

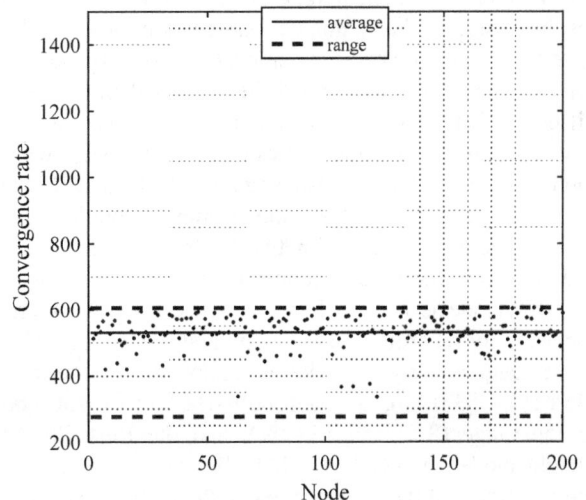

Fig. 13. Metropolis-Hasting weight model – range of convergence rates.

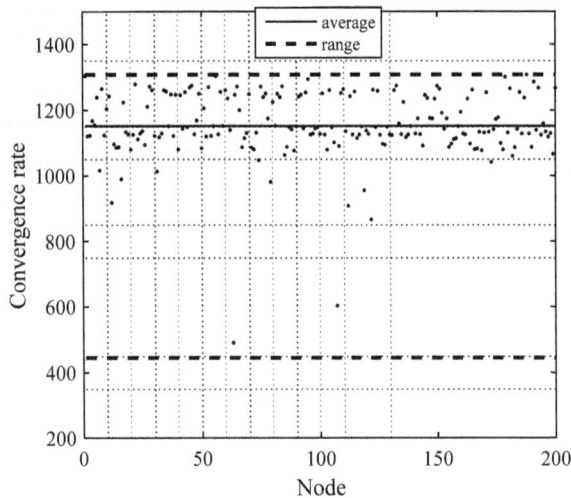

Fig. 14. Maximum Degree weight model – range of convergence rates.

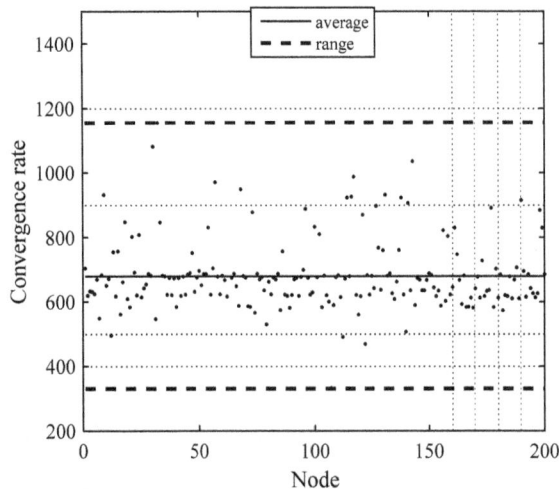

Fig. 15. Best Constant weight model – range of convergence rates.

In Fig. 12 (Biphasically configured Metropolis-Hasting weight model), Fig. 13 (Metropolis-Hasting weight model), Fig. 14 (Maximum Degree weight model) and Fig. 15 (Best Constant weight model), we have shown the results of one of the examined topologies in order to illustrate this problem. We have depicted the convergence rates for each node and highlighted the range calculated as the difference between the slowest and the fastest convergence rate. We can see from the figures that our optimized weight model achieves the smallest range and therefore, it is the best also in this aspect for the examined topology. Regarding the results from 30 networks, the best optimization is achieved in the weakly connected networks, meanwhile, the worst one is observed in the network of the average connectivity regardless of the examined model. The Maximum Degree weight model is optimized in the range (45.61% – 53.08%), the Metropolis-Hasting weight model in the range (8.51% – 15.41%), and the Best Constant weight model in (46.80% – 52.02%). Thus, as seen from the results of all the executed experiments, the best optimization within all three optimized aspects is achieved for the optimization of the impact of the choice of the leader.

6. Conclusion

In this paper, we present an optimized version of the Metropolis-Hasting weight model called the Biphasically configured Metropolis-Hasting weight model. We examined the achieved optimization compared with other constant concurrent weight models (Maximum degree weight model, Metropolis-Hasting weight model, Best Constant weight model). In our analysis, we focused on the estimation of the average value, the estimation of the network size and the impact of the choice of the leader. The improvement ensured by our optimized weight model was demonstrated in randomly generated networks with a weak, an average, and a strong connectivity. According to the depicted results, the optimization achieved by our optimized weight model poses a significant improvement of the computation process of the average consensus algorithm.

Acknowledgments

The research described in this paper was financed by the Czech Ministry of Education in frame of the National Sustainability Program under grant LO1401. For research, infrastructure of the SIX Center was used.

References

[1] BUTUN, I., MORGERA, S. D., SANKAR, R. A survey of intrusion detection systems in wireless sensor networks. *IEEE Communications Survey & Tutorials,* 2014, vol. 16, no. 1, p. 266–282. DOI: 10.1109/SURV.2013.050113.00191

[2] BAHREPOUR, M., MERATNIA, N., POEL, M., et al. Distributed event detection in wireless sensor networks for disaster management. In *Proceeding of the 2nd International Conference on Intelligent Networking and Collaborative Systems (INCOS).* Thessaloniki (Greece), 2010, p. 507–512. DOI: 10.1109/INCOS.2010.24

[3] OTHMAN, M. F., SHAZALI, K. Wireless sensor network applications: A study in environment monitoring system. *Procedia Engineering,* 2012, vol. 41, p. 1204–1210. DOI: 10.1016/j.proeng.2012.07.302

[4] YICK, J., MUKHERJEE, B., GHOSAL, D. Wireless sensor network survey. *Computer Networks,* 2008, vol. 52, no. 12, p. 2292–2330. DOI: 10.1016/j.comnet.2008.04.002

[5] SHERLY PUSPHA, L., MURUGAN, K. Energy-efficient quorum-based MAC protocol for wireless sensor networks. *ETRI Journal,* 2015, vol. 37, no. 3, p. 480–490. DOI: 10.4218/etrij.15.0114.0688

[6] HATAMIAN, M., ALMASI BARDMILY, M., ASADBOLAND, M., el al. Congestion-aware Routing and fuzzy-based rate controller for wireless sensor networks. *Radioengineering,* 2016, vol. 25, no. 1, p. 114–123. DOI: 10.13164/re.2016.0114

[7] XIE, L., SHI, Y., HOU, Y.T., et al. Wireless power transfer and applications to sensor networks. *IEEE Wireless Communications,* 2013, vol. 20, no. 4, p. 140–145. DOI: 10.1109/MWC.2013.6590061

[8] KUILA, P., JANA, P. K. Energy efficient clustering and routing algorithms for wireless sensor networks: Particle swarm

optimization approach. *Engineering Applications of Artificial Intelligence,* 2014, vol. 33, p. 127–140. DOI: 10.1016/j.engappai.2014.04.009

[9] RAULT, T., BOUABDALLAH, A., CHALLAL, Y. Energy efficiency in wireless sensor networks: A top-down survey. *Computer Networks,* 2014, vol. 67, p. 104–122. DOI: 10.1016/j.comnet.2014.03.027

[10] VISSER, H. J., VULLERS, R. J. RF energy harvesting and transport for wireless sensor network applications: Principles and requirements. *Proceedings of the IEEE,* 2013, vol. 101, no. 6, p. 1410–1423. DOI: 10.1109/JPROC.2013.2250891

[11] SHELTAMI, T. An enhanced energy saving approach for WSNs. *Procedia Computer Science,* 2013, vol. 21, p. 199–206. DOI: 10.1016/j.procs.2013.09.027

[12] KARABOGA, D., OKDEM, S., OZTURK, C. Cluster based wireless sensor network routing using artificial bee colony algorithm. *Wireless Networks,* 2012, vol. 18, no. 7, p. 847–860. DOI: 10.1007/s11276-012-0438-z

[13] KAR, S., TANDON, R., POOR, H. V., et al. Distributed detection in noisy sensor networks. In *Proceeding of the IEEE International Symposium on Information Theory Proceedings (ISIT).* St. Petersburg (Russia), 2011, p. 2856–2860. DOI: 10.1109/ISIT.2011.6034097

[14] KEMPE, D., DOBRA, A., GEHRKE, J. Gossip-based computation of aggregate information. In *Proceeding of the 44th Annual IEEE Symposium on Foundations of Computer Science.* Cambridge (Massachusetts, USA), 2003, p. 482–491. DOI: 10.1109/SFCS.2003.1238221

[15] BRACA, P., MARANO, S., MATTA, V. Enforcing consensus while monitoring the environment in wireless sensor networks. *IEEE Transactions on Signal Processing,* 2008, vol. 56, no. 7, p. 3375–3380. DOI: 10.1109/TSP.2008.917855

[16] SLUCIAK, O., RUPP, M. Reaching consensus in asynchronous WSNs: Algebraic approach. In *Proceeding of the IEEE International Conference on Acoustics Speech and Signal Processing (ICASSP).* Prague (Czech Republic), 2011, p. 3300 to 3303. DOI: 10.1109/ICASSP.2011.5946727

[17] XIONG, G., KISHORE, S. Linear high-order distributed average consensus algorithm in wireless sensor networks. *EURASIP Journal on Advances in Signal Processing,* 2010, vol. 1, no. 1, p. 1–6. DOI: 10.1155/2010/373604

[18] JAFARIZADEH, S., JAMALIPOUR, A. Weight optimization for distributed average consensus algorithm in symmetric, CCS & KCS star networks. *arXiv preprint arXiv:1001.4278.*

[19] SLUCIAK, O., RUPP, M. Network size estimation using distributed orthogonalization. *IEEE Signal Processing Letters,* 2013, vol. 20, no. 4, p. 347–350. DOI: 10.1109/lsp.2013.2247756

[20] OLFATI-SABER, R., FAX, J. A., MURRAY, R. M. Consensus and cooperation in networked multi-agent systems. *Proceedings of the IEEE,* 2007, vol. 95, no. 1, p. 215–233. DOI: 10.1109/JPROC.2006.887293

[21] SHAMES, I., CHARALAMBOUS, T., HADJICOSTIS, C.N., et al. Distributed network size estimation and average degree estimation and control in networks isomorphic to directed graphs. In *Proceeding of the 50th Annual Allerton Conference on Communication, Control, and Computing (Allerton).* Monticello (Illinois, USA), 2012, p. 1885–1892. DOI: 10.1109/Allerton.2012.6483452

[22] BIANCHIN, G., CENEDESE, A., LUVISOTTO, M., et al. Distributed faults detection in sensor networks via clustering and consensus. In *Proceeding of the IEEE 54th Conference on Decision and Control (CDC).* Osaka (Japan), 2015, p. 3828–3833. DOI: 10.1109/cdc.2015.7402814

[23] KIM, B. Y., HUR, H., AHN, H. S. Coordination and control for freeway traffic network. In *Proceeding of the IEEE International*

Systems Conference (SysCon). Orlando (Florida, USA), 2013, p. 159–163. DOI: 10.1109/syscon.2013.6549875

[24] ZENG, F., LI, C., TIAN, Z. Distributed compressive spectrum sensing in cooperative multihop cognitive networks. *IEEE Journal of Selected Topics in Signal Processing,* 2011, vol. 5, no. 1, p. 37 to 48. DOI: 10.1109/vtcfall.2013.6692163

[25] TAHBAZ-SALEHI, A., JADBABAIE, A. A one-parameter family of distributed consensus algorithms with boundary: From shortest paths to mean hitting times. In *45th IEEE Conference on Decision and Control.* San Diego (CA, USA), 2006, p. 4664–4669. DOI: 10.1109/cdc.2006.377308

[26] CHIB, S., GREENBERG, E. Understanding the Metropolis-Hastings algorithm. *The American Statistician,* 1995, vol. 49, no. 4, p. 327–335. DOI: 10.2307/2684568

[27] KIA, S. S., CORTES, J., MARTINEZ, S. Distributed event-triggered communication for dynamic average consensus in networked systems. *Automatica,* 2015, vol. 59, p. 112–119. DOI: 10.1016/j.automatica.2015.06.011

[28] WANG, X., LI, J., XING, J., et al. A novel finite-time average consensus protocol for multi-agent systems with switching topology. *Transaction of the Institute of Measurement and Control,* 2016, p. 1–9. DOI: 10.1177/0142331216663617

[29] KENYERES, M., KENYERES, J., SKORPIL, V. Split distributed computing in wireless sensor networks. *Radioengineering,* 2015, vol. 24, no. 3, p. 749–756. DOI: 10.13164/re.2015.0749

[30] KIA, S. S., CORTES, J., MARTINEZ, S. Dynamic average consensus under limited control authority and privacy requirements. *International Journal of Robust and Nonlinear Control,* 2015, vol. 25, no. 13. p. 1941–1966. DOI: 10.1002/rnc.3178

[31] PRIOLO, A., GASPARRI, A., MONTIJANO, E., et al. A distributed algorithm for average consensus on strongly connected weighted digraphs. *Automatica,* 2014, vol. 50, no. 3, p. 946–951. DOI: 10.1016/j.automatica.2013.12.026

[32] YUAN, D., XU, S., ZHAO, H., et al. Accelerating distributed average consensus by exploring the information of second-order neighbors. *Physics Letters A,* 2010, vol. 374, no. 24. p. 2438–2445. DOI: 10.1016/j.physleta.2010.03.053

[33] AYSAL, T. C., ORESHKIN, B. N., COATES, M. J. Accelerated distributed average consensus via localized node state prediction. *IEEE Transactions on Signal Processing,* 2009, vol. 57, no. 3. p. 1563–1576. DOI: 10.1109/TSP.2008.2010376

[34] LIU, W., GU, W., XU, Y., et al. Improved average consensus algorithm based distributed cost optimization for loading shedding of autonomous microgrids. *International Journal of Electrical Power & Energy Systems,* 2015, vol. 73, p. 89–96. DOI: 10.1016/j.ijepes.2015.04.006

[35] CHEN, Y., TRON, R., TERZIS, A., et al. Corrective consensus: Converging to the exact average. In *Proceeding of the 49th IEEE Conference on Decision and Control (CDC).* Atlanta (Georgia, USA), 2010, p. 1221–1228. DOI: 10.1109/CDC.2010.5717925

[36] LI, W., JIA, Y. Consensus-based distributed multiple model UKF for jump Markov nonlinear systems. *IEEE Transactions on Automatic Control,* 2012, vol. 57, no. 1, p. 227–233. DOI: 10.1109/TAC.2011.2161838

[37] LI, Z., YU, F.R., HUANG, M. A distributed consensus-based cooperative spectrum-sensing scheme in cognitive radios. *IEEE Transactions on Vehicular Technology,* 2010, vol. 59, no. 1, p. 383–393. DOI: 10.1109/TVT.2009.2031181

[38] FRASCA, P., CARLI, R., FAGNANI, F., et al. Average consensus on networks with quantized communication. *International Journal of Robust and Nonlinear Control,* 2009, vol. 19, no. 16, p. 1797 to 1816. DOI: 10.1109/TVT.2009.2031181

[39] KENYERES, M., KENYERES, J., SKORPIL, V. The distributed convergence classifier using the finite difference. *Radioengineering*, 2016, vol. 25, no. 1, p. 148–155. DOI: 10.13164/re.2016.0148

[40] TOULOUSE, M., MINH, B. Q., CURTIS, P. A consensus based network intrusion detection system. In *Proceeding of the 5th International Conference on IT Convergence and Security (ICITCS)*. Kuala Lumpur (Malaysia), 2015, p. 1–6. DOI: 10.1109/ICITCS.2015.7292913

[41] BENJAMIN, A., CHARTRAND, G., ZHANG, P. *The Fascinating World of Graph Theory*. Princeton (NJ, USA): Princeton University Press, 2015. ISBN: 9780691163819

[42] SCHWARZ, V., MATZ, G. Average consensus in wireless sensor networks: Will it blend? In *Proceeding of the IEEE International Conference on Acoustics, Speech and Signal Processing (ICASSP)*. Vancouver (British Columbia, Canada), 2013, p. 4584–4588. DOI: 10.1109/ICASSP.2013.6638528

[43] ANDERSON, W. N., MORLEY, T. D. Eigenvalues of the Laplacian of a graph. *Linear and Multilinear Algebra*, vol. 18, no. 2, 1985, p. 141–145. DOI: 10.1080/03081088508817681

[44] SPIELMAN, D. A. *The Laplacian (Lecture 2)*. 6 pages. [Online] Cited 2016-10-20. Available at: http://www.cs.yale.edu/homes/spielman/561/2009/lect02-09.pdf

[45] XIAO, L., BOYD, S., KIM, S. J. Distributed average consensus with least-mean-square deviation. *Journal of Parallel and Distributed Computing*, 2007, vol. 67, no. 1. p. 33–46. DOI: 10.1016/j.jpdc.2006.08.010

[46] XIAO, L., BOYD, S. Fast linear iterations for distributed averaging. *Systems & Control Letters*, 2004, vol. 53, no. 1, p. 65 to 78. DOI: 10.1016/j.sysconle.2004.02.022

[47] MACUA, S. V., LEON, C. M., ROMERO, J. S., et al. How to implement doubly-stochastic matrices for consensus-based distributed algorithms. In *Proceeding of the IEEE 8th Sensor Array and Multichannel Signal Processing Workshop (SAM)*. A Coruña, (Spain), 2014, p. 333–336. DOI: 10.1109/SAM.2014.6882409

[48] BAPAT, R. B. *Graphs and Matrices*. New York (NY): Springer, 2010. DOI: 10.1007/978-1-84882-981-7

[49] LI, Z., REN, W., LIU, X., et al. Consensus of multi-agent systems with general linear and Lipschitz nonlinear dynamics using distributed adaptive protocols. *IEEE Transactions on Automatic Control*, 2013, vol. 58, no. 7, p. 1786–1791. DOI: 10.1109/TAC.2012.2235715

[50] NOWZARI, C., CORTÉS, J. Zeno-free, distributed event-triggered communication and control for multi-agent average consensus. In *Proceeding of the American Control Conference (ACC)*. Portland (Oregon, USA), 2014, p. 2148–2153. DOI: 10.1109/ACC.2014.6859495

[51] AL-NAKHALA, N., RILEY, R. ELFOULY, T. Distributed algorithms in wireless sensor networks: An approach for applying binary consensus in a real testbed. *Computer Networks*, vol. 79, p. 30–38. DOI: 10.1016/j.comnet.2014.12.011

[52] SCHWARZ, V., HANNAK, G., MATZ, G. On the convergence of the average consensus with generalized Metropolis-Hasting weights. In *Proceeding of the IEEE International Conference on Acoustics, Speech and Signal Processing (ICASSP)*. Florence (Italy), 2014, p. 5442–5446, DOI: 10.1109/icassp.2014.6854643

[53] VARGA, R. S. *Gersgorin and His Circles*. Berlin (GE): Springer-Verlag, 2004. DOI: 10.1007/978-3-642-17798-9

About the Authors ...

Martin KENYERES was born in Bratislava, Slovakia. He received his M.Sc. from the Slovak University of Technology in Bratislava in 2013. His research interests include distributed computing and wireless sensor networks. In 2013, he was with the Vienna University of Technology, Austria, where he participated in NFN SISE project under Professor Markus Rupp's supervision. He dealt with the implementation of distributed algorithms for an estimation of aggregate functions into wireless sensor networks. Since 2014, he has been with Brno University of Technology (BUT), where he works towards his PhD thesis on an analysis and an optimization of distributed systems.

Jozef KENYERES was born in Bratislava, Slovakia. He received his Ph.D. from the Slovak University of Technology in Bratislava in 2014. His research interests include embedded systems, wireless sensor networks and VoIP. From 2006 to 2009, he worked as a technician at Slovak Telecom, a. s., Bratislava, Slovakia. From 2009 to 2013, he was a project assistant at the Vienna University of Technology, Austria and from 2014 to 2015 he was with Zelisko GmbH, where he worked as a software developer. Since 2015, he has been working as a software developer at Sipwise GmbH, Austria.

Vladislav ŠKORPIL was born in Brno, Czech Republic. He graduated from the BUT, Faculty of Electrical Engineering, Dept. of Telecommunications in 1980. From 1985 to 1989 he was a doctoral student in the same Department. From 1980 to 1982 he worked as a designer for the telecommunication design office. He again entered the Dept. of Telecommunications, BUT in 1982 as a university teacher and he has been working there since that time (1984 Associate Professor). From 1994 to 2013, he was a vice-head of this department. He takes a keen interest in modern telecommunication systems. He has taught in courses on transmission systems from analogue through all categories of digital up to special applications. He is the author of more than 100 international scientific papers and some manuals. He has complemented his theoretical knowledge by co-operation with a lot of firms and institutions. He has co-operated on telecommunication projects such as digital transmission and switching systems, telecommunication broadband networks, ISDN, ATM, data networks LAN and MAN, on structured cabling design, neural networks, wavelet transform, Quality of Service QoS, data bit rate compression, etc. He is a member of international organizations IEEE and WSEAS.

Radim BURGET is an Associated Professor (2014) at the Dept. of Telecommunications, Faculty of Electrical Engineering, BUT, Czech Republic. He obtained his MSc. in 2006 (Information Systems) and finished his Ph.D. in 2010. He is interested in image processing, data mining, genetic programming and optimization.

Appendix A

The convergence rates of the average estimation							
		Maximum Degree		Metropolis-Hasting		Best Constant		BMH
		CR	OPT [%]	CR / OPT [%]		CR / OPT [%]		CR
Weakly connected	Network #1	6928	53.18	3713	12.65	4366	25.71	3243.47
	Network #2	4259	54.20	2288	14.75	2431	19.76	1950.56
	Network #3	5128	55.98	2591	12.88	2775	18.66	2257.28
	Network #4	9554	55.98	4975	15.47	6268	32.91	4205.35
	Network #5	7011	53.39	3798	13.95	4502	27.41	3268.15
	Network #6	6172	47.38	3794	14.40	3560	8.77	3247.65
	Network #7	13059	60.23	6597	21.28	7854	33.88	5193.18
	Network #8	2881	49.93	1701	15.20	1643	12.20	1442.52
	Network #9	6837	64.60	3011	19.61	3779	35.95	2420.40
	Network #10	3335	51.09	1903	14.29	1970	17.20	1631.07
Averagely connected	Network #1	1064	42.29	680	9.70	706	13.03	614.02
	Network #2	1370	45.11	845	11.01	842	10.69	751.97
	Network #3	1226	47.64	700	8.29	869	26.13	641.95
	Network #4	2085	52.94	1097	10.56	1114	11.92	981.20
	Network #5	1181	44.02	727	9.06	675	2.05	661.14
	Network #6	2864	60.04	1315	12.98	1794	36.21	1144.37
	Network #7	1309	46.37	824	14.80	808	13.11	702.06
	Network #8	1568	50.02	929	15.64	1014	22.71	783.75
	Network #9	951	39.71	649	11.66	685	16.30	573.35
	Network #10	979	40.29	634	7.80	593	1.42	584.55
Strongly connected	Network #1	520	35.56	380	11.82	629	46.72	335.10
	Network #2	1870	69.30	691	16.93	1130	49.20	574.01
	Network #3	412	31.47	297	4.9400	293	3.64	282.33
	Network #4	610	46.06	355	7.3100	389	15.41	329.05
	Network #5	497	33.20	352	5.6800	336	1.19	332.00
	Network #6	622	34.82	435	6.8000	448	9.51	405.40
	Network #7	492	35.41	332	4.2900	351	9.47	317.77
	Network #8	620	42.09	394	8.8800	413	13.07	359,03
	Network #9	502	32.84	363	7.1200	326	-3.42	337.15
	Network #10	607	40.03	396	8.0800	392	7.15	363.99

The number of the messages necessary for the average estimation					
		Maximum Degree	Metropolis-Hasting	Best Constant	BMH
		MN	MN	MN	MN
Weakly connected	Network #1	1378473	738688	868635	645251.53
	Network #2	847342	455113	483570	387962.44
	Network #3	1020273	515410	552026	448999.72
	Network #4	1901047	989826	1247133	836665.65
	Network #5	1378473	738688	868635	644988.85
	Network #6	1228029	754807	708241	646083.35
	Network #7	2598542	1312604	1562747	1033243.82
	Network #8	573120	338300	326758	286862.48
	Network #9	1360364	598990	751822	481460.60
	Network #10	663466	378498	391831	324383.93
Averagely connected	Network #1	211537	135121	140295	121990.98
	Network #2	272431	167956	167359	149443.03
	Network #3	243775	139101	172732	127549.05
	Network #4	414716	218104	221487	195059.80
	Network #5	234820	144474	134126	131367.86
	Network #6	569737	261486	356807	227530.63
	Network #7	260292	163777	160593	139510.94
	Network #8	311833	184672	201587	155767.25
	Network #9	189050	128952	136116	113897.65
	Network #10	194622	125967	117808	116126.45
Strongly connected	Network #1	103480	75620	125171	66485.90
	Network #2	372130	137509	224870	114028.99
	Network #3	81988	59103	58307	55984.67
	Network #4	121390	70645	77411	65281.95
	Network #5	98903	70048	66864	65869.00
	Network #6	123778	86565	89152	80475.60
	Network #7	97908	66068	69849	63037.23
	Network #8	123380	78406	82187	71247.97
	Network #9	99898	72237	64874	66893.85
	Network #10	120793	78804	78008	72235.01

Appendix B

The convergence rates of the network size estimation								
		Maximum Degree		**Metropolis-Hasting**		**Best Constant**		**BMH**
		CR	OPT [%]	CR	OPT [%]	CR	OPT [%]	CR
Weakly connected	Network #1	2930.50	56.19	1553.11	17.35	1733.61	25.95	1283.72
	Network #2	1733.00	54.28	928.87	14.70	955.4	17.08	792.30
	Network #3	2399.74	56.36	1250.58	16.26	1334.13	21.50	1047.27
	Network #4	3435.97	56.15	1775.21	15.13	2159.40	30.23	1506.69
	Network #5	3104.41	56.44	1605.87	15.79	1812.49	25.39	1352.31
	Network #6	2278.28	47.75	1402.14	15.10	1383.30	13.94	1190.48
	Network #7	4828.82	58.99	2502.52	20.86	3117.06	36.47	1980.42
	Network #8	1206.68	49.27	740.11	17.29	726.44	15.73	612.15
	Network #9	2788.35	64.46	1234.56	19.74	1516.88	34.67	990.91
	Network #10	1498.38	50.58	860.51	13.95	861.25	14.03	740.43
Averagely connected	Network #1	486.48	43.85	304.95	10.43	290.33	5.92	273.14
	Network #2	563.86	44.62	350.46	10.90	314.70	0.78	312.26
	Network #3	591.57	47.23	342.56	8.87	321.78	2.98	312.18
	Network #4	815.70	51.73	439.52	10.41	475.28	17.15	393.76
	Network #5	481.54	45.82	288.13	9.44	273.42	4.57	260.92
	Network #6	1151.97	60.00	530.98	13.21	679.12	32.14	460.84
	Network #7	562.69	42.26	365.90	11.20	341.12	4.75	324.91
	Network #8	682.59	45.68	426.31	13.03	413.60	10.36	370.76
	Network #9	444.69	36.90	311.02	9.78	286.20	1.96	280.60
	Network #10	421.22	41.85	267.08	8.29	248.19	1.31	244.94
Strongly connected	Network #1	339.23	53.51	175.28	10.02	231.73	31.94	157.72
	Network #2	643.01	68.16	241.43	15.21	389.26	47.41	204.72
	Network #3	208.51	36.87	138.93	5.25	136.07	3.26	131.64
	Network #4	295.72	44.93	175.35	7.13	159.99	-1.78	162.84
	Network #5	240.14	32.02	172.34	5.27	167.13	2.32	163.25
	Network #6	284.81	39.61	183.94	6.49	174.17	1.25	172.00
	Network #7	248.03	34.88	173.06	6.67	182.39	11.44	161.52
	Network #8	279.66	40.23	182.12	8.21	177.24	5.69	167.16
	Network #9	251.15	30.76	187.15	7.09	156.52	-11.1	173.89
	Network #10	268.61	36.99	182.65	7.34	172.51	1.90	169.24

The number of the messages necessary for the network size estimation					
		Maximum Degree	**Metropolis-Hasting**	**Best Constant**	**BMH**
		MN	MN	MN	MN
Weakly connected	Network #1	583169.50	309068.89	344988.39	255460.28
	Network #2	344872.97	184845.13	190140.52	157667.70
	Network #3	477548.26	248865.42	265491.87	208406.73
	Network #4	683758.03	353266.79	429720.60	299831.31
	Network #5	617777.59	319568.13	360685.51	269109.69
	Network #6	453377.72	279025.86	275276.70	236905.52
	Network #7	960935.18	498001.48	620294.94	394103.58
	Network #8	240129.32	147281.89	144561.56	121817.85
	Network #9	554881.65	245677.44	301859.12	197191.09
	Network #10	298177.62	171241.49	171388.75	147345.57
Averagely connected	Network #1	96809.52	60685.05	57775.67	54354.86
	Network #2	112208.14	69741.54	62625.30	62139.74
	Network #3	117722.43	68169.44	64034.22	62123.82
	Network #4	162324.30	87464.48	94580.72	78358.24
	Network #5	95826.46	57337.87	54410.58	51923.08
	Network #6	229242.03	105665,02	135144.88	91707.16
	Network #7	111975.31	72814.10	67882.88	64657.09
	Network #8	135835.41	84835.69	82306.40	73781.24
	Network #9	88493.31	61892.98	56953.80	55839.40
	Network #10	83822.78	53148.92	49389.81	48743.06
Strongly connected	Network #1	67506.77	34880.72	46114.27	31386.28
	Network #2	127958.99	48044.57	77462.74	40739.28
	Network #3	41493.49	27647.07	27077.93	26196.36
	Network #4	58848.28	34894.65	31838.01	32405.16
	Network #5	47787.86	34295.66	33258.87	32486.75
	Network #6	56677.19	36604.06	34659.83	34228.00
	Network #7	49357.97	34438.94	36295.61	32142.48
	Network #8	55652.34	36241.88	35270.76	33264.84
	Network #9	49978.85	37242.85	31147.48	34604.11
	Network #10	53453.39	36347.35	34329.49	33678.76

Appendix C

		Maximum Degree		Metropolis-Hasting		Best Constant		BMH
		CR	**OPT [%]**	**CR**	**OPT [%]**	**CR**	**OPT [%]**	**CR**
	Network #1	2700	50.77	1574	15.54	2518	47.21	1329.34
	Network #2	1441	54.44	767	14.40	1343	51.11	656.53
	Network #3	1560	60.75	714	14.24	1638	62.62	612.36
	Network #4	3637	58.69	1822	17.54	3255	53.84	1502.48
Weakly connected	Network #5	2814	49.48	1612	11.81	2625	45.84	1421.58
	Network #6	1949	40.85	1267	9.01	1874	38.48	1152.87
	Network #7	4996	65.26	2429	28.55	4795	63.80	1735.61
	Network #8	512	49.63	271	4.84	630	59.07	257.88
	Network #9	2158	62.78	1028	21.86	2036	60.55	803.26
	Network #10	937	38.19	692	16.31	929	37.66	579.16
	Network #1	345	43.09	205	4.22	267	26.46	196.34
	Network #2	261	34.31	181	5.28	348	50.74	171.44
	Network #3	335	48.96	185	7.58	386	55.70	170.98
	Network #4	522	56.52	253	10.28	486	53.29	226.99
Averagely connected	Network #5	379	57.87	176	9.28	273	41.51	159.67
	Network #6	863	68.08	329	16.28	824	66.57	275.44
	Network #7	248	50.77	138	11.54	259	52.86	122.08
	Network #8	239	25.21	199	10.17	444	59.74	178.76
	Network #9	207	31.83	147	4.00	258	45.30	141.12
	Network #10	289	39.45	187	6.42	208	15.87	175.00
	Network #1	560	83.19	136	30.79	283	66.74	94.13
	Network #2	834	74.68	293	27.93	615	65.67	211.16
	Network #3	119	52.16	60	5.12	92	38.12	56.93
	Network #4	141	51.09	75	8.05	158	56.35	68.96
Strongly connected	Network #5	863	68.08	329	16.28	824	66.57	275.44
	Network #6	245	49.68	130	5.17	199	38.05	123.28
	Network #7	196	41.24	126	8.60	143	19.47	115.16
	Network #8	130	47.85	78	13.09	168	59.65	67.79
	Network #9	111	22.72	88	2.52	144	40.43	85.78
	Network #10	92	34.78	61	1.64	160	62.50	60.00

The range of the convergence rates caused by the change of the leader

12

Mixed-Mode Third-Order Quadrature Oscillator Based on Single MCCFTA

Khachen KHAW-NGAM[1], Montree KUMNGERN[1], Fabian KHATEB[2,3]

[1] Faculty of Engineering, King Mongkut's Institute of Technology Ladkrabang, Bangkok 10520, Thailand
[2] Department of Microelectronics, Brno University of Technology, Technická 10, Brno, Czech Republic
[3] Faculty of Biomedical Engineering, Czech Technical University in Prague, nám. Sítná 3105, Kladno, Czech Republic

kkmontre@gmail.com, khateb@feec.vutbr.cz

Abstract. *This paper presents a new mixed-mode third-order quadrature oscillator based on new modified current-controlled current follower transconductance amplifier (MCCFTA). The proposed circuit employs one MCCFTA as active element and three grounded capacitors as passive components which is highly suitable for integrated circuit implementation. The condition and frequency of oscillations can be controlled orthogonally and electronically by adjusting the bias currents of the active device. The circuit provides four quadrature current outputs and two quadrature voltage outputs into one single topology, which can be classified as mixed-mode oscillator. In addition, four quadrature current output terminals possess high-impedance level which can be directly connected to the next stage without additional buffer circuits. The performance of the proposed structure has been verified through PSPICE simulators using 0.25 μm CMOS process from TSMC and experimental results are also investigated.*

Keywords

Third-order quadrature oscillator, mixed-mode oscillator, modified current-controlled current follower transconductance amplifier (MCCFTA)

1. Introduction

At present, current-mode technique is very interesting approach due to the fact that it is easy as operating of arithmetic such as addition and subtraction signals, multiplication and division signals by a constant signal and potential to operate at lower supply voltage compared their voltage-mode circuits [1]. Regarding to a current-mode building block, the current differencing transconductance amplifier (CDTA) [2] is a really current-mode active building block, due to its input and output signals are current forms. The structure of this device is consisted of a unity-gain current source controlled by the difference of two inputs and an output transconductance amplifier providing electronic tuning capability through its transconductance gain (g_m). Therefore, CDTA is highly suitable for realizing current-mode analog signal processing and CDTA-based circuits can also reduce number of passive resistors.

A quadrature oscillator (QO) usually provides sinusoids having a phase difference of 90° that is useful in communication and measurement systems such as quadrature mixers and single sideband generators for communication system [3], vector generators and selective voltmeters for measurements system [4]. Several QOs using CDTAs as active element have been proposed in the literature; see, for example [5–12]. However, some structures do not exploit the full capability of the CDTA where typically one of two input terminals of the CDTA is floated and not used [8–12]. Unfortunately, floating terminal may increase the area of chip when these CDTA-based QOs build as IC's forms and also may cause some noise injection into the monolithic circuit.

Recently, a new concept of active building block with one current input and two kinds of current outputs, the so-called "current follower transconductance amplifier (CFTA)", has been introduced [13]. This device is modified from an original CDTA by removing the negative terminal. When subtraction current circuit as input stage is absent, the structure of CFTA is simple. Compared with CDTA, the number of transistor used for CFTA is lesser. Therefore, several CFTA-based analog signal processing circuits are reported [14–19]. The CFTA has been already used to realize QOs [20–26]. They, in conjunction, exhibit high potential for bring drown the number of components. However, all structures are second-order QOs. This work focuses on the third-order QOs, which enjoys the requirement of QOs such as orthogonal adjustability and electronic tunability of the condition of oscillation (CO) and frequency of oscillation (FO), and uses grounded capacitors and minimum number of active elements. Because of high-order network the circuit provides better frequency response and quality, compared with lower-order network [27]. This mention has been expressed in [28] by mathematical formulation, especially, to confirm phase noise reduction. Compared with a second-order oscillator, the third-order oscillator (three-stage oscillator) also offers lower phase noise [29]. As a result, a number of third-order

QOs based on different active building blocks have been proposed [30–52]. The early system using operational transconductance amplifier (OTA) [30] enjoys electronic tuning capability, but its CO and FO are not decoupled and difficult to control. The QOs using active building blocks such as operational amplifier [31], second-generation current conveyor (CCII) [32–34], differential voltage current conveyor (DVCC) [35], operational transresistance amplifier (OTRA) [36], [37], [50], offer orthogonal control of CO and FO, but these structures lack the electronic tuning capability. A number of electronic-controlled third-order QOs have been reported using active building blocks such as current-controlled CCII (CCCII) [38], [39], current difference transconductance amplifier (CDTA) [40], [41], current-controlled CDTA (CCDTA) [42], current-controlled current conveyor transconductance amplifier (CCCTA) [43], [44], OTRA and MOS-C [45], DDCC and VDTA [46], [51], [52], differential voltage current conveyor transconductance amplifier (DVCCTA) [47]. They exhibit high potential for enjoying up the electronic tunability. However, they require an excessive number of active components and the structures are not compact. The third-order oscillator QO based on log-domain technique is proposed in [48]. However, this structure is suitable only in bipolar technology. The structure in [49] proposed third-order QO using a single current-controlled current conveyor transconductance amplifier (CCCCTA), but the circuit does not exploit the full capability of the CCTA when y-terminal of CCTA is not used and attached to ground. Until now, there is no CCCFTA-third-order QO available in open literature.

Therefore, this paper presents a new mixed-mode third-order QO employing a modified current-controlled current follower transconductance amplifier (MCCFTA) and three grounded capacitors. The concept of MCCFTA is similar to conventional CFTA [13], except f-terminal provides parasitic resistance R_f that can be controlled by the bias current [53]. Identical z-copy terminal can be obtained using current-mirrors [13] and connecting transconductance amplifiers in parallels connection are available [54], [55]. Thus, MCCFCTA is an active building block that provides the possibility for utilizing its resistance simulating element and its transconductance gains that can be electronically controlled through adjusting the bias currents. Therefore we have input parasitic resistance and transconductance gains (TAs) into a single MCCFTA for realizing a third-order QO with orthogonal control of the CO and FO. The proposed structure provides four quadrature current outputs and two quadrature voltage outputs. PSPICE simulation results are given to confirm the performance of the proposed structure. The comparison between the proposed third-order QO and previously third-order QOs is expressed as Tab. 1.

2. Proposed Circuit

The circuit symbol and the equivalent circuit of the MCCFTA are shown in Fig. 1 and CMOS implementation

of MCCFTA is shown Fig. 2. The ideal port relations of Fig. 1 can be expressed by

$$\begin{pmatrix} I_{z1} \\ I_{z2} \\ I_{zc} \\ V_f \\ I_{x1} \\ I_{x2} \end{pmatrix} = \begin{pmatrix} 1 & 0 & 0 & 0 & 0 & 0 \\ 1 & 0 & 0 & 0 & 0 & 0 \\ 1 & 0 & 0 & 0 & 0 & 0 \\ R_f & 0 & 0 & 0 & 0 & 0 \\ 0 & \pm g_{m1} & 0 & 0 & 0 & 0 \\ 0 & 0 & \pm g_{m2} & 0 & 0 & 0 \end{pmatrix} \begin{pmatrix} I_f \\ V_{z1} \\ V_{z2} \\ V_{zc} \\ V_{x1} \\ V_{x2} \end{pmatrix} \quad (1)$$

where R_f is the parasitic resistance at f-terminal and g_{m1} and g_{m2} are two TAs. The z_1-, x_1-terminals and g_{m1} are existing terminals and TA of conventional CCFTA [50]. The z_2- and z_c-terminals are the z-copy terminals [13] of CCFTA that can be obtained by adding current mirrors in conventional CCFTA. By cascading g_{m2} in parallel connection, x_2-terminal in CCFTA can be obtained and hence the name MCCFTA. It should be noted that connecting n TAs in parallel connection of MCCFTA is also possible. From the CMOS implementation of MCCFTA in Fig. 2, assuming that transistors M_1 to M_4 are matched and operated in saturation regions, the parasitic resistance at f-terminal (R_f) can be given as

$$R_f \cong \frac{1}{\sqrt{8\mu C_{ox}\left(\frac{W}{L}\right)I_{b1}}} \quad (2)$$

where μ is the carrier mobility, C_{ox} is the gate oxide capacitance per unit area, W and L are the channel width and length, respectively, of MOS transistor. From (2) and Fig. 2, the parasitic resistance R_f can be controlled by adjusting the bias current I_{b1}. This property makes it different from a conventional CFTA [13].

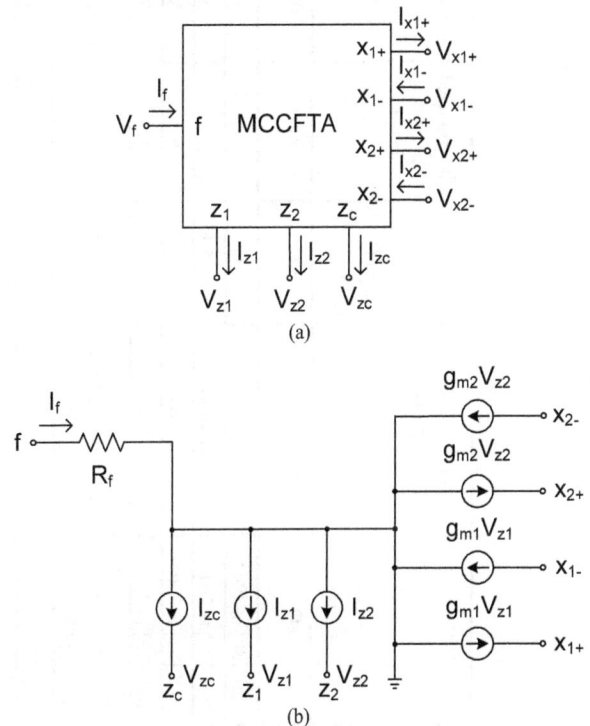

Fig. 1. MCCFTA: (a) electrical symbol, (b) equivalent circuit.

	Number of active elements	Number of resistor (R) & capacitor (C)	All-grounded passive element	Orthogonal control of CO and FO	Offer current and voltage outputs	Offer electronically tunable	Multi-phase generation	Type of dependence of FO on control/bias current/voltage	Availability of constant of output waveforms in dep. On tuning of FO	Range of output voltage/current (peak-to-peak) (Voltage) (Current)	Phased error of generate waveform (%) (Voltage) (Current)
Proposed QO	1-MCCFTA	3-C	Yes	Yes	Yes	Yes	Yes	Nonlinear	No	106-700 mV / 20-160 µA	0.3-1.89 / 0.5-5.78
Ref. [30] in 2002	3-OTA	3-C	Yes	Yes	No	Yes	Yes	Nonlinear	No	-	-
Ref. [31] in 2011	3-op-amp	3-C, 5-R	No	Yes	No	No	No	Nonlinear	No	-	-
Ref. [32] in 2005	3-CCIIs	3-C, 5-R	Yes	Yes	No	No	No	Nonlinear	No	-	-
Ref. [33] in 2011	2-CCIIs	3-C, 3-R	No	Yes	Yes	No	No	Nonlinear	No	-	-
Ref. [34] in 2012	2-CCII, 1-UVC	3-C, 3-R	Yes	Yes	Yes	No	Yes	Nonlinear	No	-	-
Ref. [35] in 2013	3-DVCC	3-C, 3-R	No	Yes	Yes	No	Yes	Nonlinear	No	-	-
Ref. [36] in 2014	3-OTRA	3-C, 5-R	No	Yes	No	No	No	Nonlinear	No	-	-
Ref. [37] in 2014	3-OTRA	3-C, 5-R	No	Yes	No	No	No	Nonlinear	No	-	-
Ref. [38] in 2005	4-CCII	3-C	Yes	Yes	No	Yes	Yes	Nonlinear	No	-	-
Ref. [39] in 2010	3-CCII	3-C	Yes	Yes	Yes	Yes	Yes	Nonlinear	No	-	-
Ref. [40] in 2009	3-CDTA	3-C	Yes	Yes	No	Yes	No	Nonlinear	No	-	-
Ref. [41] in 2010	3-CDTA	3-C	Yes	Yes	Yes	Yes	No	Nonlinear	No	-	-
Ref. [42] in 2011	1-CCCDTA, 1-OTA	3-C	Yes	Yes	Yes	Yes	No	Nonlinear	No	-	-
Ref. [43] in 2009	2-CCCCTA	3-C	Yes	Yes	No	Yes	No	Nonlinear	No	-	-
Ref. [44] in 2011	1-CCCCTA, 1-OTA	3-C	Yes	Yes	No	Yes	No	Nonlinear	No	-	-
Ref. [45] in 2012	2-OTRA	3-C, 4-R	No	No	No	Yes	No	Nonlinear	No	-	-
Ref. [46] in 2013	1-DDCC, 1-VDTA	3-C, 1-R	Yes	Yes	No	Yes	No	Nonlinear	No	-	-
Ref. [47] in 2015	2-DVCCTA	3-C, 2-R	Yes	Yes	Yes	No	Yes	Nonlinear	No	-	~0-4
Ref. [48] in 2011	18-BJT, 18-current source	3-C	Yes	Yes	No	Yes	Yes	Nonlinear	No	-	-
Ref. [49] in 2012	1 MCCCCTA	3-C	Yes	Yes	Yes	Yes	Yes	Nonlinear	No	-	-
Ref. [50] in 2014	3-OTRA	3-C, 4-R	No	Yes	No	No	No	Nonlinear	No	-	-
Ref. [51] in 2014	2-DVTA	3-C	Yes	Yes	No	Yes	No	Nonlinear	No	-	-
Ref. [52] in 2014	2-DVTA	3-C	Yes	Yes	Yes	Yes	No	Nonlinear	No	-	-

Tab. 1. Comparison of the proposed third-order QO with those of previous third-order QOs.

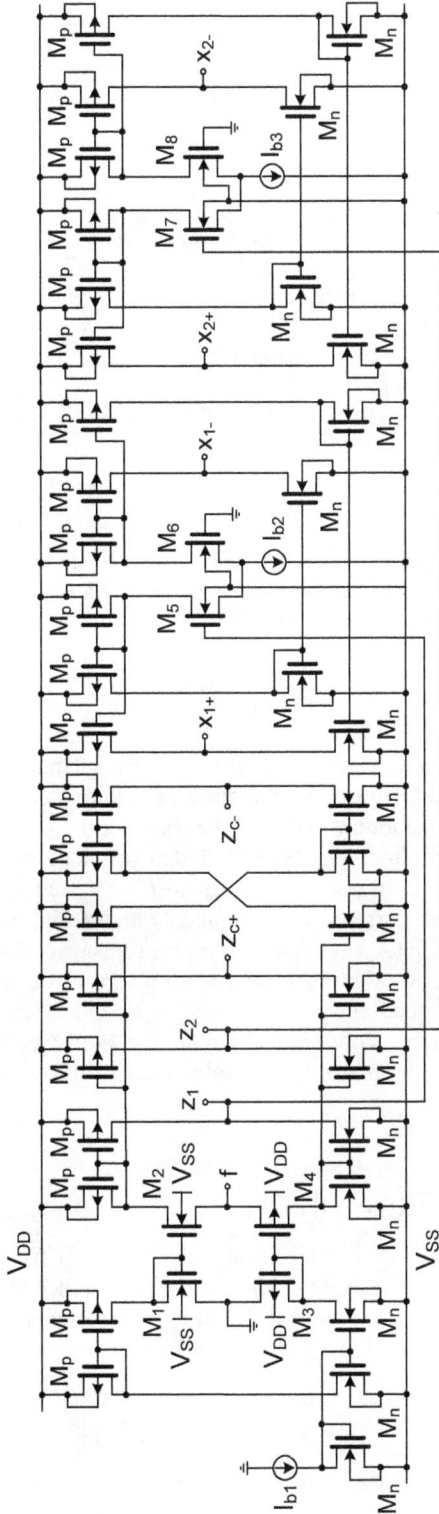

Fig. 2. CMOS implementation for MCCFTA.

The transconductance gains g_{m1} and g_{m2} can be obtained by assuming the transistors M_5-M_6 and M_7-M_8 are matched and operated in saturation region, g_{m1} and g_{m2} of MCCFTA can be expressed, respectively, as

$$g_{m1} = \sqrt{\mu C_{ox}\left(\frac{W}{L}\right) I_{b2}}, \qquad (3)$$

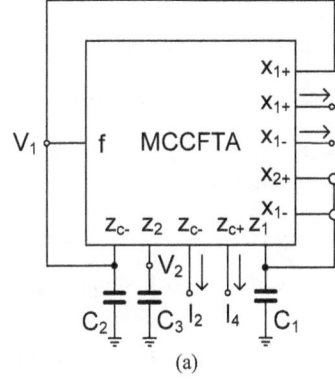

(a)

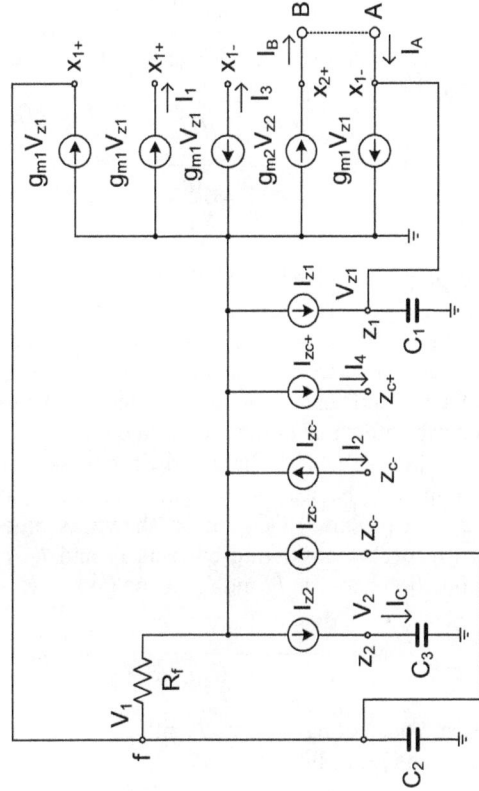

(b)

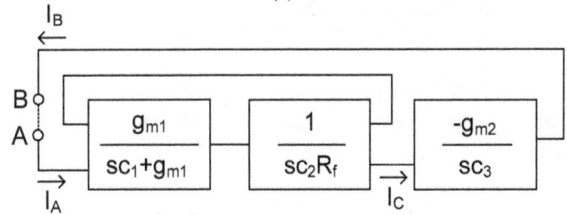

(c)

Fig. 3. Proposed third-order QO: (a) circuit symbol, (b) equivalent circuit, (c) block diagram.

$$g_{m2} = \sqrt{\mu C_{ox}\left(\frac{W}{L}\right) I_{b3}}. \qquad (4)$$

From (3) and (4), transconductance gains g_{m1} and g_{m2} can be controlled by adjusting the bias currents I_{b2} and I_{b3}, respectively. The z_{c+}- and z_{c-}-terminals (z-copy CFTA) can be obtained by adding current mirrors and cross couple current mirrors in CCFTA as shown in Fig. 2.

The third-order oscillator can be realized from the third-order polynomial equation as

$$N(s) = a_0 s^3 + a_1 s^3 + a_2 s + a_3 . \tag{5}$$

Letting $N(s) = 0$ and $s = j\omega$, (5) becomes the third-order oscillator characteristic as

$$0 = -j\omega^3 a_0 - \omega^2 a_1 + j\omega\, a_2 + a_3 . \tag{6}$$

Considering the coefficient of real and imaginary parts, we have:

$$a_3 - \omega^2 a_1 = 0 , \tag{7}$$

$$\omega a_2 - \omega^3 a_0 = 0. \tag{8}$$

From (7) and (8), the CO and FO can be given respectively [30] by

$$a_0 a_3 = a_1 a_2 , \tag{9}$$

$$\omega = \sqrt{\frac{a_3}{a_1}} = \sqrt{\frac{a_2}{a_0}} . \tag{10}$$

The proposed mixed-mode third-order QO is shown in Fig. 3. It is continuously developed from the circuit proposed in [56] by expanding text, simulation results and adding experimental results. The circuit is consisted of only one MCCFTA and three grounded capacitors. The use of grounded capacitors makes the proposed circuit ideal for IC implementation [57]. The characteristic equation of proposed circuit can be expressed using Fig. 3(b). To easy following, the proposed QO can be shown as block diagram in Fig. 3(c). Considering currents I_A and I_C, current transfer function between I_C and I_A can be given by

$$\frac{I_C}{I_A} = \frac{g_{m1}}{s^2 C_1 C_2 R_f + s C_2 g_{m1} R_f + g_{m1}} . \tag{11}$$

In addition, considering currents I_B and I_A, current transfer function can be given by

$$\frac{I_B}{I_A} = \frac{-g_{m1} g_{m2}}{s^3 C_1 C_2 C_3 R_f + s^2 C_2 C_3 g_{m1} R_f + s C_3 g_{m1}} . \tag{12}$$

Letting $I_B/I_A = 1$ (nodes A and B are closed), the characteristic equation of proposed QO in Fig. 3 can be expressed as

$$s^3 + \frac{g_{m1}}{C_1} s^2 + s \frac{g_{m1}}{C_1 C_2 R_f} + \frac{g_{m1} g_{m2}}{C_1 C_2 C_3 R_f} = 0. \tag{13}$$

Compared with (5), the coefficients can be expressed as:

$$a_0 = 1 ,$$

$$a_1 = \frac{g_{m1}}{C_1} ,$$

$$a_2 = \frac{g_{m1}}{C_1 C_2 R_f} ,$$

$$a_3 = \frac{g_{m1} g_{m2}}{C_1 C_2 C_3 R_f} .$$

According to (9) and (10), the CO and FO are obtained, respectively, by

$$g_{m2} = \frac{g_{m1} C_3}{C_1} , \tag{14}$$

$$\omega_o = \sqrt{\frac{g_{m1}}{C_1 C_2 R_f}} . \tag{15}$$

It is evident from (14) and (15) that the CO can be controlled using g_{m2} by keeping $C_1 = C_3$ and g_{m1} constant and the FO can be controlled by R_f by keeping $C_1 = C_2$ ($C_1 = C_2 = C_3$) and g_{m1} constant. Therefore, the CO and FO can be orthogonally controlled. From Fig. 3, the relationship between I_2 and I_1 can be expressed by

$$\frac{I_2}{I_1} = \frac{1}{s C_2 R_f} \tag{16}$$

while the relationship between voltages V_1 and V_2 can be expressed as

$$\frac{V_2}{V_1} = \frac{1}{s C_3 g_{m2}} \tag{17}$$

where the phase difference is $\phi = \pi/2$. This guarantees that the proposed QO provides the quadrature output currents I_1 and I_2 and quadrature output voltages V_1 and V_2. Also the uses of multiple-output MCCFTA that provides inversion of the output currents, thus it leads to $I_3 = -I_1$ and $I_4 = -I_2$. Moreover, all current output terminals are at high impedance of MCCFTA, thus ensuring insensitive current outputs that require no additional current followers to be sensed. However, if quadrature output voltages V_1 and V_2 are used, loads cannot be connected directly, the buffer circuit is needed. This problem can be solved easily using voltage follower circuit.

3. Non-Ideal Analysis

To consider the non-ideal effect of a MCCFTA by taking the non-idealities of the MCCFTA into account, the relationship of the terminal voltages and currents can be rewritten as

$$
\begin{pmatrix} I_{z1} \\ I_{z2} \\ I_{zc} \\ V_f \\ I_{x1} \\ I_{x2} \end{pmatrix}
=
\begin{pmatrix}
\alpha_1 & 0 & 0 & 0 & 0 & 0 \\
\alpha_2 & 0 & 0 & 0 & 0 & 0 \\
\alpha_c & 0 & 0 & 0 & 0 & 0 \\
R_f & 0 & 0 & 0 & 0 & 0 \\
0 & \pm g_{m1} & 0 & 0 & 0 & 0 \\
0 & 0 & \pm g_{m2} & 0 & 0 & 0
\end{pmatrix}
\begin{pmatrix} I_f \\ V_{z1} \\ V_{z2} \\ V_{zc} \\ V_{x1} \\ V_{x2} \end{pmatrix}
\tag{18}
$$

where $\alpha_1 = 1 - \varepsilon_{i1}$ and $\varepsilon_{i1}(\varepsilon_{i1} \ll 1)$ is the current tracking error from f-terminal to z_1-terminal, $\alpha_2 = 1 - \varepsilon_{i2}$ and $\varepsilon_{i2}(\varepsilon_{i2} \ll 1)$ is the current tracking error from f-terminal to z_2-terminal and $\alpha_c = 1 - \varepsilon_{ic}$ and $\varepsilon_{ic}(\varepsilon_{ic} \ll 1)$ is the current tracking error from f-terminal to z_c-terminal, R_f is the parasitic resistances at f-terminal. At high-frequency operating,

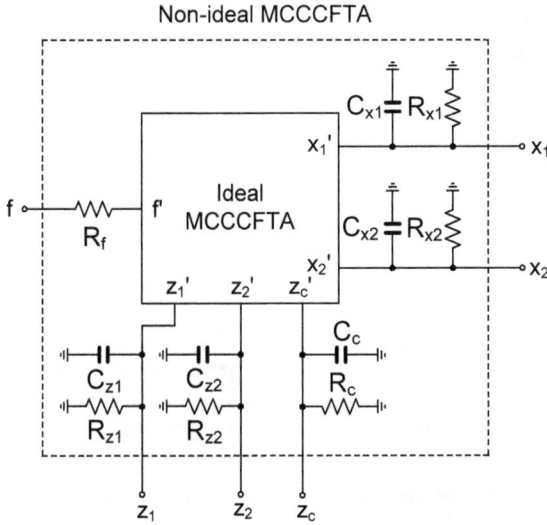

Fig. 4. Non-ideal of MCCFTA at high operating frequency.

non-ideal model of MCCFTA can be shown in Fig. 4. R_{z1}, R_{z2} and R_{zc} are respectively the high parasitic resistances at z_1-, z_2- and z_c-terminals, R_{x1} and R_{x2} are respectively the high parasitic resistances at x_1- and x_2-terminals, C_{z1}, C_{z2} and C_{zc} are respectively the low parasitic capacitances at z_1-, z_2- and z_c-terminals and C_{x1} and C_{x2} are respectively the low parasitic capacitances at x_1- and x_2-terminals.

Taking into account the non-ideal MCCFTA characteristics, current transfer function between I_B and I_A can be rewritten as

$$\frac{I_B}{I_A} =$$

$$\frac{-Z_1 Z_2 Z_3 g_{m1} g_{m2} \alpha_1 \alpha_2 \alpha_c}{\left(Z_1 Z_2 g_{m1} + Z_1 R_f g_{m1} + Z_2 + R_f + Z_1 Z_2 g_{m1} \alpha_1 \alpha_c - Z_1 Z_2 g_{m1} \alpha_1 - Z_2 \alpha_1\right)} \quad (19)$$

where

$$Z_1 = \frac{R_{p1}}{sC_1' R_{p1} + 1},$$

$$Z_2 = \frac{R_{p2}}{sC_2' R_{p2} + 1},$$

$$Z_3 = \frac{R_{p3}}{sC_3' R_{p3} + 1},$$

$$C_1' = C_1 \| C_{x1-} \| C_{x2+} \| C_z,$$

$$C_2' = C_2 \| C_{x1+} \| C_{zc-},$$

$$C_3' = C_3 \| C_{z2},$$

$$R_{p1} = R_{x1-} \| R_{x2+} \| R_z,$$

$$R_{p2} = R_{x1+} \| R_{zc-},$$

$$R_{p3} = R_{z2}.$$

The modified characteristic equation of Fig. 3 can be rewritten as

$$s^3 + s^2 \frac{g_{m1}}{C_1'} \left(1 + \frac{1}{g_{m1} R_f} + \frac{1}{g_{m1} R_{p2}} + \frac{1}{g_{m1} R_{p1}} + \frac{C_1'}{C_3' g_{m1} R_{p3}} - \frac{\alpha_1}{g_{m1} R_f} \right)$$

$$+ s \frac{g_{m1}}{C_1' C_2' R_f} \left(1 + \alpha_1 \alpha_c + \frac{1}{g_{m1} R_{p1}} + \frac{R_f}{R_{p2}} + \frac{C_2' R_f}{C_3' R_{p3}} + \frac{C_1'}{C_3' g_{m1} R_{p3}} + \frac{R_f}{g_{m1} R_{p1} R_{p2}} + \frac{C_1' R_f}{C_3' g_{m1} R_{p2} R_{p3}} + \frac{C_2' R_f}{C_3' g_{m1} R_{p1} R_{p3}} - \frac{\alpha_1}{g_{m1} R_{p1}} - \frac{\alpha_1}{g_{m1} R_{p1}} - \alpha_1 \right)$$

$$+ \frac{g_{m1} g_{m2} \alpha_1 \alpha_2 \alpha_c}{C_1' C_2' C_3' R_f} \left(1 + \frac{1}{g_{m1} g_{m2} R_{p1} R_{p3} \alpha_1 \alpha_2 \alpha_c} + \frac{1}{g_{m2} R_{p3} \alpha_2} + \frac{1}{g_{m2} R_{p3} \alpha_1 \alpha_2 \alpha_c} + \frac{R_f}{g_{m2} R_{p2} R_{p3} \alpha_1 \alpha_2 \alpha_c} + \frac{R_f}{g_{m1} g_{m2} R_{p1} R_{p2} R_{p3} \alpha_1 \alpha_2 \alpha_c} - \frac{1}{g_{m1} g_{m2} R_{p1} R_{p3} \alpha_2 \alpha_c} - \frac{1}{g_{m2} R_{p3} \alpha_2 \alpha_c} \right) \quad (20)$$

Letting, parasitic parameters R_z, R_c, R_x are very high resistance values and C_z, C_c, C_c are very low capacitance value, (20) can be approximated as

$$s^3 + s^2 \frac{g_{m1}}{C_1'} + s \frac{g_{m1}}{C_1' C_2' R_f} + \frac{g_{m1} g_{m2} \alpha_1 \alpha_2 \alpha_c}{C_1' C_2' C_3' R_f} = 0 . \quad (21)$$

In this case, the CO and FO are modified, respectively, as:

$$g_{m2} = \frac{g_{m1} C_3'}{C_1'} \alpha_1 \alpha_2 \alpha_c , \quad (22)$$

$$\omega_o = \sqrt{\frac{g_{m1} g_{m2} \alpha_1 \alpha_2 \alpha_c}{C_1' C_2' C_3' R_f}} . \quad (23)$$

The various passive sensitivities of ω_o of the proposed QO can be obtained as

$$\left. \begin{array}{l} S_{g_{m1}}^{\omega_0} = S_{g_{m2}}^{\omega_0} = S_{\alpha_1}^{\omega_0} = S_{\alpha_2}^{\omega_0} = S_{\alpha_3}^{\omega_0} = 0.5 \\ S_{R_f}^{\omega_0} = S_{C_1'}^{\omega_0} = S_{C_2'}^{\omega_0} = S_{C_3'}^{\omega_0} = -0.5 \end{array} \right\} \quad (24)$$

Thus, the proposed QO has low active and passive sensitivities.

4. Simulation Results

To verify the theoretical prediction of the proposed third-order QO, the circuit in Fig. 3 was simulated using PSPICE simulators and the MCCFTA in Fig. 2 was used. The model parameters for nMOS and pMOS transistors are taken from 0.25 μm CMOS process from TSMC. The power supply was given as ±1 V. From Fig. 2, I_{b1} was initially designed for 100 μA while I_{b2} and I_{b3} were initially designed for 150 μA. Transistors around trans-linear loop (M_1 to M_4) were designed to obtain $g_{m(nMOS)} = g_{m(pMOS)}$ and differential transistor pairs (M_5-M_6 and M_7-M_8) were designed to obtain $g_{m(total)} = (g_{m(M5)} + g_{m(M6)})/2$ or $g_{m(total)} = (g_{m(M7)} + g_{m(M8)})/2$. For 0.25 μm CMOS process used in this case, $(\mu_o C_{ox})_{nMOS} = 242.2$ μA/V^2 and $(\mu_o C_{ox})_{pMOS} = 51.8$ μA/V^2, $V_{th0(nMOS)} = 0.37$ V and $V_{th0(pMOS)} = -0.49$ V were given. Therefore, the aspect ratios of MOS transistors for MCCFTA in Fig. 3 were $W/L = 5$ μm/0.5 μm for M_1-M_2, $W/L = 15$ μm/0.5 μm for M_3-M_4, $W/L = 5$ μm/0.5 μm for M_5-M_8, $W/L = 5$ μm/0.5 μm for all M_n transistors and $W/L = 15$ μm/0.5 μm for all M_p transistors. PSPICE simulations have verified that when I_{b1} was varied from 1 to 100 μA, achieved R_f was in the range of 12 to 2.19 kΩ as shown in Fig. 5 whereas I_{b2} and I_{b3} were varied from 5 to 150 μA, achieved g_{m1} and g_{m2} were in the range of 107 to 472 μA/V as shown in Fig. 6. Both simulated R_f- and g_m-value were compared with (2) and (3), respectively. Simulated performances of MCCFTA were summarized in Tab. 2.

The voltage outputs and current outputs were shown respectively in Figs. 7 and 8, when Fig. 3 was designed with the following values: $C_1 = C_2 = C_3 = 20$ pF, $I_{b1} = 20$ μA, $I_{b2} = 80$ μA and $I_{b3} = 75$ μA. The I_{b3}-value was varied for g_{m2} so as to satisfy the CO in (14). In this case, ($R_f \approx 1/g_{m1}$), each current output and each voltage output were almost equal of magnitude. The magnitudes of the quadrature signals will not be equaled in case of R_f was varied ($R_f \neq 1/g_{m1}$) in (15). For applications requiring equal magnitude quadrature outputs, additional amplifying circuits were needed. From Fig. 7, the simulated THD of V_1 and V_2 were about 0.26 % and 0.17 %, respectively, the FO was 2.7 MHz. From Fig. 8, the simulated the FO was found to be 2.7 MHz, the results of the I_1, I_2, I_3 and I_4 total harmonic distortion (THD) analysis were obtained as 0.56 %, 0.37 %, 0.85 % and 0.38 %, respectively, and the power consumption was about 1.43 mW. This low THD was obtained for the output signals V_1, V_2, I_1, I_2, I_3 and I_4 of 204 mV$_{P-P}$, 198 mV$_{P-P}$, 70 μA$_{P-P}$, 70 μA$_{P-P}$, 70 μA$_{P-P}$ and

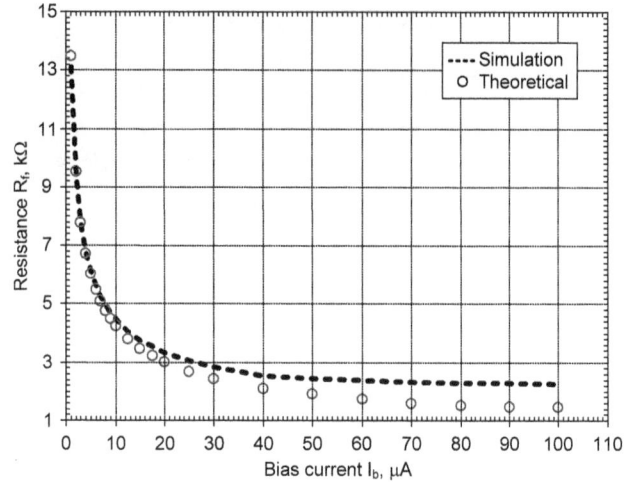

Fig. 5. Simulated resistance R_f when bias current is varied.

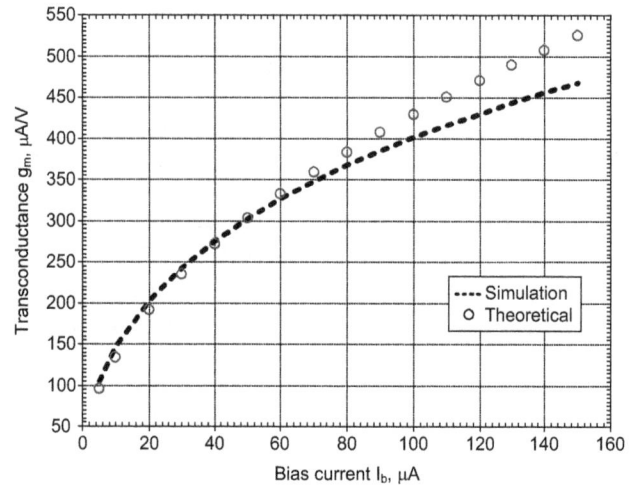

Fig. 6. Simulated transconductance g_m when bias current is varied.

66 μA$_{P-P}$, respectively. THD will be increased if output signal level was increased which can be obtained by increased CO. However, increasing CO may be increased nonlinear behavior of the system which is results in high THD output signal that should be avoided.

The electronic tuning of the oscillator was shown in Fig. 9. The result gives a variation of the FO from 1.1 to 3.25 MHz with I_{b1} in the range of 1 to 50 μA that was confirmed by (15). This FO was achieved without adjusting the CO. It shows that the simulated FO was consistent with the theoretical values. At $I_{b1} = 50$ μA, the error of FO between the simulated and theoretical values was 4.9 %. The bias current I_{b1} can be continuously varied high up to 100 μA, but the g_{m2} must be adjusted via I_{b3} to satisfy the CO. However, it was also found that there was a deviation between the theoretical and simulated values in large bias current value region over the value of 50 μA. From Fig. 9, output signal levels of V_1, V_2, I_1, I_2, I_3, and I_4 versus f_o have been obtained as shown in Fig. 10. It should be noted that the output signal level is not equal and constant when I_{b1} was varied. This problem can be solved by using automatic gain control (AGC) circuit. There are oscillators with AGC available in literature [58–61]. These techniques use AGC

Parameters	Value
Technology	0.25 μm CMOS
Power supply	± 1 V
R_f (I_b = 1-100 μA)	12 to 2.19 kΩ
g_m (I_b = 5-150μA)	107 to 472 μA/V
Input and output range (I_b = 5-150 μA)	± 40 to ±250 mV
Bandwidth (-3dB) @ I_{b1} = 50 μA, $I_{b2} = I_{b3}$ = 100 μA Current follower (I_z/I_f) Current follower (I_x/I_f) (R_z = 2.23 kΩ)	42.84 MHz 39.9 MHz
Current gain @ I_{b1} = 50 μA, $I_{b2} = I_{b3}$ = 100 μA - I_z/I_f	1.034
Parasitic parameters @ I_{b1} = 50 μA, $I_{b2} = I_{b3}$ = 100 μA - R_z, C_z - R_x, C_x	116 kΩ, 25 fF 96 kΩ, 25 fF
Power consumption @ I_{b1}= 50 μA, $I_{b2} = I_{b3}$ = 100 μA	2.12 mW

Tab. 2. Simulated specifications of MCCFTA.

circuit including into the loop of the system for controlling the amplitude of signal. Usually, AGC is suitable for applying to linear control of CO and FO oscillator, if nonlinear control of CO and FO such as this work (adjusting CO and FO in square-root domain), current squaring circuits [62], [63] are required to compensate the square-root form to obtain linear current control of CO and FO.

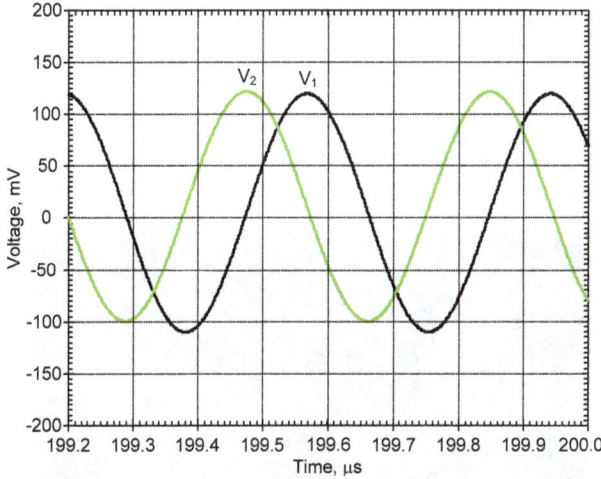

Fig. 7. Simulated two voltage output waveforms.

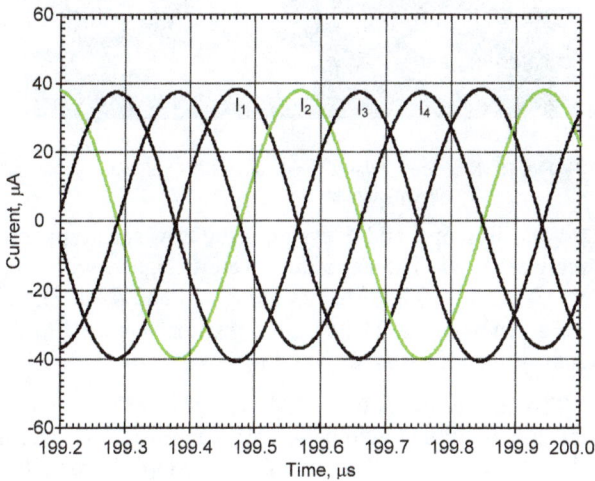

Fig. 8. Simulated four current output waveforms.

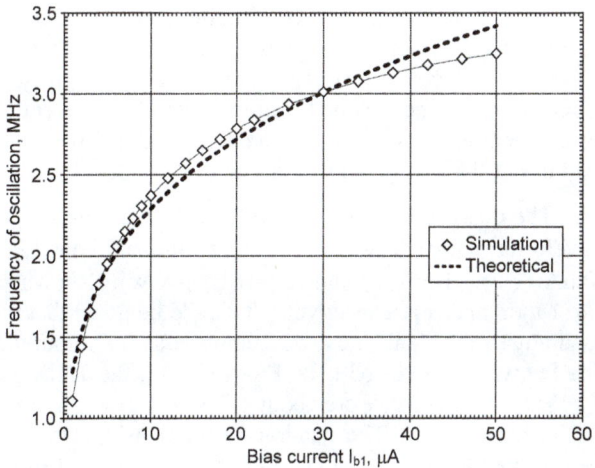

Fig. 9. Electronic frequency tuning with the bias current I_{b1}.

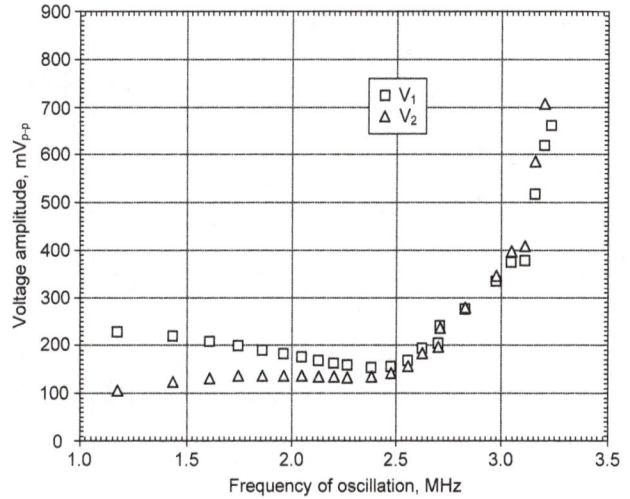

(a)

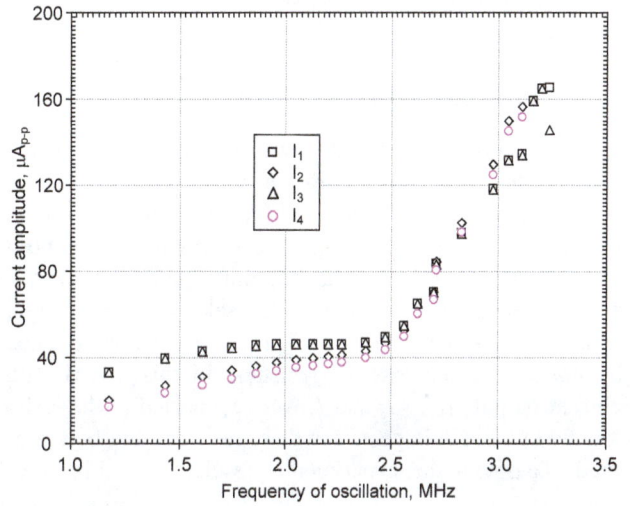

(b)

Fig. 10. Output level versus f_o: (a) V_1 and V_2, (b) I_1, I_2, I_3 and I_4.

Total harmonic distortion (THD) and phase error were shown in Figs 11 and 12, respectively. From Fig. 11, THD was increased when I_{b1} (R_f) was varied far from satisfying condition.

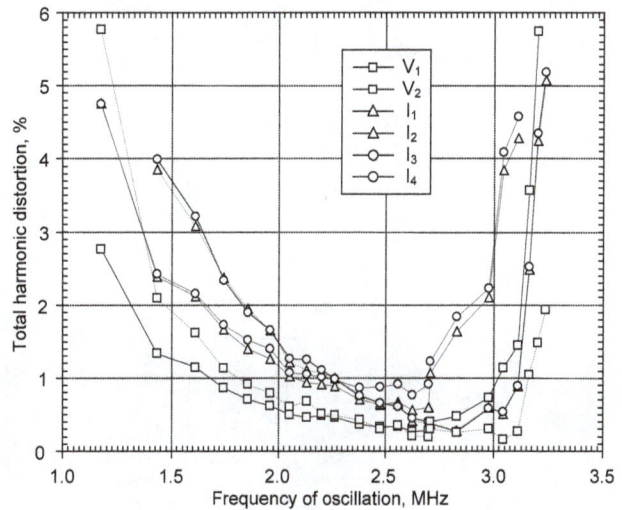

Fig. 11. Simulated THD versus f_o.

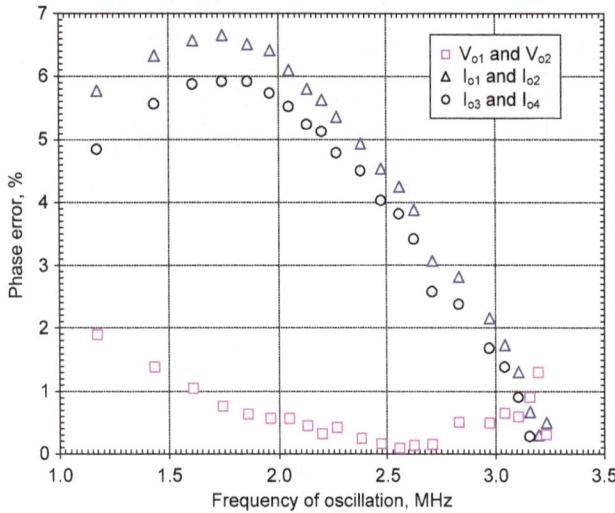

Fig. 12. Simulated phase error versus f_o.

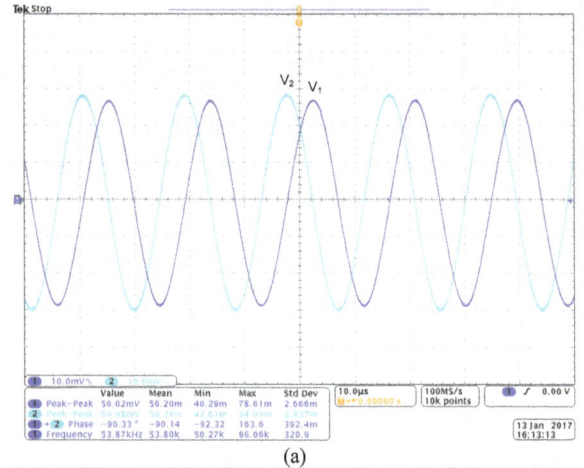

(a)

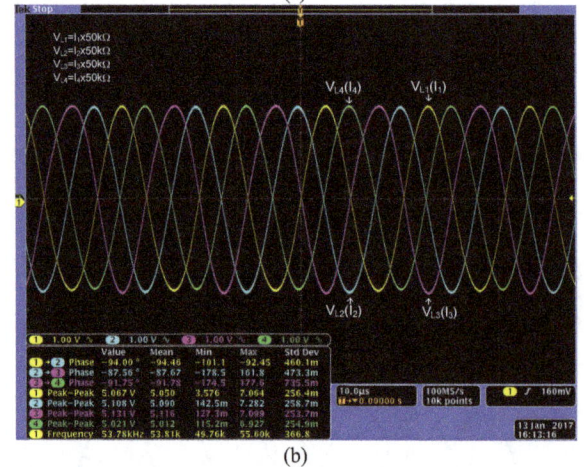

(b)

Fig. 15. Measured waveforms for: (a) voltage outputs, (b) current outputs.

5. Experimental Results

The proposed oscillator was experimentally tested wherein the functionality of the MCCFTA was implemented using CFOA AD844 and OTA LM13600 as shown in Fig. 13. The supply voltage ±10 V was used. The non-instrument of oscilloscope as measurable currents is available to the authors. However, for sake of experimental results, the current outputs, I_1, I_2, I_3 and I_4, can be obtained by connecting the external resistors. In this experiment, current outputs I_1, I_2, I_3 and I_4 were connected to the resistors 50 kΩ and voltage across these resistors will be measured. To obtain the experimental results, Tektronix MSO

4034 oscilloscope and Keysight N9030A spectrum analyzer were used. The capacitors were designed with $C_1 = C_2 = C_3 = 5$ nF . The resistors R_1, R_2, R_3 and R_4 in Fig. 13 were given as 1 kΩ. The variable resistor was used to obtain the effect of variable g_m and R_f.

To control the output amplitude of signals to be constant, the amplitude-automatic gain control (AGC) circuit in Fig. 14 was used. This circuit was adopted from [61]. From our pretest on the proposed circuit, it was found that V_1 (I_1 and I_3) was dependent on tuning of FO, thus V_1 will be used for the input of AGC circuit ($V_{in(AGC)}$). CFOA AD844 and R_{p2} were used to work as voltage-to-current converter (V-I converter). The current output $I_{out(AGC)}$ of AGC circuit will be supplied additionally to bias current I_{b3} for compensating CO. The active device and passive-value used in Fig. 14 were tabulated in Tab. 3.

The quadrature sinusoidal voltage and current output waveforms with resistor $R_f = 500$ Ω, bias currents $I_{b2} = 50$ µA ($g_{m2} = 1$ mA/V) and $I_{b3} = 61.8$ µA were shown in Fig. 15, (a) and (b), respectively. It should be noted that the quadrature sinusoidal signal was almost equal of amplitude. The FO was found as 53 kHz. From Fig. 15, the THDs for V_1, V_2, I_1, I_2, I_3 and I_4 were about 1.2, 0.9, 1.2, 1.1, 1.3 and 1.1 %, respectively. The quadrature relationship was further verified through the X-Y plots of the two output forms in Fig. 15 as shown in Fig. 16, (a) and (b), respectively.

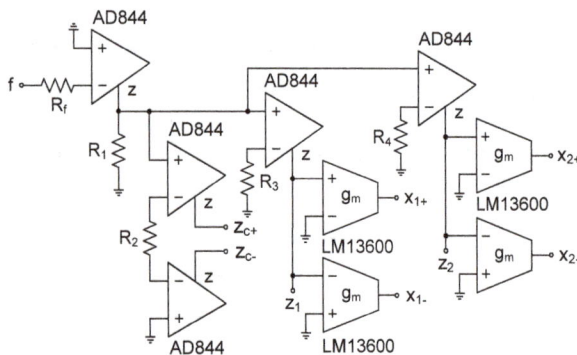

Fig. 13. Possible realization of the MCCFTA using commercial active devices (AD844 and LM13600).

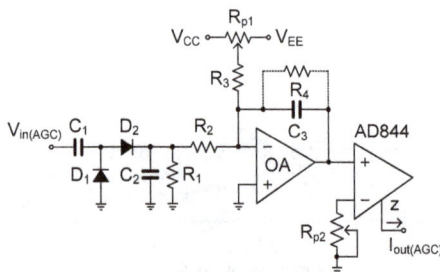

Fig. 14. Amplitude-automatic gain control circuit [61].

(a)

(b)

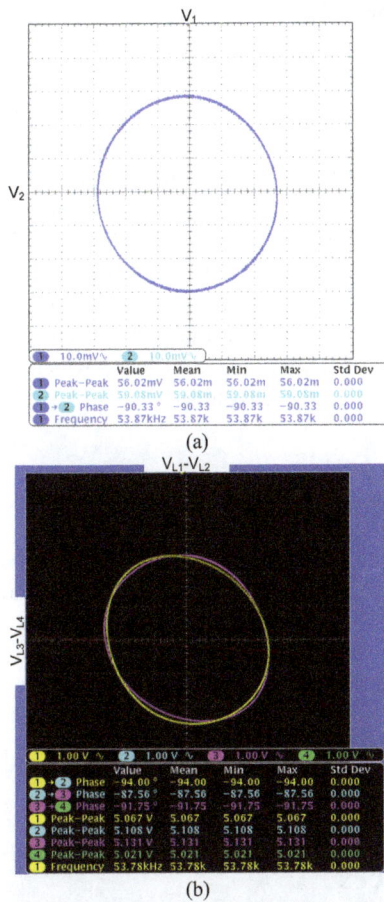

Fig. 16. X-Y plots for: (a) voltage output, (b) current output.

Device	Value
OA	OPA2650
D_1, D_2	2xBAT42
C_1, C_2	1 µF
C_3	10 nF
R_1, R_4	1 MΩ
R_2, R_3	100 kΩ
R_5	200 Ω
R_{p1} and R_{p2}	100 kΩ (variable resistor)

Tab. 3. Active and passive components used in Fig. 14.

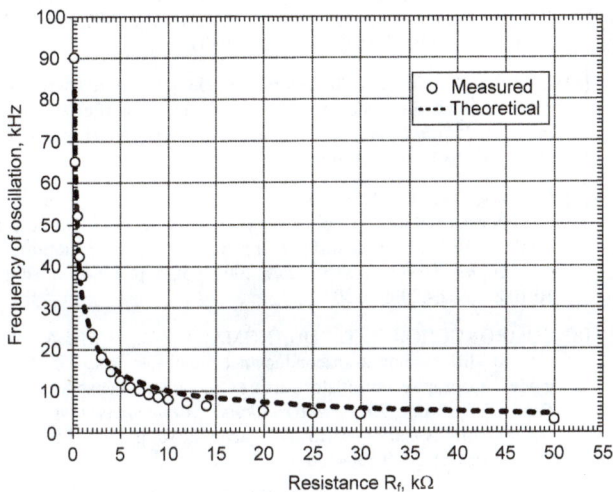

Fig. 17. FO tuning with R_f.

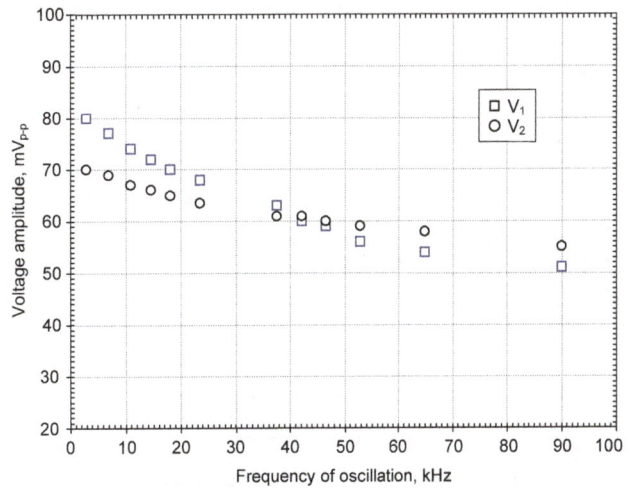

Fig. 18. Output level of V_1 and V_2 versus f_o.

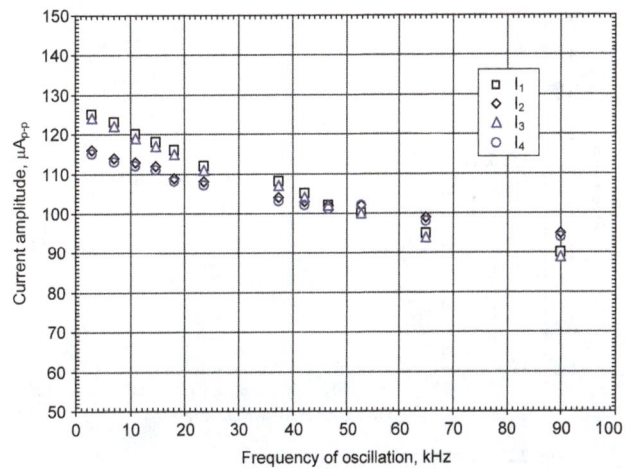

Fig. 19. Output level of I_1, I_2, I_3 and I_4 versus f_o.

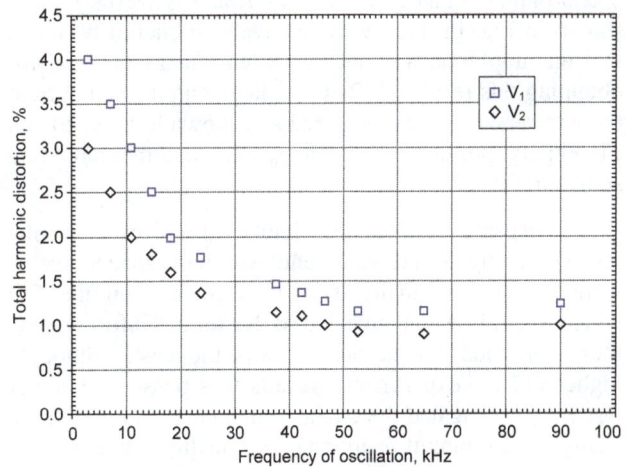

Fig. 20. Measured THD of voltage outputs versus f_o.

Figure 17 shows the experimental results of the FO by changing the value of the resistor R_f with $C_1 = C_2 = C_3 = 5$ nF, $I_{b1} = 50$ µA and $I_{b3} = 61.8$ µA. The tuning of the oscillator gives a variation of the FO from 2.9 to 90 kHz when R_f-value was decreased from 50 to 0.15 kΩ. In this case, FO was obtained by adjusting CO via AGC. The plots for theoretical value were also included for comparison.

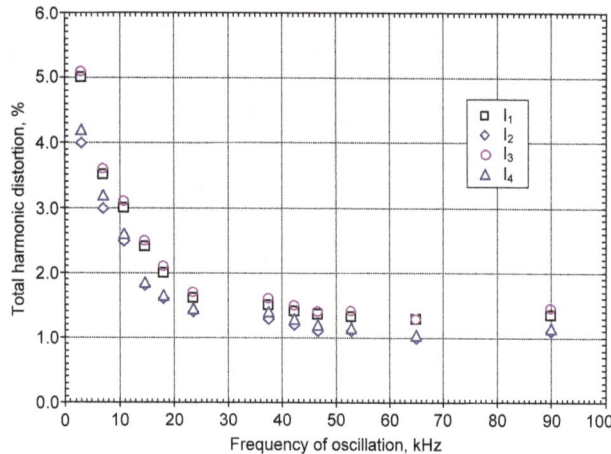

Fig. 21. Measured THD of current outputs versus f_o.

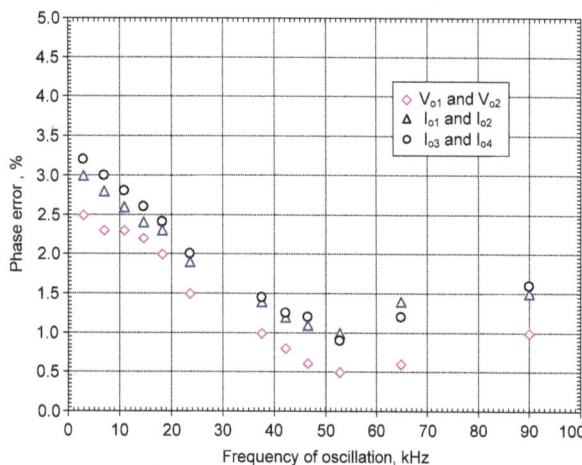

Fig. 22. Measured phase error versus f_o.

From Fig. 17, output signal level of V_1 and V_2 versus f_o and output signal level of I_1, I_2, I_3 and I_4 versus f_o were shown in Fig. 18 and 19, respectively. It should be noted that the amplitude was not fluctuated; thanks to AGC for obtaining this result. THDs for voltage outputs and current outputs versus f_o were measured and shown in Figs. 20 and 21. Finally, phase errors versus f_o were also measured and shown in Fig. 22.

Compared between the simulation and experimental results, setting of FO-value and working capacitor-value were different. Actually, this work focuses on the QO which should be implemented as IC form. Unfortunately, the obstruction for fabricating was the cost. Although, higher FO for experimental results was possible, but our test was investigated the circuit which builds on perfboard. Using printed circuit board and high quality active devices the operation band of the circuit [59–61] will be increased. Therefore, this experimental result was used only to confirm the workability of the proposed structure.

6. Conclusions

In this paper, a mixed-mode third-order QO based on new MCCFTA has been presented. The proposed circuit uses only one MCCFTA and three grounded capacitors. The use of grounded capacitors is ideally interesting from an integration point of view. The proposed structure provides four high output impedance current sources with 90° phase difference, thus these output terminals can be directly connected to the loads without additional follower circuits. In addition, voltage with 90° phase difference can be obtained without changing any topology. Also the CO and FO can be controlled orthogonally and electronically by adjusting the bias currents of MCCFTA. The active and passive sensitivities are no more than unity of magnitude. Simulation and experimental results, which confirm the theoretical analysis, are included.

Acknowledgments

The authors would like to express sincere appreciation to the anonymous reviewers and editor for their valuable comments. The research described in this paper was financed by the Czech Ministry of Education in frame of the National Sustainability Program under grant LO1401 and by the Czech Science Foundation under grant No. P102-15-21942S. For research, infrastructure of the SIX Center was used.

References

[1] TOUMAZOU, C., LIDGEY, F., MAKRIS, C. A. Extending voltage-mode op amps to current-mode performance, *IEE Proceeding Part-G*, 1990, vol. 137, p. 116–130. DOI: 10.1049/ip-g-2.1990.0020

[2] BIOLEK, D. CDTA-building block for current-mode analog signal processing, In *Proceedings of European Conference on Circuit Theory and Design (ECCTD)*. Poland, 2003, p. 397–400, vol. III.

[3] HAYKIN, S., MOHER, M. *An Introduction to Analog and Digital Communications*. New York: John Wiley & Sons, 2007. ISBN-13: 978-0471432227

[4] BOLTON, W. *Measurement and Instrumentation Systems*. Oxford: Newnes, 1996. ISBN-9780128011324

[5] KESKIN, A. U., BIOLEK, D. Current mode quadrature oscillator using current differencing transconductance amplifiers (CDTA). *IEE Proceeding of Circuits Devices and Systems*, 2006, vol. 153, p. 214–218: DOI: 10.1049/ip-cds:20050304

[6] LAHIRI, A. Novel voltage/current-mode quadrature oscillator using current differencing transconductance amplifier. *Analog Integrated Circuits and Signal Processing*, 2009, vol. 61, p. 199 to 203. DOI: 10.1007/s10470-009-9291-0

[7] BIOLEK, D., KESKIN, A. U., BIOLKOVA, V. Grounded capacitor current mode single resistance-controlled oscillator using single modified current differencing transconductance amplifier. *IET Circuits, Devices and Systems*, 2009, vol. 4, p. 496–520. DOI: 10.1049/iet-cds.2009.0330

[8] BUMRONGCHOKE, T., DUANGMALAI, D., JAIKLA, W. Current differencing transconductance amplifier based current-mode quadrature oscillator using grounded capacitors. In *Proceedings of 2010 International Communications and Information Technologies (ISCIT)*. Japan, 2010, p. 192–195. DOI: 10.1109/ISCIT.2010.5664834

[9] JAIKLA, W., SIRIPRUCHYANUN, M., BAJER, J., BIOLEK, D.

A simple current-mode quadrature oscillator using single CDTA. *Radioengineering*, 2008, vol. 17, no. 4, p. 33–40.

[10] KUMNGERN, M., LAMUN, P., DEJHAN, K. Current-mode quadrature oscillator using current differencing transconductance amplifiers. *International Journal of Electronics*, 2012, vol. 99, p. 971–986. DOI: 10.1080/00207217.2011.651693

[11] JAIKLA, W., LAHIRI, A. Resistor-less current-mode four-phase quadrature oscillator using CCCDTAs and grounded capacitors. *International Journal of Electronics and Communications*, 2012, vol. 66, p. 214–218. DOI: 10.1016/j.aeue.2011.07.001

[12] JIN, J., WANG, C. Single CDTA-based current-mode quadrature oscillator. *International Journal of Electronics and Communications*, 2012, vol. 66, p. 933–936. DOI: 10.1016/j.aeue.2012.03.018

[13] BIOLEK, D., SENANI, R., BIOLKOVA, V., KOLKA, Z. Active elements for analog signal processing: classification: Review and new proposals. *Radioengineering*, 2008, vol. 17, no. 4, p. 15–32.

[14] HERENCSAR, N. KOTON, J. VRBA, K., LAHIRI, A. Floating simulators based on current follower transconductance amplifiers (CFTAs). In *Proceedings of the European Conference of Circuits Technology and Devices*. Tenerife (Spain), 2010, p. 23–26.

[15] HERENCSAR, N., KOTON, J., VRBA, K., CICEKOGLU, O. New active-C grounded positive inductance simulator based on CFTAs. In *Proceedings of 33rd International Conference on Telecommunications and Signal Processing*. Czech Republic, 2010, p. 35–37.

[16] HERENCSAR, N., KOTON, J., VRBA, K. Realization of current-mode KHN-equivalent biquad using current follower transconductance amplifiers (CFTAs). *IEICE Transactions on Fundamentals of Electronics, Communications and Computer Sciences*, 2010, vol. E93.A, p. 1816–1819. DOI: 10.1587/transfun.E93.A.1816

[17] HERENCSAR, N., KOTON, J., VRBA, K., LATTENBERG, I. Current follower transconductance amplifier (CFTA)-a useful building block for analog signal processing. *Journal of Active and Passive Electronic Devices*, 2011, vol. 6, p. 217–229.

[18] INTAWICHAI, K., TANGSRIRAT, W. Signal flow graph realization of nth-order current-mode allpass filters using CFTAs. In *Proceedings of 2013 10th International Conference on Electrical Engineering/Electronics, Computer, Telecommunications and Information Technology (ECTI-CON)*. Thailand, 2013, p. 1–6. DOI: 10.1109/ECTICon.2013.6559519

[19] TORTEANCHAI, U., KUMNGERN, M. First-order allpass network using CFTA. In *Proceedings of 2014 4th Joint International Conference on Information and Communication Technology, Electronic and Electrical Engineering (JICTEE)*. Thailand, 2014, p. 1–4. DOI: 10.1109/JICTEE.2014.6804120

[20] HERENCSAR, N., VRBA, K., KOTON, J., LAHIRI, A. Realisations of single-resistance-controlled quadrature oscillators using a generalised current follower transconductance amplifier and a unity-gain voltage-follower. *International Journal of Electronics*, 2010, vol. 97, p. 897–906. DOI: 10.1080/00207211003733320

[21] TANGSRIRAT, W., MONGKOLWAI, P., PUKKALANUN, T. Current-mode high-Q bandpass filter and mixed-mode quadrature oscillator using ZC-CFTAs and grounded capacitors. *Indian Journal of Pure & Applied Physics*, 2012, vol. 50, p. 600–607. URI: http://hdl.handle.net/123456789/14480

[22] PRASERTSOM, D., TANGSRIRAT, W. Current gain controlled CFTA and its application to resistorless quadrature oscillator. In *Proceedings of 2012 9th International Conference on Electrical Engineering/Electronics, Computer, Telecommunications and Information Technology (ECTI-CON)*. Thailand, 2012, p. 1–4. DOI: 10.1109/ECTICon.2012.6254271

[23] LAMUN, P. KUMNGERN, M., TORTEANCHAI, U.,

SARSITTHITHUM, K. Tunable current-mode quadrature sinusoidal oscillator using CCCFTAs and grounded capacitors. In *Proceedings of 2013 4th International Conference on Intelligent Systems, Modelling and Simulation (ISMS)*. Thailand, 2013, p. 665–668. DOI: 10.1109/ISMS.2013.138

[24] KUMNGERN, M., LAMUN, P., JUNNAPIYA, S. CFTA-based electronically tunable quadrature sinusoidal oscillator. In *Proceedings of 2014 International Electrical Engineering Congress (iEECON)*. Thailand, 2014, p. 1–4. DOI: 10.1109/iEECON.2014.6925891

[25] PHATSORNSIRI, P., LAMUN, P. Tunable current-mode quadrature oscillator using CFTAs and grounded capacitors. In *Proceedings of 2015 12th International Conference on Electrical Engineering/Electronics, Computer, Telecommunications and Information Technology (ECTI-CON)*. Thailand, 2015, p. 1–4. DOI: 10.1109/ECTICon.2015.7207104

[26] SRISAKULTIEW, S., SIRIPRUCHYANUN, M., JAIKLA, W. Single-resistance-controlled current-mode quadrature sinusoidal oscillator using single CCCFTA with grounded elements. In *Proceedings of 2013 36th International Conference on Telecommunications and Signal Processing (TSP)*. Rome (Italy), 2013, p. 436-439. DOI: 10.1109/TSP.2013.6613969

[27] VAN VALKENBURG, M. E. *Analog Filter Design*. Holt Sounders International Edition, 1987.

[28] RUBIOLA, E. *Phase Noise and Frequency Stability in Oscillators*. Cambridge University Press, 2010. ISBN-13: 9780521153287

[29] RAZAVI, B. A study of phase noise in CMOS oscillators. *IEEE Journal of Solid-State Circuits*, 1996, vol. 31, p. 331–343. DOI: 10.1109/9780470545492.ch19

[30] PROMMEE, P. DEJHAN, K. An integrable electronic-controlled quadrature sinusoidal oscillator using CMOS operational transconductance amplifier. *International Journal of Electronics*, 2002, vol. 89, p. 365–379. DOI: 10.1080/713810385

[31] HORNG, J.-W. Quadrature oscillators using operational amplifiers. *Active and Passive Electronic Components*, vol. 2011, Article ID 320367. DOI: 10.1155/2011/320367

[32] HORNG, J.-W., HOU, C.-L., CHANG, C.-M., CHUNG, W.-Y., TANG, H.-I., WEN, Y.-I. Quadrature oscillator using CCIIs. *International Journal of Electronics*, 2005, vol. 92, p. 21–31. DOI: 10.1080/00207210412331332899

[33] HORNG, J.-W. Current/voltage-mode third order quadrature oscillator employing two multiple outputs CCIIs and grounded capacitors. *Indian Journal of Pure & Applied Physics*, 2011, vol. 49, p. 494–498. URI: http://hdl.handle.net/123456789/12012

[34] KOTON, J., HERENCSAR, N., VRBA, K., METIN, B. Current- and voltage-mode third-order quadrature oscillator. In *Proceedings of 2012 13th International Conference on Optimization of Electrical and Electronic Equipment (OPTIM)*. Romania, 2012, p. 1203–1206. DOI: 10.1109/OPTIM.2012.6231795

[35] CHATURVEDI, B., MAHESHWARI, S. Third-order quadrature oscillator circuit with current and voltage outputs. *ISRN Electronics*, vol. 2013, Article ID 385062, DOI: 10.1155/2013/385062

[36] PANDEY, R., PANDEY, N., KOMANAPALLI, G., ANURAG, R. OTRA based voltage mode third order quadrature oscillator. *ISRN Electronics*, vol. 2014, Article ID: 126471, p. 1–5. DOI: 10.1155/2014/126471

[37] KUMNGERN, M., KANSIRI, I. Single-element control third-order quadrature oscillator using OTRAs. In *Proceedings 2014 12th ICT and Knowledge Engineering (ICT&KE)*. Thailand, 2014, p. 24–27. DOI: 10.1109/ICTKE.2014.7001529

[38] MAHESHWARI, S., KHAN, I. A. Current controlled third order quadrature oscillator. *IEE Proceeding of Circuits Devices and System*, 2005, vol. 152, p. 605–607. DOI: 10.1049/ip-cds.20045185

[39] MAHESHWARI, S. Current-mode third-order quadrature oscillator. *IET Circuits, Devices and Systems*, 2010, vol. 4, p. 188 to 195. DOI: 10.1049/iet-cds.2009.0259

[40] HORNG, J.-W. Current-mode third-order quadrature oscillator using CDTAs. *Active and Passive Electronic Components*, vol. 2009, Article ID 789171. DOI: 10.1155/2009/789171

[41] HORNG, J.-W., LEE, H., WU, J.-Y. Electronically tunable third-order quadrature ocillator using CDTAs. *Radioengineering*, 2010, vol. 19, p. 326–330.

[42] KUMNGERN, M., JUNNAPIYA, S. Current-mode third-order quadrature oscillator using minimum elements. In *Proceedings of International Conference on Electrical Engineering and Informatics (ICEEI)*. Indonesia, 2011, p. 1–4. DOI: 10.1109/ICEEI.2011.6021799

[43] LAWANWISUT, S., SIRIPRUCHAYANUN, M. High output-impedance current-mode third-order quadrature oscillator based on CCCCTAs. In *Proceedings of 2009 IEEE Region 10 Conference (TENCON)*. Singapore, 2009, p. 1–4. DOI: 10.1109/TENCON.2009.5395961

[44] DUANGMALAI, D., JAIKLA, W. Realization of current-mode quadrature oscillator based on third order technique. *ACEEE International Journal on Electrical and Power Engineering*, 2011, vol. 2, p. 46–49. Available at: http://agritech.pcru.ac.th/new/doc/3rd%20oscillator-20120222-093615.pdf

[45] PANDEY, R., PANDEY, N., PAUL, S. K. MOS-C third order quadrature oscillator using OTRA. In *Proceedings of 2012 Third International Conference on Computer and Communication Technology (ICCCT)*. India, 2012, p. 77–80. DOI: 10.1109/ICCCT.2012.24

[46] PHANRUTTANACHAI, K., JAIKLA, W. Third order current-mode quadrature sinusoidal oscillator with high output impedances. *World Academy of Science, Engineering and Technology*, 2013, vol. 7, p. 472–475.

[47] PANDEY, N., PANDEY, R. Approach for third order quadrature oscillator realization. *IET Circuits, Devices & Systems*, 2015, vol. 9, p. 161–171. DOI: 10.1049/iet-cds.2014.0170

[48] PROMMEE, P., ANGKEAW, K. Log-domain current-mode third-order sinusoidal oscillator. In *Proceedings of 2011 IEEE 54th International Midwest Symposium on Circuits and Systems (MWSCAS)*. Korea, 2011, p. 1–4.

[49] KUMNGERN, M., CHANWUTITUM, J. Single MCCCCTA-based mixed-mode third-order quadrature oscillator. In *Proceedings of Fourth International Conference on Communications and Electronics (ICCE)*. Vietnam, 2012, p. 426 to 429. DOI: 10.1109/CCE.2012.6315943

[50] PANDEY, R., PANDEY, N., KOMANAPALLI, G., ANURAG, R. OTRA based voltage mode third order quadrature oscillator. *ISRN Electronics*, vol. 2014, Article ID 126471.

[51] PHATSORNSIRI, P. LAMUN, P. KUMNGERN, M. TORTEANCHAI, U. Current-mode third-order quadrature oscillator using VDTAs and grounded capacitors. in *Proceeding of The 4th Joint International Conference on Information and Communication Technology, Electronic and Electrical Engineering (JICTEE)*. Thailand, 2014, p. 1–4. DOI: 10.1109/JICTEE.2014.6804103

[52] CHANNUMSIN, O. JANTAKUN, A. Third-order sinusoidal oscillator using VDTAs and grounded capacitors with amplitude controllability. In *Proceeding of The 4th Joint International Conference on Information and Communication Technology, Electronic and Electrical Engineering (JICTEE)*. Thailand, 2014, p. 1–4. DOI: 10.1109/JICTEE.2014.6804103

[53] HERENCSAR, N., KOTON, J., VRBA, K. LAHIRI, A., CICEKOGLU, O. Current-controlled CFTA-based current-mode SITO universal filter and quadrature oscillator. In *Proceedings of*

2010 International Conference on Applied Electronics (AE), Czech Republic, 2010, p. 1–4.

[54] LI, Y. A modified CDTA (MCDTA) and its applications: designing current-mode sixth-order elliptic band-pass filter. *Circuits, Systems, and Signal Processing*, 2011, vol. 30, p. 1383 to 1390. DOI: 10.1007/s00034-011-9329-2

[55] JAIKLA, W., SIRIPRUCHYANUN, M., LAHIRI, A. Resistorless dual-mode quadrature sinusoidal oscillator using a single active building block. *Microelectronics Journal*, 2011, vol. 42, p. 135 to 104. DOI: 10.1016/j.mejo.2010.08.017

[56] KUMNGERN, M., TORTEANCHAI, U. A current-mode four-phase third-order quadrature oscillator using a MCCCFTA. In *Proceedings of 2012 IEEE International Conference on Cyber Technology in Automation, Control, and Intelligent Systems (CYBER)*. Thailand, 2012, p. 156–159. DOI: 10.1109/CYBER.2012.6392545

[57] BHUSAN, M., NEWCOMB, R. W. Grounding of capacitors in integrated circuits. *Electronics Letters*, 1967, vol. 3, p. 148–149. DOI: 10.1049/el:19670114

[58] BIOLEK, B., LAHIRI, A., JAIKLA, W., SIRIPRUCHYANUN, M., BIOLKOVA, V. Realization of electronically tunable voltage-mode/current-mode quadrature sinusoidal oscillator using ZC-CG-CDBA. *Microelectronics Journal*, 2011, vol. 42, p. 1116–1123. DOI: 10.1016/j.mejo.2011.07.004

[59] SOTNER, R., JERABEK, J., HERENCSAR, N., PETRZELA, J., VRBA, K., KINCL, Z. Linear tunable quadrature oscillator derived from LC Colpitts structure using voltage differencing transconductance amplifier and adjustable current amplifier. *Analog Integrated Circuits and Signal Processing*, 2014, vol. 81, p. 121–136. DOI: 10.1007/s10470-014-0353-6

[60] SOTNER, R., HRUBOS, Z., HERENCSAR, N., JERABEK, J., DOSTAL, T., VRBA, K. Precise electronically adjustable oscillator suitable for quadrature signal generation employing active elements with current and voltage gain control. *Circuits, Systems, and Signal Processing*, 2014, vol. 33, p. 1–35. DOI: 10.1007/s00034-013-9623-2

[61] SOTNER, R., JERABEK, J., LANGHAMMER, L., POLAK, J., HERENCSAR, N., PROKOP, R., PETRZELA, J., JAIKLA, W. Comparison of two solutions of quadrature oscillators with linear control of frequency of oscillation employing modern commercially available devices. *Circuits, Systems, and Signal Processing*, 2015, vol. 34, p. 3449–3469. DOI: 10.1007/s00034-015-0015-7

[62] BULT, K., WALLINGA, H. A class of analog CMOS circuits based on the square-law characteristic of an MOS transistor in saturation. *IEEE Journal of Solid-State Circuits*, 1987, vol. 22, p. 357–365. DOI: 10.1109/JSSC.1987.1052733

[63] KUMNGERN, M., DEJHAN, K. Versatile dual-mode class-AB four-quadrant analog multiplier. *International Journal of Electrical, Computer, Energetic, Electronic and Communication Engineering*, 2008, vol. 2, p. 215–221. Available at: www.waset.org/publications/401

About the Authors ...

Khachen KHAW-NGAM was born in Surin, Thailand. He received his B.Eng. from Suranaree University of Technology, Nakornratchasima, Thailand, in 1996, and the M.Eng. from King Mongkut's Institute of Technology Ladkrabang (KMITL), Bangkok, Thailand in 2006. He is now pursuing a doctoral degree in Electrical Engineering, KMITL. His research interests include analog signal processing circuit design.

Montree KUMNGERN received the B.S.Ind.Ed. degree from King Mongkut's University of Technology Thonburi (KMUTT), Bangkok, Thailand, in 1998, the M.Eng. and D.Eng. degrees from King Mongkut's Institute of Technology Ladkrabang (KMITL), Bangkok, Thailand, in 2002 and 2006, respectively, all in major of electrical engineering. He is currently an Assistant Professor at Faculty of Engineering, KMITL. His research interests include analog electronics, analog and digital VLSI circuits and nonlinear electronic circuits. He is author or co-author of more than 150 publications in journals and proceedings of international conferences.

Fabian KHATEB was born in 1976. He received the Ing. and Ph.D. degrees in Electrical Engineering and Communication and also in Business and Management from Brno University of Technology (BUT), Czech Republic in 2002, 2005, 2003 and 2007, respectively. He is currently working as an Associate Professor at the Dept. of Microelectronics BUT and also at the Czech Technical University in Prague, Faculty of Biomedical Engineering, Joint Centre for Biomedical Engineering of the Czech Technical University and Charles University in Prague. He has expertise in new principles of designing low-voltage low-power analog circuits, particularly biomedical applications. He has acted as a reviewer for numerous scientific international journals. He is an author or co-author of more than 100 publications in journals and proceedings of international conferences. He is an Associate Editor for Circuits, Systems and Signal Processing and International Journal of Electronics. He is a member of the Editorial Board of IET Circuits, Devices & Systems and Microelectronics Journal. He holds four Czech national patents.

13

Digital Color Images Ownership Authentication via Efficient and Robust Watermarking in a Hybrid Domain

Manuel CEDILLO-HERNANDEZ [1], Antonio CEDILLO-HERNANDEZ [1], Francisco GARCIA-UGALDE [2], Mariko NAKANO-MIYATAKE [1], Hector PEREZ-MEANA [1]

[1] Instituto Politecnico Nacional SEPI ESIME Culhuacan, Avenida Santa Ana 1000, San Francisco Culhuacan Coyoacan, Ciudad de Mexico, Mexico
[2] Universidad Nacional Autonoma de Mexico, Facultad de Ingenieria, Avenida Universidad 3000 Ciudad Universitaria Coyoacan, Ciudad de Mexico, Mexico

{ mcedilloh, mnakano, hmperezm }@ipn.mx, antoniochz@hotmail.com, fgarciau@unam.mx

Abstract. *We propose an efficient, imperceptible and highly robust digital watermarking scheme applied to color images for ownership authentication purposes. A hybrid domain for embedding the same watermark is used in this algorithm, which is composed by a couple of watermarking techniques based on spread spectrum and frequency domain. The visual quality is measured by three metrics called Peak Signal to Noise Ratio (PSNR), Structural Similarity Index (SSIM) and Visual Information Fidelity (VIF). The difference color between the original and watermarked image is computed using the Normalized Color Difference (NCD) measure. Experimentation shows that the proposed method provides high robustness against several geometric distortions including large image cropping, removal attacks, image replacement and affine transformation; signal processing operations including several image filtering, JPEG lossy compression, visual watermark added and noisy image, as well as combined distortions between all of them. Also, we present a comparison with some previously published methods which reported outstanding results and have a similar purpose as our proposal, i.e. they are focused in robust watermarking.*

Keywords

Robust digital watermarking; ownership authentication, spread spectrum, discrete Fourier transform, discrete Contourlet transform

1. Introduction

During the recent years, digital multimedia technologies associated mainly with image, video and audio, are widely consumed by the end users within personal computers and mobile devices through networks, which is a common practice that growing dramatically. This practice allows that digital multimedia data may be easily edited and/or re-distributed without any control type. This behavior requires the necessity of developing efficient tools to solve the problems associated with the infringing of the intellectual property of the multimedia's owner. In the context of digital images, watermarking is considered as a suitable solution for ownership authentication purposes. In this, commonly a small signal called "watermark" is embedded using the information from the spatial or frequency domain of the image, without affecting their visual quality and at the same time it can be detected using a detection algorithm [1], [2]. According to the different applications and requirements, digital image watermarking is classified into two types: visible and invisible. In the invisible context, watermarking is classified into two types: fragile and robust as well. Fragile watermarking modality is used for content protection, authentication, and detection tamper applications while the robust watermarking is used for copyright protection and ownership authentication. Thus, in robust watermarking, according to the detection procedure, the methods are classified into two types: blind and non-blind. In blind watermarking, the original image is not needed to detect the presence of the watermark signal while into the non-blind watermarking the original image is required. In robust watermarking with blind detection, the synchronization loss between embedding-detection stages commonly causes watermark detection errors. Geometric operations such as cropping, removal, rotation, scaling or affine transformation are the principal reasons of this desynchronization. In the literature, several works are related to robust image watermarking with geometric invariance feature [3–7]. These plans show robustness against rotation and scaling geometric distortions as well as against signal processing operations such as filtering, JPEG compression and among others; because these methods embed the watermark into invariant geometric domains, however, may be typically weak to cropping and removal attacks, affine transformations, and other aggressive distortions. Additionally, while several watermarking algorithms have been proposed to watermark gray-scale images [3–7], until nowadays only a few have been designed specifically for color images [8]. The use of color information has become an essential property to steganography and watermarking of image and video [8], [9]. In this respect, several robust color image watermarking methods have been proposed in the literature, and some of them are based on the frequency domain transform [10], [11], [12], spatial domain [13], [14], histogram modification [15], [16], [17] and Singular Value Decomposition (SVD) [18], [19]. In a particular way, the discrete Contourlet Transform (CT) has been used in the literature as a frequency alternative domain to develop robust color watermarking methods [20], [21]. In general terms, CT has been developed as an accurate bi-dimensional representation that can efficiently represent images containing contours and textures, the CT can capture the directional edges superior to wavelets [22].

In this respect, authors in [20] proposed a robust color watermarking method based on Support Vector Regression (SVR) and Non-Subsampled Contourlet Transform (NSCT), together with an image normalization procedure, to obtain geometric invariance against general affine transformation. Here, the color image is decomposed into three RGB color model components and a region of interest is obtained from the normalized components using the invariant centroid theory. Then, the NSCT is performed on the G channel of the important region. Finally, the watermark is embedded into the color original image by modifying the low-frequency NSCT coefficients, in which a Human Visual System (HVS)-based masking is used to control the watermark embedding strength. According to the high correlation among different channels of the color image, the digital watermark can be recovered using the SVR technique. This algorithm presents robustness against several geometric and signal processing distortions, including cropping attacks. However, the method presents an important drawback: high computation time is needed for SVR training, performing NSCT as well as image normalization process.

Meanwhile, authors in [21] present a blind and highly robust color watermarking scheme method by combining of information from spatial and frequency domain. The watermark signal is generated for each channel RGB of the color image by extracting spatial domain features using gray level co-occurrence matrix as well as a unique identification number. The watermark is embedded in Principal Component Analysis (PCA) less correlated between the low and high frequency of the CT sub-bands to preserve the perceptual quality of the image. This algorithm presents high imperceptibility and at same time robustness against several geometric and signal processing distortions, including cropping attacks and combined distortions; however, the algorithm is not robust against affine general transformation.

To boost the robustness without diminishing the imperceptibility, a very auspicious research direction consists in developing hybrid watermarking algorithms. These algorithms may combine, e.g., the frequency and color image information in conjunction with a geometric correction procedure [20], or the frequency and color image informa-

tion in conjunction with a frequency analysis procedure [21]. In this context, our paper proposes a highly robust digital watermarking applied to color images for ownership authentication purposes. A hybrid domain for embedding the same watermark is used in this algorithm, which is composed by a pair of watermarking algorithms. In the first one, the luminance channel is used to embed the watermark into the spectrum of the middle frequencies of the Discrete Fourier Transform (DFT) via Direct Sequence Code Division Multiple Access (DS-CDMA). In the second one, the chrominance blue-difference channel is used to embed the watermark into the Contourlet Transform (CT) domain coefficients using an Improved Spread Spectrum (ISS) method. The quality of the watermarked image is measured using the following three well-known indices Peak Signal to Noise Ratio (PSNR), Structural Similarity Index (SSIM) and Visual Information Fidelity (VIF). The difference color between the original and watermarked image is computed using the Normalized Color Difference (NCD) measure. Experimentation shows that the proposed method provides high robustness against several geometric distortions including image cropping, removal attacks, image replacement and affine transformation; signal processing operations including several image filtering, JPEG lossy compression, visual watermark added and noisy image, as well as combined distortions. Also, we present a comparison with some previously published methods which reported outstanding results and have a similar purpose as our proposal, i.e. they are focused in robust watermarking.

The rest of the paper is organized as follows: Section 2 describes the embedding and detection process of the proposed algorithm, and experimental results including comparison with previously reported watermarking algorithms are presented in Sec. 3. Finally, Sec. 4 concludes this work.

2. Proposed Method

The proposed watermarking method consists of the embedding and detection process, which are explained in detail as follows.

2.1 Discrete Fourier Transform Embedding Process

Embedding process is carried out through two stages: the first one operates on DFT domain and the second one on CT domain, respectively. Moreover, the embedding algorithm is designed to avert one embedding process interfering in the other. Watermark embedding in the DFT domain has robust properties respect to rotation, scaling and translation (RST) distortions as well as robustness against common signal processing such as compression, filtering, and noise contamination, among others. The DFT domain embedding algorithm is described as follows:

1) Rescale the color image I into a size of $N_1 \times N_2$, these dimensions will be stored and considered as a secret key K_1 in the detection stage.

2) Since the RGB has the most correlated components while the YCbCr are the less correlated as well as the forward and backward transformations between RGB and YCbCr color models are linear [8], [9], using the information of the image I converts the RGB to YCbCr color model representation and isolates the luminance component $Y(x,y)$ from YCbCr.

3) The watermark is a zero mean 1-D binary pseudo-random pattern formed by {1, 0} values achieved by a secret key K_2, $W = \{w_i| i=1, ...,L\}$, where L is the length of the watermark.

4) Apply the 2D DFT transform to the original luminance component $Y(x,y)$. The 2D DFT transform of $Y(x,y)$ of size $N_1 \times N_2$ is given by (1):

$$F(u,v) =$$
$$\sum_{x=1}^{N_1}\sum_{y=1}^{N_2} Y(x,y)\exp(-j2\pi(ux/N_1 + vy/N_2)). \tag{1}$$

5) Get the magnitude $M(u,v) = |F(u,v)|$ and phase $P(u,v)$ of the $F(u,v)$. By DFT properties [1], the translation in the spatial domain does not affect the magnitude of the DFT transform, as shown in (2):

$$|DFT[Y(x+t_x, y+t_y)]| = M(u,v) \tag{2}$$

where t_x and t_y are the translation parameters in x and y directions, respectively. Meanwhile, the scaling in the spatial domain causes an inverse scaling in the DFT domain, as shown in (3):

$$DFT[Y(s_f x, s_f y)] = \frac{1}{s_f} F(\frac{u}{s_f}, \frac{v}{s_f}) \tag{3}$$

where s_f is the scaling factor. Concerning the rotation in the spatial domain causes the same rotation in the DFT domain, as shown in (4):

$$DFT[Y(x\cos\theta - y\sin\theta, x\sin\theta + y\cos\theta)] = \tag{4}$$
$$F(u\cos\theta - v\sin\theta, u\sin\theta + v\cos\theta)$$

where θ is the rotation angle. Then motivation to selecting the DFT domain to embed the watermark W is due to a certain number of advantages for rotation, scaling and translation (RST) invariance as well as robustness against common signal processing. However, the DFT domain presents weak robustness against other aggressive distortions mainly cropping and image corruption by Gaussian noise. Thus, to increase the robustness without decreasing the watermark imperceptibility, in our method, the technique based on CT domain is designed to complement and improve the robustness against the above weakness and is explained later.

6) Select a pair of radiuses r_1 and r_2 in $M(u,v)$ and the annular area $A = \pi(r_2^2 - r_1^2)$ between r_1 and r_2 should cover the middle frequencies coefficients in $M(u,v)$ around the zero frequency term. Because modifications in the lower frequencies of $M(u,v)$ will cause visible distortion in the spatial domain of the image. On the other hand, the coefficients of the higher frequencies are vulnerable to the JPEG compression. Thus, the watermark W should be embedded in the band of the middle frequencies because, in this spectral region, it will be robust against JPEG compression and at the same time imperceptible. The pair of radiuses r_1 and r_2 will be stored and considered as a secret key K_4 in the detection stage.

7) Scramble the watermark data bits to guarantee their security using a secret key K_3.

8) For each watermark data bit w_i a pseudorandom $\{-1,1\}$ g_i pattern with length $A/2$ is assigned according to a predefined secret key K_5. Each g_i value is dependent on w_i in the following way:

$$\begin{cases} +g_i & \text{if} \quad w_i = 0, \\ -g_i & \text{if} \quad w_i = 1. \end{cases} \tag{5}$$

After that, the sum of all random patterns g_i defines the encoded watermark W_e as follows:

$$W_e = \sum_{i=1}^{L} \pm g_i \tag{6}$$

where the sign of each g_i is dependent of w_i value as defined in (5).

9) Considering a linear version of the DS-CDMA, embed the encoded watermark W_e into the magnitude coefficients of the annular area $A/2$ corresponding to the upper half of the original magnitude M that cover the middle frequency, in an additive form:

$$M' = M + \alpha W_e \tag{7}$$

where α is the watermark strength and M, M', are the original and the watermarked magnitude coefficients into the middle-frequency band, respectively. A larger value of α would boost the robustness of the watermark, on the other hand, the watermark imperceptibility is less altered by a small value of α. Hence, there is a tradeoff between robustness and imperceptibility. According to DFT symmetrical properties to produce real values after the DFT magnitude M modification, the watermark was embedded into the upper half part of middle frequencies of the DFT magnitude coefficients, and subsequently, the lower half part of the middle-frequency band should be modified symmetrically.

10) Finally, the watermarked luminance component $Y_w(x,y)$ is obtained applying the inverse DFT (IDFT) to the watermarked magnitude $M'(u,v)$ and the corresponding initial phase $P(u,v)$ as shown follows:

$$Y_w(x,y) = IDFT(F'(u,v)), F' = \tag{8}$$
$$(M'(u,v), P(u,v)).$$

2.2 Discrete Contourlet Transform Embedding Process

Once the watermarked luminance channel Y_w is acquired, the watermark embedding procedure starts the second method into CT domain and thus getting the watermarked color image, which is explained as follows. Watermark embedding into the chrominance information using the CT domain has robust properties respect to high image cropping, image replacement, rotation with cropping, as well as robustness against common signal processing such as filtering and Gaussian noise contamination, among others. The CT domain embedding process is described as follows:

1) Isolate the blue difference chrominance component $Cb(x,y)$ from YCbCr color model representation. According to the human color vision, color information is detected at normal (daylight) levels of illumination by the three types of photoreceptors denoted as cones, named L, M, S, corresponding to the light sensitive pigments at long, medium, and short wavelengths, respectively [9]. In a global manner and considering that the amount of S-cones is scarce compared with the number of L-M-cones into the human eye, the human color vision is less sensitive to the blue color than it is to the red and green colors.

2) Apply the 2D CT transform to the original blue-difference chrominance component $Cb(x,y)$ with three levels of decomposition.

3) For each watermark data bit w_i a pseudorandom $\{-1,1\}$ pattern h_i is assigned according to a predefined secret key K_6.

4) Using a linear version of the improved spread spectrum watermarking technique [23], [24] embeds the watermark data bits w_i as follows:

$$c_s' = c_s + (\gamma w_i - \lambda z) h_i \qquad (9)$$

where c_s and c_s' are the original and watermarked eight CT directional sub-bands of the third decomposition level respectively. Meanwhile, w_i is the i-th watermark data bit, γ is the watermark strength, λ is a distortion control parameter, h_i the i-th pseudorandom sequence and $z \equiv \langle c_s | h_i \rangle / \langle h_i | h_i \rangle$, the operator $\langle A | B \rangle$ denotes inner product and is defined as:

$$\langle A | B \rangle \doteq \frac{1}{N} \sum_{j=1}^{N} A_j B_j \qquad (10)$$

where N is the length of some given vectors A and B. From (9), in the conventional spread spectrum watermarking scheme $\lambda = 0$. To simplify the analysis to determinate an optimal value to the distortion control parameter λ, considering only a single watermark data bit w with a given pseudorandom sequence h, as well as the information channel is modeling as additive noise, we get:

$$s = c_s' + n \;, \qquad (11)$$

with the channel noise modeled as in (11), the receiver sufficient statistics is:

$$r = \frac{\langle s | h \rangle}{\langle h | h \rangle} = \frac{\langle c_s + (\gamma w - \lambda z) h + n | h \rangle}{\langle h | h \rangle}$$
$$= \gamma w + (1 - \lambda) z + n \qquad (12)$$

where $n \equiv \langle n | h \rangle / \langle h | h \rangle$. Therefore, from (12) we can see that the closer we make λ to 1, the more the influence of z is removed from r. The optimum value of λ can be computed as in [23] and is given by:

$$\lambda_{optimum} = 0.5(Q1 - Q2),$$
$$Q1 = \left(1 + \frac{\sigma_n^2}{\sigma_{c_s}^2} + \frac{N\sigma_h^2}{\sigma_{c_s}^2}\right),$$
$$Q2 = \left(\sqrt{\left(1 + \frac{\sigma_n^2}{\sigma_{c_s}^2} + \frac{N\sigma_h^2}{\sigma_{c_s}^2}\right)^2 - 4\frac{N\sigma_h^2}{\sigma_{c_s}^2}}\right) \qquad (13)$$

where N is the length of n, h and c_s. Variables $\sigma_{c_s}^2$, σ_n^2 and σ_h^2 denote the variances of c_s, n, and h respectively. From (13) we can see that to N large enough, the value of $\lambda_{optimum} \rightarrow 1$ and the signal to noise ratio SNR$\rightarrow\infty$. As we can compute the optimum value to λ from (13), we can vary γ to find the best performance of the trade-off imperceptibility- robustness.

5) Then, the watermarked component $Cb_w(x,y)$ is obtained by CT image reconstruction. Thus, the watermarked image I_w is assembled using the watermarked luminance component $Y_w(x,y)$, the watermarked blue difference chrominance component $Cb_w(x,y)$ and the original red difference chrominance component $C_r(x,y)$; restoring the Y_wCb_wCr watermarked components to RGB color model representation. Rescale the watermarked image I_w to the dimensions of the original image I. The diagram of the embedding process is shown in Fig. 1. The secret keys K_1, K_2, K_3, K_4, K_5 and K_6 shown in Fig. 1 are also known by the watermark detector.

2.3 Detection Process

The detection process diagram is shown in Fig. 2, and it is described as follows:

1) Rescale the color watermarked image I_w into a size of $N_1 \times N_2$ using the secret key K_1.

2) Using the information of the image I_w, converts the RGB to YCbCr color model representation and obtain the watermarked components Y_w and Cb_w respectively. If I_w was distorted by a general affine transformation, then, from luminance information Y_w and supported by our resynchronization method previously reported in the literature, we can restore geometrically the attacked image detecting the watermark

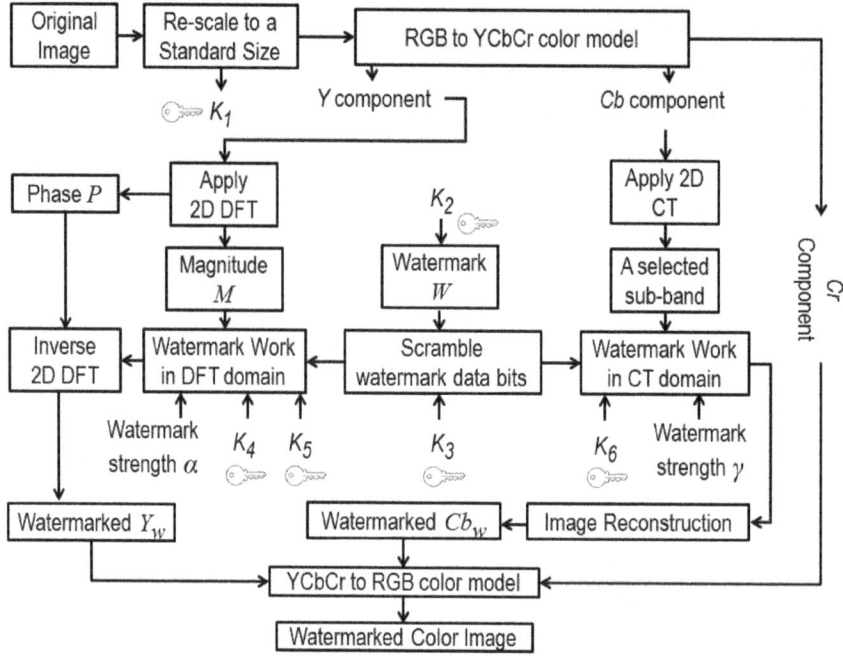

Fig. 1. Flowchart of watermark embedding procedure.

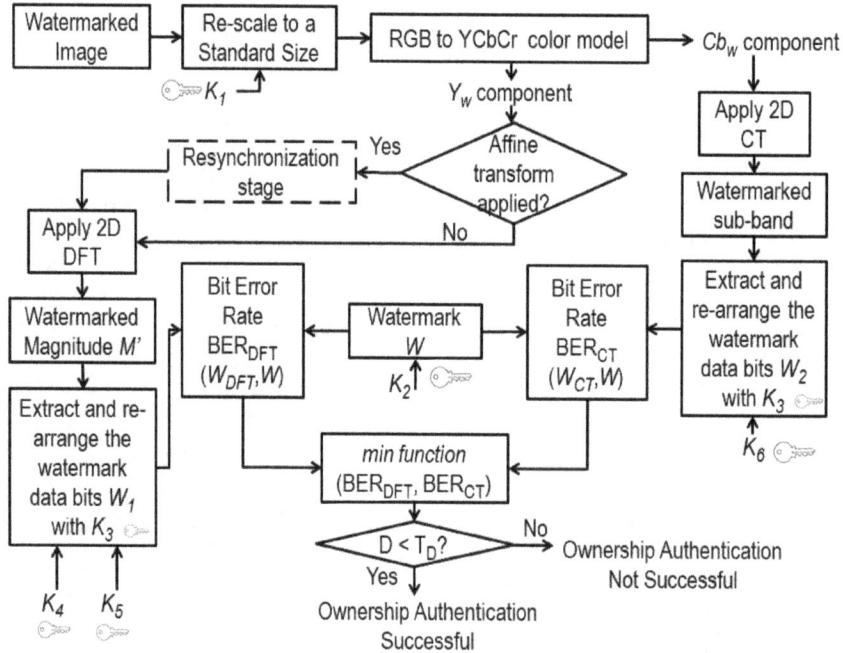

Fig. 2. Flowchart of watermark detection procedure.

correctly. To more details of the resynchronization technique, interested readers can refer to [29].

3) Compute the bi-dimensional DFT transform $F'(u,v)$ of the watermarked luminance component $Y_w(x,y)$. Then from $F'(u,v)$ get the watermarked magnitude $M'(u,v) = |F'(u,v)|$.

4) The annular area A is computed using the secret key K_4 that contains the values of radiuses r_1 and r_2 used in the embedding process.

5) Split the DFT watermarked magnitude $M'(u,v)$ in two parts, the upper half, and the lower half respectively.

6) By symmetrical DFT properties, using only information from the upper half part of watermarked magnitude M', the embedded watermark can be extracted one bit at a time by calculating the correlation between the normalized watermarked magnitude coefficients M'_{norm} and the i-th pseudorandom pattern g_i. Thus, using the secret key K_5, compute the linear correlation C_i^{DFT} between the normalized watermarked magnitude coefficients M'_{norm} and the i-th pseudorandom pattern g_i as follows:

$$C_i^{DFT} = \sum_{i=1}^{L} ((g_i - \hat{g}_i) \cdot M'_{norm}) \qquad (14)$$

where $\hat{g}_i$ is the average of all values in g_i and $M'_{\text{norm}} = M' - M'_{\text{av}}$, where M'_{av} is the average of all values in M'.

7) Decode the watermark pattern $W_{\text{DFT}}\{w'_i \mid i=1, ..., L\}$ using the sign function as follows: if $\text{sign}(C_i^{\text{DFT}})$ is '+' then $w'_i= 0$, otherwise $w'_i= 1$. Re-arrange W_{DFT} using the secret key K_3.

8) Using the watermarked blue-difference chrominance component $Cb_w(x,y)$, apply the 2D CT transform with three levels of decomposition.

9) Using only information from the eight sub-bands that compose the third decomposition level, the embedded watermark can be extracted one bit at a time by calculating the linear correlation C_i^{CT} between the watermarked directional sub-band coefficients c_s' of the third CT decomposition level and the pseudo-random sequences h_i as follows:

$$C_i^{\text{CT}} = \sum_{i=1}^{L}(c_s' \cdot h_i) \cdot \qquad (15)$$

10) Decode the watermark pattern $W_{\text{CT}}\{w'_i \mid i=1, ..., L\}$ using the sign function as follows: if $\text{sign}(C_i^{\text{CT}})$ is '+' then $w'_i= 0$, otherwise $w'_i= 1$. Re-arrange W_{CT} using the secret key K_3.

11) Reorganize the original watermark pattern W with the secret key K_2 and compute the bit error rate (BER) between (W, W_{DFT}) and (W, W_{CT}) denoted by BER_{DFT} and BER_{CT} respectively.

12) Compare and select the minimum value between BER_{DFT} and BER_{CT} using a min function. The result is indicated as a decision value D.

13) Adopting ergodicity, the BER is defined as the ratio between the number of incorrectly decoded bits and the total number of embedded bits. A decision threshold value T_{D} must be set to determine if the watermark W is present or not into the color image. In this concern, considering a binomial distribution with success probability equal to 0.5, the false alarm probability P_{fa} for L bits embedded watermark data is given by (16), and a threshold value T must be set to ensure that P_{fa} is smaller than a predetermined value.

$$P_{\text{fa}} = \sum_{q=T}^{L}(0.5)^L \cdot \left(\frac{L!}{q!(L-q)!}\right) \qquad (16)$$

where L is the total number of watermark data bits, whose value in our experiments is empirically set to 32. The false alarm probability must be less than $P_{\text{fa}} = 5.6537 \times 10^{-5}$, which is to be able to satisfy the requirements of most watermarking applications for a reliable detection. Then an adequate decision threshold value $T_{\text{D}} (= 1 - (T/L) = 1 - (27/32))$ is equal to 0.1563, according to the fact that the bit error rate (BER) + the bit correct rate (BCR) must be equal to 1. If $D > T_{\text{D}}$ (more than five error bits) the watermark detection is failed, else if $D < T_{\text{D}}$ the watermark detection is successful and the detection process is terminated.

3. Results and Discussion

In this section, the performance of the proposed algorithm is evaluated considering watermark imperceptibility and robustness properties and using a variety of digital color images. We have used 1000 images with different content among which are Goldhill, Barbara, Lena, Airplane, Baboon, Peppers, among others, all of sizing 512×512 and color resolution of 24bits/pixel. Our experiments were carried out on a personal computer running Microsoft Windows 7© with an Intel© Xeon processor (2.4 GHz) and 16 GB RAM while the embedding and extracting procedures were implemented on Matlab© 8.1. In our system, the average computing time for the embedding process has been 1.64 seconds while an average of 1.13 seconds was needed for the detection procedure. A 1D binary pseudorandom sequence of size $L = 32$ bits is used as the watermark pattern W, which is embedded in a redundant manner as explained, getting a watermark payload of 64. For the Contourlet transform as suggested in [22], we use the 9–7 biorthogonal filters with three levels of pyramidal decomposition for the multi-scale decomposition stage and the 'dmaxflat7' filters for the multidirectional decomposition stage. We partition the finest scale to eight directional sub-bands. The false alarm probability is $P_{\text{fa}} = 5.6537 \times 10^{-5}$ when the decision threshold $T_{\text{D}} = 0.1563$. The values $N_1 = N_2 = 768$ composes the secret key K_1 used. The secret key K_4 is formed by the pair of radiuses employed in the DFT domain embedding process and were $r_1 = 50$ and $r_2 = 150$. The watermark strengths used in the embedding are equal to $\alpha = 1.5$ and $\gamma = 0.3$. The watermarked image quality is measured using the following well-known indices Peak Signal to Noise Ratio (PSNR), Visual Information Fidelity (VIF) [25] and Structural Similarity Index (SSIM) [26]. The difference color of the watermarked image is obtained using the Normalized Color Difference (NCD) measure [27]. Finally, we present a comparison with some previously published methods which reported outstanding results and have a similar purpose as our proposal.

3.1 Setting Parameters r_1, r_2 and Directional Sub-bands c_s

Considering the DFT domain embedding process into the luminance component (Y) from YCbCr color model of the original color image, a watermark strength $\alpha = 1.5$ and $\gamma = 0.3$, a pair of experimental radiuses $r_1 = 5$, $r_2 = 105$ for low, $r_1 = 50$, $r_2 = 150$ for middle and $r_1 = 150$, $r_2 = 250$ for high DFT magnitude frequency respectively, and a value of $L = 32$, in Tab. 1 we show the average VIF after the watermark embedding in each spectral region, obtaining 0.7536 for low, 0.9283 for middle and 0.9633 for high DFT

Visual Information Fidelity		
Low Frequency $[r_1=5, r_2=105]$	Middle Frequency $[r_1=50, r_2=150]$	High Frequency $[r_1=150, r_2=250]$
VIF=0.7536	VIF=0.9283	VIF=0.9633

Tab. 1. Average VIF after the watermark embedding in each different spectral region.

magnitude frequency respectively. The range of VIF is [0, 1] and the closer value to 1 represents the better fidelity respect to the original image. Then according to the VIF results in Tab. 1, we can see that from the imperceptibility point of view, the modifications in the magnitude of lower frequencies of the DFT will produce visible distortion in the spatial domain of the image.

However, although the magnitude coefficients of the high frequency offer the high watermark imperceptibility, but on the other hand are susceptible to the JPEG compression. Considering the same parameters used in the above experiment, and applying a JPEG lossy compression to the watermarked color image with quality factor equal to 20; in Fig. 3 (a) we show the average BER after the watermark embedding in each spectral region, obtaining 0 for low, 0.0313 for middle and 0.3438 for high DFT magnitude frequency respectively. BER values of the low and middle frequencies are less than the decision threshold value $T_D=0.1563$. However, BER value of the high frequency is greater than $T_D = 0.1563$, affirming the susceptibility of the high frequency against JPEG compression. Thus, the watermark should be embedded in the range of the middle frequencies $r_1 = 50$, $r_2 = 150$ because, in this spectral region, it will be robust against JPEG compression and at the same time imperceptible. Once that the pair of radiuses $r_1 = 50$ and $r_2 = 150$ are set, we consider the CT domain embedding process, a watermark strength $\alpha = 1.5$, $\gamma = 0.3$ and a value of $L = 32$. Then, use the four, eight and sixteen directional subbands that compose the second, third and fourth CT decomposition levels respectively.

Table 2 shows the average PSNR after the watermark embedding in each decomposition level, obtaining 57.6391 dB for the second, 53.7513 dB for the third and 48.5229 dB for the four decomposition level respectively. According to the PSNR results in Tab. 2, we can see that from the imperceptibility point of view, embedding the watermark into the directional sub-bands of the fourth decomposition level will cause a decreasing of the quality image since PSNR value is less than 49 dB. However, although the embedding into the second decomposition level provides high watermark imperceptibility, it is vulnerable to the image corruption by Gaussian noise. Considering the same parameters used in the above experiment, and applying Gaussian noise contamination to the watermarked color image with mean $\mu = 0$ and variance $\sigma^2 = 0.05$; in Fig. 3(b) we show the average BER after the watermark embedding in each decomposition level, obtaining 0.1875 for the second, 0.0313 for the third and 0 for the four decomposition level, respectively. BER values of the third and fourth decomposition level are less than the decision threshold value $T_D = 0.1563$. But, BER value of the second decomposition

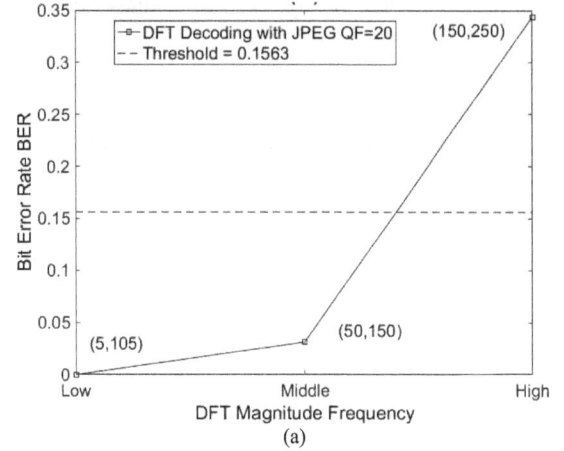

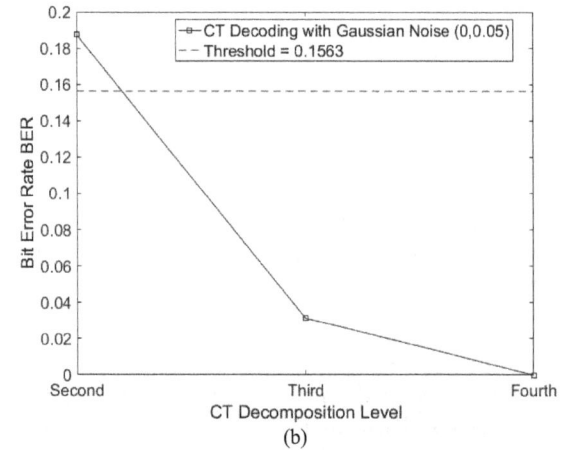

Fig. 3. (a) Average BER after DFT decoding in each spectral region: BER = 0 for low, BER = 0.0313 for middle and BER = 0.3438 for high DFT magnitude frequency respectively. (b) Average BER after CT decoding in each decomposition level: BER = 0 for the 4th, BER = 0.0313 for the 3rd and BER = 0.1875 for the 2nd, respectively.

Peak Signal to Noise Ratio PSNR		
2nd Decomposition Level [4 directional sub-bands]	3rd Decomposition Level [8 directional sub-bands]	4th Decomposition Level [16 directional sub-bands]
57.6391 dB	53.7513 dB	48.5229 dB

Tab. 2. Average PSNR after the watermark embedding in each CT decomposition level.

level is greater than $T_D = 0.1563$, confirming the vulnerability of the embedding into the second decomposition level against Gaussian noise. Thus, in our proposed method, the watermark should be embedded in the directional sub-bands of the third decomposition level because, in this spectral region, it will be robust against Gaussian noise and at the same time imperceptible.

3.2 Watermark Imperceptibility: Setting Watermark Strength α and γ

As explained in Sec. 3.1 the proposed algorithm embeds a watermark sequence twice using two different frequency domains, i.e., DFT and CT respectively. In this

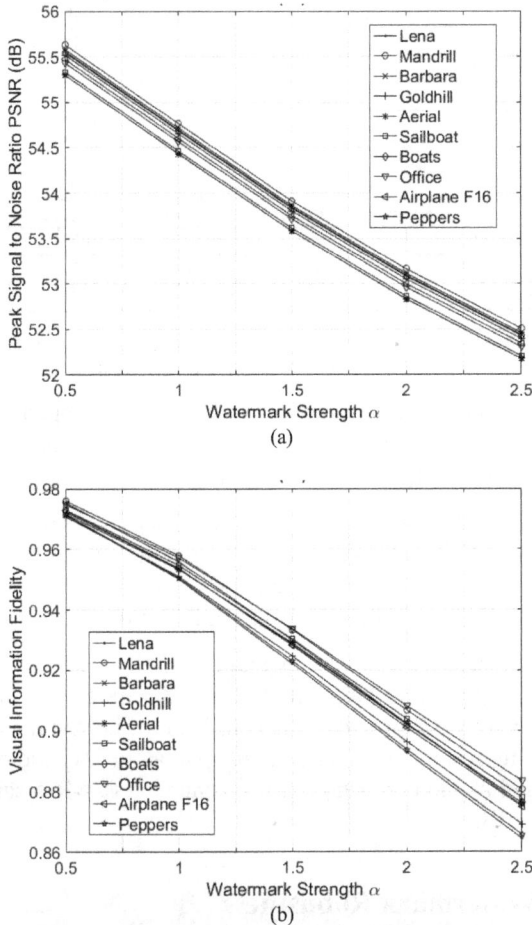

Fig. 4. Average (a) PSNR and (b) VIF with variable α.

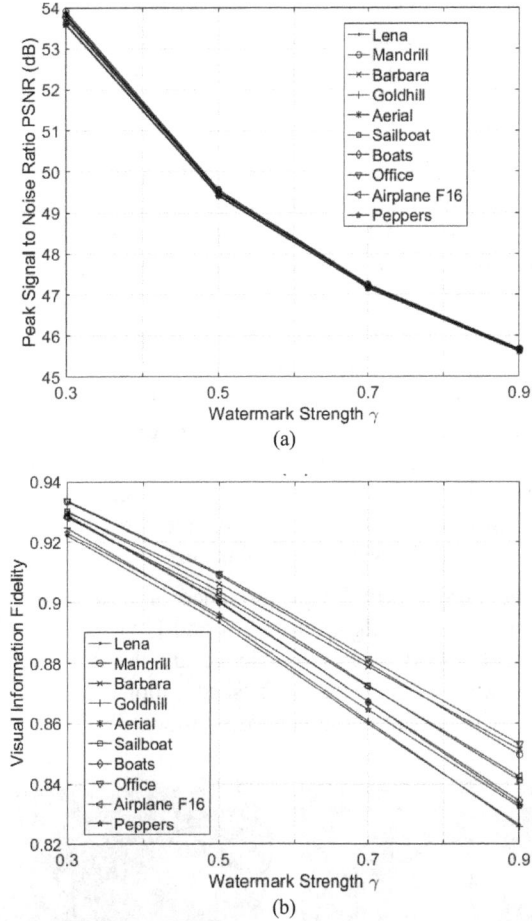

Fig. 5. Average (a) PSNR and (b) VIF with variable γ.

way, a careful watermark imperceptibility evaluation is required. To set the watermark strength α, using a pair of radiuses $r_1 = 50$ and $r_2 = 150$ in DFT domain, watermark length $L = 32$, variable α from 0.5 to 2.5, and a set of ten test color images. The watermark imperceptibility is evaluated regarding the PSNR and VIF image quality metrics. As it is known in the literature, the VIF value reflects perceptual distortions more precisely than PSNR. In Fig. 4, the average PSNR and VIF are plotted with variable watermark strength α ranging from 0.5 to 2.5 respectively.

As shown in Fig. 4(a) and (b), a larger value of α would boost the robustness of the watermark, but the watermark imperceptibility is decreased. Hence, there is a trade-off between robustness and imperceptibility. To preserve the trade-off between robustness and imperceptibility, based on the experimentation, we considered a watermark strength of $\alpha = 1.5$ as a suitable value. To set the watermark strength γ, using the eight directional sub-bands of the third CT decomposition level, watermark length $L = 32$, variable watermark strength γ from 0.3 to 0.9, and a set of ten test color images; the watermark imperceptibility was evaluated regarding the PSNR and VIF image quality metrics. In Fig. 5, the average PSNR and VIF are plotted with variable watermark strength γ ranging from 0.3 to 0.9 respectively. As shown in Fig. 5(a) and (b), a larger value of γ would boost the robustness of the watermark, but the watermark imperceptibility is declined.

Image	PSNR (dB)	VIF	SSIM	NCD
Lena	53.8638	0.9222	0.9872	0.0240
Baboon	53.9135	0.9334	0.9947	0.0248
Barbara	53.8461	0.9300	0.9882	0.0318
Goldhill	53.8047	0.9247	0.9886	0.0364
Sailboat	53.6154	0.9303	0.9899	0.0274
Boats	53.8335	0.9286	0.9865	0.0305
Office	53.7169	0.9337	0.9891	0.0270
Airplane	53.7516	0.9282	0.986	0.0202
Peppers	53.5839	0.9231	0.9875	0.0213
Aerial	53.5838	0.929	0.9931	0.0320

Tab. 3. Watermark imperceptibility measured regarding PSNR, VIF, SSIM and NCD metrics.

Hence, once again there is a trade-off between robustness and imperceptibility. To preserve the trade-off between robustness and imperceptibility, based on our experiments, we considered a watermark strength of $\gamma = 0.3$ as a suitable value.

According to the results of Figs. 4 and 5, establishing the watermark strength $\alpha = 1.5$ and $\gamma = 0.3$ we obtain a PSNR greater than 53 dB and the VIF value near to 1, it follows that the proposed technique preserves the trade-off between robustness and imperceptibility.

In order to complement the watermark imperceptibility evaluation, using $r_1 = 50$ and $r_2 = 150$, $\alpha = 1.5$, $\gamma = 0.3$, the eight directional sub-bands of the third CT decomposition level and a watermark with $L = 32$, in Tab. 3 we show the values of PSNR, VIF, SSIM and NCD of watermarked test images respect to the original ones, and in Fig. 6, some original images (a-c) together with their respective watermarked versions (d-f) are shown.

From Tab. 3 and Fig. 6, it follows that the proposed watermarking algorithm provides a sufficiently good fidelity of the watermarked color image, and also the color difference provided by NCD metric, between the watermarked image and the original one is insignificant [27], i.e., is near to 0.

From Tab. 3 we show that the average PSNR is greater than 53 dB, and the SSIM, as well as VIF values obtained, are near to 1. The range of SSIM is [0, 1], and the closer value to 1 represents the better quality respect to the original image, a value SSIM = 1 indicates that the original and the reference image are the same. In this manner, it follows that the proposed scheme provides a fairly good fidelity of the watermarked image.

The imperceptibility performance is compared with results reported by algorithms [20] and [21] respectively, which to the best of our knowledge are the most robust watermarking algorithms published applied to color images, with similar purposes as our proposed scheme. To

Image	Proposed Method	Pan-Pan et al. [20]
Lena	53.87 dB	40.57 dB
Baboon	53.91 dB	41.67 dB
Barbara	53.85 dB	40.71 dB

Tab. 4. Comparison of watermark imperceptibility in terms of PSNR between our method and Pan-Pan et al. [20].

Image	Proposed Method	Prathap et al. [21]
Lena	53.87 dB	54.68 dB
Baboon	53.91 dB	53.55 dB
Peppers	53.58 dB	58.32 dB

Tab. 5. Comparison of watermark imperceptibility in terms of PSNR between our method and Prathap et al. [21].

get a proper comparison, we consider a homogeneous format of color images of $512 \times 512 \times 24$ bits. The comparison results are shown in Tab. 4 and 5.

From Tab. 4 and 5 it follows that our proposed method provides a reasonably good fidelity of the watermarked color image, achieving a PSNR greater than 53 dB, avoiding the perceptual distortions in the color images.

Comparison results show that the PSNR results of the method reported by Pan-Pan et al. in [20] are clearly outperformed by our proposed method. Meanwhile, the imperceptibility results obtained by Prathap et al. in [21] and our proposed method are very similar, achieving PSNR greater than 53 dB.

3.3 Watermark Robustness

To evaluate the watermark robustness of the proposed algorithm, several geometrical, signal processing, and combined distortions are applied to watermarked color images. In the flowchart showed in Fig. 2 and described in detail in Sec. 2.3, the watermark detector makes a decision based on two calculated BER values that correspond in turn to each watermark embedding process introduced in this proposal.

To have a clear perception of robustness achieved by each watermark decoding against performed distortions, the output of each detector is displayed separately in a form CT/DFT linked to the Contourlet Transform/Discrete Fourier Transform decoding respectively. In this way, the strengths and weakness of each embedding method can be precisely determined. Tab. 6, 7 and 8 show the BER obtained after applying the distortions mentioned above to a set of six test watermarked images. In Tab. 6, 7 and 8 italic characters indicate failure detection against the respective distortion.

From Tab. 6 and considering the decision value D criterion described in Sec. 2.3, we can observe that the embedded watermark signal in our proposed method is sufficiently robust to most common signal processing distortions. These distortions including JPEG lossy compression with quality factor until 10, Gaussian and median filtering with different size windowing, sharpening, brightness, and image corruption by the determined amount of Gaussian and impulsive noise respectively, histogram

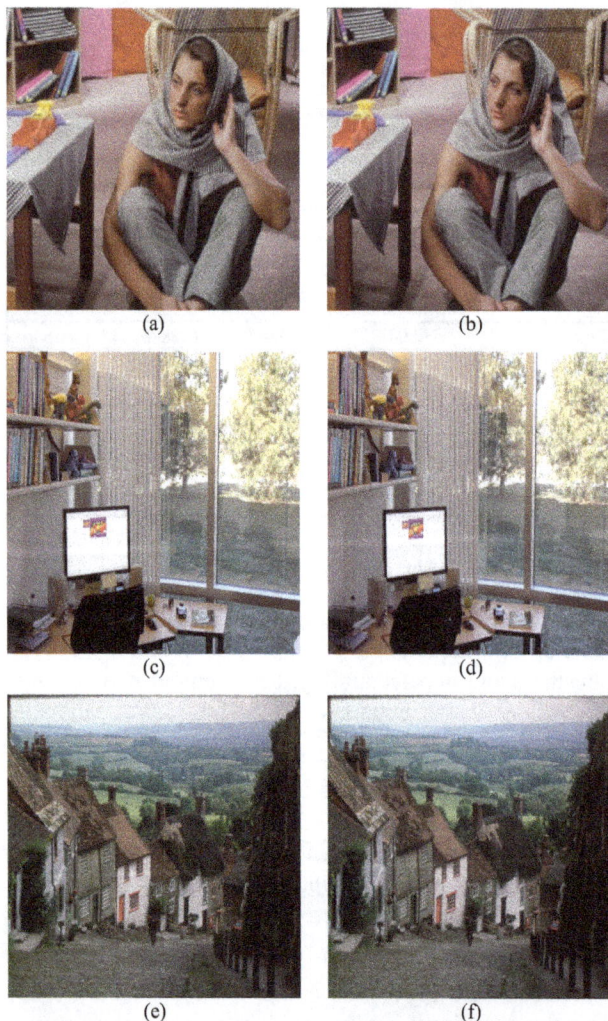

Fig. 6. Original (a), (c), (e), Watermarked versions (b), (d), (f).

equalization, motion blurring, gamma correction and visual watermark added into RGB channels. Obtaining BER values less than the decision threshold $T_D = 0.1563$, calculated as mentioned in Sec. 2.3, and used to determine if the watermark W is present or not in the watermarked color image.

From Tab. 7 we can observe that our proposed method is sufficiently robust to geometric attacks. These distortions including all rotation angles with and without cropping, image scaling with several scale factors, dynamic image cropping until 95%, centered cropping, image replacement, translation with removal columns and rows, general affine transformations including shearing in x-direction and aspect ratio changes. In all cases, using the decision value D criterion, we obtained BER values less than the decision threshold $T_D = 0.1563$.

To complement the robustness testing, we design a set of combined distortions composed by JPEG lossy compression with quality factor 50 in conjunction with several common signal processing and geometric distortions shown in Tab. 6 and 7 respectively. According to the experimental results, from Tab. 8 we demonstrate that the proposed method is robust against this kind of combined distortions, obtaining BER values less than $T_D = 0.1563$.

With illustrative purposes, in Fig. 7 we show the Airplane watermarked image after being processed by six of the most aggressive distortions. In all cases, the BER value is less than the decision threshold $T_D = 0.1563$.

The robustness performance is compared with that reported by the algorithms [20] and [21] respectively. Again, to get a proper comparison, we consider a homogeneous format of color images of $512 \times 512 \times 24$ bits. To design a compact robustness testing, the set of distortions discussed in the comparative include only the most aggressive distortions reported in the literature. Tab. 9 and 10 show the robustness relative in BER terms with that reported by the algorithms [20] and [21] respectively.

From Tab. 9 we show that the algorithm of Pan-Pan et al. [20] and our proposed watermarking method are robust against several geometric distortions including rotation, scaling, translation, cropping, affine transformation and aspect ratio changes. Both proposals are robust against signal processing including JPEG compression, median, and Gaussian filtering, sharpening, impulsive and Gaussian noise. Moreover, both methods are robust to the combined distortions composed by operations of the same type, i.e., geometric/geometric or signal processing/signal processing respectively. However, the method of Pan-Pan et al. [20] is outperformed by our proposed method because in almost all test our method get BER values close to 0. Moreover, the tolerance of Pan-Pan et al. [20] against several distortions is weak compared with the tolerance of our proposed method, which was previously shown in Tab. 6, 7 and 8. Furthermore, our proposal considers a broader range of distortions compared with the reported by Pan-Pan [20].

From Tab. 10 we show that the algorithm of Prathap et al. [21] and our proposed watermarking method are robust against several geometric distortions including a rotation with and without cropping, scaling, translation, and cropping. Meanwhile, both approaches are robust against signal processing including JPEG compression, median, and Gaussian filtering, sharpening, impulsive and Gaussian noise. Moreover, both approaches are robust to the combined distortions composed JPEG lossy compression with quality factor 50 in conjunction with signal processing or geometric distortion. However, the method of Prathap et al. [21] is outperformed by our proposed method

Distortion	Lena CT/DFT	Baboon CT/DFT	Barbara CT/DFT	Goldhill CT/DFT	Peppers CT/DFT	Airplane CT/DFT
Without attack	0/0	0/0	0/0	0/0	0/0	0/0
JPEG 90	0/0	0/0	0/0	0/0	0/0	0/0
JPEG 70	0.1875/0	0.125/0	0.2813/0	0.2188/0	0.4063/0	0.2188/0
JPEG 50	0.2813/0	0.2188/0	0.4063/0	0.2188/0	0.4375/0	0.375/0
JPEG 20	0.4063/0.0313	0.25/0	0.5938/0	0.4063/0	0.5313/0.0313	0.4063/0.0625
JPEG 10	0.5/0.0313	0.3125/0	0.6563/0.0625	0.3438/0.125	0.4063/0.0625	0.5313/0.125
Gaussian filter 5x5	0/0	0/0	0/0	0/0	0/0	0/0
Gaussian filter 7x7	0/0	0/0	0/0	0/0	0/0	0/0
Sharpen	0/0	0/0	0/0	0/0	0/0	0/0
Median filter 3x3	0/0	0/0	0/0	0/0	0/0	0/0
Median filter 5x5	0/0.1875	0.0938/0.25	0/0.0938	0/0.1563	0.0313/0.1875	0.0625/0.2813
Brightness	0/0	0/0	0/0	0/0	0/0	0/0.0313
Gaussian noise (0,0.06)	0.0313/0.1563	0.0313/0.125	0.0313/0.125	0.0313/0.0625	0.0313/0.0938	0.0313/0.0313
Gaussian noise (0,0.07)	0.0313/0.2188	0.0313/0.1563	0.0313/0.1875	0.0625/0.1563	0.0625/0.25	0.0313/0.0938
Impulsive noise density 0.08	0/0	0/0	0/0	0/0.0938	0/0.0313	0/0.0313
Impulsive noise density 0.09	0/0.0313	0/0	0/0	0/0.0625	0/0.0625	0/0
Histogram equalization	0/0	0/0	0/0	0/0	0/0	0/0
Motion blurring	0/0	0/0.0313	0/0	0/0	0/0	0/0
Gamma correction	0/0	0/0	0/0	0/0	0/0	0/0
Visual watermark added	0/0	0/0	0/0	0/0	0/0	0/0

Tab. 6. BER of CT/DFT decoding respectively obtained from six test watermarked images after signal processing distortions. Decision threshold value $T_D = 0.1563$.

Distortion	Lena CT/DFT	Baboon CT/DFT	Barbara CT/DFT	Goldhill CT/DFT	Peppers CT/DFT	Airplane CT/DFT
Rotation 35° with crop	0/0	0/0	0/0	0/0	0/0	0/0
Rotation 75° with crop	0/0	0/0	0/0	0/0	0/0	0/0
Rotation 195° with crop	0/0	0/0	0/0	0/0	0/0	0/0
Scaling 0.3	0/0.25	0.0313/0.25	0/0.1875	0/0.1875	0.0313/0.125	0/0.3125
Scaling 0.5	0/0	0/0	0/0	0/0	0/0	0/0
Scaling 0.7	0/0	0/0	0/0	0/0	0/0	0/0
Scaling 1.5	0/0	0/0	0/0	0/0	0/0	0/0
Scaling 2.0	0/0	0/0	0/0	0/0	0/0	0/0
Cropping 65%	0/0	0/0.1563	0/0	0/0.0313	0/0.0625	0/0.0938
Cropping 95%	0/0.4563	0/0.4875	0/0.4938	0/0.5000	0/0.5000	0/0.5000
Centered cropping 100x100	0/0	0/0	0/0	0/0	0/0	0/0
Image replacement	0/0.0625	0/0.0313	0/0.0625	0/0.125	0/0.125	0/0.125
Rotation 45° without crop	0/0	0/0	0/0	0/0	0/0	0/0
Rotation 105° without crop	0/0	0/0	0/0	0/0	0/0	0/0
Rotation 285° without crop	0/0	0/0	0/0	0/0	0/0	0/0
Translation $x=30, y=30$	0.6563/0	0.6563/0	0.5313/0	0.625/0	0.4063/0	0.4063/0
Translation $x=70, y=70$	0.4375/0.0313	0.5625/0.0313	0.4375/0	0.5/0	0.5/0	0.5625/0.0313
Aspect ratio (1.2:1)	0/0	0/0	0/0	0/0	0/0	0/0
Aspect ratio (0.7:1.2)	0/0	0/0	0/0	0/0	0/0	0/0
Shearing 0.2x	0.2813/0	0.2813/0	0.2500/0	0.4688/0	0.5313/0	0.3125/0
Affine [0.9,0.2,0;0.1,1.2,0;0,0,1]	0.375/0	0.3125/0	0.2813/0	0.4375/0	0.4688/0	0.3438/0
Affine [1.01,0.1,0;0.1,0.9,0;0,0,1]	0.4375/0	0.4688/0	0.25/0	0.25/0	0.3438/0	0.4375/0

Tab. 7. BER of CT/DFT decoding respectively obtained from six test watermarked images after geometric distortions. Decision threshold value $T_D = 0.1563$.

Combined distortions composed by JPEG compression 50 + *distortion*	Lena CT/DFT	Baboon CT/DFT	Barbara CT/DFT	Goldhill CT/DFT	Peppers CT/DFT	Airplane CT/DFT
Gaussian filter 7x7	0.25/0	0.25/0	0.375/0	0.3125/0	0.4063/0	0.3438/0
Sharpen	0.3125/0	0.3438/0	0.375/0	0.25/0	0.3438/0	0.3438/0
Brightness	0.375/0	0.2188/0	0.25/0	0.25/0	0.5/0.0313	0.3438/0
Gaussian noise (0,0.02)	0.5/0	0.4688/0.0313	0.5/0.0313	0.4688/0	0.5/0.0938	0.5/0.0313
Impulsive noise density 0.05	0.4688/0.0313	0.4375/0	0.5/0.0313	0.5/0.0938	0.5625/0.0313	0.5/0.0313
Median filter 3x3	0.3438/0	0.25/0	0.3125/0	0.2813/0	0.375/0	0.375/0
Histogram equalization	0.375/0	0.2813/0	0.4375/0	0.2188/0	0.375/0	0.4063/0
Gamma correction	0.2813/0	0.1875/0	0.25/0	0.2188/0	0.4375/0	0.3125/0
Visual watermark added	0.3125/0.0313	0.2188/0	0.3438/0	0.2813/0	0.3125/0	0.4375/0.0313
Rotation 35° with crop	0.4375/0	0.4688/0.0313	0.4375/0	0.4375/0	0.4375/0	0.4688/0.0313
Rotation 145° with crop	0.375/0.0313	0.4375/0	0.4063/0	0.4063/0	0.4063/0	0.4063/0.0625
Scaling 0.5	0.2813/0	0.2188/0	0.4063/0	0.25/0	0.375/0	0.375/0
Scaling 2.0	0.3438/0	0.2188/0	0.375/0	0.2813/0	0.3438/0	0.3438/0
Cropping 40%	0.3125/0	0.2188/0	0.3125/0	0.1875/0	0.3438/0.0313	0.25/0
Centered cropping 100x100	0.375/0.0313	0.2813/0	0.4063/0	0.3125/0	0.4688/0.0313	0.3438/0
Rotation 15° without crop	0.3438/0	0.2813/0	0.4688/0	0.2188/0	0.4063/0	0.375/0
Rotation 125° without crop	0.375/0	0.2813/0	0.4688/0	0.1875/0	0.4063/0	0.4375/0
Translation $x=30, y=30$	0.4375/0	0.4688/0.0313	0.5/0	0.625/0	0.375/0.0313	0.5/0.0313
Aspect ratio (1.2:1)	0.3438/0	0.25/0	0.5313/0	0.2188/0	0.375/0	0.375/0
Aspect ratio (0.7:1.2)	0.375/0	0.3125/0	0.4688/0	0.2813/0	0.3438/0	0.375/0
Shearing 0.2x	0.6875/x	0.4688/0	0.625/0.0625	0.4688/0	0.375/0	0.4375/0.0313
Affine [0.9,0.2,0;0.1,1.2,0;0,0,1]	0.5625/0	0.375/0	0.4688/0.0313	0.5/0	0.5313/0	0.375/0

Tab. 8. BER of CT/DFT decoding respectively obtained from six test watermarked images after combined distortions. Decision threshold value $T_D = 0.1563$.

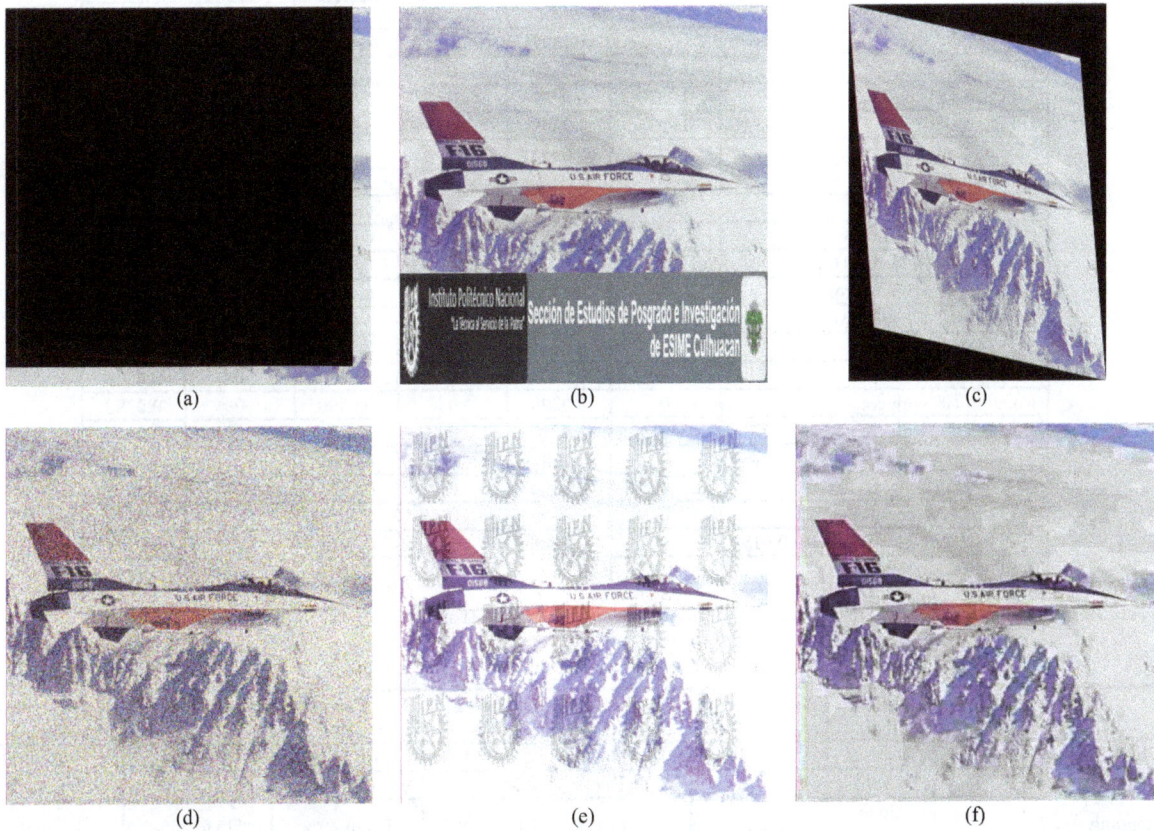

Fig. 7. Aggressive geometric and signal processing distortions in Airplane watermarked image. (a) Cropping with 95%, BER = 0. (b) Image replacement, BER = 0. (c) Affine transformation, BER = 0. (d) Gaussian noise (0,0.07), BER = 0.0313. (e) Visual watermark added, BER = 0. (f) JPEG with QF = 10, BER = 0.125.

Distortion	Lena		Baboon		Barbara	
	Proposed	Ref.[20]	Proposed	Ref.[20]	Proposed	Ref.[20]
JPEG 50	0	0.0334	0	0.0293	0	0.0244
JPEG 30	0	0.0400	0	0.0322	0	0.0283
Median filter 3x3	0	0.0303	0	0.0049	0	0.0234
Gaussian filter 3x3	0	0.0313	0	0.0107	0	0.0225
Sharpen	0	0.0225	0	0.0449	0	0.0273
Gaussian noise (0,0.006)	0	0.0273	0	0.0234	0	0.0215
Impulsive noise density 0.003	0	0.0234	0	0.0205	0	0.0164
Median filter 3x3 + Gaussian Noise (0,0.006)	0	0.0244	0	0.0137	0	0.0186
Gaussian Noise (0,0.006) + Sharpen	0	0.0449	0	0.0811	0	0.0547
JPEG 70 + Gaussian filter 3x3	0	0.0381	0	0.0234	0	0.0303
JPEG 70 + Median filter 3x3	0	0.0264	0.0313	0.0195	0	0.0196
Rotation 45° without crop	0	0.0342	0	0.0164	0	0.0244
Scaling 2	0	0.0273	0	0.0137	0	0.0303
Translation x=20,y=20	0	0.1240	0	0.0605	0	0.1201
Cropping 50%	0	0.1250	0	0.1240	0	0.1250
Aspect ratio (1.2,1.0)	0	0.0244	0	0.0166	0	0.0244
Affine transformation [10; 1.0, 1.0; 0.5, 0.2]	0	0.0225	0	0.0137	0.0313	0.0195
Scaling 2 + Translation x=5,y=0	0	0.0596	0	0.0273	0	0.0713
Rotation 5° + Scaling 2	0	0.0332	0	0.0234	0	0.0254
Rotation 5° + Translation x=5, y=15	0	0.0498	0	0.0479	0	0.1240
Rotation 45° + Scaling 2 + Translation x=20, y=20	0	0.0709	0	0.1318	0	0.1221

Tab. 9. Comparison of BER of extracted watermark for our proposed method and Pan-Pan et al. [20].

Distortion	Lena		Baboon		Peppers	
	Proposed	Ref.[21]	Proposed	Ref.[21]	Proposed	Ref.[21]
JPEG 50	0	0.0256	0	0.0359	0	0.0417
JPEG 20	0.0313	0.0369	0	0.0381	0.0313	0.0396
JPEG 10	0.0313	0.0359	0	0.0379	0.0625	0.0336
Median filter 5x5	0	0.0435	0.0938	0.0401	0.0313	0.0372
Gaussian filter 7x7	0	0	0	0	0	0
Sharpen	0	0.0241	0	0.0412	0	0.0464
Gaussian noise (0,0.05)	0	0.0485	0	0.0407	0	0.0487
Impulsive noise density 0.08	0	0.0393	0	0.0320	0.0313	0.0355
Rotation 10° without crop	0	0.0370	0	0.0610	0	0.0410
Rotation 45° without crop	0	0.0660	0	0.0510	0	0.0770
Scaling 0.3	0	0.0463	0.0313	0.0534	0.0313	0.0478
Scaling 0.5	0	0.0523	0	0.0623	0	0.0701
Rotation 10° with crop	0	0.0290	0	0.0590	0	0.0280
Rotation 60° with crop	0	0.0510	0	0.0690	0	0.0460
Translation x=40,y=40	0	0.05	0	0.0560	0	0.0380
Cropping 25%	0	0.0410	0	0.0435	0	0.0523
JPEG 50 + Median Filter 3x3	0	0.0429	0	0.0443	0	0.0471
JPEG 50 + Gaussian Noise (0,0.01)	0	0.0448	0	0.0261	0	0.0322
JPEG 50 + Scaling 0.2	0.0625	0.0625	0.0938	0.0436	0.0939	0.0666

Tab. 10. Comparison of BER of extracted watermark for our method and Prathap et al. [21].

Comparison	Najih, et al. [6]	Xiang-Yang, et al. [7]	Chrysochos et al. [16]	Shao-Li. [18]	Pan-Pan et al. [20]	Prathap et al. [21]	Proposed Method
JPEG (Quality Factor)	Detected	20 − 80	25 − 100	10 − 100	30 − 100	5 − 100	10 − 100
Scaling	0.5 − 1	0.5 − 1.5	Detected	0.5 − 2.5	0.5 − 2	0.2 − 1	0.3 − 2
Cropping	Up to 25%	Up to 25%	Up to 20%	Up to 50%	Up to 20%	Up to 25%	Up to 95%
Affine Transformation	-	-	-	-	Detected	-	Detected
Rotation	Detected	0° − 45°	0° − 360°	0° − 30°	0° − 45°	0° − 90°	0° − 360°
Visual Watermark Added	-	-	-	-	-	-	Detected
Image Replacement	-	-	-	-	-	-	Detected
Gaussian Noise	(0, 0.01)	(0, 0.01)	(0, 0.95)	(0, 0.25)	(0, 0.006)	(0, 0.01)	(0, 0.07)
Combined Distortions - Geometric (G) - Signal Processing (SP)	-	a) JPEG50 + (G) or (SP)	-	-	a) JPEG70 + (SP) b) (G) + (G) c) (SP) + (SP)	a) JPEG50 + (SP) or (G) b) (G) + (G)	a) JPEG50 + (G) or (SP) b) (G) + (G) c) (SP) + (SP)
Watermark Length (bits)	Not Provided	Not provided	30	1024	1024	200	64
Image Quality Metrics	Average PSNR 61dB	Not measured	Average: wPSNR=50dB PSNR=37dB	Average SSIM 0.9887	Average PSNR 40.98dB	Average PSNR 53.55dB	Average: PSNR=53.75dB SSIM=0.989 VIF=0.928 NCD=0.02
Image kind	Grayscale	Grayscale	Grayscale	Color	Color	Color	Color

Tab. 11. Performance comparison.

because in almost all test our method get BER values close to 0. Moreover, the method of Prathap et al. [21] is not robust to affine transformations and its tolerance against image cropping attacks is weak compared with the tolerance of our proposed method, which was previously shown in Tab. 6, 7 and 8. Furthermore, our proposal considers a broader range of distortions compared with [21].

3.4 Robustness against Geometric Distortions

According to the experimental results, our proposed watermarking method presents a high robustness against a broader range of distortions. Focusing on the geometric distortions, the robustness against rotations with and without cropping is obtained through exhaustive search from 0° to 180° rotation degrees to DFT decoding (by symmetrical properties) and 0° to 360° to CT decoding. On the other hand, the use of the secret key K_1 that re-scales the color image to a standard size allows robustness against scaling and aspect ratio changes. Moreover, the method is robust against aggressive cropping, which is considered as a correlated noise, because the DS-CDMA and ISS spread spectrum techniques preserve the second Shannon's theorem [30]. Finally, our method presents robustness against general affine transformations because when a watermarked color image is deformed with an affine operation, from luminance information and supported by our resynchronization method previously reported in the literature, we can restore geometrically the attacked image detecting the watermark correctly. To more details of the resynchronization technique, interested readers can refer to [29].

3.5 Payload

Since our design implies an ownership authentication application, to preserve the trade-off between imperceptibility and robustness we consider a watermark length $L = 32$ as optimal value to determine the presence or absence of watermark with a false alarm probability $P_{fa} = 5.6537 \times 10^{-5}$, which is to be able to satisfy the requirements of ownership authentication applications. Because our method embeds the watermark by duplicate, the total payload of our proposed method is 64 watermark data bits.

3.6 Security

In addition to robustness and imperceptibility, the security of our scheme is another important aspect to consider. Then, the security level is defined by the number of observations the opponent needs to estimate the secret keys [28], [31] accurately. It is ensured by the set of six secret keys K_1, K_2, K_3, K_4, K_5 and K_6, which additionally could be renewed periodically by the ownership to keep the security level and avoid the watermark removal.

3.7 Performance Comparison

Finally, this investigation compares the performance of the proposed method with the algorithm based on angle quantization in discrete Contourlet transform developed by Najih, et al. [6] in 2016, the algorithm based on the exponent moments invariants in non-subsampled Contourlet transform domain proposed by Xiang-Yang, et al. [7] in 2014, the hybrid watermarking based on chaos and histogram modification proposed by Chrysochos et al. [16] in 2014, the watermarking to color images based on Singular Value Decomposition (SVD) developed by Shao-Li. [18] in 2014, the color image watermarking scheme in non-sampled Contourlet-domain proposed by Pan-Pan et al. [20] in 2011, and the hybrid robust watermarking for color images proposed by Prathap et al. [21] in 2014, under JPEG lossy compression, scaling, cropping, affine transformation, rotation, visual watermark added, image replacement, Gaussian noise and combined distortions. Table 11 compares the performance of the watermark detector outputs, the watermark data length, image quality metrics and the kind of image associated with each algorithm. Table 11 presents also the tolerance under distortions, and designates the capacity to resist as 'detected', when the tolerance is not given in detail by the other six methods above mentioned. A grid-cell is marked with a dash for attack simulations not mentioned in the literature. These results show better performance of the proposed method compared with principal methods reported previously in terms of imperceptibility and robustness against most common geometric, signal processing and combined attacks.

4. Conclusions

In this paper, we have designed a high robust, blind, color image watermarking algorithm which employs DS-CDMA and ISS watermark embedding in both DFT and CT domain respectively. This method is applicable for ownership authentication of color pictures. The proposed scheme can tolerate a broader range of distortions, particularly signal processing, geometric and combined distortions. Authenticity is achieved by the thresholding criterion regarding bit error rate. Our proposed method satisfies the primary watermarking requirements such as imperceptibility, security, and robustness. Algorithm is very robust against geometric manipulations including rotation by several angles with and without cropping, affine transformation, image replacement, scaling, aspect ratio change, aggressive cropping attacks among others. Also, the method is robust against several common signal processing distortions such as JPEG lossy compression, median and Gaussian filtering, impulsive and Gaussian noise perturbation, brightness, contrast, visual watermark added, sharpening, and histogram equalization among others. The method presents good robustness against combined distortions composed by several geometric and signal processing attacks. The comparison of the proposed method with other existing schemes shows the improved performance in terms of imperceptibility and robustness, in the context of robust watermarking techniques.

Acknowledgments

Authors thank the Instituto Politecnico Nacional (IPN), the Consejo Nacional de Ciencia y Tecnologia de Mexico (CONACyT) as well as the Post-Doctorate Scholarships program and the PAPIIT IN106816 project from DGAPA in Universidad Nacional Autonoma de Mexico (UNAM) by the support provided during the realization of this research.

References

[1] BARNI, M., BARTOLINI, F. *Watermarking Systems Engineering: Enabling Digital Assets Security and Other Applications*. CRC Press, 2004. ISBN: 9780824750916

[2] LANGELAAR, G. C., SETYAWAN, I., LAGENDIJK, R. L. Watermarking digital image and video data. A state-of-the-art overview. *IEEE Signal Processing Magazine*, 2000, vol. 17, p. 20 to 46. DOI: 10.1109/79.879337

[3] WÓJTOWICZ, W., OGIELA, M. R. Digital images authentication scheme based on bimodal biometric watermarking in

an independent domain. *Journal of Visual Communication and Image Representation,* 2016, vol. 38, p. 1–10. DOI: 10.1016/j.jvcir.2016.02.006

[4] RABIZADEH, M., AMIRMAZLAGHANI, M., AHMADIAN-ATTARI, M. A new detector for contourlet domain multiplicative image watermarking using Bessel K form distribution. *Journal of Visual Communication and Image Representation,* 2016, vol. 40, Part A, p. 324–334. DOI: 10.1016/j.jvcir.2016.07.001

[5] BUM-SOO, K., CHOI, J. G., PARK, C. H., et al. Robust digital image watermarking method against geometrical attacks. *Real-Time Imaging,* 2003, vol. 9, p. 139–149. DOI: 10.1016/s1077-2014(03)00020-2

[6] NAJIH, A., AL-HADDAD, S.A.R., RAMLI, A. R., et al. Digital image watermarking based on angle quantization in discrete contourlet transform. *Journal of King Saud University - Computer and Information Sciences.* [Online] Cited April 4, 2016. DOI: 10.1016/j.jksuci.2016.02.005

[7] XIANG-YANG WANG, AI-LONG WANG, HONG-JING YANG, et al. A new robust digital watermarking based on exponent moments invariants in nonsubsampled contourlet transform domain. *Computers & Electrical Engineering,* April 2014, vol. 40, no. 3, p. 942–955. DOI: 10.1016/j.compeleceng.2013.12.017

[8] CHAREYRON, G., DA RUGNA, J., TRÉMEAU, A. Chapter 3: Color in Image Watermarking. *Advanced Techniques in Multimedia Watermarking: Image, Video and Audio Applications.* IGI Global Press, p. 36–56, 2010. ISBN: 1615209042

[9] TRÉMEAU, A., TOMINAGA, S., PLATANIOTIS, K. Color in image and video processing: Most recent trends and future research directions. *EURASIP Journal on Image and Video Processing,* 2008. DOI: 10.1155/2008/581371

[10] CHEMAK, C., LAPAYRE J. C., BOUHLEL, M. S. New watermarking scheme for security and transmission of medical images for pocket neuro project. *Radioengineering,* 2007, vol. 16, no. 4, p. 58–63. ISSN 1210-2512

[11] KUO-CHENG, L. Wavelet-based watermarking for color images through visual masking. *AEU - International Journal of Electronics and Communications,* 2010, vol. 64, no. 2, p. 112–124. DOI: 10.1016/j.aeue.2008.11.006

[12] CEDILLO-HERNANDEZ, M., CEDILLO-HERNANDEZ, A., GARCIA-UGALDE, F., NAKANO-MIYATAKE, M., PEREZ-MEANA, H. Copyright protection of color imaging using robust-encoded watermarking. *Radioengineering,* 2015, vol. 24, no. 1, p. 240–251. DOI: 10.13164/re.2015.0240

[13] BATTIATO, S., CATALANO, D., GALLO, G., GENNARO, R. Robust watermarking for images based on color manipulation. In *Proceedings of the Third International Workshop on Information Hiding.* Dresden (Germany), 1999. Springer-Verlag, 2000, vol. 1768, p. 302–317. DOI: 10.1007/10719724_21

[14] CHAREYRON, G., COLTUC, D., TREMEAU, A. Watermarking and authentication of color images based on segmentation of the XYZ color space. *Journal of Imaging Science and Technology,* 2006, vol. 50, no. 5, p. 411–423. DOI: 10.2352/J.ImagingSci.Technol.(2006)50:5(411)

[15] JIA, X., QI, Y., SHAO, L., JIA, X. An anti-geometric digital watermark algorithm based on histogram grouping and fault-tolerance channel. *Intelligent Science and Intelligent Data Engineering. Lecture Notes in Computer Science,* 2012, vol. 7202, p. 753–760. DOI: 10.1007/978-3-642-31919-8_96

[16] CHRYSOCHOS, E., FOTOPOULOS, V., XENOS, M., SKODRAS, A. N. Hybrid watermarking based on chaos and histogram modification. *Signal, Image and Video Processing,* 2014, vol. 8, no.5, p. 843–857. DOI 10.1007/s11760-012-0307-3

[17] CEDILLO-HERNANDEZ, M., GARCIA-UGALDE, F., NAKANO-MIYATAKE, M., PEREZ-MEANA H. Robust hybrid color image watermarking method based on DFT domain and 2D histogram modification. *Signal, Image and Video Processing,* 2014, vol. 8, no. 1, p. 49–63. DOI: 10.1007/s11760-013-0459-9

[18] SHAO-LI JIA. A novel blind color images watermarking based on SVD. *Optik - International Journal for Light and Electron Optics,* 2014, vol. 125, no. 12, p. 2868–2874. DOI: 10.1016/j.ijleo.2014.01.002

[19] LAUR, L., RASTI, P., AGOYI, M., ANBARJAFARI, G. A robust color image watermarking scheme using entropy and QR decomposition. *Radioengineering,* 2015, vol. 24, no. 4, p. 1025 to 1032. DOI: 10.13164/re.2015.1025

[20] PAN-PAN NIU, XIANG-YANG WANG, YI-PING YANG, MING-YU LU. A novel color image watermarking scheme in nonsampled contourlet-domain. *Expert Systems with Applications,* 2011, vol. 38, no. 3, p. 2081–2098. DOI: 10.1016/j.eswa.2010.07.147

[21] PRATHAP, I., NATARAJAN, V., ANITHA, R. Hybrid robust watermarking for color images. *Computers & Electrical Engineering,* 2014, vol. 40, no. 3, p. 920–930. DOI: 10.1016/j.compeleceng.2014.01.006

[22] DO, M. N., VETTERLI, M. The Contourlet transform: an efficient directional multiresolution image representation. *IEEE Transactions on Image Processing,* 2005, vol. 14, no. 12, p. 2091 to 2106. DOI: 10.1109/TIP.2005.859376

[23] MALVAR, H., FLORÊNCIO, D. Improved spread spectrum: a new modulation technique for robust watermarking. *IEEE Transactions on Image Processing,* 2003, vol. 51, no. 4, p. 898 to 905. DOI: 10.1109/TSP.2003.809385

[24] JIMENEZ-SALINAS, M., GARCIA-UGALDE, F. Improved spread spectrum image watermarking in contourlet domain. In *2008 23rd International Conference Image and Vision Computing.* Christchurch (New Zealand), 2008, 6 p. DOI: 10.1109/IVCNZ.2008.4762090

[25] SHEIKH, H. R., BOVIK, A. C. Image information and visual quality. *IEEE Transactions on Image Processing,* 2006, vol. 15, no. 2, p. 430–444. DOI: 10.1109/TIP.2005.859378

[26] WANG, Z., BOVIK, A. C., SHEIKH, H. R., SIMONCELLI, E. P. Image quality assessment: From error measurement to structural similarity. *IEEE Transactions on Image Processing,* 2004, vol. 13, no. 4, p. 600–612. DOI: 10.1109/TIP.2003.819861

[27] CHANG, H., CHEN, H. H. Stochastic color interpolation for digital cameras. *IEEE Transactions on Circuits and Systems for Video Technology,* 2007, vol. 17, no. 8, p. 964–973. DOI: 10.1109/TCSVT.2007.897471

[28] CAYRE, F., FONTAINE, C., FURON, T. Watermarking security: theory and practice. *IEEE Transactions on Signal Processing,* 2005, vol. 53, no. 10, p. 3976–3987. DOI: 10.1109/TSP.2005.855418

[29] CEDILLO-HERNANDEZ, M., GARCIA-UGALDE, F., NAKANO-MIYATAKE, M., PEREZ-MEANA, H. Robust object-based watermarking using SURF feature matching and DFT domain. *Radioengineering,* 2013, vol. 22, no. 4, p. 1057–1071. ISSN 1210-2512

[30] PROAKIS, J. G., SALEHI, M. *Digital Communications.* 5th ed. McGraw-Hill, 2008. p. 336, 351, 361. ISB: 0072321113

[31] KALKER, T. Considerations on watermarking security. In *Proceedings of 2001 IEEE Fourth Workshop on Multimedia Signal Processing MMSP.* Cannes (France), 2001, p. 201–206. DOI: 10.1109/MMSP.2001.962734

About the Authors ...

Manuel CEDILLO-HERNANDEZ was born in Mexico. He received the B.S. degree in Computer Engineering, M.S. degree in Microelectronics Engineering and his Ph.D. in Communications and Electronic in the National Polytechnic Institute of Mexico (IPN) in 2003, 2006 and 2011, respectively. Currently, he is a full-time researcher at Seccion de Estudios de Posgrado e Investigacion of ESIME Culhuacan in the Instituto Politecnico Nacional de Mexico. His principal research interests are image and video processing, watermarking, software development and related fields.

Antonio CEDILLO-HERNANDEZ was born in Mexico. He received the B.S. degree in Computer Engineering, M.S. degree in Microelectronics Engineering and his Ph.D. in Communications and Electronic in IPN in 2005, 2007 and 2013, respectively. His principal research interests are video processing, watermarking, software development and related fields.

Francisco GARCIA-UGALDE was born in Mexico. He obtained his bachelor in 1977 in Electronics and Electrical System Engineering from National Autonomous University of Mexico, his Diplôme d'Ingénieur in 1980 from SUPELEC France, and his Ph.D. in 1982 in Information Processing from Université de Rennes I, France. Since 1983 is a full-time professor at UNAM, Engineering Faculty. His current interest fields are: Image and video coding, image analysis, watermarking, theory and applications of error control coding, turbo coding, and applications of cryptography, parallel processing and data bases.

Mariko NAKANO-MIYATAKE was born in Japan. She received the M.E. degree in Electrical Engineering from the University of Electro-Communications, Tokyo, Japan in 1985, and her Ph.D. in Electrical Engineering from The Universidad Autonoma Metropolitana (UAM), Mexico City, in 1998. In February 1997, she joined the Graduate Department of The Mechanical and Electrical Engineering School, IPN, where she is now a Professor. Her research interests are in information security, image processing, pattern recognition and related field.

Hector PEREZ-MEANA was born in Mexico. He received his M.S. Degree in Electrical Engineering from the Electro-Communications University of Tokyo, Japan in 1986 and his Ph.D. degree in Electrical Engineering from the Tokyo Institute of Technology, Tokyo, Japan, in 1989. In February 1997, he joined the Graduate Studies and Research Section of The Mechanical and Electrical Engineering School, IPN, where he is now a Professor. His principal research interests are adaptive systems, image processing, pattern recognition, watermarking and related fields.

14

Adaptive Measurement Partitioning Algorithm for a Gaussian Inverse Wishart PHD Filter that Tracks Closely Spaced Extended Targets

Peng LI[1,2], Hongwei GE[1,2], Jinlong YANG[1,2]

[1] Key Laboratory of Advanced Process Control for Light Industry (Jiangnan University), Ministry of Education, Wuxi, 214122, China
[2] School of Internet of Things Engineering, Jiangnan University, Wuxi, 214122, China

lipengjiangnan@163.com, ghw8601@163.com, lptiancaiyesss@163.com

Abstract. *Use of the Gaussian inverse Wishart probability hypothesis density (GIW-PHD) filter has demonstrated promise as an approach to track an unknown number of extended targets. However, when targets of various sizes are spaced closely together and performing maneuvers, estimation errors will occur because measurement partitioning algorithms fail to provide the correct partitions. Specifically, the sub-partitioning algorithm fails to handle cases in which targets are of different sizes, while other partitioning approaches are sensitive to target maneuvers. This paper presents an improved partitioning algorithm for a GIW-PHD filter in order to solve the above problems. The sub-partitioning algorithm is improved by considering target extension information and by employing Mahalanobis distances to distinguish among measurement cells of different sizes. Thus, the improved approach is not sensitive to either differences in target sizes or target maneuvering. Simulation results show that the use of the proposed partitioning approach can improve the tracking performance of a GIW-PHD filter when target are spaced closely together.*

Keywords

Target tracking, extended target, filtering, GIW-PHD filter, measurement partition.

1. Introduction

Multi-target tracking (MTT) typically assumes that each target can produce at most one measurement per scan. Many studies have been done on MTT based on random finite sets (RFSs), such as [1], [2], and this kind of tracking has been used in several fields, such as those in [3], [4]. However, along with the development of modern and high-resolution sensors, a target may have a larger volume than the sensor resolution cell has and produce more than one measurement per scan. Such a target is called an extended target. Extended target tracking (ETT) is a vibrant area of research and has received increasing attention in recent years [5–10]. This is especially true for multiple extended target tracking (METT) [11–17].

An RFS-based observation model for MTT based on finite set statistics (FISST) was first introduced by Mahler. One RFS implementation is the probability hypothesis density (PHD) filter [1], which can estimate target states by computing the function of the first order moment over the target state space. The Gaussian mixture PHD (GM-PHD) filter, presented in [2], is an effective approach to MTT in which the target state is approximated with a Gaussian mixture method.

In contrast to typical single-point targets, target extension is considered by ETT approaches as an extended state. Random hypersurface models (RHMs) for ETT are introduced in [5–7]. These methods assume that target measurements are produced from a certain hypersurface by random selection. Thus, the targets can be tracked and their shapes can be estimated by computing the parameters of a given hypersurface function. Another typical ETT approach is the random matrix (RM) model [8–10]. Because target extension is usually treated as elliptical extension, RM models assume that it follows a Wishart distribution. Therefore, the target extension state can be estimated with an inverse Wishart distribution.

To track an unknown number of extended targets, a METT PHD framework is presented in [11] by Mahler. Granström et al. employ both this framework and the Gaussian mixture method, presenting an extended target GM-PHD (ET-GM-PHD) filter for METT [12], [13]. This filter uses the distance partitioning (DP) and sub-partitioning (SP) methods to provide partitions, and then the partitions are used to update the Gaussian components by computing the likelihood function. The ET-GM-PHD filter can track an unknown number of extended targets effectively, but it cannot estimate target extension states (i.e., the geometrical target extension must be a predetermined value). To solve this problem, a GIW-PHD filter is presented in [14]. This filter assumes that "the last estimated PHD is approximated with an unnormalized mixture of Gaussian inverse Wishart distributions". Therefore, the RM approach can be employed in ETT to estimate the target extension state of a METT PHD filter. More discussion on determining the parameters of a GIW-PHD filter can be found in [15], [16], and the implementation of a GIW-PHD filter in X-band marine radar is introduced in [17]. However, when targets are spatially close and performing maneuvers, the target states will be incorrectly estimated. The reason for this is that the partitioning approaches applied fail to provide the correct partitions, which leads to the failure of the GIW-PHD filter in successive scans.

This paper presents an extension of the GIW-PHD filter, and its main contributions are as follows:

1) We present an adaptive sub-partitioning (ASP) approach to a GIW-PHD filter in order to solve the partitioning problems that occur when targets are spaced closely together. First, the candidate extension information is computed by using the GIW components that are around the measurements. Then the extension information is used to improve SP by employing Mahalanobis distances. As a result, the proposed ASP will not be sensitive to either the differences in target extensions or target maneuvers.

2) Since the GIW-PHD filter cannot provide the identities of the estimated targets, we present a labeling approach and introduce label evolvement.

The paper is organized as follows. Section 2 introduces the GIW-PHD filter. Section 3 presents the ASP algorithm. The labeling approach for the GIW-PHD filter is presented in Sec. 4. Simulation results are presented in Sec. 5. Section 6 contains our conclusions.

2. Background: The Gaussian Inverse Wishart PHD Filter

The GIW-PHD filter is an important implementation of the extended target PHD framework. The target states at time k are defined as

$$\mathbf{X}_k = \left\{ \xi_k^{(i)} \right\}_{i=1}^{N_{x,k}}, \quad \xi_k^{(i)} = \left(x_k^{(i)}, \mathbf{X}_k^{(i)} \right) \tag{1}$$

where $N_{x,k}$ is the unknown number of targets, and $x_k^{(i)}$ and $\mathbf{X}_k^{(i)}$ are the kinematic and the extension states, respectively.

Update: The detection intensity is

$$D_{k|k}\left(\xi_k \right) = D_{k|k}^{ND}\left(\xi_k \right) + \sum_{p \angle Z_k} \sum_{W \in p} D_{k|k}^{D}\left(\xi_k, W \right) \tag{2}$$

where the notation $p \angle Z_k$ indicates that the summation is taken of all partitions p in the current set of measurements Z_k. $D_{k|k}^{ND}$ is the missed detection PHD, and the detection PHD $D_{k|k}^{D}$ can be approximated by a mixture of Gaussian inverse Wishart distributions as follows:

$$D_{k|k}^{D}\left(\xi_k, W \right) = \sum_{j=1}^{J_k} w_{k|k}^{(j)} \mathcal{N}\left(x_k; m_{k|k}^{(j,W)}, \mathbf{P}_{k|k}^{(j,W)} \otimes \mathbf{X}_k \right)$$
$$\times IW\left(\mathbf{X}_k; v_{k|k}^{(j,W)}, \mathbf{V}_{k|k}^{(j,W)} \right) \tag{3}$$

where $m_{k|k}^{(j,W)}$ and $\mathbf{P}_{k|k}^{(j,W)} \otimes \mathbf{X}_k$ are the mean and covariance of the jth Gaussian distribution, respectively. Here, $\otimes$ denotes the Kronecker product of the matrices. $v_{k|k}^{(j,W)}$ and $\mathbf{V}_{k|k}^{(j,W)}$ are the degrees of freedom and the scale matrix of the jth GIW distribution, respectively. The updated component weight is given by

$$w_{k|k}^{(j)} = \frac{w_p}{d_W} \exp\left(-\gamma^{(j)} \right) \left(\frac{\gamma^{(j)}}{\beta_{FA,k}} \right)^{|W|} p_D L_k^{(j,W)} w_{k|k-1}^{(j)}, \tag{4a}$$

$$d_W = \delta_{|W|,1} + \sum_{\ell=1}^{J_{k|k-1}} \exp\left(-\gamma^{(\eta)} \right) \left(\frac{\gamma^{(\ell)}}{\beta_{FA,k}} \right)^{|W|} p_D L_k^{(\ell,W)} w_{k|k-1}^{(\ell)}, \tag{4b}$$

$$w_p = \frac{\prod_{W \in p} d_W}{\sum_{p' \angle Z'} \prod_{W' \in p'} d_{W'}}, \tag{4c}$$

$$\delta_{|W|,1} = \begin{cases} 1 & \text{if } |W| = 1 \\ 0 & \text{if } |W| \neq 1 \end{cases} \tag{4d}$$

where d_W denotes the probability intensity of cell W, and w_p is the weight of a partition p. $\gamma^{(j)}$ is the mean number of the measurements produced by a target, and $\beta_{FA,k}$ is a rate parameter to determine the clutter measurements per surveillance volume per scan. $\delta_{|W|,1}$ is the Kronecker delta and p_D is the detection probability. $L_k^{(j,W)}$ denotes the likelihood between the jth components and the cell W.

Since we have not made original contributions in this section, only the known basic methods and functions are introduced. The details of the GIW-PHD filter can be found in [14], and the measurement partitioning methods are discussed in Sec. 3.

3. Adaptive Sub-partitioning Algorithm

Measurement partitioning is an integral part of an extended target PHD filter, because incorrect partitions will lead directly to estimation error. However, the difficulty of partitioning closely spaced targets is still an unresolved issue that leads to serious errors in GIW PHD filtering (see [14], section VI). This section proposes an ASP algorithm to improve the estimation performance of a GIW PHD filter when targets are closely spaced.

3.1 Problems and Key Methods

Several partitioning approaches to extended target PHD filtering have been proposed, such as distance partitioning (DP) [12], SP [13], mixture partitioning (MP) [18], expectation maximum partitioning (EMP), and prediction partitioning (PP) [14]. However, MP, EMP, and PP are all sensitive to target maneuvers because of their uses of pre-dicted target positions. DP is not sensitive to target maneu-vers, but it fails to handle closely spaced targets correctly. SP is an additional partitioning that follows DP to divide the measurement cells produced by multiple targets. Thus, SP is not sensitive to target maneuvers and can handle cases in which targets are closely spaced. However, when targets are closely spaced and their extensions are different, SP will provide incorrect partitions because a clustering error occurs in K–means++ (an algorithm employed in SP) [19]. Therefore, the typical partitioning approaches en-counter the following problems: 1) no partitioning ap-proach can handle cases in which targets are both closely spaced and performing maneuvers; 2) the use of additional partitioning approaches can indeed improve the perfor-mance of the GIW-PHD filter (DP-SP, EMP and PP are used in [14]), but doing so will require more computations.

The SP algorithm is comprised of two steps: detecting and dividing. Suppose $W = \{z_\ell\}_{\ell=1}^n$ denotes a measurement cell of a partition obtained by DP. First, SP will detect whether W should be divided into sub-cells with

$$\hat{N} = \arg\max_n p\left(|W| \middle| N = n\right), \qquad (5a)$$

$$p\left(|W| \middle| N_j = n\right) = \text{Pois}\left(|W|, \gamma n\right) \qquad (5b)$$

where $\hat{N}$ denotes the possible number of targets. $\text{Pois}(\cdot)$ and $|\cdot|$ denote the Poisson distribution and the number of elements in a set, respectively.

If $\hat{N} \geq 2$, the cell W should be divided into $\hat{N}$ sub-cells by the K–means++ algorithm as follows:

Step 1: Choose the initial centers $C = \{c_j\}_{j=1}^{\hat{N}}$ using the method in [19].

Step 2: For each $j \in \{1, \ldots, \hat{N}\}$, assign each z_ℓ to the cluster with the shortest distance $D_{\ell,j}$ between z_ℓ and c_j. Thus, $\hat{N}$ sub–cells $\widetilde{W}_j$ are obtained.

Step 3: Update the centers with $c_j = \frac{1}{|\widetilde{W}_j|}\sum_{z_\ell \in \widetilde{W}_j} z_\ell$.

Step 4: Repeat Steps 2 and 3 until no $\widetilde{W}_j$ values change any longer.

The distance $D_{\ell,j}$ is generally considered a Euclidean distance, meaning

$$D_{\ell,j} = \sqrt{\left(z_\ell - c_j\right)^{\mathrm{T}}\left(z_\ell - c_j\right)}. \qquad (6)$$

Using a Euclidean distance is similar to using a straight line to divide the measurements. However, if target extensions differ sufficiently, such a line may con-verge to the incorrect position, as shown in Fig. 1.

In contrast to Euclidean distances, Mahalanobis dis-tances are scale-invariant distances that can employ covari-ance matrices to eliminate the influence differences in target sizes. A key innovation of the proposed approach is

employing Mahalanobis distances to improve the K–means++ algorithm. When this is done, (6) becomes

$$D_{\ell,j} = \sqrt{\left(z_\ell - c_j\right)^{\mathrm{T}} \mathbf{S}_j^{-1}\left(z_\ell - c_j\right)} \qquad (7)$$

where $\mathbf{S}_j$ denotes the corresponding covariance matrix of $\widetilde{W}_j$. Note that the parameter $\mathbf{S}_j$ is not given, but it can be calculated using the inverse scale matrix provided by the GIW-PHD filter. The primary method of determining $\mathbf{S}_j$ can be summarized as follows:

Find the predicted components around the cell W. The extension matrices $\{\tilde{\mathbf{X}}_i\}$ of these components can be em-ployed to calculate the candidates for $\{\mathbf{S}_j\}$. For each sub-cell $\widetilde{W}_j$, a corresponding $\mathbf{S}_j$ can be initialized using these candidates.

Do Step 2 and Step 3 of the K–means++ algorithm.

Calculate the covariance matrix $\boldsymbol{\Psi}_j$ for each sub-cell $\widetilde{W}_j$ and assign each $\hat{\mathbf{X}}_i$ to its most similar $\boldsymbol{\Psi}_j$. For example, suppose $\hat{\mathbf{X}}_1$ and $\hat{\mathbf{X}}_2$ are the covariance matrices of the large and small targets, respectively, in Fig. 1; $\boldsymbol{\Psi}_1$ and $\boldsymbol{\Psi}_2$ denote the covariance matrices of cluster 1 and cluster 2, respec-tively, in Fig. 2 (c). $\{\hat{\mathbf{X}}_1, \hat{\mathbf{X}}_2\}$ will be assigned to $\{\boldsymbol{\Psi}_1, \boldsymbol{\Psi}_2\}$ (i.e., $\mathbf{S}_2 = \hat{\mathbf{X}}_1$ and $\mathbf{S}_1 = \hat{\mathbf{X}}_2$) because of the similarity of their extensions (the details are shown in Sec. 3.2). Thus, the partitioning result will converge to the extension similarity maximum.

Figure 2 shows an example the partitioning process:

- (a) shows the result of the first loop. The triangles are the initial centers, which are obviously inaccurate. The original K–means++ result may converge to the result shown in Fig. 1, because the use of (6) is like using a straight line to divide measurements, and the situation in Fig. 1 demonstrates the best convergence result.

- However, as shown in (b), Equation (7) ensures that the proposed approach converges to an elliptical ex-tension. Thus, the result in (b) is more similar to two elliptical extensions than that of (a).

- (c) shows the results of the third loop. As we would intuitively expect, the measurements are divided al-most correctly with the exceptions of several points. The extension of cluster 2 is similar to that found in $\hat{\mathbf{X}}_1$, and the extension of cluster 1 is similar to that found in $\hat{\mathbf{X}}_2$. Hence, the covariance matrices in (7) be-come $\mathbf{S}_1 = \hat{\mathbf{X}}_2$ and $\mathbf{S}_2 = \hat{\mathbf{X}}_1$.

- Finally, the proposed algorithm converges to the re-sult shown in (d). The proposed method can ensure the coverage of results whose clusters are similar to those of $\hat{\mathbf{X}}_1$ and $\hat{\mathbf{X}}_2$. This is why the proposed method can achieve a better performance than the algorithm originally used in the GIW-PHD filter can.

Remark: the proposed method applies only to the situation that the number of elements in $\{\hat{\mathbf{X}}_j\}$ is equal to $\hat{N}$. If this condition are not established, the number of targets may be estimated incorrectly in the last time step (i.e., the

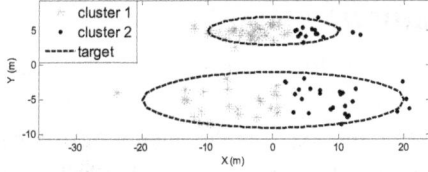

Fig. 1. The partitioning result of SP.

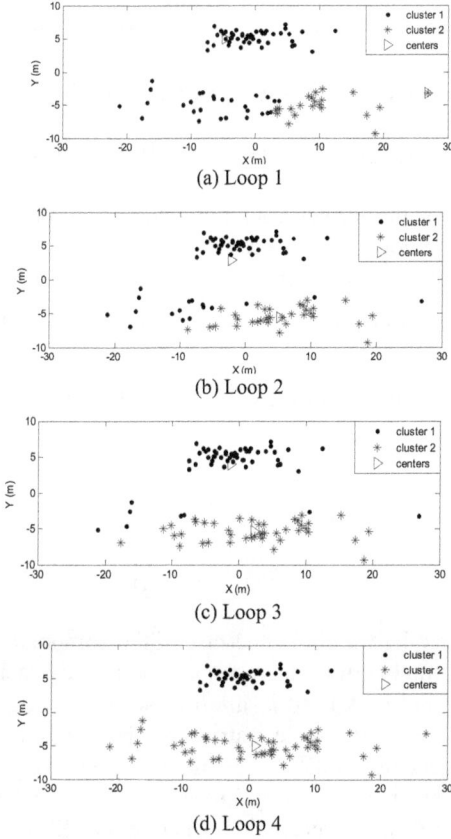

(a) Loop 1

(b) Loop 2

(c) Loop 3

(d) Loop 4

Fig. 2. Illustration of ASP.

extension information of the predicted components may be incorrect). Thus, the original K–means++ algorithm should be used in this step.

3.2 Implementation

3.2.1 Initialization

Suppose $\left\{\xi^i_{k|k-1}\right\}^{N_k}_i$ denotes the set of predicted components and $\left\{m^{i,d}_{k|k-1}\right\}^{N_k}_i$ denotes the corresponding d–dimensional set of predicted positions. The states of the predicted components around W can be defined as

$$M = \left\{m^{i,d}_{k|k-1} \mid \sigma_i = 1\right\}^{N_k}_i , \tag{8a}$$

$$\aleph = \left\{\hat{\mathbf{X}}^i_{k|k-1} \mid \sigma_i = 1\right\}^{N_k}_i , \tag{8b}$$

$$\sigma_i = \begin{cases} 1, & \left\{\left(m^{i,d}_{k|k-1} - z_\ell\right)^T \left(m^{i,d}_{k|k-1} - z_\ell\right) < \vartheta\right\}_{z_\ell \in W} \\ 0, & \text{others} \end{cases} \tag{8c}$$

where ϑ is the maximum threshold of DP. M includes the predicted positions around the cell W, and $\aleph$ includes the corresponding extension matrices. $\hat{\mathbf{X}}^i_{k|k-1}$ is the extension matrix and can be calculated by

$$\hat{\mathbf{X}}^i_{k|k-1} = \frac{\mathbf{V}^i_{k|k-1}}{v^i_{k|k-1} - d - 1} \tag{9}$$

where d denotes the number of dimensions of the physical space [9]. $\mathbf{V}^i_{k|k-1}$ and $v^i_{k|k-1}$ are the predicted inverse scale matrix and the inverse Wishart degrees of freedom, respectively.

ASP is only used when $|M| = \hat{N}$. The calculation of the initial centers can give results equal to those of the original K–means++. In addition, to improve the accuracy of the initialization, the centers can be obtained by translating the corresponding $m^{i,d}_{k|k-1}$ by

$$C = \left\{c_j\right\}^{|M|}_{j=1} , \tag{10a}$$

$$c_j = m^{i,d}_{k|k-1} - \overline{m}_{k|k-1} + \overline{z} , \tag{10b}$$

$$\overline{m}_{k|k-1} = \frac{\sum_{m^{i,d}_{k|k-1} \in \aleph} m^{i,d}_{k|k-1}}{|M|} , \tag{10c}$$

$$\overline{z} = \frac{\sum_{z_\ell \in W} z_\ell}{|W|} \tag{10d}$$

where $\overline{m}_{k|k-1}$ and $\overline{z}$ are the mean positions of M and W, respectively. c_j indicates the relative position and removes the influence of target maneuvers. Thus, the initialization of the covariance matrix is

$$\left\{\mathbf{S}_j\right\}^{\hat{N}}_{j=1} = \aleph . \tag{11}$$

3.2.2 Modification of K–means++

In the modified K–means ++ algorithm, the dividing process remains the same as in Step 2 and Step 3 in Sec. 3.1, but the distance $D_{\ell,j}$ should be calculated by (7) instead. Then the covariance matrix $\mathbf{\Psi}_j$ of each sub-cell $\tilde{W}_j$ can be calculated by

$$\mathbf{\Psi}_j = \frac{1}{|\tilde{W}_j|} \sum_{z_\ell \in \tilde{W}_j} \left(z_\ell - \overline{z}^{(\tilde{W}_j)}\right)\left(z_\ell - \overline{z}^{(\tilde{W}_j)}\right)^T , \tag{12a}$$

$$\overline{z}^{(\tilde{W}_j)} = \frac{1}{|\tilde{W}_j|} \sum_{z_\ell \in \tilde{W}_j} z_\ell . \tag{12b}$$

$\mathbf{S}_j$ is updated based on the similarity of $\mathbf{\Psi}_j$ and $\hat{\mathbf{X}}^i_{k|k-1}$. Suppose $\aleph'_\kappa = \left\{\hat{\mathbf{X}}'^j_{\kappa,k|k-1}\right\}^{\hat{N}}_{i=1}$ denotes a permutation of the elements of $\aleph$ (e.g., $\left\{\hat{\mathbf{X}}'^j_{1,k|k-1}\right\}^2_{i=1} = \left\{\hat{\mathbf{X}}^1_{k|k-1}, \hat{\mathbf{X}}^2_{k|k-1}\right\}$) and $\left\{\hat{\mathbf{X}}'^j_{2,k|k-1}\right\}^2_{i=1} = \left\{\hat{\mathbf{X}}^2_{k|k-1}, \hat{\mathbf{X}}^1_{k|k-1}\right\}$ when $\hat{N} = 2$. The difference of matrices is defined as

$$\omega_\kappa = \frac{1}{\hat{N}} \sum_{j=1}^{\hat{N}} \mathrm{tr}\left[\left(\Psi_j - \hat{\mathbf{X}}_{\kappa,k|k-1}^{\prime j} \right)^2 \right] \quad (13)$$

where $\mathrm{tr}[\cdot]$ denotes the trace of the matrix. ω_κ denotes the error of $\{\Psi_j\}_{j=1}^{\hat{N}}$ and $\aleph_\kappa'$. Supposing that $\omega_{\kappa'}$ is the smallest of all ω_κ values, $\{\mathbf{S}_j\}_{j=1}^{\hat{N}}$ is updated by

$$\{\mathbf{S}_j\}_{j=1}^{\hat{N}} = \aleph_{\kappa'}'. \quad (14)$$

4. Track Maintenance

The GIW-PHD filter cannot provide the trajectories of individual extended targets. This section proposes a track maintenance approach to a GIW-PHD filter based on a component labeling technique [20]. The proposed approach consists of the following steps:

1) Prediction:

Suppose that at time $k-1$ ($k \geq 2$), the labels of the GIW components are denoted by

$$L_{k-1} = \left\{ l_{k-1}^{(j)} \right\}_{j=1}^{J_{k-1}} = \left\{ l_{k-1}^{(1)}, l_{k-1}^{(2)} \cdots l_{k-1}^{(J_{k-1})} \right\} \quad (15)$$

where J_{k-1} is the number of GIW components at time $k-1$ and that $l_{k-1}^{(j)}$ denotes the label of the jth GIW component. The labels of predicted GIW components can be written as

$$L_{k|k-1} = L_{k-1} \bigcup L_\beta, \quad (16a)$$

$$L_\beta = \left\{ l_\beta^{(i)} \right\}_{i=1}^{J_\beta} = \left\{ l_\beta^{(1)}, l_\beta^{(2)} \cdots l_\beta^{(J_\beta)} \right\} \quad (16b)$$

where $l_\beta^{(i)}$ denotes the labels of the birth GIW components and J_β is the number of birth components.

2) Update:

Suppose that $\tilde{N}_k$ denotes the total number of all cells in all partitions (i.e., $\tilde{N}_k = \sum_{i=1}^{\tilde{N}_{p,k}} \tilde{N}_{W,k}^{(i)}$, where $\tilde{N}_{p,k}$ is the number of partitions at time k, and $\tilde{N}_{W,k}^{(i)}$ is the number of the cells of the ith partition). According to (2), there would now be $\tilde{N}_k + 1$ times the number of predicted GIW components. Thus, each predicted component gives rise to $\tilde{N}_k + 1$ corresponding updated components. The corresponding labels can be expressed as

$$L_{k|k} = L_{k|k-1} \bigcup \left\{ L_{k|k-1}^{(i)} \right\}_{i=1}^{\tilde{N}_k} \quad (17)$$

where $L_{k|k-1}^{(i)} = L_{k|k-1}$.

3) Pruning & merging:

When the GIW components with low weights are pruned, the corresponding labels and their attributes should be also pruned. Note that if several components $\left\{ \cdots, \xi_{k|k}^{(n)}, \cdots \right\}$ are merged and $\xi_{k|k}^{(n)}$ has the largest weight, the label of the merged component is equal to the label of $\xi_{k|k}^{(n)}$.

5. Simulation Results

The extended targets are modeled by $U\left(\mathbf{A}\mathbf{X}_k\mathbf{A}^{\mathrm{T}}\right) + \mathcal{N}(\mathbf{R})$ [9], where $\mathbf{X}_k$ is a symmetric positive definite (SPD) matrix, $\mathbf{A}$ is the rotation matrix determined by the motion model, and $\mathbf{R}$ is the Gaussian measurement noise. The number of measurements of each target follows the Poisson distribution.

The parameters of simulated scenarios are given as

$$T_\mathrm{s} = 1\,\mathrm{s},\ \mathcal{S}{=}4000 \times 4000\ \mathrm{m}^2,\ \beta_{FA,k} = 6.25 \times 10^{-7}, \quad (18a)$$

$$\mathbf{R}_k = \mathrm{diag}\left([1,1]\right),\quad \mathbf{Q}_k = \mathrm{diag}\left([0.5, 0.5, 0, 0]\right) \quad (18b)$$

where T_s is the sensor-scanning interval. $\mathcal{S}$ denotes the surveillance volume with a rate parameter (i.e., the Poisson mean of clutter measurements is $\mathcal{S} \times \beta_{FA,k}{=}10$). $\mathbf{R}_k$ and $\mathbf{Q}_k$ are the covariances of process noise and measurement noise, respectively.

The parameters of the GIW-PHD filter are

$$w_0 = 0.1,\ v_0 = 7,\ \tau = 5,\ \gamma^{(j)}{=}15, \quad (19a)$$

$$\mathbf{V}_0 = \mathrm{diag}\left([50, 50]\right), \quad (19b)$$

$$\mathbf{P}_0 = \mathrm{diag}\left([25, 100]\right) \quad (19c)$$

where w_0 is the weight of the birth GIW component. $\gamma^{(j)}$ is the mean number of measurements produced by each target each time and v_0, $\mathbf{V}_0$ and $\mathbf{P}_0$ are inverse Wishart degrees of freedom, the inverse scale matrix, and the Gaussian covariance of generated GIW components, respectively. Measurements are partitioned by DP with $d = \{1,3,5,10,15,\ldots, 30\}$ as the distance thresholds. Then, either SP or ASP is used to divide cells containing more than one target into sub-cells. Thus, the partitioning approaches are called DP-SP and DP-ASP.

5.1 Extended Target OSPA (ET–OSPA)

According to [21], the Optimal SubPattern Assignment (OSPA) distance is employed to evaluate the performance of a PHD filter. For METT, the ET-OSPA is defined as

$$\bar{d}_p^{(c)}(X,Y) = \left(\frac{1}{n} \left(\min_{\pi \in \prod_n} \sum_{i=1}^{m} d^{(c)} \cdots \times \left(x_i, \tilde{X}_i, y_{\pi(i)}, \tilde{Y}_{\pi(i)} \right)^p + c^p(n-m) \right) \right)^{1/p}, \quad (20a)$$

$$d^{(c)}\left(x_i, \tilde{X}_i, y_{\pi(i)}, \tilde{Y}_{\pi(i)} \right)^p = \left(x_i - y_{\pi(i)} \right)^p + RMSE_{\tilde{X}_i, \tilde{Y}_{\pi(i)}}^p, \quad (20b)$$

$$RMSE_{\tilde{X}_i, \tilde{Y}_{\pi(i)}}^p = \left(\mathrm{tr}\left[\left(\tilde{\mathbf{X}}_i - \tilde{\mathbf{Y}}_{\pi(i)} \right)^2 \right] \right)^p \quad (20c)$$

where $RMSE_{\tilde{X}_i, \tilde{Y}_{\pi(i)}}^p$ denotes the extension error and $\{\tilde{\mathbf{X}}_i\}$ and $\{\tilde{\mathbf{Y}}_{\pi(i)}\}$ are the estimated and true extension matrices, respectively. The first term on the right hand side of (20b), namely the localization error, is equivalent to original function of the OSPA algorithm. The second term is the added extension error. Therefore, the error function $d^{(c)}$ involves both localization and extension errors.

5.2 Results

The target tracks simulated in this study are shown in Fig. 3. The targets moved closer together from 0–20 s and then moved linearly together from 21–40 s. They began circular motion at 41 s and moved uniformly again from 61–80 s. Finally, they separated from each other at 81 s. The edges of the targets separated from each other by 2 m. The shape matrices were $\mathbf{X}^{(1)} = \text{diag}([20, 4])$ and $\mathbf{X}^{(2)} = \text{diag}([10, 2])$.

Figure 4 shows the average results of 100 Monte Carlo (MC) runs. The ET-OSPA distance of each approach is shown in (a). The ET-OSPA values from 20–80 s are higher when DP-SP is used than they are when other approaches are used. This means that DP-SP cannot provide the correct partitions when targets with different extensions are closely spaced. The results of using DP-SP with EMP and PP are better than those of using just DP-SP because EMP and PP consider the target extension information and can provide the correct partitions. However, the ET-OSPA values increase significantly during the times 20–25 s and 40–60 s. The reason for this is that PP and EMP are sensitive to target maneuvers. The ET-OSPA values of the proposed DP-ASP were clearly lower than those of DP-SP with EMP and PP when the targets were both closely spaced and performing maneuvers. Equation (10) insured that DP-ASP was not sensitive to target maneuvers, while the use of target extension information made DP-ASP insensitive to differences in target extensions. Therefore, DP-ASP could provide more correct partitions than other partitioning approaches could when the targets were closely spaced. This conclusion is also evident in Fig. 4(b), which shows the sums of the weights of each approach.

Figure 5 shows the average time costs of making 100 MC runs for each partitioning approach. The use of more partitioning approaches required a greater number of computations. Thus, the values of DP-SP with EMP and PP were significantly larger than those of the other two approaches. The time costs of DP-ASP and DP-SP were roughly the same because the two approaches provided the same number of partitions. Note that providing a greater number of correct partitions can reduce the number of GIW components. Thus, the time costs of DP-ASP are sometimes lower than those of DP-SP.

Figure 6 shows the association results of a single trial, and Figure 6(a) shows the results of the GIW-PHD filter using DP-SP with EMP and PP. When the targets were maneuvering, many incorrect and excrescent tracks occurred. The reason for this is that the partitioning approaches failed to provide accurate partitions. Then, the inaccurate partitions led to estimation errors in the filter, which ultimately resulted in association problems. The

association results when using DP-ASP, shown in (b), were relatively better than those in (a) (i.e., most of the tracks were correct, and the number of excrescent tracks was lower than that of (a)). This means that even though an error may occur during an individual scan, DP-ASP can generally provide more accurate partitions than other approaches when targets are closely spaced.

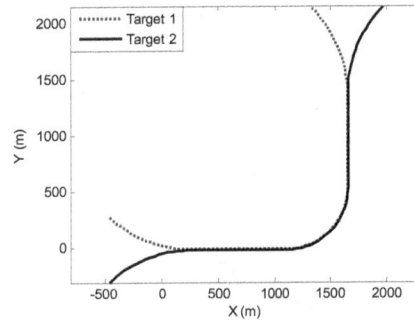

Fig. 3. True tracks.

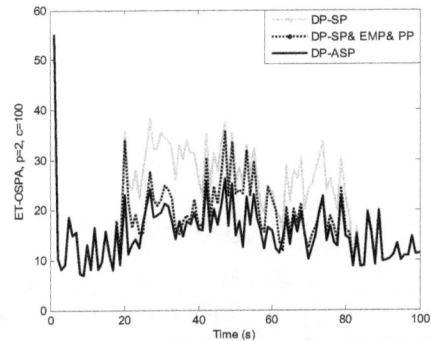

(a) ET-OSPA.

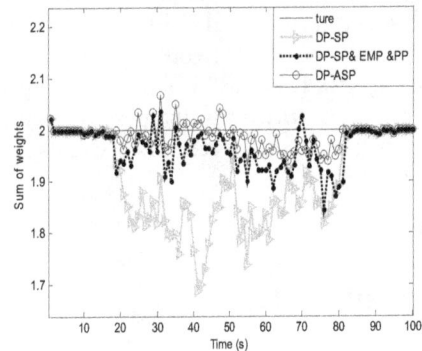

(b) Sums of weights.

Fig. 4. Results of 100 MC runs.

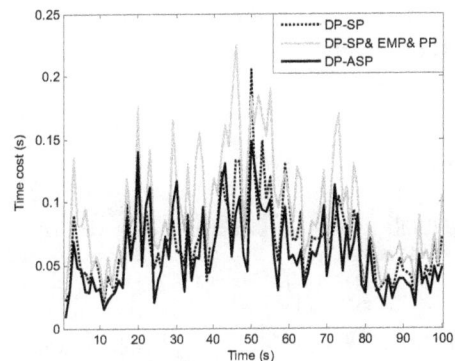

Fig. 5. Time cost.

(a) Results of DP-SP with EMP and PP.

(b) Results of DP-ASP.

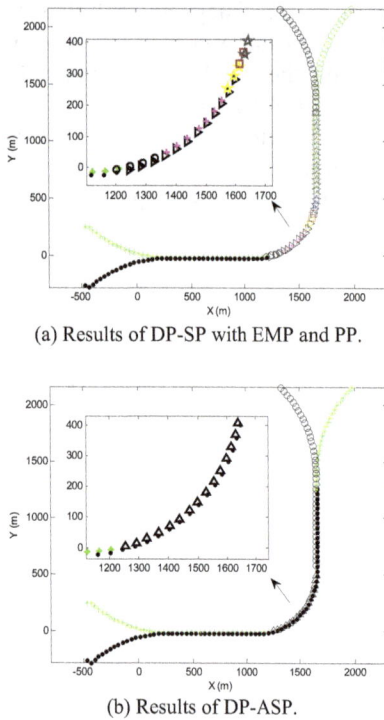

Fig. 6. Track association results.

6. Conclusions

This paper proposes a measurement partitioning algorithm for a GIW PHD filter, called ASP, which can provide better partitioning results than other approaches can when targets are closely spaced. In addition, a track maintenance approach is included in the GIW PHD filter. In the future works, we plan to apply ASP to other METT particle filters as was done in [22], [23]. ASP is a promising approach for dividing particle swarms. Thus, the performance of METT particle filters can be improved when targets are closely spaced.

Acknowledgments

This paper is supported by the National Natural Science Foundation of China (No. 61305017), the Graduate Innovation Foundation of Jiangsu Province (No. KYLX16_0782), the 111 Project under Grant No. B12018, and PAPD of Jiangsu Higher Education Institutions.

References

[1] MAHLER, R. Multi-target Bayes filtering via first-order multi-target moments. *IEEE Transactions on Aerospace and Electronic Systems*, 2003, vol. 39, no. 4, p. 1152–1178. DOI: 10.1109/TAES.2003.1261119

[2] VO, B. N., MA, W. K. The Gaussian mixture probability hypothesis density filter. *IEEE Transactions on Signal Processing*, 2010, vol. 54, no. 11, p. 4091–4104. DOI: 10.1109/TSP.2006.881190

[3] MULLANE, J., VO, B. N., ADAMS, M. D., et al. A random-finite-set approach to Bayesian SLAM. *IEEE Transactions on Robotics*, 2011, vol. 27, no. 2, p. 268–282. DOI: 10.1109/TRO.2010.2101370

[4] DRALLE, K., RUDEMO, M. Automatic estimation of individual tree positions from aerial photos. *Canadian Journal of Forest Research*, 1997, vol. 27, no. 27, p. 1728–1736. DOI: 10.1109/LGRS.2013.2242044

[5] BAUM, M., HANEBECK, U. D. Random hypersurface models for extended object tracking. In *Proceedings of IEEE International Symposium on Signal Processing and Information Technology*. Ajman (United Arab. Emirates), December 2009, p. 178–183. DOI: 10.1109/ISSPIT.2009.5407526

[6] BAUM, M., HANEBECK, U. D. Shape tracking of extended objects and group targets with star-convex RHMs. In *Proceedings of the 14th International Conference on Information Fusion*. Chicago (Illinois, USA), July 2011, p. 1–8. ISBN: 9780982443828

[7] ZEA, A., FAION, F., BAUM, M., et al. Level-set random hypersurface models for tracking non-convex extended objects. In *Proceedings of the 16th International Conference on Information Fusion*. Istanbul (Turkey), July 2013, p. 1760–1767. ISBN: 9781479902842

[8] KOCH, W. Bayesian approach to extended object and cluster tracking using random matrices. *IEEE Transactions on Aerospace and Electronic Systems*, 2008, vol. 44, no. 3, p. 1042–1059. DOI: 10.1109/TAES.2008.4655362

[9] FELDMANN, M., FRÄNKEN, D. Tracking of extended objects and group targets using random matrices. *IEEE Transactions on Signal Processing*, 2011, vol. 59, no. 4, p. 1409–1420. DOI: 10.1109/TSP.2010.2101064

[10] LAN, J., LI, X. R. Tracking of maneuvering non-ellipsoidal extended object or target group using random matrix. *IEEE Transactions on Signal Processing*, 2014, vol. 62, no. 9, p. 2450 to 2463. DOI: 10.1109/TSP.2014.2309561

[11] MAHLER, R. PHD filters for nonstandard targets, I: Extended targets. In *Proceedings of the 12th International Conference on Information Fusion*. Seattle (WA, USA), July 2009, p. 915–921. ISBN: 9780982443804

[12] GRANSTRÖM, K., LUNDQUIST, C., ORGUNER, U. A Gaussian mixture PHD filter for extended target tracking. In *Proceedings of the 13th International Conference on Information Fusion*. Edinburgh (Scotland, UK), July 2010, p. 1–8. DOI: 10.1109/ICIF.2010.5711885

[13] GRANSTRÖM, K., LUNDQUIST, C., ORGUNER, U. Extended target tracking using a Gaussian-mixture PHD filter. *IEEE Transactions on Aerospace and Electronic Systems*, 2012, vol. 48, no. 4, p. 3268–3286. DOI: 10.1109/TAES.2012.6324703

[14] GRANSTRÖM, K., ORGUNER, U. A PHD filter for tracking multiple extended targets using random matrices. *IEEE Transactions on Signal Processing*, 2012, vol. 60, no. 1, p. 5657 to 5671. DOI: 10.1109/TSP.2012.2212888

[15] GRANSTRÖM, K., ORGUNER, U. Estimation and maintenance of measurement rates for multiple extended target tracking. In *Proceedings of the 15th Conference on Information Fusion*. Singapore, September 2012, p. 2170–2176. ISBN: 978-0-9824438-5-9

[16] GRANSTRÖM, K., ORGUNER, U. On the reduction of Gaussian inverse Wishart mixtures. In *Proceedings of the 15th Conference on Information Fusion*. Singapore, September 2012, p. 2162–2169. ISBN: 978-0-9824438-5-9

[17] GRANSTRÖM, K., NATALE, A., BRACA, P., et al. Gamma Gaussian inverse Wishart probability hypothesis density for

extended target tracking using X-band marine radar data. *IEEE Transactions on Geoscience & Remote Sensing*, 2015, vol. 53, p. 1 to 15. DOI: 10.1109/TGRS.2015.2444794

[18] HAN Y., ZHU H., HAN C. Z. Maintaining track continuity for extended targets using Gaussian-mixture probability hypothesis density filter. *Mathematical Problems in Engineering*, 2015, no. 3, p. 1–16. DOI: 10.1155/2015/501915

[19] ARTHUR, D., VASSILVITSKII, S. k-means++: The advantages of careful seeding. In *Proceedings of the 18th Annual ACM-SIAM Symposium on Discrete Algorithms*. New Orleans (USA), Jan. 2007, p. 1027–1035. DOI: 10.1145/1283383.1283494

[20] PANTA, K., CLARK, D., VO, B. N. Data association and track management for the Gaussian mixture probability hypothesis density filter. *IEEE Transactions on Aerospace and Electronic Systems*, 2009, vol. 45, no. 3, p. 1003–1016. DOI: 10.1109/TAES.2009.5259179

[21] SCHUHMACHER, D., VO, B. T., VO, B. N. A consistent metric for performance evaluation of multi-object filters. *IEEE Transactions on Signal Processing*, 2008, vol. 56, no. 8, p. 3447–3457. DOI: 10.1109/TSP.2008.920469

[22] SONG, L. P., LIANG, M., JI, H. B. Box-particle implementation and comparison of cardinalized probability hypothesis density filter. *Radioengineering*, 2016, vol. 25, no. 1, p. 177–186. DOI: 10.13164/re.2016.0177

[23] LI, M., LIN, Z., AN, W., et al. Box-particle labeled multi-Bernoulli filter for multiple extended target tracking. *Radioengineering*, 2016, vol. 25, no. 3, p. 527–535. DOI: 10.13164/re.2016.0527

About the Authors ...

Peng LI was born in 1989. He received his B.S. degree in Mathematics and Applied Mathematics from Langfang Normal University. He is currently a Ph.D. candidate at the School of Internet of Things Engineering in Jiangnan University. His research interests include target tracking and signal processing.

Hongwei GE (corresponding author) was born in 1967. He received the M.S. degree in Computer Science from Nanjing University of Aeronautics and Astronautics, China, in 1992 and the Ph.D. degree in Control Engineering from Jiangnan University, China, in 2008. Currently, he is a professor and Ph.D. supervisor in the School of Internet of Things of Jiangnan University. His research interests include target tracking, artificial intelligence, machine learning, pattern recognition and their applications.

Jinlong YANG was born in 1981. He received his M.S. degree in Circuit and System from the Northwest Normal University, China in 2009, and his Ph.D. degree in Pattern Recognition and Intelligent System from Xidian University, China in 2012. Currently he is an associate professor in Jiangnan University. His research interests include target tracking, signal processing and pattern recognition.

15

Modeling and Link Performance Analysis of Busbar Distribution Systems for Narrowband PLC

Zeynep HASIRCI, Ismail Hakki CAVDAR, Mehmet OZTURK

Dept. of Electrical and Electronics Engineering, Karadeniz Technical University, 61080, Trabzon, Turkey

{zhasirci, cavdar, mehmetozturk}@ktu.edu.tr

Abstract. *Busbar distribution system is used as a modular infrastructure to carry electrical energy in low voltage grid. Due to the widespread usage in industrial areas, the power line communication possibilities should be investigated in terms of smart grid concept. This paper addresses modeling of the busbar distribution system as a transmission line and gives some suggestions on the link performance for narrowband power line communication for the first time in literature. Firstly, S-parameters of different current level busbars were measured up to 500 kHz for all possible two-port signal paths. The utilization of the frequency-dependent RLCG(f) model was proposed to extract transmission line characteristics to eliminate the unwanted measurement effects. Particle swarm algorithm was used to optimize the model parameters with a good agreement between measured and simulated S-parameters. Additionally, link performance of busbar distribution system as a power line communication channel at 3 kHz–148.5 kHz band was examined for frequency shift keying and phase shift keying modulations under different network configurations such as varying busbar type, the line length between transmitter and receiver, branch number, and terminating load impedance. Obtained results were presented as bit-error-rate vs. signal-to-noise ratio graphs.*

Keywords

Bit-error-rate, busbar, channel capacity, channel modeling, M2M, narrow band, parameter optimization, power line communication, smart grid, S-parameters

1. Introduction

Creating alternative energy sources and increasing only production, may not be sufficient to ensure the production-consumption balance. What really matters is using energy effectively and efficiently. Smart grid disperses control of the network infrastructure. It ensures transfer and monitoring of data with a protection of various technologies. Power line communication (PLC) is one of the most powerful communication alternatives for smart grid

applications that use existing grid infrastructure as the communication medium. Low voltage (LV) network is more complex than the medium/high voltage grid due to the necessity of multi-point communication network. Due to many advantages, busbar is an electrical distribution system element which has a modular structure that carries electrical energy in the buildings. They are used mostly in industrial areas which have high power consumption. Starting from the fact that solution to some problems - monitoring energy from the point of production to consumption, loss-efficiency analysis, machine to machine communication (M2M) - should be more significant in the industrial field, busbars emerge as mediums that should be examined in terms of smart grids.

PLC studies on cables in the literature do not provide a complete and accurate identification for busbars mainly due to the electrical installation differences. The subject of this study has emerged from the importance of this issue in terms of inadequate scientific work conducted on. It is necessary to know the characteristics of the busbars for providing PLC communication. Accurate transmission line models are required for accurate simulation of signal paths implemented in communication systems. Such transmission line models are typically given in per-unit-length (p.u.l.) parameters (RLGC parameters) to derive the echo transfer function of the power line [1]. Whereas lossy transmission line models may equally be described in terms of the characteristic impedance (Z_c) and propagation constant (γ), designers are generally focused on RLGC parameters. Judging from here, busbar system can also be described as a transmission line [2], [3] and can be represented with RLGC parameters as the other power cables. In the literature, there are many studies about power cables and extraction of the p.u.l. parameters. Many of these studies are based on S-parameters measurements [1], [4–8]. Besides, others are based on time-domain measurements [9–11]. Some methods support the calculation of the RLCG parameters from Z_c and γ [1, 4, 12]. Almost other methods estimate Z_c from γ extracted from measurements by some assumptions [13–17]. Additionally, line parameters extraction is made with some optimization algorithms such as genetic algorithm etc. [18–20].

There is a few study about busbars for PLC. [2] reports scattering parameters of the copper conductor series

1000 A busbar system by an EM analysis simulation tool for CENELEC (*European Committee for Electrotechnical Standardization*), FCC (*Federal Communications Commission*) and Broadband, [3] presented propagation characteristics of different current levels copper conductor series busbar systems in 1-50 MHz. [2] and [3] are simulation based studies for copper conductor series busbar system and need experimental validation. Then, more preferable type aluminum conductor series busbar system (630 A) are modeled by Sonnet Suites 13.52 and results are validated with measured S-parameters [21]. [22] presented a parameter extraction approach for 630 A busbar.

In this study, S-parameters of different current level busbar distribution systems (630 A, 1250 A, 2000 A) were measured with a Vector Network Analyzer (VNA) up to 500 kHz for all possible signal paths. RLCG(*f*) modeling procedure was used to estimate p.u.l. parameters. After particle swarm optimization algorithm (PSO), the measured and simulated S-parameters (from the model) show a good agreement. With the utilization of the chain-scattering matrix method, different combinations of single-branch networks were simulated for N-branch busbar distribution network. After that, the effects of terminating load impedance, busbar line length, and branch number on the system performance in terms of the bit-error-rate (BER) were calculated for FSK and PSK modulations. BER vs. signal-to-noise ratio (SNR) graphics were presented for different busbar network configurations.

2. Modeling of Busbar

2.1 Measurements and Estimation of Model Parameters

Aluminum busbars have more widespread usage than copper ones due to their low cost. Thus, in this study, aluminum busbars which have 630 A, 1250 A, and 2000 A current levels were used for measurements and modeling.

Different current levels change the physical characteristics of the busbar system such as the cross-sectional area of the conductor. Physical characteristics of used busbar systems for different current levels at 20°C are listed in Tab. 1 and detailed information can be found in [23].

As it is mentioned before, busbar system can also be described as a transmission line [2, 3, 21, 22] and can be represented with Z_c and γ as other power cables to find the

transfer function of the communication channel. It is a three-phase system (L1, L2, L3) with a neutral (N) and it can be analyzed as six different two-port networks for L1-N, L2-N, L3-N, L2-L1, L3-L1, and L3-L2 port connections (signal paths), separately. It is commonly known that a two-port device can be described by some parameter sets such as impedance, admittance, hybrid and voltage/current transmission matrices at each of the two ports. For these type measurements, ideal short/open circuit terminations are required. However, frequency dependence of these terminations can have caused measurement issues. Thus, traveling waves as variables, S-parameters, can be a good alternative to overcome extra measurement faults. S-parameters are defined with respect to traveling waves, unlike terminal voltages and currents and can be measured on a device located at some distance from the instrument [24]. These measurements are carried out by terminating one or the other port with normalizing impedance Z_0 (generally 50 Ω) [1], [25]. Two-port S-parameter measurements are made via VNA, Agilent Technologies N9913A Field Fox RF Analyzer. Measurement setup is shown in Fig. 1. Two 1.5 m length M17/75-RG214 type coaxial cables with 50 Ω characteristic impedance are attached to the ports for the connection between VNA and DUT (unit length-3 m busbar) to measure the S-parameters. Two port calibration is made to eliminate the extra connection effects on measurements. However, calibration does not guarantee to eliminate the busbar S-measurement faults, the preferred method for parameter extraction plays an important role in eliminating these unwanted effects. These experimental studies are made without extra load connection to find the transmission line parameters.

S-parameters represent the relationship between normalized incident (a_1, a_2) and reflected (b_1, b_2) voltage waves as shown in (1).

$$\begin{bmatrix} b_1 \\ b_2 \end{bmatrix} = \begin{bmatrix} S_{11} & S_{12} \\ S_{21} & S_{22} \end{bmatrix} \begin{bmatrix} a_1 \\ a_2 \end{bmatrix}. \tag{1}$$

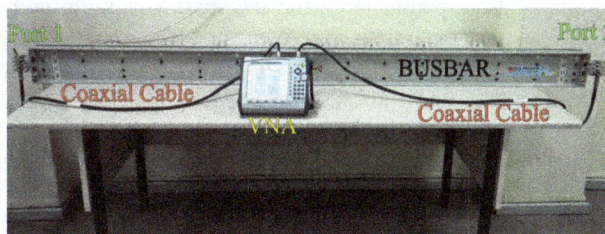

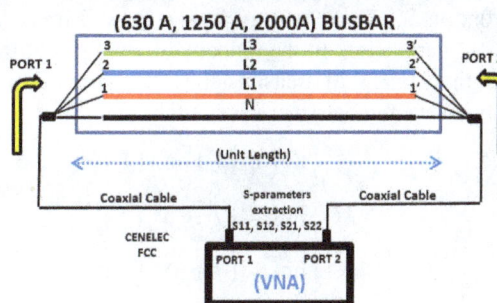

Fig. 1. Measurement setup for S-parameters.

Current levels R [A]	Resistance R [mΩ/m]	Reactance X [mΩ/m]	Impedance Z [mΩ/m]	Weight [kg/m]	Conductor size [mm × mm]
630	0.121	0.027	0.124	7.9	6 × 40
1250	0.044	0.013	0.046	13.9	6 × 110
2000	0.026	0.008	0.027	21.7	6 × 200

Tab. 1. Physical characteristics of used busbar systems for different current levels at 50 Hz [23].

A network that consists entirely of linear passive components, such as transmission lines, is reciprocal. Thus, the full S-parameters matrix of the reciprocal circuit must be symmetric. That means $S_{11} = S_{22}$ (symmetric) and $S_{21} = S_{12}$ (reciprocity).

Incident and reflected voltage waves are also used to define scattering transfer parameters (T-parameters) as well S-parameters. However, T-parameters represent the relation between the waves at port 1 to the waves at port 2 as given in (2).

$$\begin{bmatrix} a_1 \\ b_1 \end{bmatrix} = \begin{bmatrix} T_{11} & T_{12} \\ T_{21} & T_{22} \end{bmatrix} \begin{bmatrix} b_2 \\ a_2 \end{bmatrix}. \quad (2)$$

The direct measurement of T-parameters is not as easy as S-parameters. Thus, S-parameters were measured and used for busbar modeling. Then, due to the ease of calculating cascaded networks by matrix multiplication with T-parameters, S-matrix of each single-branch network was converted to T-matrix as explained in Sec. 3.

Necessary parameters to find the transfer function of a transmission line (Z_c and γ) can be calculated as (3) and (4) if p.u.l. parameters are known.

$$Z_c = \sqrt{\frac{R + j\omega L}{G + j\omega C}}, \quad (3)$$

$$\gamma = \alpha + j\beta = \sqrt{(R + j\omega L)(G + j\omega C)}. \quad (4)$$

There are some conventional and modified transmission line characterization methods from VNA measurements [25], [26–28]. This paper uses frequency-dependent $RLGC(f)$ model method for p.u.l. parameters extraction from measured S-parameters. The results have been more accurate and efficient in a large frequency band due to they eliminate the discontinuity caused by the hyperbolic functions [18–20], [22]. Additionally, the limitations of the conventional methods such as specific line lengths, frequencies, the number of lines, etc. will be exceeded. Estimation of p.u.l. parameters from measurements with modeling includes the minimization of an objective function which can generally be non-linear. The error between measured data and the data from the model defines the objective function. In this study, the objective function consists of mean squared errors between real (Re) and imaginary (Im) parts of only S_{11} and S_{21} as given in (5) due to the reciprocity. The modeling experiments using the only magnitude of S_{21} gives relatively significant estimation errors on S_{11}. Thus, with including Re and Im parts of S_{11} and S_{21} parameters in the objective function, optimized unit length element of transmission line parameters will reveal the measured data more accurately.

$$F_{obj} = \frac{1}{M} \sum_{i=1}^{M} \begin{Bmatrix} \mathrm{Re}\left(S_{11}^{m}(f_i) - S_{11}^{e}(f_i)\right)^2 + \\ \mathrm{Re}\left(S_{21}^{m}(f_i) - S_{21}^{e}(f_i)\right)^2 + \\ \mathrm{Im}\left(S_{11}^{m}(f_i) - S_{11}^{e}(f_i)\right)^2 + \\ \mathrm{Im}\left(S_{21}^{m}(f_i) - S_{21}^{e}(f_i)\right)^2 \end{Bmatrix} \quad (5)$$

In (5), M is the total number of measurement points, $S^m(f_i)$ and $S^e(f_i)$ correspond to measured and estimated S-parameters, respectively. PSO [29] was selected to minimize F_{obj} due to its superior performance to find global minimum in case of a large parameter search range. In our previous study [17], a trial and error method was utilized. This method needs some well-selected initial values, unlike PSO which only needs parameter search bounds. Thus, in [22], trial and error procedure started at the point that is inspired from extracted parameters of a parallel plate transmission line. When modeling a different type of busbar (different current level), a different set of initial values has to be selected for optimization. With PSO, this difficulty is removed, and the algorithm is generalized in this paper.

A frequency-dependent $RLGC(f)$ model was used to extract the p.u.l. parameters of the busbar transmission line as in (6)

$$\begin{aligned} R(f) &= R_1 + R_2\sqrt{f}, \\ L(f) &= L_1 + R_2/\left(2\pi\sqrt{f}\right), \\ G(f) &= G_1 + G_2 f, \\ C(f) &= C_1 \end{aligned} \quad (6)$$

where R_1, R_2, L_1, G_1, G_2, and C_1 are constant unknowns that should be estimated. R_1, R_2, L_1, G_1, G_2, and C_1 refer to DC resistance, skin effect loss, inductance at high frequencies (generally constant), shunt current due to free electrons in an imperfect dielectric, power loss due to dielectric polarization, and geometry-related capacitance constant, respectively. For all different type busbar modeling procedures, the search range set given in Tab. 2 was used.

The conversion from RLGC parameters to S-parameters starts with calculation of Z_c and γ with (3) and (4), respectively. Then, for a specific line length of busbar ($l = 3$ m), ABCD matrix is created with known Z_c and γ parameters as in (7)

$$\begin{bmatrix} A & B \\ C & D \end{bmatrix} = \begin{bmatrix} Z_c \cosh(\gamma l) Z_c^{-1} & Z_c \sinh(\gamma l) \\ \sinh(\gamma l) Z_c^{-1} & \cosh(\gamma l) \end{bmatrix}. \quad (7)$$

Finally, the S-parameters for busbar S^e are obtained from the ABCD matrix as in (8) [30].

$$S^e = \begin{bmatrix} S_{11}^e & S_{12}^e \\ S_{21}^e & S_{22}^e \end{bmatrix} = $$
$$\begin{bmatrix} \dfrac{AZ_0 + B - CZ_0^2 - DZ_0}{X} & \dfrac{2(AD - BC)Z_0}{X} \\ \dfrac{2Z_0}{X} & \dfrac{-AZ_0 + B - CZ_0^2 + DZ_0}{X} \end{bmatrix}$$
$$X = AZ_0 + B + CZ_0^2 + DZ_0. \quad (8)$$

The overall parameter estimation process with PSO is given as a flowchart in Fig. 2.

R_1 [Ω/m]	R_2 [Ω/(m√Hz)]	L_1 [H/m]	G_1 [S/m]	G_2 [Ω/(mHz)]	C_1 [F/m]
0~10	0~1	0~10^{-5}	0~1	10^{-11}~10^{-8}	10^{-11}~10^{-8}

Tab. 2. Lower and upper bounds of parameters used in PSO.

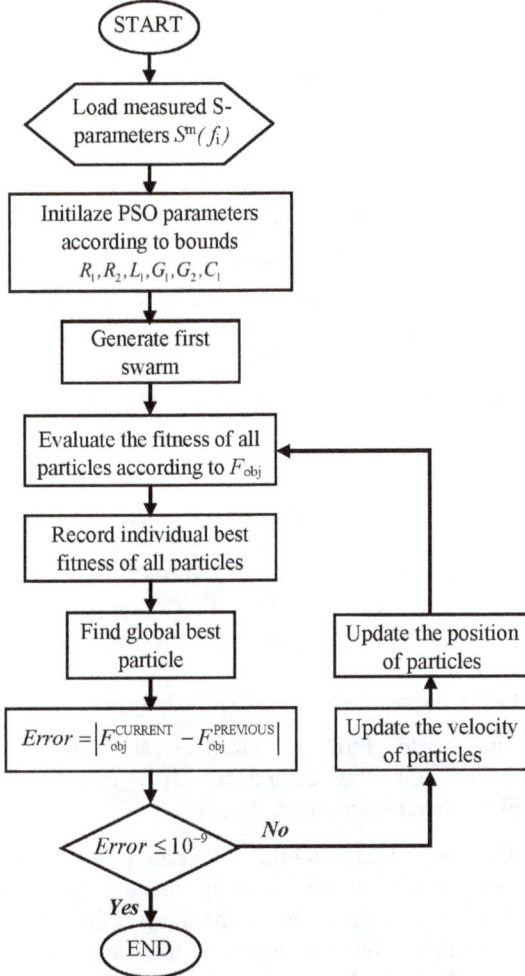

Fig. 2. Optimization process flowchart of the model parameter extraction.

2.2 Modeling Results

In this study, all of the signal paths were modeled for each current level busbar and the results were analyzed. As there are 18 different estimation results, it is not feasible to draw all of the combinations in this paper. Thus, a comparison of measured and simulated S-parameters for only some phase to neutral (L1-N, L2-L1, L3-N) and phase to phase (L2-L1 L3-L1, L3-L2) signal paths of 630 A, 1250 A, and 2000 A current level busbars are given in Figs. 3, 4, and 5, respectively. As the objective function uses measured S-parameters, visual presentations of results are given as a qualitative analysis. On the other hand, since the voltage transfer function of a two-port network is S_{21}, the goodness of fit value was calculated via amplitude and phase of S_{21}. The R-square value [31] was used as quantitative error measure. The averaged R-square values are 0.93 and 0.98 for amplitude and phase of S_{21}, respectively.

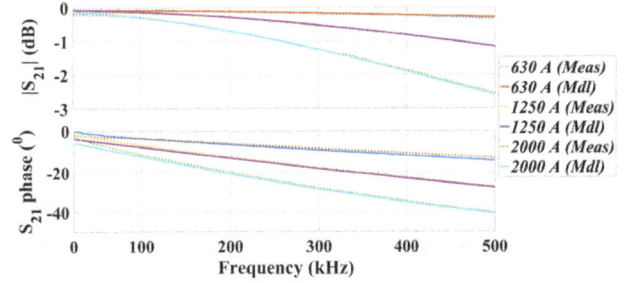

Fig. 3. Measured (*Meas*) and Estimated (*Mdl*) S_{21} parameters for L1-N signal path (3 m).

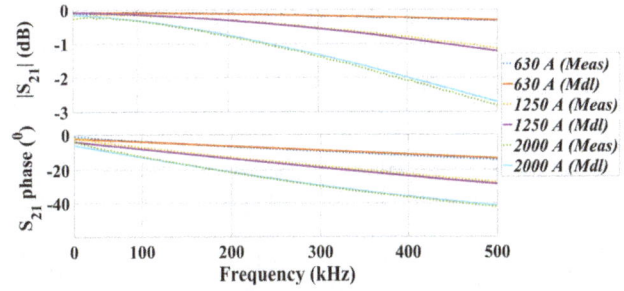

Fig. 4. Measured (*Meas*) and Estimated (*Mdl*) S_{21} parameters for L2-L1 signal path (3 m).

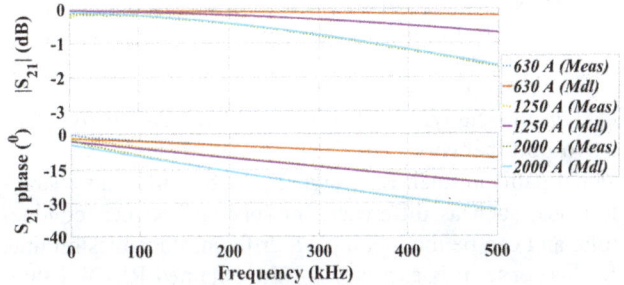

Fig. 5. Measured (*Meas*) and Estimated (*Mdl*) S_{21} parameters for L3-N signal path (3 m).

Because the R-square value of 1 means an exact match, the obtained error values show that the model results can be considered as a very close estimation.

Calculated frequency-dependent RLGC values are presented in Figs. 6, 7 and 8 for different current level busbars for some signal paths. When the figures are examined, it is easily seen that resistance and inductance are decreased with the increasing cross-section area as expected for all signal paths. Additionally, conductance and capacitance are increased.

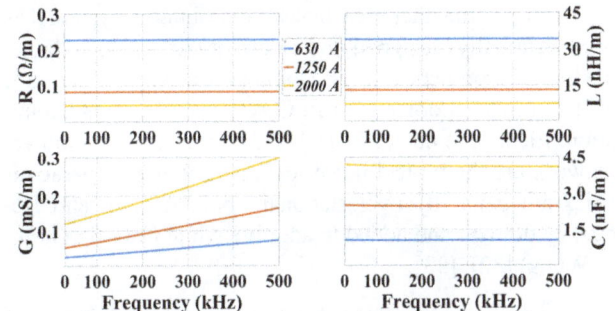

Fig. 6. Estimated RLGC parameters of different current level busbars from measured S-parameters for L1-N signal path.

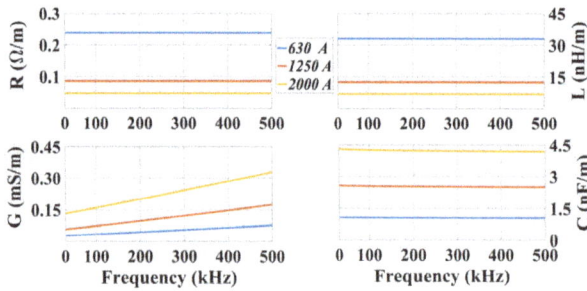

Fig. 7. Estimated RLGC parameters of different current level busbars from measured S-parameters for L2-L1 signal path.

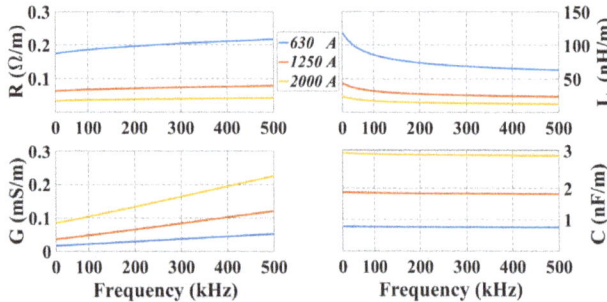

Fig. 8. Estimated RLGC parameters of different current level busbars from measured S-parameters for L3-N signal path.

In this study, busbar was considered and analyzed as a two-port network. So, there are six different two-port signal paths in total. Each signal path has different characteristics, such as differences in spacing between conductors, and can be thought of as a different transmission line. In this sense, it is expected for the obtained RLGC values to be different. Additionally, L1-N and L2-L1 signal paths have physically similar conditions (such as inter-conductor distances and insulating layer thickness). For this reason, similar RLGC values are estimated, as can be seen from Figs. 6 and 7. On the other hand, L3-N has some differences such as distance between conductors and an increased insulation thickness. This is the main reason for the differences between the estimated RLGC values.

3. Bit-Error-Rate Performance Analysis

Due to the fact that busbar distribution system has been originally designed for electricity transfer, it has a hostile environment for data communication. Noise is one of the primary problems for PLC systems as well as attenuation. Electronic devices which are connected to the power network are the typical noise sources. The noise characteristics are generally cyclostationary both in time and frequency domain and added to additive white Gaussian noise (AWGN) [32], [33].

The averaged power spectral density (PSD) for the typical colored background noise in narrowband PLC can be seen in Fig. 9 [34]. In this study, we used an assumption

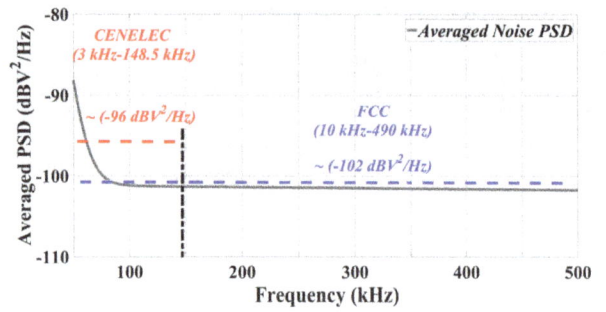

Fig. 9. Averaged PSD of background noise [27].

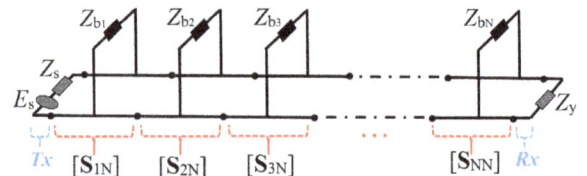

Fig. 10. An example of a busbar distribution network.

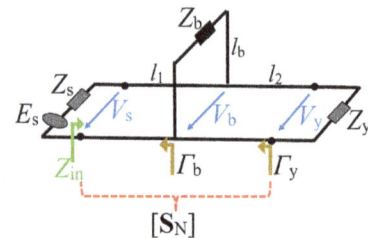

Fig. 11. Representation of a single-branch network.

that the noise PSD is constant as approximately 96 dBV^2/Hz for the theoretical BER calculations at CENELEC band according to Fig. 9.

The attenuation is another important parameter for the PLC and it is directly related to the channel transfer function. When Z_c and γ are known, the transfer function of the PLC channel can be determined. In this study, bottom-up frequency-domain channel modeling approach was preferred. Z_c and γ parameters were calculated from (2) and (3) for all busbar types and each signal paths with the help of estimated RLGC parameters. Basically, the busbar PLC channel can be defined as separated N-cascaded single-branch networks [35] as shown in Fig. 10.

Firstly, the S-matrix of each single-branch network as shown in Fig. 11 was found with the transmission line theory. Then, the chain-scattering matrix method was used to determine the whole busbar PLC network with the different combinations of single-branch networks. The S_{21} term of the scattering matrix of the whole PLC network gives the desired transfer function [30]. When the configuration of the busbar PLC network is changed, the transfer function of the entire network is affected. The configuration of network changes with busbar type, signal paths, busbar line length between the transmitter (Tx) and the receiver (Rx), branch number, branch length and connected load impedance.

In Fig. 11, Z_s, Z_y, and Z_b are source impedance (50 Ω), load impedance (50 Ω), and branch load impedance, respectively. Z_{in} corresponds to the input impedance of the whole single-branch network. Γ_y and Γ_b refer to the reflection coefficients from the load and from the branch tap, respectively. l is the line lengths in m for the relevant part of the network. S_{11} and S_{21} parameters of the single-branch network can be calculated with (9) [36]

$$S_{11} = \frac{Z_{in} - 50}{Z_{in} + 50}, \quad S_{21} = 2\frac{V_y}{E_s} = 2\frac{V_y}{V_b}\frac{V_b}{V_s}\frac{V_s}{E_s}. \quad (9)$$

S_{21} is obtained indirectly with a well-known approach by applying shifting in the reference planes (10) [30].

$$\frac{V_s}{E_s} = \frac{Z_s}{Z_s + Z_{in}}, \quad \frac{V_y}{V_b} = \frac{(1+\Gamma_y)e^{-\gamma_2 l_2}}{1+\Gamma_y e^{-2\gamma_2 l_2}}, \quad \frac{V_b}{V_s} = \frac{(1+\Gamma_b)e^{-\gamma_1 l_1}}{1+\Gamma_b e^{-2\gamma_1 l_1}}$$
$$(10)$$

Then, the T-matrix for the each single-branch network can be calculated from S-matrix. For this calculation, the conversion equations are used as shown in (11)

$$\mathbf{T} = \begin{bmatrix} \dfrac{1}{S_{21}} & -\dfrac{S_{22}}{S_{21}} \\ \dfrac{S_{11}}{S_{21}} & S_{12} - \dfrac{S_{11}S_{22}}{S_{21}} \end{bmatrix}, \quad \mathbf{S} = \begin{bmatrix} \dfrac{T_{21}}{T_{11}} & T_{22} - \dfrac{T_{21}T_{12}}{T_{11}} \\ \dfrac{1}{T_{11}} & -\dfrac{T_{12}}{T_{11}} \end{bmatrix}. \quad (11)$$

Total T-matrix for the N-branch network ($\mathbf{T}_N$) can be calculated as given in (20) where $\mathbf{T}_{iN}$ represents the T-matrix of the ith single-branch network. Then, the T-matrix of the whole network can be converted to S-matrix using the (12). The S_{21} term of the S-matrix gives the transfer function of the whole PLC network, $H(f)$.

$$\mathbf{T}_N = \prod_{i=1}^{N}\left[\mathbf{T}_{iN}\right] \quad (12)$$

For a specific channel bandwidth and a certain channel capacity, the communication link performance can be calculated as a function of energy per bit to noise density ratio E_b/N_0 as given in (13)

$$\frac{E_b}{N_0} = \frac{C}{N}\frac{B_w}{f_b} \quad (13)$$

where C is the received signal power (W), N is the total noise power (W), B_w is the channel bandwidth (Hz), and f_b is channel data rate (bit/s) or capacity. In this study, the communication link performance of the busbar PLC system for different network configurations was investigated at CENELEC band ($B_w = 145.5$ kHz between 3 kHz–148.5 kHz). In Europe, CENELEC has formed the standard EN-50 065-1, in which the frequency bands, signaling levels and procedures are specified [37]. The transmitted signaling level (V_t) is limited to 134 dBµV for industrial areas according to the standard [37] and this value is used for the calculations. Transmitted signal power S_t (W) according to the well-known maximum power transfer theorem and received signal power C (W) was calculated using (14) where Z_{in} is input impedance of the network and $H(f)$

is channel transfer function. Z_{in} is a frequency dependent value. Thus, the mean value of Z_{in} was utilized for the calculations

$$S_t = \frac{V_t^2}{4Z_{in}}, \quad C = S_t|H(f)|^2. \quad (14)$$

Channel transfer function shows also a frequency-dependent characteristic and the attenuation increases with increasing frequency. In (14), the maximum attenuation value of $H(f)$ at this frequency band (worst case) for each network configuration was used for calculations. Finally, the E_b/N_0 values were calculated with these assumptions for a certain data rate. BER performance analysis vs. SNR of the busbar systems for different network configurations was made for FSK and PSK modulations. Theoretical bit error probabilities (P_e) for coherent FSK and PSK are given in (15)

$$P_e^{(FSK)} = \frac{1}{2}\mathrm{erfc}\sqrt{\frac{E_b}{2N_0}}, \quad P_e^{(PSK)} = \frac{1}{2}\mathrm{erfc}\sqrt{\frac{E_b}{N_0}} \quad (15)$$

where erfc(*) is the complementary error function.

4. Numerical Results

To show the effect of the different network topologies on BER performances, some simulations were conducted for 630 A, 1250 A, and 2000 A current level busbars at CENELEC band. Instead of giving results for all signal paths, the worst and best cases in terms of channel attenuation were selected according to the transfer functions as given in Fig. 12.

Thus, L2-L1 signal path for the 630 A current level busbar was determined as the worst case scenario due to maximum attenuation behavior while L3-N signal path for the 2000 A current level busbar was the best case. Then, the achievable channel capacities were determined for the best and worst busbar network topology cases using the Shannon's Theorem as shown in Fig. 13. It shows a rough variation range of channel capacity (Mbps) against PSD of transmitted signal power (dBm/Hz). For an average PSD of transmitted power –60 dBm/Hz, the mean of achievable channel capacity is approximately $f_b = 1$ Mbps from Fig. 13.

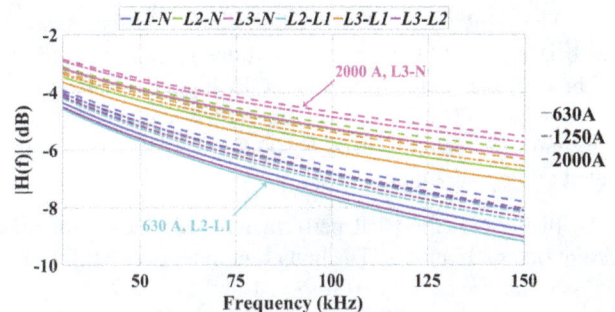

Fig. 12. Transfer functions for all signal paths of a length 100 m without branch.

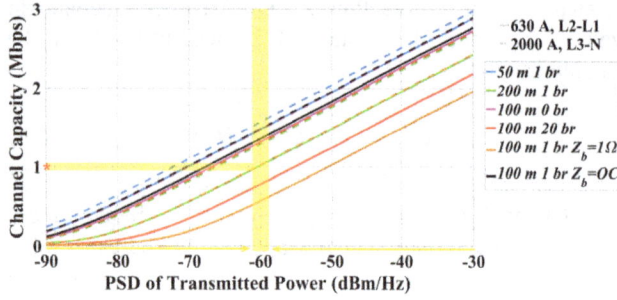

Fig. 13. Channel capacities for different busbar network topologies.

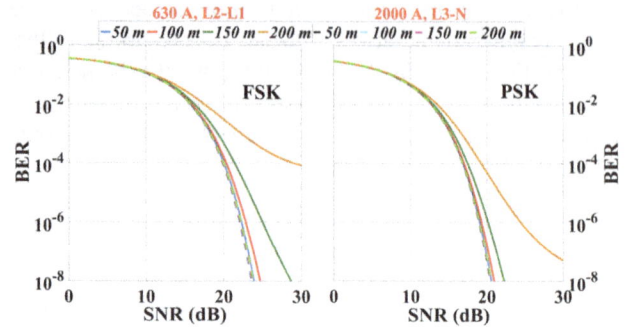

Fig. 14. FSK-PSK bit error probabilities of different busbars for different line lengths between *Tx* and *Rx*.

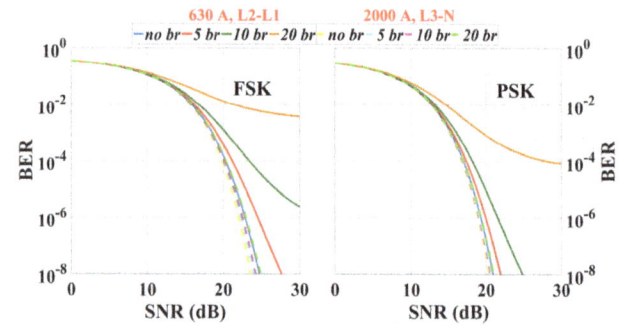

Fig. 15. FSK-PSK bit error probabilities of different busbars for different branch numbers between *Tx* and *Rx*.

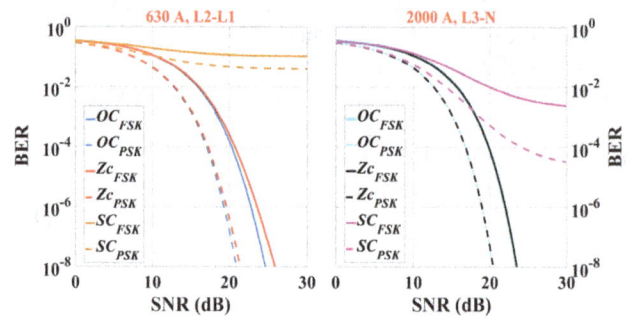

Fig. 16. FSK-PSK bit error probabilities of different busbars for different terminating load impedances.

To support the 1 Mbps value, [38] can be referred for typical narrowband channel capacities. For these calculations, load impedances and branch lengths were taken 100 Ω and 3 m, respectively.

Since the largest voltage drop permitted in the LV distribution system is about 3%, the maximum line length (l_T) between *Tx* and *Rx* for a 630 A current level busbar can be calculated with provided technical characteristics [23] [18] for $\cos\varphi = 0.8$, and $\alpha_y = 0.5$ as approximately 150 m with a well-known formula as shown in (16)

$$\Delta V = \alpha_y \sqrt{3}\left[l_T I \left(R\cos\varphi + X\sin\varphi\right)10^{-3}\right] \quad (16)$$

where $\cos\varphi$ is power factor of the line; α is the distribution factor of the loads (or the utilization factor), which depends on the type and number of loads; l_T is the line length (m); I is the line current (A); R is the line resistance (mΩ/m); and X is the line reactance (mΩ/m). At the same conditions, the maximum l_T will be approximately 195 and 205 m for 1250 A and 2000 A busbars, respectively. For this reason, in the simulation studies, a line length of 200 m was taken as the maximum length between *Tx* and *Rx*.

BER performances vs. SNR are shown in Fig. 14, Fig. 15, and Fig. 16. These figures illustrate the relationship between bit error probability and busbar type, busbar length between *Tx* and *Rx* (l_T in m), branch number (*br*), branch length (l_b in m), and branch terminating load (Z_b) at CENELEC band. The branch terminating load refers to the load impedance at the end of the branch. These results can give an idea for busbar PLC channel especially for remote control applications such as automatic meter reading and remote control systems design.

The BER vs. SNR of a single-branch busbar system for best case and worst case conditions to show the effect of busbar line length are shown in Fig. 14. The l_T was taken as 50, 100, 150, and 200 m, branch length was set as 3 m and terminated with $Z_b = 100$ Ω. The branch was located at the midpoint of *Tx-Rx*.

In Fig. 15, the BER performances are shown for different branch numbers. The branch number between *Tx* and *Rx* was increased from 0 to 5, 10 and 20 for $l_T = 100$ m. They also were positioned at equal intervals along the l_T. Branches were terminated with $Z_b = 100$ Ω, and their lengths were taken $l_b = 3$ m.

The load impedance effect on the BER performance of a single-branch busbar was examined for varying Z_b as short circuit (*SC*), characteristic impedance, and open circuit (*OC*) as shown in Fig. 16. These examinations were made for $l_T = 100$ m and $l_b = 3$ m and the branch was located at the midpoint of *Tx-Rx*.

The effect of branch length, on the other hand, was also investigated and concluded that there are no significant changes over BER performances. Thus, they were not included in this paper.

Obtained findings are very important in solving the communication link design and performance analysis of the narrowband PLC channel for busbar distribution systems. The results of this study can be summarized as follows:

- Different current level busbars (630 A, 1250 A, and 2000 A) were examined for six different signal paths (L1-N, L2-N, L3-N, L2-L1, L3-L1, L3-L2). A frequency-dependent RLGC(f) model was used to represent the busbar transmission line and model parameters were extracted from measured S-parameters with PSO. The results are given in Figs. 3–8 with 0.93 and 0.98 averaged R-square values for amplitude and phase of S_{21}, respectively.

- A bottom-up frequency-domain channel modeling approach was preferred to obtain N-branch busbar distribution network and chain-scattering matrix method was used to determine the whole busbar PLC network. For this aim, a GUI was created to simulate the different busbar network topologies by varying busbar type, signal path, line length, branch number, and load impedances.

- 630 A, L2-L1 was determined as the worst case scenario due to maximum attenuation behavior of $H(f)$ while 2000 A, L3-N was the best channel (Fig. 12).

- The possible line length (l_T) is limited according to the permitted voltage drop (3%) at LV grid. Calculated l_T values for 630 A, 1250 A and 2000 A current level busbars are 150, 195 and 205 m, respectively using (16). In this study, maximum line length was selected as 200 m.

- For the studied network topologies in terms of different busbar line lengths, different branch numbers and different load impedances, Z_{in} and attenuation (*Atten.*) ranges can be given in Tab. 3. These values show the used Z_{in} and *Atten.* ranges for the calculations.

- Capacity (data rate) was calculated according to Shannon's Theorem. Bandwidth (B_w) was taken 145.5 kHz (all CENELEC band). It means that the calculated capacity is the total capacity of the communication link. Capacity varies from 300 kbps to 3 Mbps for the different network topologies as shown in Fig. 13. For an average PSD of transmitted power, the capacity was taken $f_b = 1$ Mbps for calculations.

- BER vs. SNR performances were obtained for the basic modulations, FSK and PSK. In these calculations, V_t and averaged PSD of background noise (from Fig. 9) were taken 134 dBμV and −96 dBV^2/Hz, respectively.

- Figs. 14–16 show the BER vs. SNR performances for different channel conditions at both modulation types: FSK and PSK. 200 m line length has the worst performance (Fig. 14). Additionally, increasing branch number decreases BER performance (Fig. 15). Another important disruptive effect may be defined as load impedance. When the terminating load impedance at the branches decreases, the BER performance of the system decreases (Fig. 16).

As a result, if the line length, branch number, and the terminating load impedances are assumed within a meaningful range, the system performance is an acceptable level for a busbar PLC link. Obtained results from Figs. 14–16 conclude that the BER is smaller than 10^{-5} for a 20 dB SNR value.

Different network topologies					
Line Length 50 m < l_T < 200 m		Branch Number 0 < br < 20		Load Impedance SC < Z_b < OC	
Z_{in} (Ω)	Atten. (dB)	Z_{in} (Ω)	Atten. (dB)	Z_{in} (Ω)	Atten. (dB)
630 A, L2-L1					
25-30	4.5-18	27-33.5	8.8-20	12.5-29.5	8.2-31
2000 A, L3-N					
6.5-14.5	2.7-10.5	9.6-30	5-8.8	2-10.3	5-26.3

Tab. 3. The observed Z_{in} and *Atten.* ranges for studied different network configurations.

5. Conclusion

Busbar is a key element of modern distribution systems in LV grid especially in industrial areas to carry high power. As one of the main targets of the smart grid is to increase the energy quality, existing grid undoubtedly needs some information and communication technologies (ICT). PLC is one of the most powerful communication technique in the smart grid concept. Thus, busbar should be examined in terms of PLC possibilities to provide an integrated communication network in the LV grid. On the other hand, M2M is one of the important backbones in the smart grid. Because busbar is generally used in the industrial areas, it provides a natural link between machines. If PLC over busbar is succeeded, no extra cabling and RF link are needed for an integrated M2M communication. Thus, it is essential to know the characteristics of the busbar distribution systems for providing PLC communication to contribute to the existing literature.

While cable models have been proposed for narrowband PLC applications, no modeling of busbars as transmission lines have been offered in terms of PLC and the smart-grid concept to the best of the authors' knowledge. The results gained from this study show that discussed busbar PLC system can be realized and applied in existing power grids. On the other hand, the worst-case BER performances can be improved using advanced modulation, coding and error correction techniques. As a result, PLC system designers will be aware of the busbar's characteristics and possible link performances with the help of this study.

Acknowledgments

This work was supported by the TUBITAK (The Scientific and Technological Research Council of Turkey), 1003- Primary Subjects R&D Funding Program, Project No: EEEAG-115E137. The authors would like to thank also EAE Company for their support providing the busbar distribution systems.

References

[1] PAPAZYAN, R. PETTERSON, P., EDIN, H., ERIKSSON, R., GAFVERT, U. Extraction of high frequency power cable

characteristics from S-parameter measurements. *IEEE Transactions on Dielectrics and Insulation*, 2004, vol. 11, no. 3, p. 461–470. DOI: 10.1109/TDEI.2004.1306724

[2] HASIRCI, Z., CAVDAR, I. H. Modeling of high power busbar systems for power line communications. In *IEEE International Energy Conference (ENERGYCON 2014)*. Dubrovnik (Croatia), 2014, p. 1515–1519. DOI: 10.1109/ENERGYCON.2014.6850623

[3] HASIRCI, Z., CAVDAR, I. H., SULJANOVIC, N., MUJCIC, A. Investigation of current variation effect on PLC channel characteristics of LV high power busbar systems. In *The 5th IEEE PES European 2014 Conference on Innovative Smart Grid Technologies (ISGT)*. Istanbul (Turkey), 2014, 5 p. DOI: 10.1109/ISGTEurope.2014.7028778

[4] MARKS, R. B., WILLIAMS, D. F. Characteristic impedance determination using propagation constant measurement. *IEEE Microwave Guided Wave Letters*, 1991, vol. 1, no. 6, p. 141–143. DOI: 10.1109/75.91092 10.1109/75.91092

[5] GOLDBERG, S. B., STEER, M. B., FRANZON, P. D. Experimental electrical characterization of interconnects and discontinuities in high-speed digital systems. *IEEE Transactions on Components, Hybrids, and Manufacturing Technology*, 1991, vol. 14, no. 4, p. 761–765. DOI: 10.1109/33.105130

[6] WILLIAMS, D. F., ROGERS, J. E., HOLLOWAY, C. L. Multi-conductor transmission - line characterization: Representations, approximations, and accuracy. *IEEE Transactions on Microwave Theory and Techniques*, 1999, vol. 47, no. 4, p. 403–409. DOI: 10.1109/22.754872

[7] CHEN, G., ZHU, L., MELDE, K. Extraction of frequency dependent RLCG parameters of the packaging interconnects on low-loss substrates from frequency domain measurements. In *14th Topical Meeting on Electrical Performance of Electronic Packaging (IEEE-EPEP 2005)*. 2005, p. 25–28. DOI: 10.1109/EPEP.2005.1563691

[8] KIM, J., HAN, D. Hybrid method for frequency-dependent lossy coupled transmission line characterization and modeling. In *Electrical Performance of Electronic Packaging (EPEP2003)*. Princeton (NJ, USA), 2003, p. 239–242. DOI: 10.1109/EPEP.2003.1250040

[9] DEUTSCH, A., ARJAVALINGAM, G., KOPCSAY, G. V. Characterization of resistive transmission lines by short-pulse propagation. *IEEE Microwave and Guided Wave Letters*, 1992 (current ver. 2002), vol. 2, no. 1, p. 25–27. DOI: 10.1109/75.109132

[10] FERRARI, P., FLECHET, B., ANGENIEUX, G. Time domain characterization of lossy arbitrary characteristic impedance transmission lines. *IEEE Microwave and Guided Wave Letters*, 1994 (current ver. 2002), vol. 4, no. 6, p. 177–179. DOI: 10.1109/75.294284

[11] KIM, W., LEE, S., SEO, M., SWAMINATHAN, M., TUMMALA, R. Determination of propagation constants of transmission lines using 1-Port TDR measurements. In *59th ARFTG Conference Digest*. Seattle (WA, USA), 2002, p. 119–126. DOI: 10.1109/ARFTGS.2002.1214689

[12] DEGERSTROM, M. J., GILBERT, B. K., DANIEL, E. S. Accurate resistance, inductance, capacitance, and conductance (RLCG) from uniform transmission line measurements. In *Electrical Performance of Electronic Packaging (IEEE-EPEP2008)*. San Jose (CA, USA), 2008, p. 77–80. DOI: 10.1109/EPEP.2008.4675881

[13] ZUNIGA-JUAREZ, J. E., REYNOSO-HERNANDEZ, J. A., MAYA-SANCHEZ, M. C., MURPHY-ARTEAGA, R. S. A new analytical method to calculate the characteristic impedance Zc of uniform transmission line. *Computación y Sistemas*, 2012, vol. 16, no. 3, p. 277–285.

[14] REYNOSO-HERNANDEZ, J. A. Unified method for determining the complex propagation constant of reflecting and nonreflecting transmission lines. *IEEE Microwave and Wireless Components Letters*, 2003, vol. 13, no. 8, p. 351–353. DOI: 10.1109/LMWC.2003.815695

[15] BIANCO, B., PARODI, R. M. Determination of the propagation constant of uniform microstrip lines. *Alta Frequence*, 1976, vol. 45, no. 2, p.107–110.

[16] JANEZIC, M. D., JARGON, J. A. Complex permittivity determination from propagation constant measurements. *IEEE Microwave and Guided Wave Letters*, 1999, vol. 9, no. 2, p. 76–78. DOI: 10.1109/75.755052

[17] REYNOSO-HERNÁNDEZ, J. A., ESTRADA-MALDONADO, C. F., PARRA, T., GRENIER, K., GRAFFEUIL, J. An improved method for the wave propagation constant estimation in broadband uniform millimeter-wave transmission line. *Microwave and Optical Technology Letters*, 1999, vol. 22, no. 4, p. 268–271. DOI: 10.1002/(SICI)1098-2760(19990820)22:4<268::AID-MOP16>3.0.CO;2-6

[18] ZHANG, J., CHEN, Q. B., QIU, Z., DREWNIAK, J. L., ORLANDI, A. Extraction of causal RLGC models from measurements for signal link path analysis. In *2008 International Symposium on Electromagnetic Compatibility - EMC Europe*. Hamburg (Germany), 2008, 6 p. ISSN: 2325-0356. DOI: 10.1109/EMCEUROPE.2008.4786839

[19] ZHANG, J., DREWNIAK, J. L., POMMERENKE, D. J., KOLEDINTSEVA, M. Y., DUBROFF, R. E., CHENG, W., YANG, Z., CHEN, Q. B., ORLANDI, A. Causal RLGC(f) models for transmission lines from measured S-parameters. *IEEE Transactions on Electromagnetic Compatibility*, 2010, vol. 52, no. 1, p. 189–198. DOI: 10.1109/TEMC.2009.2035055

[20] ZHANG, J., KOLEDINTSEVA, M. Y., DREWNIAK, J. L., ANTONINI, G., ORLANDI, A. Extracting R, L, G, C parameters of dispersive planar transmission lines from measured S-parameters using a genetic algorithm. In *International Symposium on Electromagnetic Compatibility, EMC 2004*. Eindhoven (Netherlands), 2004, vol. 2, p. 572–576. DOI: 10.1109/ISEMC.2004.1349861

[21] HASIRCI, Z., CAVDAR, I. H. Extraction of narrowband propagation properties of a 630 A current level busbar. In *39th International Conference on Telecommunications and Signal Processing, TSP 2016*. Vienna (Austria), 2016, p. 203–206. DOI: 10.1109/TSP.2016.7760860

[22] HASIRCI, Z., CAVDAR, I. H., OZTURK, M. Estimation of propagation parameters for aluminum busbar up to 500 kHz. In *2016 International Symposium on Innovations in Intelligent Systems and Applications (INISTA)*. Sinaia (Romania), 2016. DOI: 10.1109/INISTA.2016.7571824

[23] EAE COMPANY, TURKEY. *E-Line KX Busbar Power Distribution System (datasheet)*. 57 pages. [Online] Cited 2016-05-03. Available at: http://eae.com.tr/EAE-ENG/upload/E-Line%20KX_eng.pdf

[24] GUPTA, K. C., GARG, R., CHADHA, R. *Computer Aided Design of Microwave Circuits*. Dedham (MA): Artech House, p. 2543, 1981. ISBN: 9780890061053

[25] BAKHOUM, E. G. S-parameters model for data communications over 3-phase transmission lines. *IEEE Transactions on Smart Grid*, 2011, vol. 2, no. 4, p. 615–623. DOI: 10.1109/TSG.2011.2168613

[26] SAMPATH, M. K. On addressing the practical issues in the extraction of RLGC parameters for lossy multi-conductor transmission lines using S-parameter models. In *Electrical Performance of Electronic Packaging (IEEE-EPEP2008)*. San Jose (CA, USA), 2008, p. 259–262. DOI: 10.1109/EPEP.2008.4675929

[27] MARKS, R. B. A multiline method of network analyzer calibration. *IEEE Transactions on Microwave Theory and Technique*, 1991 (current ver. 2002), vol. 39, no. 7, p. 1205–1215. DOI: 10.1109/22.85388

[28] KIM, J., HAN, D. H. Hybrid method for frequency-dependent lossy coupled transmission line characterization and modelling. In *12th Topical Meeting on Electrical Performance of Electronic Packaging (IEEE-EPEP 2003)*. Princeton (NJ, USA), 2003, p. 239–242. DOI: 10.1109/EPEP.2003.1250040

[29] KENNEDY, J., EBERHART, R. C. Particle swarm optimization. In *Proceedings of IEEE International Conference on Neural Networks*. Piscataway (NJ), 1995, p. 1942–1948. DOI: 10.1109/ICNN.1995.488968

[30] POZAR, D. M. *Microwave Engineering*. 4th ed. New York (USA): Wiley, 2012. ISBN: 9780470631553

[31] *Goodness-of-Fit Statistics*. [Online] Cited 2016-08-30. Available at:http://web.maths.unsw.edu.au/~adelle/Garvan/Assays/Goodness OfFit.html

[32] CORTES, J. A., DIEZ, L., CANETE, F. J., SANCHEZ-MARTINEZ, J. J. Analysis of the indoor broadband power-line noise scenario. *IEEE Transactions on Electromagnetic Compatibility*, 2010, vol. 52, no. 4, p. 849–858. DOI: 10.1109/TEMC.2010.2052463

[33] KATAYAMA, M., YAMAZATO, T., OKADA, H. A. Mathematical model of noise in narrowband power line communication systems. *IEEE Journal on Selected Areas in Communications*, 2006, vol. 24, no. 7, p. 1267–1276. DOI: 10.1109/JSAC.2006.874408

[34] LIU, W. *Emulation of Narrowband Powerline Data Transmission Channels and Evaluation of PLC Systems*. Ph.D. dissertation, p. 163–166 Elektrotechnik und Informationstechnik, Karlsruher Institut für Technologie, Karlsruhe, Germany, 2013. ISBN:978-3-7315-0071-1

[35] MENG, H., CHEN, S., GUAN, Y. L., LAW, C. L., SO, P. L., GUNAWAN, E., LIE, T. T. Modeling of transfer characteristics for the broadband power line communication channel. *IEEE Transactions on Power Delivery*, 2004, vol. 19, no. 3, p. 1057–1064. DOI: 10.1109/TPWRD.2004.824430

[36] GONZALEZ, G. *Microwave Transistor Amplifiers*. 2nd ed. Englewood Cliffs (NJ): Prentice-Hall, 1997. ISBN: 9780132543354

[37] *Standard on Low-Voltage Electrical Installations in the Frequency Range 3 kHz to 148.5 kHz, Part One*, CENELEC Std. EN50065-1:19 917A1:1992, 1992.

[38] TONELLO, A. M., PITTOLO, A. Considerations on narrowband and broadband power line communication for smart grids. In *IEEE International Conference on Smart Grid Communications (SmartGridComm)*. Miami (FL, USA), 2015. DOI: 10.1109/SmartGridComm.2015.7436269

About the Authors ...

Zeynep HASIRCI was born in Samsun, Turkey. She received B.Sc. and M.Sc. degrees from Karadeniz Technical University (KTU) in 2008 and 2011, respectively. She studied for her Ph.D. thesis in Halmstad University, Sweden for one year. She is currently a Ph.D. student at the Dept. of Electrical and Electronics Engineering, KTU. Her research interests include communication systems, mobile and satellite communication, propagation modeling and power line communication.

Ismail Hakki CAVDAR was born in Trabzon, Turkey. He received his Ph.D. degree in Electrical and Electronics Engineering from Karadeniz Technical University (KTU), in 1994. He has been a Full Professor at KTU since 1985. He was a Visiting Professor at Smart Grid Lab in the Dept. of Electrical and Computer Engineering, The University of Akron, OH USA, in 2011. His research interests include communications systems, mobile and satellite communication, power line communications, smart grids and power electronics.

Mehmet OZTURK was born in Trabzon, Turkey. He received his B.Sc degree in Electronics Engineering from Kadir Has University in 2004. He received his M. Sc. and Ph.D. degrees in Electrical and Electronics Engineering from Karadeniz Technical University (KTU) in 2007 and 2013, respectively. His research interests include biomedical, signal and image processing, point clouds.

16

Half-Mode Substrate Integrated Waveguide Yagi Array with Low Cross Polarization

Zhao ZHANG[1], Xiangyu CAO[1], Jun GAO[1], Sijia LI[1], Liming XU[2]

[1] Information and Navigation College, Air Force Engineering University, Xi'an 710077, China
[2] Science and Technology on Electronic Information Control Laboratory, Chengdu 610036, China

bjzhangzhao323@126.com, xiangyucaokdy@163.com, gjgj9694@163.com, lsj051@126.com, 10575229@qq.com

Abstract. *Low cross polarization Yagi array with mirrored arrangement is proposed. First, the half-mode substrate integrated waveguide (HMSIW) and magnetic wall are introduced to realize the miniaturization of Yagi antenna. Simulated results show that the total area of the Yagi antenna is 1.82λ × 0.57λ and the peak gain is about 6.0~7.9 dBi in the 10.5% relative bandwidth with VSWR less than 2. Then the element arrangement of Yagi array composed by the HMSIW Yagi antenna is analyzed to accomplish cross polarization elimination. It is found that the mirrored arrangement eliminates the far field cross-polar component and leads to the lower cross polarization than the other non-mirrored arrays under the condition that they have almost the same bandwidth, peak gain and beam direction. The low cross polarization four-element array with mirrored arrangement is fabricated and measured, and experimental results agree well with the simulation.*

Keywords

Half-mode substrate integrated waveguide, low cross polarization, mirrored arrangement, Yagi array

1. Introduction

With the development of satellite communications, antennas achieving compact size, low profile, high gain, specific radiation patterns and ease integration with planar circuits are highly demanded. As a quasi end-fire antenna with a tilted beam in the elevation plane, the microstrip Yagi antenna is one of the candidates for the satellite communications system [1]. The design methods and the performance of microstrip Yagi antennas have been reported since 1990s [2–4]. Based on these antennas, many microstrip Yagi arrays have been developed to extend the performance. By duplicating the microstrip Yagi antenna perpendicular to the projective direction of the tilted beam, linear arrays would be obtained [5–7]. This method of array arrangement is mainly used to achieve a high gain greater than 10 dBi. In addition, electronic scanning could be achieved by feeding each element with calibrated phase

shifters [6]. Another array arrangement is the crossed Yagi array [8–11], which usually shares a common element at the center. Combined with the switching elements, such as single pole three throw switches and pin diodes, the crossed Yagi arrays have reconfigurable radiation patterns covering the full azimuth plane. These different methods of array arrangement provide more performance features for the Yagi array.

Recently, the half-mode substrate integrated waveguide (HMSIW) and the quarter-mode substrate integrated waveguide (QMSIW) with asymmetrical configurations have been proposed by employing the fictitious quasi magnetic wall in the substrate integrated waveguide (SIW) [12–14]. The HMSIW and QMSIW are only half or quarter size of the SIW but they almost preserve the field distribution of the original SIW. Therefore, the antennas based on the HMSIW and QMSIW not only reduce the antenna size, but also provide more choices for special array arrangements. In [15], an eighth-mode SIW (EMSIW) cavity is formed by bisecting the QMSIW, and then two EMSIW cavities are orthogonally placed to generate the circularly polarized wave [16]. Another compact circularly polarized antenna is proposed by sequential rotation of the isosceles right triangular QMSIW antennas of linear polarization [17]. However, the above two antennas exhibit a different cross polarization, which indicates that the element arrangement of incomplete-mode SIW has a particular effect on the cross polarization.

In this paper, the low cross polarization HMSIW Yagi array with mirrored arrangement is proposed. Firstly, a compact Yagi antenna is designed by introducing the HMSIW and magnetic wall to realize the antenna miniaturization. The total area of the HMSIW Yagi antenna is 1.82λ × 0.57λ (λ represents the wavelength at 10 GHz). Within the 10.5% relative bandwidth with VSWR less than 2, the peak gain is about 6.0~7.9 dBi. Furthermore, three kinds of two-element array and two kinds of four-element arrays with the same distance of array elements but different arrangements are comparatively analyzed to explore the cross polarization elimination. Simulated results show that the symmetrical arrays with mirrored arrangement, in which each element antenna is the mirrored image of the neighboring one, have a much lower cross polarization than the other arrays under the condition that they have almost the same bandwidth, peak gain and beam direction. Finally, the four-element array with mirrored arrangement is fabricated and measured, and the experimental results validate the simulation.

2. Design of HMSIW Yagi Antenna

The SIW can be regarded as a special rectangular waveguide with a series of slots on the bilateral narrow walls. This feature leads to the result that only TE_{n0} modes can exist in SIW [18], [19]. According to the boundary condition on the surface of SIW when the forward wave propagates along the x direction, the surface current density J_s can be expressed as [20]

$$\boldsymbol{J}_s = \boldsymbol{e}_n \times \boldsymbol{H} = \boldsymbol{e}_x H_x - \boldsymbol{e}_y H_y = \boldsymbol{J}_x + \boldsymbol{J}_y. \quad (1)$$

As shown in Fig. 1(a), the surface current of SIW for the dominant TE_{10} mode has two components of J_x and J_y. They are symmetrically distributed along the A-A' plane and the component J_y has the opposite flowing direction at two sides of A-A'. It also can be found that the current density has its maximum at the A-A' plane. So the fictitious magnetic wall is available for the A-A' plane. After cutting the A-A' plane, the SIW becomes HMSIW which keeps half of the surface current distribution of the dominant mode as shown in Fig. 1(b). The current density has its maximum at the magnetic wall B-B'. It should be pointed out that the resonant frequencies of the HMSIW are slightly shifted up compared with the corresponding SIW one due to the fringing fields of the non-ideal magnetic wall B-B'.

In this paper, we introduce the HMSIW as the driven element of Yagi antenna. The configuration of the proposed HMSIW Yagi antenna is shown in Fig. 2(a). It consists of a driven element D, a coupling patch C and two director elements (D1 and D2). All the metal patches are printed on the upside of Rogers RT/duroid 5880 substrate

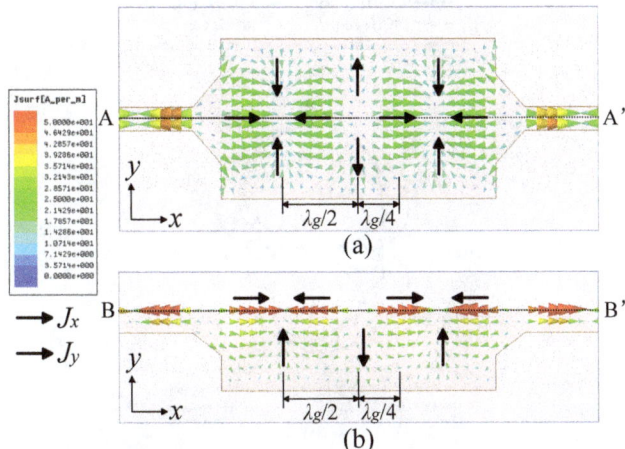

Fig. 1. Comparison of surface current distribution in (a) SIW and (b) HMSIW.

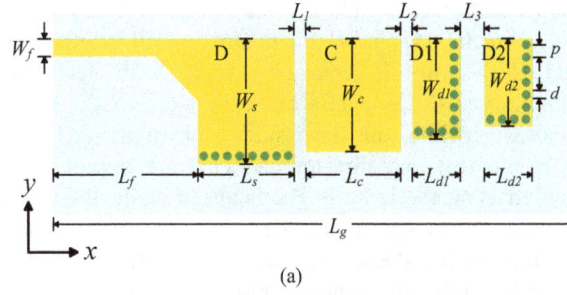

(a)

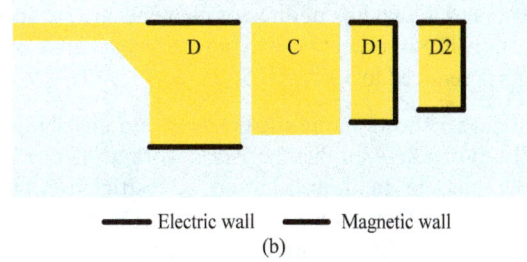

(b)

Fig. 2. Configuration of the HMSIW Yagi antenna. (a) Top view. (b) Equivalent model with electric and magnetic wall.

Parameters	Value	Parameters	Value
W_s	11.3	L_s	7.4
W_c	10.0	L_c	8.0
W_{d1}	8.8	L_{d1}	4.3
W_{d2}	8.3	L_{d2}	4.3
W_g	17	L_g	54.6
L_1	0.15	p	1
L_2	0.55	d	0.6

Tab. 1. Parameters of the proposed antenna (Unit: millimeter).

($\varepsilon_r = 2.2$ and $\tan \delta = 0.001$) with thickness h and the underside is a copper ground plane. The driven element D is a part of HMSIW which works under the dominant TE_{10} mode. Its width W_s determines the cutoff frequency of HMSIW as well as the working band of the proposed antenna. Due to the period distribution of the standing wave, the length of HMSIW can be flexible. Here, the length L_s of driven element is optimized between $\lambda_g/4$ and $\lambda_g/2$. The operating principles of the coupling patch and director elements are similar to those of the magnetic Yagi antenna in [4]. The coupling patch is utilized to improve the impedance matching when its resonant frequency is close to that of HMSIW [21]. Also, it functions as a bridge that guides the electromagnetic energy to the director element so as to radiate forward. The director elements with two edges shorted by plated through holes (PTHs) work as magnetic dipoles to increase the gain and the front-to-back ratio. All the PTHs in HMSIW and the director elements should meet the equations $0.05 < p/\lambda_c < 0.25$ where p is the distance between PTHs and λ_c is the cutoff wavelength [22], so that they can be equivalent to electric wall and leakage losses are negligible. Parametric analyses were conducted using Ansoft HFSS and the optimized dimensions are listed in Tab. 1.

What makes the proposed antenna asymmetrical and compact is the introduction of magnetic wall. Due to the symmetrical electric field distribution at both sides of the driven-director axis in [4], the magnetic wall is introduced. Figure 2(b) shows the equivalent model with electric and magnetic walls. The PTHs are equivalent to the electric walls. Meanwhile, the open sides (shown in red) of the driven element and director elements are approximately regarded as magnetic walls. By means of image theory, this arrangement has a pronounced effect on the antenna minia-turization with radiation characteristics almost preserved. The width of driven element is reduced from 1.05λ (in [4]) to 0.38λ, and the widths of director elements are reduced to 0.29λ and 0.28λ respectively. Thus, the total area of the antenna is reduced to $1.82\lambda \times 0.57\lambda$.

Figure 3 shows the inside electric field distribution at 10 GHz. It can be seen that the PTHs work as electric wall and the electric field distribution is restricted. For the driven element D and director elements (D1 and D2), the electric field reaches the maximum along the open sides, which indicates that the open sides are approximately equivalent to magnetic wall. It should be noted that the magnetic wall is strictly a quasi-one. Outside the magnetic wall, there exists fringing field which will generate influ-ence on cross polarization. It is also found that the coupling between element D2 and D1 is relatively weak. Here, ele-ment D2 has an important influence on beam steering angle and F/B ratio. And properly increasing director elements will generate lager beam steering angle and F/B ratio.

Figure 4 shows the simulated 3-D radiation pattern at 10 GHz. Obviously, the antenna has a tilted beam and the main beam points toward the low elevation direction.

The simulated results of reflection coefficient S_{11} and peak gain are shown in Fig. 5. The working band of $S_{11} < -10$ dB is 9.5~10.6 GHz (the relative bandwidth is 10.5% with VSWR less than 2). The peak gain is about 6.0~7.9 dBi in the working band.

Figure 6 presents the simulated radiation patterns at 9.5, 10 and 10.6 GHz in elevation plane and azimuth plane. The elevation plane and the azimuth plane represent the *xoz* plane and the *xoy* plane respectively according to the Cartesian coordinate in Fig. 4. In the elevation plane, the co-polarization is vertical polarization and the peak gain direction occurs around 40°, which presents a tilted beam similar to that of conventional microstrip Yagi antenna. Within the direction of 0°~90°, the cross polarization is lower than –15 dB and the cross polarization isolation (the difference between co- and cross-polarization) is greater than 14 dB. In the azimuth plane, the main beam occurs at the end-fire direction and the cross polarization isolation is greater than 10 dB. Meanwhile, both of the co-polarization and cross-polarization radiation patterns are asymmetrical along the 0°–180° in azimuth plane due to the asymmet-rical antenna structure. In addition, the radiation efficiency is simulated and results show that the efficiency is 0.964, 0.989 and 0.937 at 9.5, 10 and 10.6 GHz, respectively. So the loss is very small and the antenna has a relatively high efficiency.

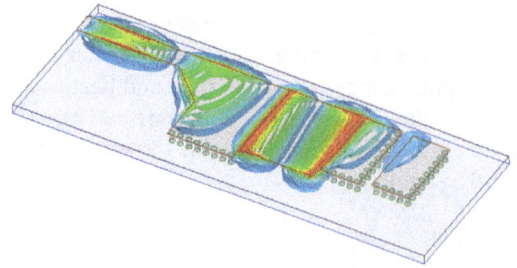

Fig. 3. Electric field distribution at 10 GHz.

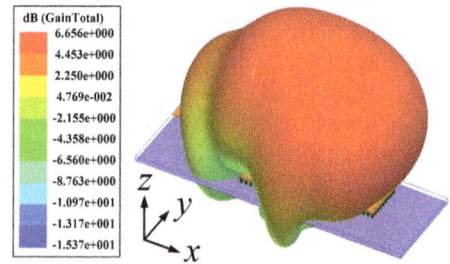

Fig. 4. Simulated 3-D radiation pattern at 10 GHz.

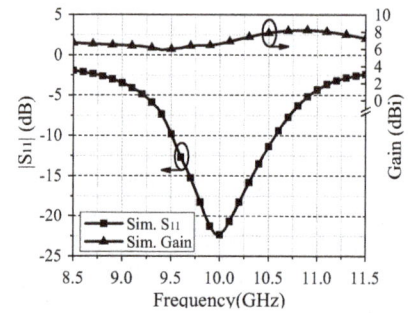

Fig. 5. Simulated S_{11} and peak gain of the proposed antenna.

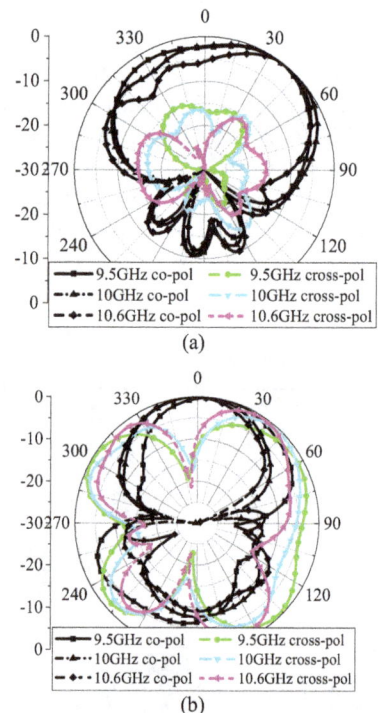

(a)

(b)

Fig. 6. Simulated radiation patterns.
(a) Elevation plane. (b) Azimuth plane.

3. Analysis of Two-element Yagi Arrays

Due to the asymmetrical structure and field distribution, the Yagi arrays composed by the HMSIW Yagi antenna with different arrangements have different radiation characteristics. We analyzed the characteristics of three kinds of two-element Yagi array. As shown in Fig. 7, the relative positions of array element are different. The array (I) is an asymmetrical structure and array elements have the same direction of the open sides. The arrays (II) and (III) are mirrored symmetrical structures, and adjacent elements have the opposite direction of the open sides. Differently, the open side direction in array (II) is contrary to the other, while the one in array (III) is facing to the other.

In view of the fringing wave around the open sides, the distances between array elements should be optimized respectively. Namely, array (II) with the contrary direction of open sides should have the minimum distance between array elements, while array (III) with the facing direction of open sides should have the maximum one and array (I) should have the middle one, so as to improve the coupling effect. Here, in order to comparatively analyze the radiation characteristics, we have selected the same distances between arrays elements for the three arrays, and they are $dx_i|_{i=1,2,3}=17$ mm (about 0.57λ).

The simulated results of reflection coefficient and peak gain are shown in Fig. 8. It can be found that the three arrays have the same curve trend of peak gain, and the peak gain increases by 2.5~3.0 dBi compared with the single Yagi antenna. As to the reflection coefficient, the array (I) and (II) has the same curve trend and bandwidth, while the array (III) has a slightly wider bandwidth.

Figure 9 shows the radiation patterns at 10 GHz. In elevation plane, the co-polarization patterns still coincide with each other and the beam directions point at 40° at the same time. In azimuth plane, the array (III) has the maximum beam width, while the array (II) has the minimum one. On the other hand, they all have similar cross polarization radiation patterns. It should be pointed out that the difference between beam widths can be eliminated by adjusting the distance between array elements. In a word, although the three arrays have completely different arrangements, they exhibit a remarkable similarity in peak gain, reflection coefficient, co- and cross-polarization radiation pattern in azimuth plane, and co-polarization radiation pattern in elevation plane.

However, what makes one array distinctly different from the others is the cross polarization in elevation plane. For the array (II) and (III), the cross polarization is below −50 dB within the direction of 0°~90° (the cross polarization isolation is greater than 45 dB), which is much lower than that of the array (I). The difference of cross polarization can be attributed to the particular element arrangement. With the mirrored arrangement, each element in array (II) and (III) has the mirrored current distribution of its neigh-

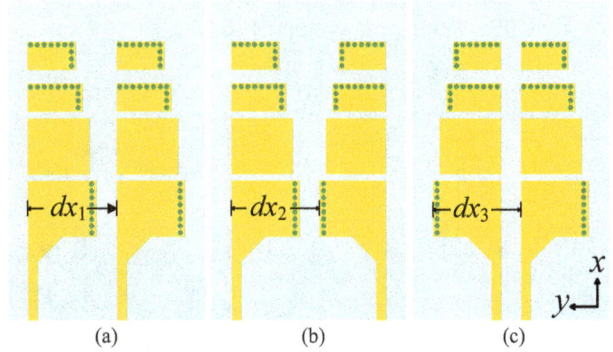

Fig. 7. Two-element arrays.
(a) Array (I). (b) Array (II). (c) Array (III).

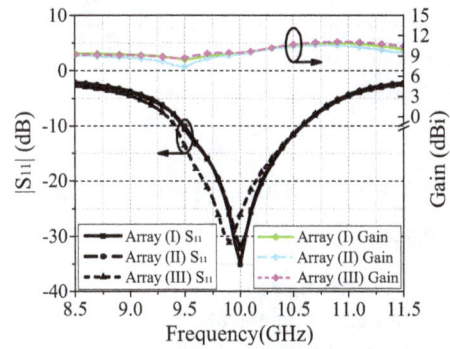

Fig. 8. Simulated reflection coefficient and peak gain.

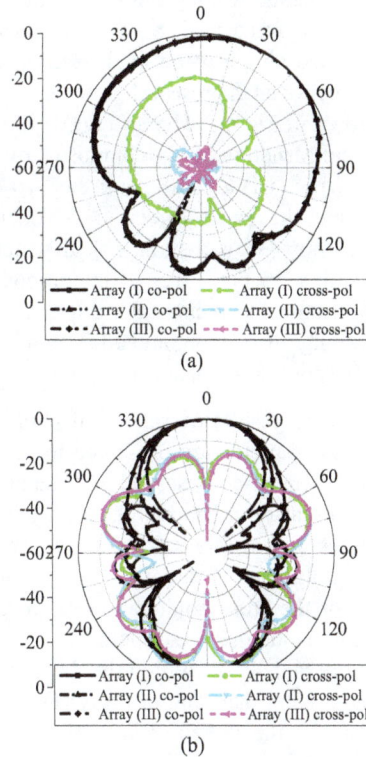

(a)

(b)

Fig. 9. Simulated radiation patterns at 10 GHz. (a) Elevation plane. (b) Azimuth plane.

bor element as shown in Fig. 10. In other words, the far-field cross-polar component would be eliminated since current components responsible for cross polarization, i.e. J_y components, are all directed in opposite directions [23].

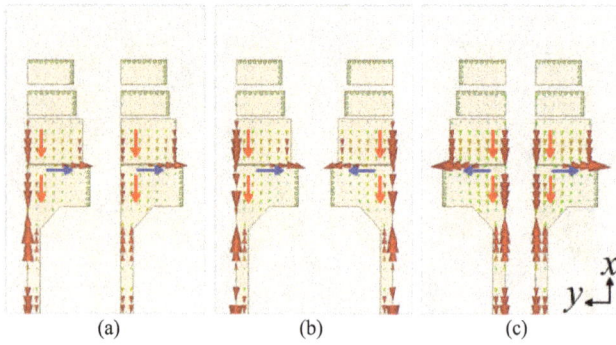

Fig. 10. Comparison of current distributions at 10 GHz. (a) Array (I). (b) Array (II). (c) Array (III).

On the contrary, current components responsible for cross polarization have the same direction in array (I), so the far field cross-polar component could not be eliminated and cross polarization still remains high.

4. Analysis of Four-element Yagi Arrays

To further investigate the radiation performance of arrays with different arrangements, four-element Yagi arrays are analyzed. Due to the asymmetrical structure, four antenna elements could make up sixteen kinds of arrays with different arrangements. Among the sixteen arrays, six couples are mutually symmetrical. So the number of arrays is reduced to ten. To simplify the analyses, we choose two representative arrays as the research objects. Their configurations are shown in Fig. 11. The array (IV) in Fig. 11(a) is an asymmetrical structure. It is composed by combining array (II) and array (III) together, and the two elements in the middle have the same direction of open sides. In Fig. 11(b), the array (V) is a symmetrical structure and its formation is shown in Fig. 11(c). It is composed by truncating four continuous elements after duplicating array (II) or array (III) along y-direction, and adjacent elements have the opposite direction of open sides.

The simulated results of reflection coefficient, peak gain and radiation patterns are presented in Figs. 12 to 15. From these plots, it can be observed that the arrays exhibit a remarkable similarity in bandwidth, peak gain, co-polarization and cross-polarization radiation pattern in azimuth plane, and co-polarization radiation pattern in elevation plane, except for the cross polarization in elevation plane. In the working band, the cross polarization isolation of array (V) is more than 47 dB within the direction of 0°~90°, which is much greater than that of array (IV). Just like the two-element arrays, the reason of low cross polarization is that each element in array (V) has the mirrored current distribution of its neighbor element.

Furthermore, a reasonable corollary could be obtained that arrays with more elements composed by the proposed HMSIW Yagi antennas should have mirrored symmetry that each element is mirrored image of the neighboring one, so as to obtain a lower cross polarization, since the radiated

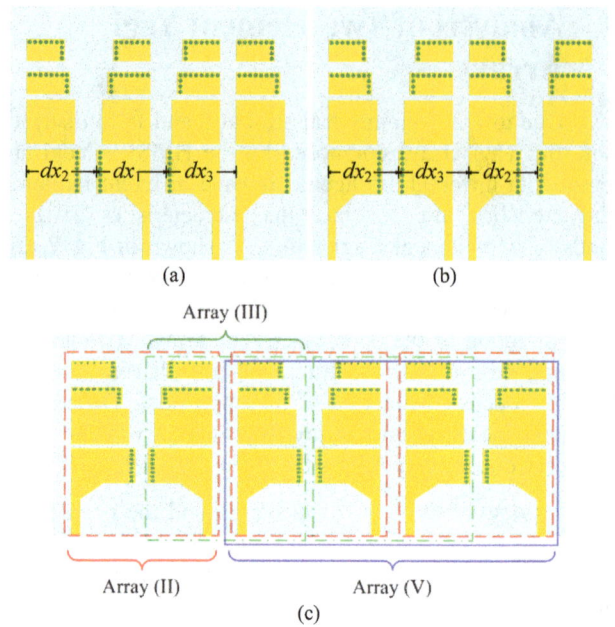

Fig. 11. Four-element arrays. (a) Array (IV). (b) Array (V). (c) Formation of array (V).

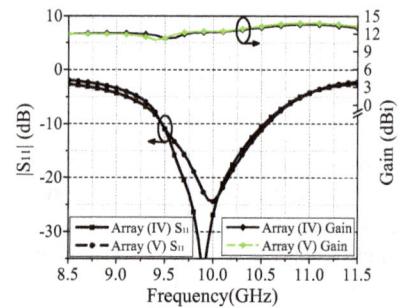

Fig. 12. Simulated reflection coefficient and peak gain.

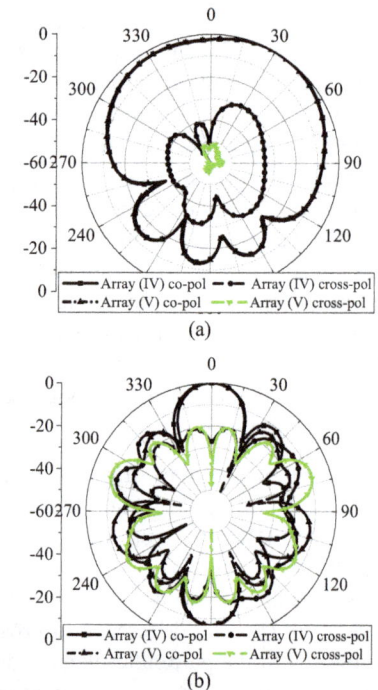

Fig. 13. Simulated radiation patterns at 9.5 GHz. (a) Elevation plane. (b) Azimuth plane.

fields of every two adjacent elements viewed as a pair will eliminate the far field cross-polar component.

In addition, we should clarify some issues about the radiation performance. Like other Yagi antennas and arrays [1–10], all the presented Yagi antenna and arrays in our work have the tilted beam. Usually, an effective method to adjust the beam steering angle is to change the number of

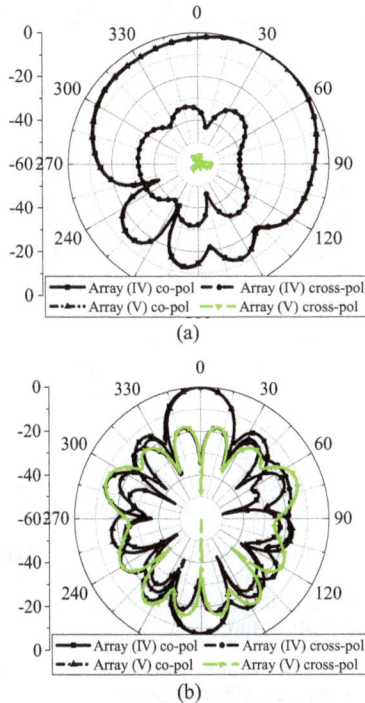

Fig. 14. Simulated radiation patterns at 10 GHz. (a) Elevation plane. (b) Azimuth plane.

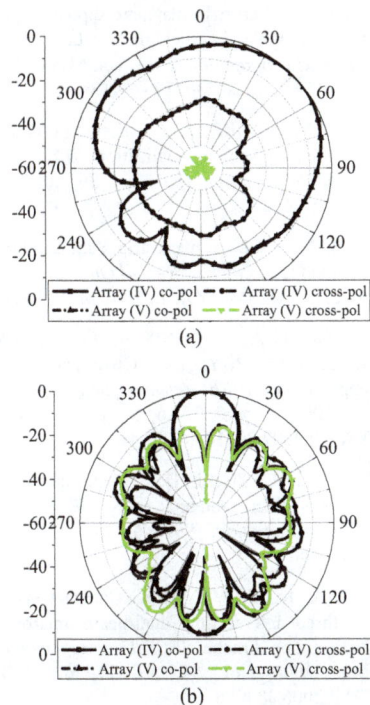

Fig. 15. Simulated radiation patterns at 10.5 GHz. (a) Elevation plane. (b) Azimuth plane.

director element according to the principle of field super-position. Another applicative method is to adjust the size of antenna ground plane. Researchers have proved that the Yagi antenna produces a beam radiating certain angle for finite ground plane, while it produces a beam radiating at exactly end-fire for infinite ground plane [4]. The above methods will not change the mirrored arrangement, such as the array (II), array (III) and array (V). So the cross-polarization will still remain at a fairly low level.

To validate the simulation, the four-element array (V) was fabricated and measured in a microwave anechoic chamber as shown in Fig. 16. A vector network analyzer (Agilent N5230C) and a linearly polarized standard-gain horn antenna (2~18 GHz) were used to constitute a test system. A microstrip 1 × 4 Wilkinson power divider was used to feed the array. The power divider works in X-band of 8~12 GHz. The loss and attenuation performance of the power divider is measured to calibrate the measured reflection coefficient of the array. The comparison between simulated and measured results of reflection coefficient and peak gain is shown in Fig. 17. The simulated and measured radiation patterns at 10 GHz are shown in Fig. 18. From these plots, a good agreement between simulation and measurement can be observed. The minor deviation between simulation and measurement is caused by fabrication and measurement.

Finally, performance comparison between different microstrip Yagi antennas and arrays is listed in Tab. 2. Obviously, the intrinsic advantages of the proposed Yagi arrays with mirrored arrangement include not only that they exhibit low profile, high gain and wide bandwidth, but also that the cross-polarization is much lower than others.

5. Conclusion

In this paper, low cross polarization Yagi array composed by compact HMSIW Yagi antenna is proposed. Firstly, the HMSIW and magnetic wall are introduced to design the compact Yagi antenna. Results show that the size of driven and director elements are all reduced and the total area is reduced to $1.82\lambda \times 0.27\lambda$. Meanwhile the peak gain is about 6.0~7.9 dBi within the 10.5% relative bandwidth with VSWR less than 2. Furthermore, two-element and four-element arrays with different arrangements are

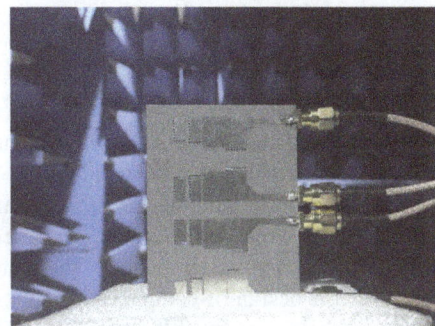

Fig. 16. Measurement in microwave anechoic chamber.

Antenna	Cross-polarization [dB]	Gain [dB]	Fractional Bandwidth	Area [λ^2]	Profile [λ]
Proposed Yagi antenna	−15	6.0~7.9	10.5%	1.04	0.033
Array (II)	−50	8.2~10.5	10%	2.20	0.033
Array (V)	−54	11.2~13.4	10%	4.26	0.033
Ref. [3]	−20	9~11.5	10%	3.80	0.027
Ref. [4]	−25	about 10.0	13.1%	3.82	0.027
Ref. [7]	−20	15.3~16.0	5.0%	8.20	0.027
Ref. [10]	−10	6.4~8.8	3.0%	1.11	0.013

Tab. 2. Comparison between Yagi antennas and arrays.

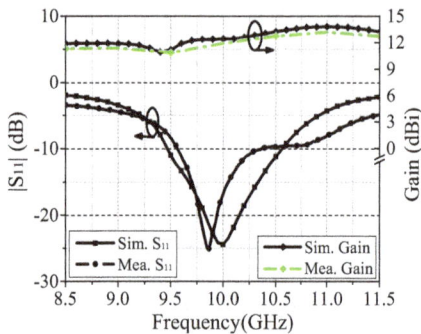

Fig. 17. Simulated and measured reflection coefficient and peak gain.

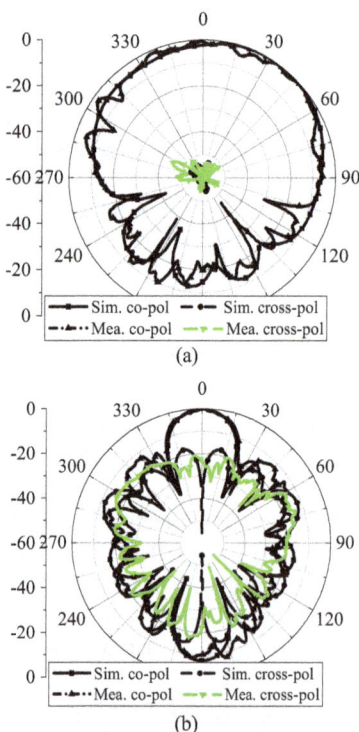

Fig. 18. Measured radiation patterns.
(a) Elevation plane. (b) Azimuth plane.

analyzed respectively. It has been found that symmetrical arrays with mirrored arrangement, in which each element antenna is the mirrored image of the neighboring one, exhibit a much lower cross polarization because the far field cross-polar component is eliminated. The results provide reference for the design of compact Yagi antenna and the Yagi array with low cross polarization composed by HMSIW elements.

Acknowledgments

This work is supported by the National Natural Science Foundation of China under Grant No. 61271100, No. 61471389, No. 61501494 and No. 61671464. Authors also thank reviewers for their valuable comments.

References

[1] HUANG, J. Planar microstrip Yagi array antenna. In *International Symposium Digest-Antennas and Propagation*, San Jose (USA), 1989, p. 894–897. DOI: 10.1109/APS.1989.134838

[2] PADHI, S. K., BIALKOWSKI, M. E. An X-band aperture-coupled microstrip Yagi array antenna for wireless communications. *Microwave and Optical Technology Letters*, 1998, vol. 168, no. 5, p. 331–335. DOI: 10.1002/mop.21344

[3] DEJEAN, G. R., TENTZERIS, M. M. A new high-gain microstrip Yagi array antenna with a high Front-to-Back (F/B) ratio for WLAN and millimetre-wave applications. *IEEE Transactions on Antennas and Propagation*, 2007, vol. 55, no. 2, p. 298–304. DOI: 10.1109/TAP.2006.889818

[4] LIU, J., XUE, Q. Microstrip magnetic dipole Yagi array antenna with endfire radiation and vertical polarization. *IEEE Transactions on Antennas and Propagation*, 2013, vol. 61, no. 3, p. 1140–1147. DOI: 10.1109/TAP.2012.2230239

[5] HUANG, J., DENSMORE, A. C. Microstrip Yagi array antenna for mobile satellite vehicle application. *IEEE Transactions on Antennas and Propagation*, 1991, vol. 39, no. 7, p. 1024–1030. DOI: 10.1109/8.86924

[6] VENKATESAN, J. X-band microstrip Yagi array with dual-offset aperture coupled feed. *Microwave and Optical Technology Letters*, vol. 48, no. 2, p. 341–344. DOI: 10.1002/mop.21344

[7] DEJEAN, G. R., THAI, T. T., NIKOLAOU, S., et al. Design and analysis of microstrip Bi-Yagi and Quad-Yagi antenna arrays for WLAN applications. *IEEE Antennas and Wireless Propagation Letters*, 2007, vol. 6, p. 244–248. DOI: 10.1109/LAWP.2007.893104

[8] GRAY, D., LU, J. W., THIEL, D. V. Electronically steerable Yagi-Uda microstrip patch antenna array. *IEEE Transactions on Antennas and Propagation*, 1998, vol. 46, no. 5, p. 605–608. DOI: 10.1109/8.668900

[9] TSUNEKAWA, K., SAWAYA, K. Compact six-sector antenna employing three intersecting dual-beam microstrip Yagi–Uda arrays with common director. *IEEE Transactions on Antennas and Propagation*, 2006, vol. 54, no. 11, p. 3055–3062. DOI: 10.1109/TAP.2006.883980

[10] YANG, X.-S., WANG, B.-Z., WU, W., et al. Yagi patch antenna with dual-band and pattern reconfigurable characteristics. *IEEE*

Antennas and Wireless Propagation Letters, 2007, vol. 6, p. 168 to 171. DOI: 10.1109/LAWP.2007.895292

[11] YANG, X.-S., WANG, B.-Z., YEUNG, S. H., et al. Circularly polarized reconfigurable crossed-Yagi patch antenna. *IEEE Antennas and Propagation Magazine*, 2011, vol. 53, no. 5, p. 65 to 80. DOI: 10.1109/MAP.2011.6138429

[12] DESLANDES, D., WU, K. Integrated microstrip and rectangular waveguide in planar form. *IEEE Microwave Wireless Component Letters*, 2001, vol. 11, no. 2, p. 68–70. DOI: 10.1109/7260.914305

[13] XU, J., HONG, W., TANG, H., et al. Half-mode substrate integrated waveguide (HMSIW) leaky-wave antenna for millimetre-wave applications. *IEEE Antennas and Wireless Propagation Letters*, 2008, vol. 7, p. 85–88. DOI: 10.1109/LAWP.2008.919353

[14] JIN, C., LI, R., ALPHONES, A., et al. Quarter-mode substrate integrated waveguide and its application to antennas design. *IEEE Transactions on Antennas and Propagation*, 2013, vol. 61, no. 6, p. 2921–2928. DOI: 10.1109/TAP.2013.2250238

[15] SAM, S., LIM, S. Electrically small eighth-mode substrate integrated waveguide antenna with different resonant frequencies depending on rotation of complementary split ring resonator. *IEEE Transactions on Antennas and Propagation*, 2013, vol. 61, no. 10, p. 4933–4939. DOI: 10.1109/TAP.2013.2272676

[16] SAM, S., LIM, S. Miniaturized circular polarized TE-mode substrate integrated waveguide antenna. *IEEE Antennas and Wireless Propagation Letters*, 2014, vol. 13, p. 658–661. DOI: 10.1109/LAWP.2014.2313747

[17] JIN, C., LI, R., ALPHONES, A. Compact circularly polarized antenna based on quarter-mode substrate integrated waveguide sub-array. *IEEE Transactions on Antennas and Propagation*, 2014, vol. 62, no. 2, p. 963–967. DOI: 10.1109/TAP.2013.2291574

[18] XU, F., WU, K. Guided-wave and leakage characteristics of substrate integrated waveguide. *IEEE Transactions on Microwave Theory and Techniques*, 2005, vol. 53, no. 1, p. 66–73. DOI: 10.1109/TMTT.2004.839303

[19] GARG, R., BAHL, I., BOZZI, M. *Microstrip Lines and Slotlines.* 3rd ed. Artech House, 2013. ISBN: 9781608075355.

[20] POZAR, D. M. *Microwave Engineering.* 4th ed. Wiley Publishing, 2004. ISBN: 978-0-470-63155-3

[21] ESQUIUS-MOROTE, M., FUCHS, B., ZÜRCHER, J.-F., et al. A printed transition for matching improvement of SIW horn

antennas. *IEEE Transactions on Antennas and Propagation*, 2013, vol. 61, no. 4, p. 1923–1930. DOI: 10.1109/TAP.2012.2231923

[22] DESLANDES, D., WU, K. Accurate modeling, wave mechanisms, and design considerations of a substrate integrated waveguide. *IEEE Transactions on Microwave Theory and Techniques*, 2006, vol. 54, no. 6, p. 2516–2526. DOI: 10.1109/TMTT.2006.875807

[23] HASANI, H., KAMYAB, M., MIRKAMALI, A. Low cross-polarization reflectarray antenna. *IEEE Transactions on Antennas and Propagation*, 2011, vol. 59, p. 1752–1756. DOI: 10.1109/TAP.2011.2123071

About the Authors ...

Zhao ZHANG was born in Baoji, Shaanxi province, P.R. China in 1990. He received B.S. and M.S. from the Air Force Engineering University, Xi'an China, in 2012 and 2014. He is currently working toward Ph.D. degree at the Information and Navigation College, Air Force Engineering University. His main interests include titled beam antennas, circularly polarized antennas, metamaterial design and RCS reduction.

Xiangyu CAO received the B.Sc and M.A.Sc degrees from the Air Force Missile Institute in 1986 and 1989, respectively. She joined the Air Force Missile Institute in 1989 as an assistant teacher. She became an associate professor in 1996. She received Ph.D. degree in the Missile Institute of Air Force Engineering University in 1999. From 1999 to 2002, she was engaged in postdoctoral research in Xidian University, China. She was a Senior Research Associate in the Department of Electronic Engineering, City University of Hong Kong from Jun. 2002 to Dec. 2003. She is currently a professor of the Air Force Engineering University of CPLA. Her research interests include computational electromagnetic, smart antennas, electromagnetic metamaterial and their antenna applications, and electromagnetic compatibility.

17

Compact Microstrip Triple-Mode Bandpass Filters Using Dual-Stub-Loaded Spiral Resonators

*Kai-da XU [1, 2], Meng-ze LI [1], Yan-hui LIU [*1], Jing AI [3], Ye-cheng BAI [1]*

[1] Inst. of Electromagnetics and Acoustics & Dept. of Electronic Science, Xiamen University, Xiamen 361005, China
[2] Shenzhen Research Inst. of Xiamen University, Shenzhen 518057, China
[3] EHF Key Lab of Science, University of Electronic Science and Technology of China, Chengdu 611731, China

yanhuiliu@xmu.edu.cn

Abstract. *Two new microstrip triple-mode resonators loaded with T-shaped open stubs using axially and centrally symmetric spiral structures, respectively, are presented. Spiraled for circuit size reduction, these two half-wavelength resonators can both generate three resonant modes over a wide frequency band by loading two T-stubs with different lengths. Due to the structural symmetry, they can be analyzed by odd- and even-mode method. To validate the design concept, two compact bandpass filters (BPFs) using these two novel resonators with center frequencies of 1.76 GHz and 2.44 GHz for the GSM1800 and WLAN/Zigbee applications, respectively, have been designed, fabricated and tested. The center frequencies and bandwidths can be tunable through the analysis of resonant frequency responses, fractional bandwidths and external quality factor versus the resonator parameters. The final measured results have achieved good consistence with the simulations of these two BPFs.*

Keywords

Spiral resonator, stub-loaded resonators, transmission zeros, triple-mode bandpass filter

1. Introduction

Bandpass filters (BPFs) are basic building blocks in the RF front-end and microwave wireless communication systems. Over the past decade, the BPFs with compact size [1], [2], sharp frequency selectivity [3], [4], low insertion loss and wide bandwidth have been studied and exploited extensively, and various design approaches have been proposed [5–12]. An effective way for compact filter design is that modifying the traditional resonator to generate additional resonant modes, resulting in one resonator with more fundamental resonant frequencies. Dual-mode resonators and filters are the main research topics in recent years, which have been analyzed deeply and comprehensively in numerous reports with a variety of structures, including rectangle loop resonators [5], meander loops [6], triangular

loop structures [7], and stub-loaded resonators [8]. Planar microstrip triple-mode or multi-mode filters have encountered many difficulties in the integrated process as they tend to be large in volume and are not easy for integration.

A harmonic-suppressed BPF based on a triple-mode stub-loaded resonator is proposed in [9], which has the advantage that the even-mode frequencies can be flexibly controlled whereas the odd-mode frequencies are fixed. However, it involves complex structure in the design process. In [10], a triple-mode BPF using a modified circular patch resonator is introduced, but the bad rejection in the stopband occurs. Reference [11] presents a wideband BPF with controllable bandwidth and suppression of the harmonic band, in which four same open loop resonators are adopted. However, the frequency selectivity still needs to be improved. In [12], a novel defected open-loop resonator as the slotline configuration is applied to design a compact triple-mode defected ground waveguide resonator-based BPF, which is compact and easy for integration with planar technology. Moreover, a novel triple-mode hexagonal BPF with capacitive loading stubs is introduced in [13], which is developed from a conventional hexagonal loop dual-mode resonator. In [14], a triple mode filter using a spiral resonator loaded with two short-stubs and an open-stub is presented.

In this paper, two different types of modified triple-mode spiral resonators have been proposed. The theoretical odd- and even-mode analysis is given to verify the performance of the resonators. Two BPFs using axially and centrally symmetric structures with center frequencies of 1.76 GHz and 2.44 GHz for the GSM1800 and WLAN/Zigbee applications, respectively, have been simulated, fabricated and tested. Good agreements are shown between simulated and experimental results.

2. Resonators Analysis

The schematic layouts of two proposed dual-T-stub-loaded spiral resonators are shown in Fig. 1, which share the same equivalent structure. Because of the spiral struc-

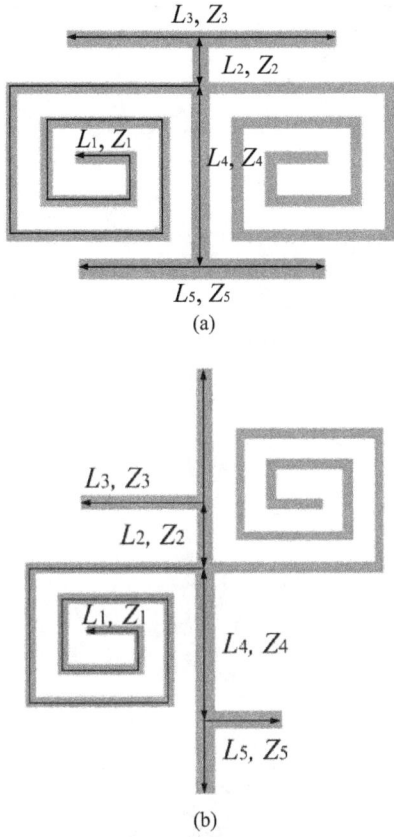

Fig. 1. Layouts of two proposed spiral resonators: (a) Axially symmetric spiral resonator. (b) Centrally symmetric spiral resonator.

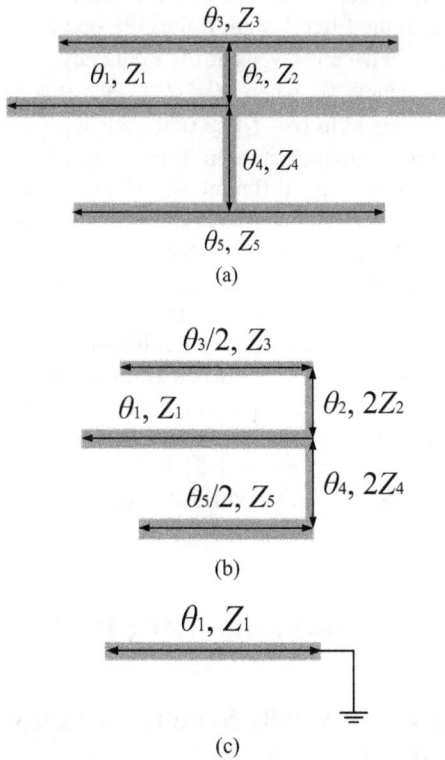

Fig. 2. Equivalent configuration and equivalent circuits of the proposed spiral resonators: (a) Equivalent configuration. (b) Even-mode equivalent circuit. (c) Odd-mode equivalent circuit.

tures, when the electrical lengths are fixed, the sizes of these two proposed resonators can be reduced greatly.

Figure 2(a) illustrates the equivalent configuration of the proposed spiral resonators, which is composed of a uniform microstrip half-wavelength resonator and a pair of T-shaped stubs with different lengths. θ_i refers to the electrical length of L_i, and Z_i denotes the characteristic impedance of the corresponding strips ($i = 1, 2, 3, 4$ and 5).

As the equivalent configuration is symmetrical in structure, apparently, we could utilize the odd- and even-mode method to further analyze this configuration. For the even-mode excitation, an approximately equivalent circuit is depicted in Fig. 2(b). We simplify the analysis by setting $Z_1 = Z_3 = 2Z_2 = 2Z_4 = Z_5$, therefore, the input impedances of the two even-mode equivalent circuits $Z_{in\_even\_1}$ and $Z_{in\_even\_2}$ can be deduced as follows:

$$Z_{in\_even\_1} = \frac{Z_1}{j\tan(\theta_1 + \theta_2 + \theta_3/2)}, \qquad (1)$$

$$Z_{in\_even\_2} = \frac{Z_1}{j\tan(\theta_1 + \theta_4 + \theta_5/2)}. \qquad (2)$$

Because of the resonance conditions that $\text{Im}(Z_{in\_even\_1}) = \infty$ and $\text{Im}(Z_{in\_even\_2}) = \infty$, we can derive the following results:

$$f_{even\_1} = \frac{nc}{2(L_1 + L_2 + L_3/2)\sqrt{\varepsilon_{eff}}}, \qquad (3)$$

$$f_{even\_2} = \frac{nc}{2(L_1 + L_4 + L_5/2)\sqrt{\varepsilon_{eff}}} \qquad (4)$$

where $n = 1, 2, 3...$, c is the velocity of light in free space, and ε_{eff} denotes the effective dielectric constant of the substrate.

For the odd-mode excitation, its equivalent circuit is depicted in Fig. 2(c). The input impedance of the odd-mode circuit Z_{in_odd} can be obtained as follows:

$$Z_{in\_odd} = jZ_1 \tan\theta_1 \qquad (5)$$

The resonance condition is $\text{Im}(Z_{in\_odd}) = \infty$. Therefore, when the odd mode is excited, the resonant frequency can be deduced as:

$$f_{odd} = \frac{(2n-1)c}{4L_1\sqrt{\varepsilon_{eff}}}. \qquad (6)$$

When $L_2 + L_3/2 < L_1 < L_4 + L_5/2$, it can be derived that $f_{even\_2} < f_{odd} < f_{even\_1}$. Following this analysis, a triple-mode resonator is presented. The above analysis shows that the resonant frequencies of two proposed triple-mode resonators could be changed by tuning the lengths of L_1, L_2, L_3, L_4, and L_5.

Figure 3 presents the variations of these three resonant frequencies with respect to L_3 and L_5 for the proposed resonators. In Fig. 3(a), by changing the length of L_3 from 16.8 to 19.2 mm, f_{odd} and f_{even_2} are rarely changed, but f_{even_1} will reduce considerably. Figure 3(b) plots the trend of these three resonant modes likewise. When L_5 increases from 13.4 to 15.8 mm, f_{odd} and f_{even_1} will almost keep unchanged, but f_{even_2} will decline significantly. These simulated results agree well with the deduced equations (3), (4) and (6). Therefore, to design a desired passband of the filter using the proposed structure, we must tailor these three resonant modes initially, which are mainly dominated by the electrical lengths of the resonator and two T-stubs.

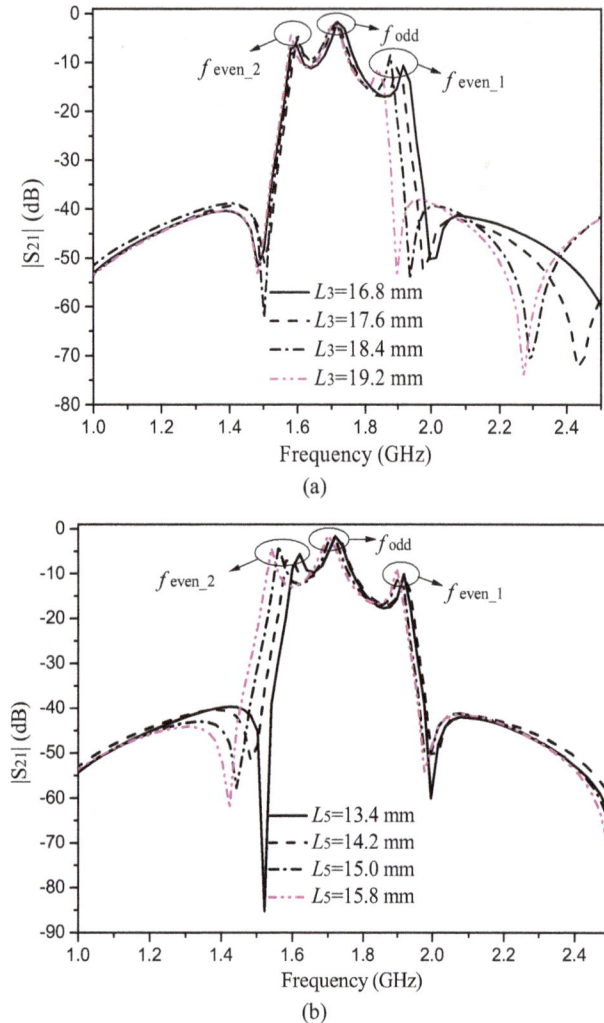

(a)

(b)

Fig. 3. Simulated frequency responses of the two proposed triple-mode resonators.

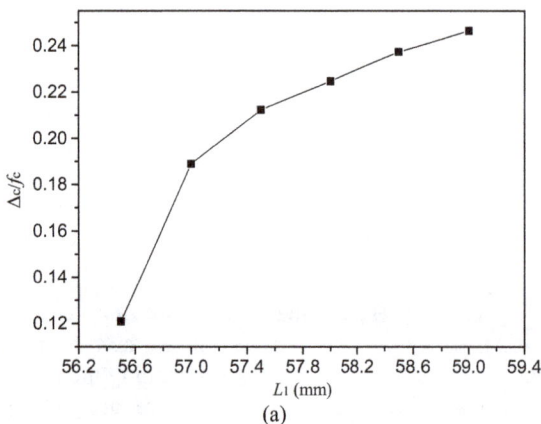

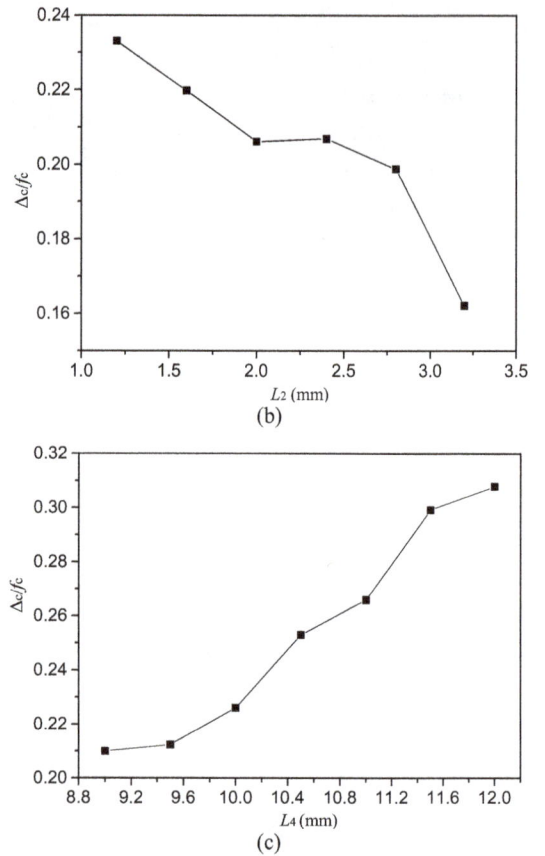

(a)

(b)

(c)

Fig. 4. Variation of fractional bandwidth $\Delta c/f_c$ with different values of (a) L_1, (b) L_2, (c) L_4.

The fractional bandwidth is a key and important technical index in the filter design. If the BPF operates at center frequency f_c between lower cut-off frequency f_1 and upper cut-off frequency f_2, where $f_c = (f_1 + f_2)/2$, the fractional bandwidth is equal to $(f_2 - f_1)/f_c$. When we apply the above-mentioned resonator structure in filter design, the fractional bandwidth $\Delta c/f_c$ with different varied parameters of the filter has been demonstrated in Fig. 4, where Δc is the 3-dB bandwidth, i.e., $\Delta c = f_2 - f_1$. As the length of L_1 increases from 56.5 to 59.0 mm, the fractional bandwidths will rise up from 12 % to 25 % as seen in Fig. 4(a). In addition, apparently that the fractional bandwidth will decrease from 23.3 % to 16 % when L_2 increases from 1.2 to 3.2 mm as shown in Fig. 4(b). For L_4 in Fig. 4(c), when it changes from 9.0 to 12.0 mm, the fractional bandwidth will also extend to 31 % from original 21 %. Therefore, the fractional bandwidth can be tuned flexibly by slightly adjusting the lengths of L_1, L_2, or L_4.

3. Two Triple-Mode BPFs Design

3.1 BPF using Axially Symmetric Spiral Resonator

Based on the proposed axially symmetric dual-T-stub-loaded spiral resonator, a compact wideband triple-mode bandpass filter is designed on the substrate with a relative

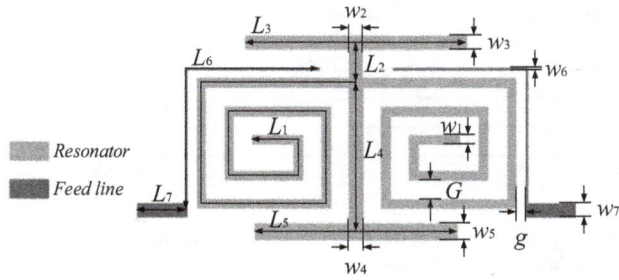

Fig. 5. Layout of the proposed axial symmetric BPF.

dielectric constant of 9.2 and a thickness of 1 mm. The layout of this filter and its dimension parameters are illustrated in Fig. 5. We also analyze the influence of the spiral gap parameter G on the resonance characteristics of the resonator. As depicted in Fig. 6, the upper even mode f_{even_1} will almost keep fixed as the value of the gap G changes but the other dimensions remain unchanged. As the parameter G decreases from 1.5 to 0.1 mm, the odd-mode frequency f_{odd} will decline fast from 1.72 to 1.42 GHz, while the lower even-mode frequency f_{even_2} will decrease from 1.6 to 1.42 GHz slowly. Eventually, these two modes overlap each other at 1.42 GHz. Thus, the gap of the spiral resonator can be used to slightly tune the distances among these three resonant frequencies, which can control the bandwidth and center frequency of the proposed filter.

In addition, the bandwidth of this filter can be affected with the coupling gap g, and Figure 7 illustrates the extracted external quality factor Q_e with varied g. When g increases from 0.1 to 1.2 mm, higher Q_e of the filter will be obtained, which causes the decrease of the bandwidth. Therefore, for filter design, we can obtain the center frequency by tuning the gap G first, and then adjust the bandwidth by changing the coupling gap g. After optimization, the parameters of this BPF are finally chosen as follows: $L_1 = 57.5$, $L_2 = 2.85$, $L_3 = 16.8$, $L_4 = 10.45$, $L_5 = 14.2$, $L_6 = 19.7$, $L_7 = 3.5$, $W_1 = 0.7$, $W_2 = 1$, $W_3 = 1$, $W_4 = 1.1$, $W_5 = 1.2$, $W_6 = 0.2$, $W_7 = 1$, $G = 1.3$, $g = 0.1$, where the units are all in millimeter.

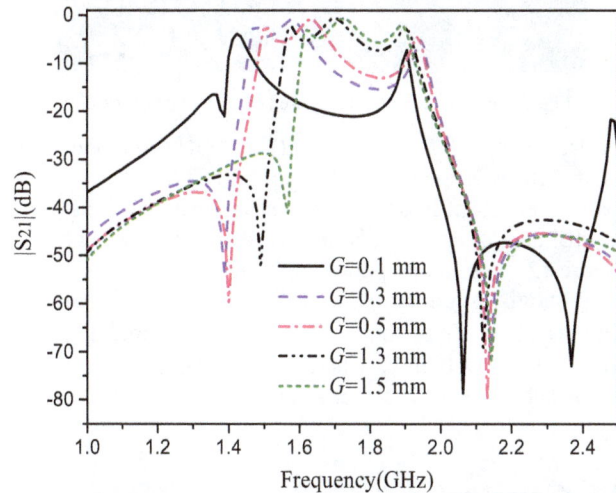

Fig. 7. Extracted external quality factor Q_e with varied g.

Figure 8 shows the measured and simulated results of the fabricated axially symmetric BPF. The size of this filter is about 13.9 mm × 23.6 mm corresponding to a size of $0.2 \lambda_g \times 0.33 \lambda_g$, where λ_g is the guided wavelength on the substrate at the center frequency. Seen from the simulated results, we will find that this filter has the center frequency of 1.79 GHz and 3-dB fractional bandwidth of about 22.5 % (1.59–1.98 GHz). Its insertion losses and return losses are less than 0.31 dB and over 19.3 dB in the passband, respectively. While the center frequency of the measured results is at 1.76 GHz and the 3-dB bandwidth is around from 1.59 to 1.92 GHz. The measured insertion losses and return losses are less than 1.5 dB and better than 19 dB in the passband, respectively. Four TZs are introduced due to this unique schematic structure, which are located at 1.03, 1.48, 2.04 and 2.61 GHz, respectively. Slight deviation between measurements and simulations could be attributed to the fabrication and measurement error. Figure 9 presents a photograph of the fabricated BPF using axially symmetric spiral resonator. Table 1 tabulates the performance comparisons among the proposed axially symmetric filter (i.e. Filter A) and several other reported triple-mode BPFs.

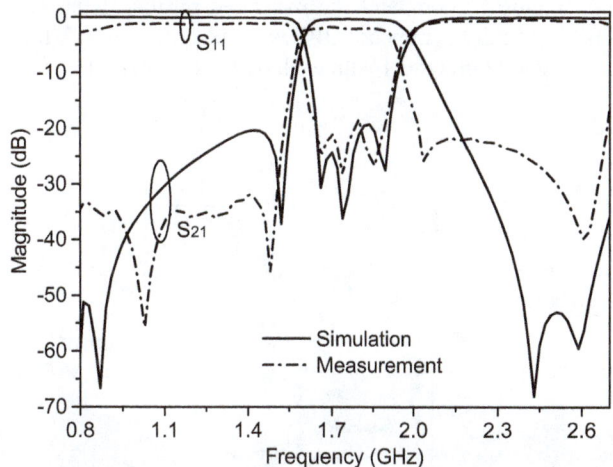

Fig. 6. Simulated frequency responses of the BPF under weak coupling with different G.

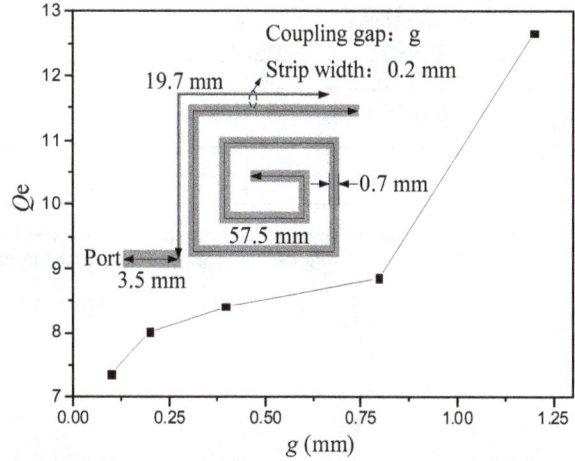

Fig. 8. Measured and simulated results of the proposed filter.

Ref.	Substrate Height (mm)/ε_r	Center Frequency (GHz)	Insertion Loss (dB)	Return Loss (dB)	3-dB FBW (%)	Circuit Size ($\lambda_g \times \lambda_g$)	Triple-mode Production Mechanism
[10]	0.625/10.2	2.4	0.5	16	29	0.34 × 0.34	DMS[1]
[11]	1.14/3.2	4.17	0.6	9	26	0.24 × 0.61	(2OLRs+2Stubs)[2]
[12]	0.8/4.5	3.5	1.9	11.5	16	0.25 × 0.39	DGW[3]
[13]	1/9.2	2.4	2.4	17	4.8	0.29 × 0.39	SPoRR[4]
Filter A	1/9.2	1.76	1.5	19	18.8	0.2 × 0.33	DSLSR[5]

[1]**DMS:** degenerate modes split; [2]**(2OLRs+2Stubs):** two open loop resonators and two stubs; [3]**DGW:** defected ground waveguide structure; [4]**SPoRR:** small perturbation on the ring resonator; [5]**DSLSR:** dual-stub-loaded spiral resonator.

Tab. 1. Performance comparisons of some reported triple-mode BPFs.

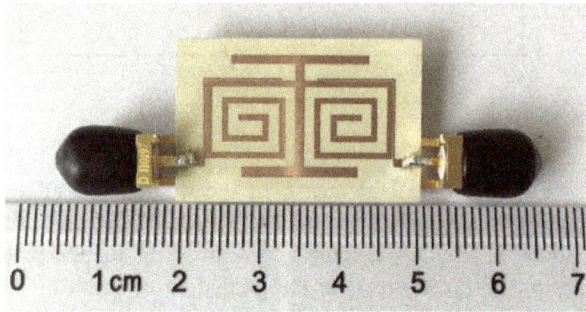

Fig. 9. Photograph of the proposed axially symmetric BPF.

3.2 BPF using Centrally Symmetric Spiral Resonator

In this sub-section, another compact triple-mode BPF is designed based on a centrally symmetric spiral resonator. Similarly to the above-mentioned axially symmetric spiral resonator structure, two T-shaped open stubs with different lengths are also loaded on this centrally symmetric spiral resonator for three resonances generation as depicted in Fig. 10. For this filter design, we use Rogers RO4350B substrate with a relative dielectric constant of 3.48 and a thickness of 0.508 mm. Accordingly, the design parameters in Fig. 10 are finally chosen as follows (unit: mm): $L_1 = 59.3$, $L_2 = 4.25$, $L_3 = 20$, $L_4 = 10$, $L_5 = 14.2$, $L_6 = 19.2$, $L_7 = 3.5$, $W_1 = 0.7$, $W_2 = 1$, $W_3 = 1$, $W_4 = 1.1$, $W_5 = 1.1$, $W_6 = 0.2$, $W_7 = 1.1$, $G = 1.3$, $g = 0.1$.

Figure 11 compares the measured results of the proposed centrally symmetric BPF with its simulations, which are in good agreement with each other. The measured cen-

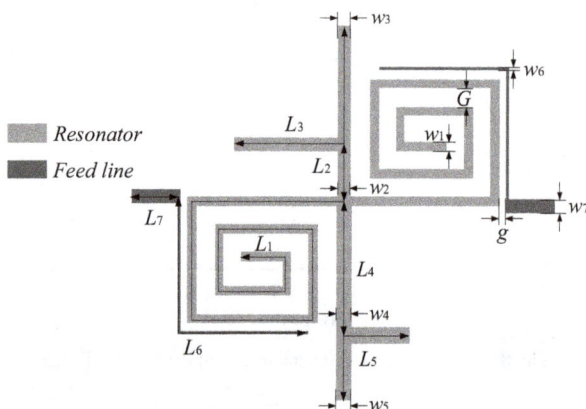

Fig. 11. Measured and simulated results of the proposed filter.

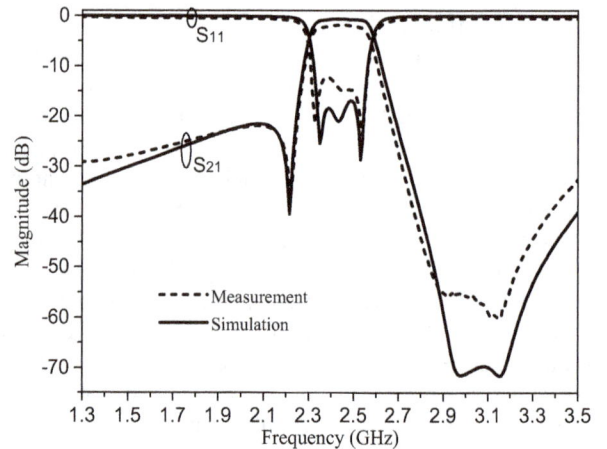

Fig. 12. Photograph of the proposed centrally symmetric BPF.

ter frequency of this BPF is at 2.44 GHz and the 3-dB bandwidth is around from 2.32 to 2.56 GHz. The insertion losses and return losses of the measurements are less than 1.5 dB and over 12.2 dB in the passband, respectively. Figure 12 presents a photograph of the fabricated BPF using axially symmetric spiral resonator. The size of this filter is about 29.25 mm × 27.2 mm corresponding to the size of 0.39 λ_g × 0.36 λ_g, where λ_g is the guided wavelength on the substrate at the center frequency.

4. Conclusion

In this paper, two compact triple-mode BPFs based on axially and centrally symmetric spiral resonators loaded with two T-shaped open stubs have been demonstrated.

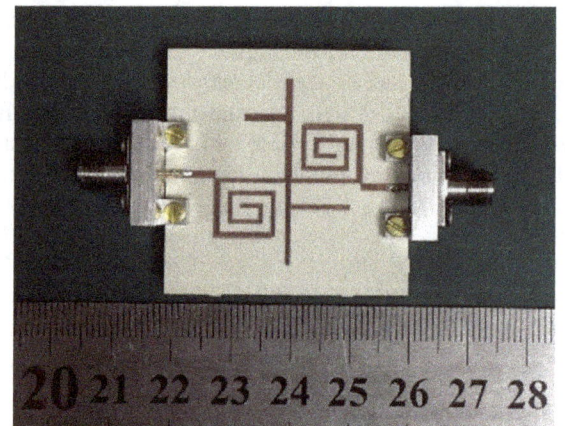

Fig. 10. Layout of the proposed centrally symmetric BPF.

The center frequencies and bandwidths of the filters are tunable through adjustment of the resonator parameters. For the fabricated axially symmetric BPF working at 1.76 GHz, the measured insertion losses are less than 1.5 dB and the return losses are better than 19 dB in the passband. While for the fabricated centrally symmetric BPF working at 2.44 GHz, the measured insertion losses and return losses are lower than 1.5 dB and over 12.2 dB in the passband, respectively. The measured results of both fabricated filters agree with their simulations. Performance comparisons of some reported triple-mode BPFs show that the proposed axially symmetric BPF has achieved miniaturization with good frequency selectivity.

Acknowledgment

This work was supported in part by the National Natural Science Foundation of China (No. 61601390), Guangdong Natural Science Foundation (No. 2016A030310375), Natural Science Foundation of Fujian Province of China (No. 2016J05164), Young and Middle-aged Teachers Education and Scientific Research Foundation of Fujian Province (No. JAT160007).

References

[1] KIM, S., KIM, N. Y. Compact bandpass filter with wide stop band response based on meandered-line stepped-impedance resonator using IPD process. *Microwave and Optical Technology Letters*, 2015, vol. 57, no. 6, p. 1466–1470. DOI: 10.1002/mop.29111

[2] XU, K. D., ZHANG, Y. H., LI, J. L. W., et al. Compact ultra-wideband bandpass filter using quad-T-stub-loaded ring structure. *Microwave and Optical Technology Letters*, 2014, vol. 56, no. 9, p. 1988–1991. DOI: 10.1002/mop.28508

[3] PAN, T., SONG, K., FAN, Y. Novel wide-stopband bandpass filter with good frequency selectivity based on composite right/left handed transmission line. *Microwave and Optical Technology Letters*, 2012, vol. 54, no. 11, p. 2494–2497. DOI: 10.1002/mop.27104

[4] WU, Y., HU, B., NAN, L., et al. Compact high-selectivity bandpass filter using a novel uniform coupled-line dual-mode resonator. *Microwave and Optical Technology Letters*, 2015, vol. 57, no. 10, p. 2355–2358. DOI: 10.1002/mop.29336

[5] ZHANG, X. Y., CHEN, J. X., XUE, Q., et al. Dual-band bandpass filters using stub-loaded resonators. *IEEE Microwave and Wireless Components Letters*, 2007, vol. 17, no. 8, p. 583–585. DOI: 10.1109/LMWC.2007.901768

[6] TORABI, A., FOROORAGHI, K. Miniature harmonic-suppressed microstrip bandpass filter using a triple-mode stub-loaded resonator and spur lines. *IEEE Microwave and Wireless Components Letters*, 2011, vol. 21, no. 5, p. 255–257. DOI: 10.1109/LMWC.2011.2122304

[7] LUGO, C., PAPAPOLYMEROU, J. Bandpass filter design using a microstrip triangular loop resonator with dual-mode operation. *IEEE Microwave and Wireless Components Letters*, 2005, vol. 15, no. 7, p. 475–477. DOI: 10.1109/LMWC.2005.851573

[8] XU, K. D., ZHANG, Y. H., ZHUGE, C. L., et al. Miniaturized dual-band bandpass filter using short stub-loaded dual-mode resonators. *Journal of Electromagnetic Waves and Applications*,
2011, vol. 25, no. 16, p. 2264–2273. DOI: 10.1163/156939311798147060

[9] DAI, G. L., ZHANG, X. Y., CHAN, C.-H., et al. An investigation of open-and short-ended resonators and their applications to bandpass filters. *IEEE Transactions on Microwave Theory and Techniques*, 2009, vol. 57, no. 9, p. 2203–2210. DOI: 10.1109/TMTT.2009.2027173

[10] SERRANO, L., CORRETA, F. S. A triple-mode bandpass filter using a modified circular patch resonator. *Microwave and Optical Technology Letters*, 2009, vol. 51, no. 1, p. 178–182. DOI: 10.1002/mop.23950

[11] MA, X. B., JIANG, T. Compact wideband bandpass filter with controllable bandwidth and suppression of the second passband using a trimode resonator. *Microwave and Optical Technology Letters*, 2015, vol. 57, no. 12, p. 2939–2943. DOI: 10.1002/mop.29475

[12] LIU, H. W., SHEN, L., JIANG, Y., et al. Triple-mode bandpass filter using defected ground waveguide. *Electronics Letters*, 2011, vol. 47, no. 6, p. 388–389. DOI: 10.1049/el.2011.0006

[13] MO, S. G., YU, Z. Y., ZHANG, L. Design of triple-mode bandpass filter using improved hexagonal loop resonator. *Progress in Electromagnetics Research*, 2009, vol. 96, no. 4, p. 117–125. DOI: 10.2528/PIER09080304

[14] XU, H., XU, K., LIU, Y., LIU, Q. H. Compact triple-mode bandpass filter using short-and open-stub loaded spiral resonator. In *2016 IEEE/ACES International Conference on Wireless Information Technology and Systems (ICWITS) and Applied Computational Electromagnetics (ACES)*. Honolulu (USA), 2016, 2 p. DOI: 10.1109/ROPACES.2016.7465476

About the Authors ...

Kai-da XU was born in Zhejiang, China. He received the B.S. and Ph.D. degrees in Electromagnetic Field and Microwave Technology from the University of Electronic Science and Technology of China (UESTC), Chengdu, China, in 2009 and 2015, respectively. From September 2012 to August 2014, he was a Visiting Researcher in the Dept. of Electrical and Computer Engineering, Duke University, Durham, NC, under the financial support from the China Scholarship Council (CSC). He received the UESTC Outstanding Graduate Awards in 2009 and 2015, respectively. He was the recipient of the National Graduate Student Scholarship in 2012, 2013, and 2014 from the Ministry of Education, China. He is now an Assistant Professor with the Inst. of Electromagnetics and Acoustics, and Dept. of Electronic Science, Xiamen University, Xiamen, China. He has authored and coauthored over 60 papers in peer-reviewed journals and conference proceedings. Since 2014, he has served as a reviewer for some journals including IEEE Transactions on Microwave Theory and Techniques, IEEE Microwave and Wireless Components Letters, IEEE Transactions on Electron Devices, IEEE Transactions on Computer-Aided Design of Integrated Circuits and Systems, IEEE Transactions on Applied Superconductivity, International Journal of RF and Microwave Computer-Aided Engineering, ACES Journal, PIER, JEMWA and so on. His research interests include RF/microwave and mm-wave circuits, antennas, and nanoscale memristors.

Meng-ze LI was born in Hunan, China. She received her B.Sc. from Hunan University, Changsha, China, in 2015, and currently she is working toward the M.S. degree in Xiamen University. Her research interests include RF/microwave components and circuits.

Yanhui LIU (corresponding author) was born in Guangxi, China. He received the B.S. and Ph.D. degrees both in Electrical Engineering from the University of Electronic Science and Technology of China (UESTC), Sichuan, China, in 2004 and 2009, respectively. From Sept. 2007 to June 2009, he was a Visiting Scholar in the Dept of Electrical Engineering, Duke University, Durham, NC. In July 2011, he joined the Dept. of Electronic Science, Xiamen University, Fujian, China, where he is now a Professor. He has authored and co-authored over 80 papers in peer-reviewed journals and conference proceedings. He received the UESTC Outstanding Graduate Award in 2004, and the Excellent Doctoral Dissertation Award of Sichuan Province of China in 2012. His research interests include antenna array design, array signal processing, and microwave imaging methods.

Jing AI was born in Sichuan, China. He received the B.Sc. degree in Electronic Science and Technology and M.S. degree in Electronic and Communication Engineering from the University of Electronic Science and Technology of China (UESTC), Chengdu, China, in 2007 and 2013, respectively, where he is currently working toward the Ph.D. degree in Electromagnetic Field and Microwave Technology. His recent research interests include microwave and millimeter-wave circuits and systems.

Ye-cheng BAI was born in Jiangsu, China. He received his B.Sc. from Hefei University of Technology, Hefei, China, in 2012, and currently he is working toward the M.S. degree in Xiamen University. His research interests include RF/microwave components and circuits.

18

Complementary Split Ring Resonator Based Triple Band Microstrip Antenna for WLAN/ WiMAX Applications

[1] *Wael ALI[1], Ehab HAMAD[2], Mohamed BASSIUNY[3], Mohamed HAMDALLAH[3]*

Dept. of Electronics & Comm. Engineering, College of Engineering, Arab Academy for Science, Technology and
[3] Maritime Transport (AASTMT), Alexandria, Egypt
[2] Dept. of Electrical Engineering, Aswan Faculty of Engineering, Aswan University, Aswan 81542, Egypt
Dept. of Electronics & Comm. Engineering, College of Engineering, Arab Academy for Science, Technology and
Maritime Transport (AASTMT), Aswan, Egypt

wael_abd_ellatif@yahoo.com, e.hamad@aswu.edu.eg, atefrana@yahoo.com, eng_zakaria_aast@yahoo.com

Abstract. *A new simple design of a triple-band microstrip antenna using metamaterial concept is presented in this paper. Multi-unit cell was the key of the multi resonance response that was obtained by etching two circular and one rectangular split ring resonator (SRR) unit cells in the ground plane of a conventional patch operating at 3.56 GHz. The circular unit cells are resonating at 5.6 GHz for the upper band of Wi-MAX, while the rectangular cell is designed to produce a resonance at 2.45 GHz for the lower band of WLAN. WiMAX's/WLAN's operating bands are covered by the triple resonances which are achieved by the proposed antenna with quite enhanced performance. A detailed parametric study of the placement for the metamaterial unit cells is introduced and the most suitable positions are chosen to be the place of the unit cells for enhanced performance. A good consistency between simulation and measurement confirms the ability of the proposed antenna to achieve an improved gain at the three different frequencies.*

Keywords

Metamaterial, metasurface, multi band antennas, CSRR, split ring resonators

1. Introduction

Microstrip patch antennas are the preferable type of antennas used for wireless communication systems, because of their attractive features such as light weight, low profile, low cost, easy fabrication, and compatibility with planar monolithic microwave integrated circuit (MMIC) components [1]. Various printed antenna topologies have been proposed by researchers for the purpose of enhancing their performance. One of these topologies can be achieved by changing the geometry of the patch itself or defecting the ground plane of the antenna as proposed in [2], where an antenna for ultra-high frequency application is designed.

Multi-circular shape of the patch in [3] was used to provide an enhanced performance of multiband antennas. Also, authors in [4] proposed a dual-band antenna realized by two different single-slotted single-band rectangular microstrip antennas.

Because of their fascinating properties, metamaterials are recently used for many applications in the field of antenna design [5]. There are a lot of antennas that have been developed based on metamaterials such as the antennas based on engineered dispersion curves (k–β diagram) [6], and the antennas based on the split-ring resonators (SRRs) and/or complementary split-ring resonators (CSRRs) [7]. Metamaterial antennas provide various techniques that can be used for improving antenna performance such as in [8] where TL-MTM technique is used to generate a multi band response. Metamaterial can also be used for enhancing the antenna parameters such as its bandwidth and gain. In [9], the design of a dual band antenna with an enhanced bandwidth is presented. In [10], metasurfaces are used to improve the gain of the antenna. In some applications, the main required feature for an antenna is its polarization and this parameter can also be controlled by using metamaterials in [11].

In this paper, a compact triple-band microstrip antenna based on CSRRs with two different geometries loaded on the ground plane is proposed. A conventional patch antenna operating at 3.6 GHz, which is operating in the middle WiMAX band (3.2 GHz to 3.8 GHz), is loaded on a ground plane with a rectangular CSRR of suitable dimensions that enables the excitation of the lower WLAN band (2.4 GHz to 2.484 GHz). Also, two circular CSRRs of suitable dimensions are loaded on the ground plane in order to resonate at the upper WiMAX frequency bands (5.25 GHz to 5.85 GHz). Finite Element Method (FEM) based software, Ansoft HFSS 13, is used for the analysis of the proposed antenna and optimizing its geometrical parameters. The main advantage of this proposed design over different multiband antenna designs is its accurate determination of all resonance frequencies i.e. every specific resonance frequency is corresponding to a specific resonator. Also, its relatively high radiation characteristics at the different achieved frequencies which make our design good candidate for WLAN/WiMAX applications.

2. Antenna Design

In this paper, a metamaterial-inspired rectangular antenna is introduced. A microstrip patch antenna operating at 3.6 GHz is chosen as the reference design. The width and length of the patch antenna are 28 mm and 31 mm, respectively with a copper thickness of 35 μm. The patch radiator is printed on Rogers RT/duroid 5880 (tm) substrate (ε_r = .2, tan δ = 0.0009) with thickness of 1.575 mm as shown in Fig. 1. A microstrip inset feed is presented to perfectly match the patch radiator to the microstrip feed line that is characterized by width and length equal to 4.9 mm and 25 mm, respectively. The feed line is inset to the patch with a value of Y_0 = 6.5 mm and the two sides of the feed with a cut of width W_0 = 5 mm.

The reference antenna has a reasonable gain of 4.8 dBi that will be affected by etching the ground plane with the unit cells. The return loss of the antenna is shown in Fig. 2 and it can be noticed that the antenna resonates at 3.56 GHz.

CSRRs, the complementary of the basic unit cells of metamaterials, are etched from the ground plane to reso-

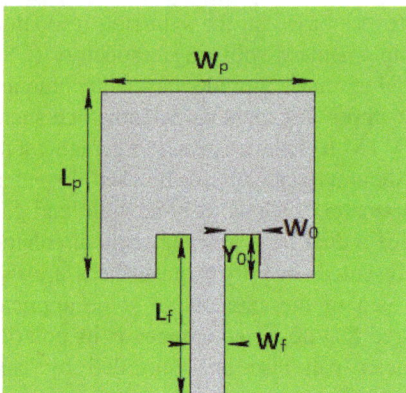

Fig. 3. Rectangular CSRR unit cell.

Fig. 4. Circular CSRR unit cell.

nate at 2.45 GHz and 5.3 GHz. The rectangular unit cell and the circular unit cell, shown in Fig. 3 and 4, are utilized to resonate at 2.45 GHz and 5.3 GHz, respectively. The bandwidth of the SRR can be increased by widening the width of the unit cell as usual [12]. Two different shapes of CSRRs are used to achieve a resonance at 5.3 GHz and 2.45 GHz because the operating bandwidth of lower WLAN (2.4 to 2.484 GHz) is narrower than the operating bandwidth of upper WiMAX (5.25 to 5.85 GHz). Hence, there is no need to utilize a circular CSRR for achieving resonance at the 2.45 GHz since it can provide a larger bandwidth than a rectangular CSRR as was mentioned in [13]. To test and extract the scattering parameters of this unit cell, the boundaries need to be adjusted as illustrated in Fig. 5. Perfect magnetic conductor (PMC) boundary condition sets on the left and right faces of the waveguide, and perfect electric conductor (PEC) boundary condition sets on the top and bottom of the waveguide. The scattering parameters are calculated over a suitable frequency range in order to determine the resonance frequency and the effective parameters caused by that unit cell. The incident TEM wave propagates in the y-axis direction. The E-field of the incident wave is polarized along the z-axis, and the H-field is polarized along the x-axis. Figure 6 shows the S-parameter characteristics of the rectangular CSRR and it can be observed that the rectangular unit cell resonates at 2.45 GHz with a return loss equal to –25 dB with narrower bandwidth, which make it suitable for lower WLAN band.

Figure 7 shows the negative permittivity characteristics of the rectangular CSRR which are extracted from the scattering parameters using the algorithm presented in [14]. In order to obtain a resonance at 5.3 GHz, a pair of circular CSRRs shown in Fig. 4 is etched from the ground plane.

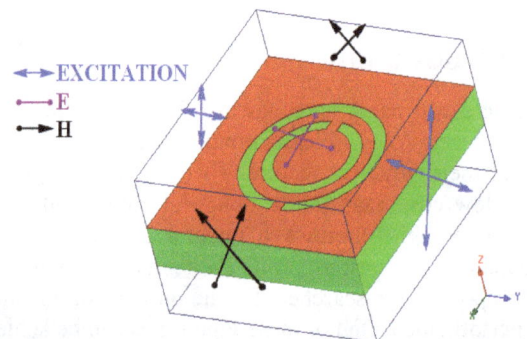

Fig. 1. Microstrip antenna operates at 3.6 GHz.

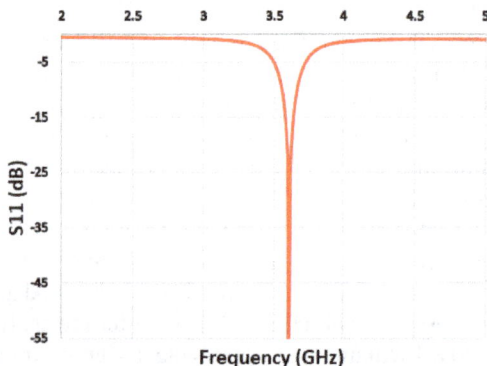

Fig. 2. S_{11} of the conventional microstrip antenna operates at 3.6 GHz.

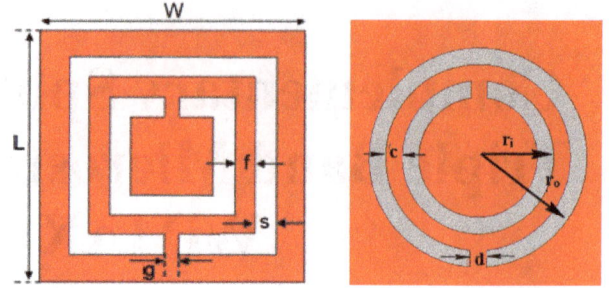

Fig. 5. The setup of the CSRR unit cell for transmission analysis.

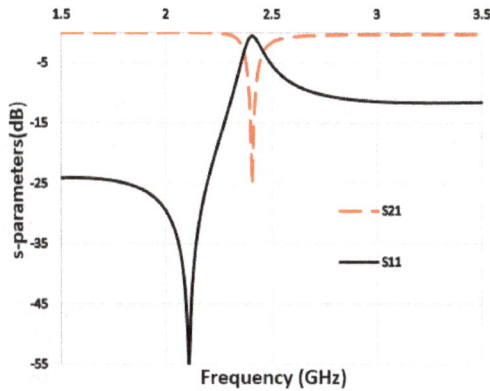

Fig. 6. S-parameters of the rectangular CSRR unit cell.

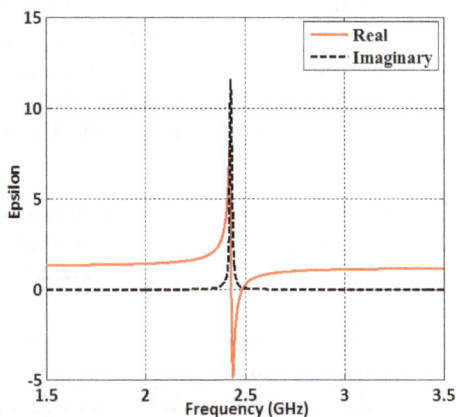

Fig. 7. Permittivity of the rectangular unit cell.

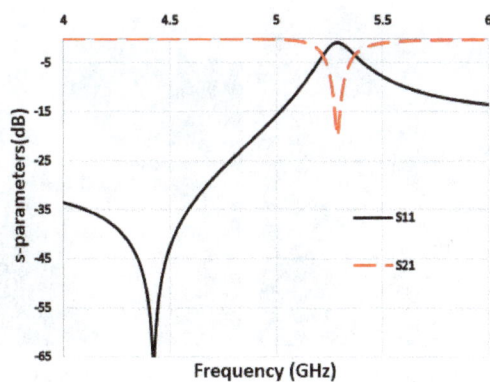

Fig. 8. S-parameters of the circular CSRR unit cell.

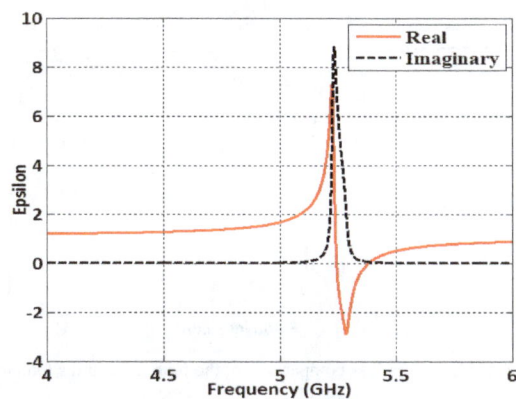

Fig. 9. Permittivity of the circular unit cell.

Parameter	L_p	W_p	Y_0	W_0	L_f	W_f	W	L
Value (mm)	28	31	6.5	5	20	4.9	18	18
parameter	G	f	s	r_o	r_i	c	d	
Value (mm)	0.5	1	2.5	3.3	2.3	0.5	0.5	

Tab. 1. Optimal values of the proposed antenna dimensions.

Figure 8 demonstrates the S-parameter characteristics of the circular CSRR and it is obvious that the return loss is less than −15 dB at 5.3 GHz with a reasonable bandwidth for further operation in the upper WiMAX band. The negative permittivity characteristics of the circular CSRR that have been extracted following the same procedure used for the rectangular unit cell are shown in Fig. 9. The permittivity of CSRR unit cells are negative at the desired frequencies as expected, and that is the main feature of the metasurfaces [15]. All the optimal dimensions of the antenna and the unit cells are listed in Tab. 1.

3. Simulation Results and Discussion

The resonant frequency of the patch antenna without any CSRRs is 3.6 GHz. Embedding a rectangular unit cell of CSRR in the ground creates a second resonance at 2.45 GHz. Etching two circular unit cells generates the third resonance which is 5.3 GHz. The configuration of the proposed triple band antenna is demonstrated in Fig. 10.

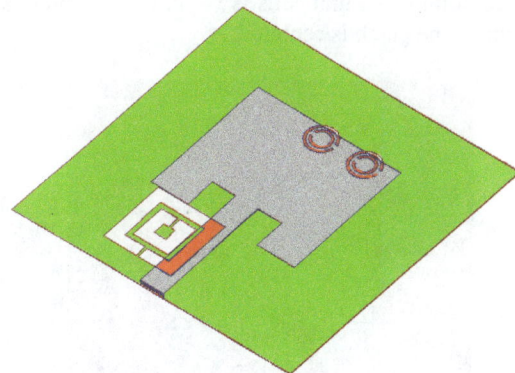

Fig. 10. Triple band microstrip antenna configuration.

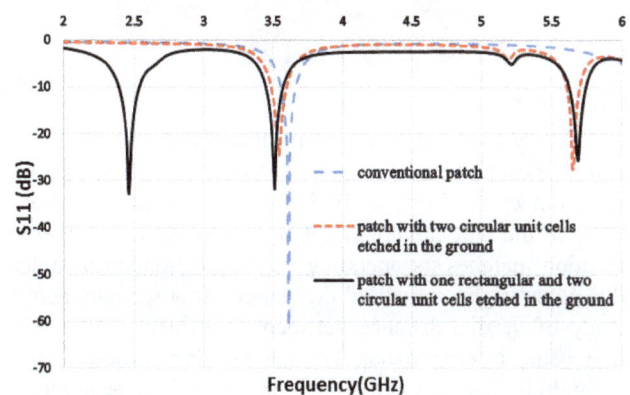

Fig. 11. S-parameters of the antenna at different configurations.

It was expected to obtain the three resonant frequencies, each one corresponding to its generating element but the overall design response suffers from frequency shift and this discrepancy attributed to the mutual coupling between the unit cells was previously noticed in [16]. As a result of coupling effect, the resonance of the circular unit cell (5.3 GHz) has been shifted to 5.6 GHz and the resonance frequency of the patch (3.6 GHz) is shifted to 3.5 GHz as shown in Fig. 11.

4. Parametric Study

The parametric study in this section shows how the position of the rectangular and circular unit cells (without changing their dimensions) will affect the resonance frequencies at 2.45 and 5.6 GHz in order to determine the definite position of the cells for proper operation.

4.1 The First Resonance Corresponding to the Circular Unit Cell (Upper Resonance 5.6 GHz)

For some locations of the CSRRs in the ground plane as shown in Fig. 12, radiation efficiencies up to 96% can be achieved for both resonances. The simulated antenna in the parametric study has the same dimensions as the fabricated one, excluding the placement of the circular CSRR inside the ground, which varies from $cx = -13$ mm to -17 mm along x-axis direction. It should be noted that the reference point of all unit cells is $x = 0$, $y = 0$, which is the point where the patch is centered.

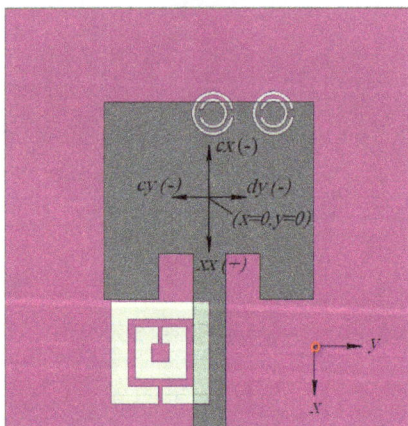

Fig. 12. CSRRs position study.

For y direction, the best position of the circular unit cell is chosen after the parametric study which was carried out from $cy = 8.5$ mm to 10.5 mm in the y-axis direction. Gaps of the unit cell are oriented in y-direction as this orientation matches its operating resonance frequency under the upper right corner of the patch. Another parametric study of dy, the distance between the centers of the two unit cells, is carried out. For simple clarification of the second circular unit cell's position, the reference point of

dy is chosen to be the center point of the first unit cell. The design of the patch with single circular unit cell can only achieve 3.7 dBi peak realized gain with 79% radiation efficiency, so the second circular unit cell is used in order to increase the gain and the bandwidth at 5.6 GHz. Table 2 shows the best position of circular unit cells of this parametric study. The structure of the antenna containing two circular unit cells in the ground plane is fabricated to evaluate its effectiveness.

The fabricated prototype of the dual band antenna is shown in Fig. 13. A comparison between simulated and measured results is demonstrated in Fig. 14. The impedance characteristics of the proposed dual band antenna have been tested using R&S ZVB 20 vector network analyzer (VNA). It can be observed from Fig. 14 that the antenna resonates at the two pre-specified frequencies but due to low fabrication accuracy, a loss of about 13 dB is presented at 3.6 GHz. Also, the resonance created by the circular unit cells at 5.6 GHz has a little shift but with very low return loss.

Position	First circular CSRR	Second circular CSRR
At x-axis (mm)	−15	−15
At y-axis (mm)	9.5	0.5

Tab. 2. Optimal positions of the proposed circular unit cells.

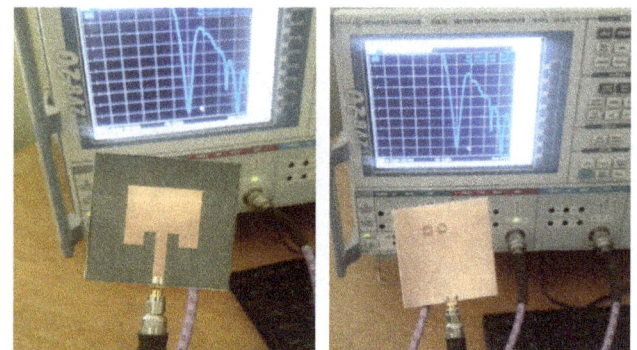

(a) Top view   (b) Bottom view

Fig. 13. Fabricated prototype of the dual band antenna.

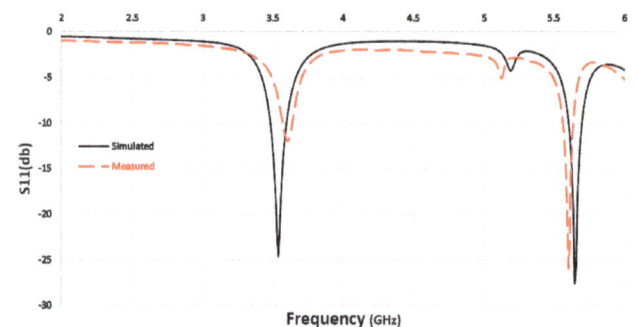

Fig. 14. Return loss comparison of the fabricated and simulated antenna.

4.2 The Second Resonance Corresponding to the Rectangular Unit Cell (Lower Resonance 2.45 GHz)

After choosing the best position of the unit cells of the first resonance (5.6 GHz), the next step is to optimize these cells with the new rectangular unit cell that will be etched from the ground plane to generate the second resonance (2.45 GHz) in order to finally achieve the triple band behavior. The position of the rectangular unit cell was a field distribution-based selection. Regions with highly concentered electric field when the patch is excited at 2.45 GHz, need to be clarified for the purpose of finding the best regions on the ground plane for etching the new rectangular unit cell. Figure 15 shows the electric field distribution over the ground plane of the dual band model at 2.45 GHz.

The orientation of the unit cell (gaps are made in x-direction) was the suitable choice, as the unit cell with this orientation matches its operating resonance frequency. The unit cell is approximately positioned underneath the lower left corner of the patch where the field is intensively confined.

Figure 16 depicts the parametric study that can easily find the best position of the unit cell in x-direction only as the unit cell position in y-direction is -7.2 mm from the center of the patch. After varying the position from $xx = 17$ mm to $xx = 19$ mm, it can be noticed from Fig. 16

that the best position of the rectangular unit cell in x-direction is $xx = 18$ mm.

5. Fabrication and Measurements

The fabricated scheme, depicted in Fig. 17, gives satisfying results that match the simulated results. The simulated and measured return losses of the proposed triple band antenna are illustrated in Fig. 18. It can be observed from Fig. 18 that the fabricated prototype resonates at 2.45, 3.56, 5.62 GHz, respectively. Reasonable values of gain and radiation efficiency are achieved for each frequency but a little discrepancy occurs for the measured radiation efficiency at 2.45 GHz, due to the small distance between the SMA connector and the rectangular unit cell, which directly affects the gain at 2.45 GHz.

A comparison between simulated and measured radiation patterns is shown in Fig. 19 for the proposed triple band antenna. As investigated in Fig. 19, the simulated and measured E and H planes of radiation pattern of the proposed antenna at the three frequencies are mostly omnidirectional patterns with some ripples because the electrical size of the ground plane is reduced as a result of etching the three unit cells on it, resulting in partial absence of the reflector. Moreover, it can be observed from the radiation pattern of the two planes at 5.62 GHz that the maximum radiation not in the broadside and this is due to the coupling effect between the two circular CSRRs. Also, the non-symmetric position of the two circular CSRRs on the ground plane may be a reason for that inclination of the

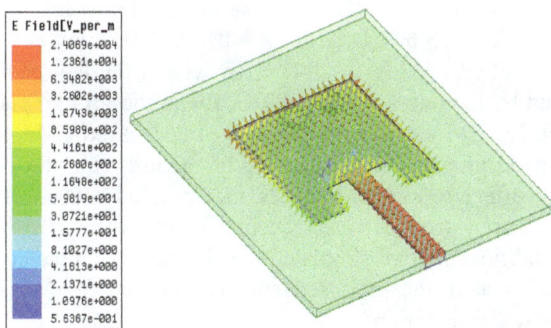

Fig. 15. Electric field distribution in the ground plane at 2.45 GHz.

Fig. 17. Top and bottom view of the fabricated antenna.

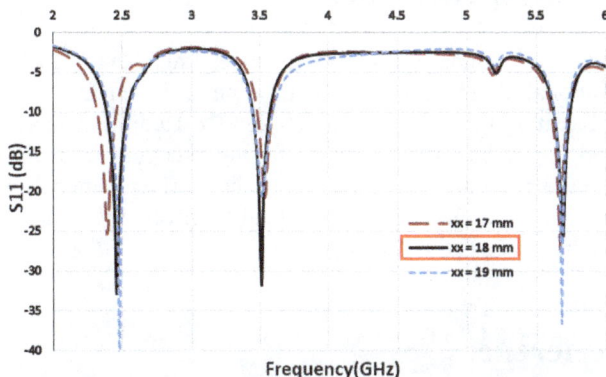

Fig. 16. Parametric study for the rectangular unit cell, operating at 2.45 GHz notched in the ground, positioned in x direction.

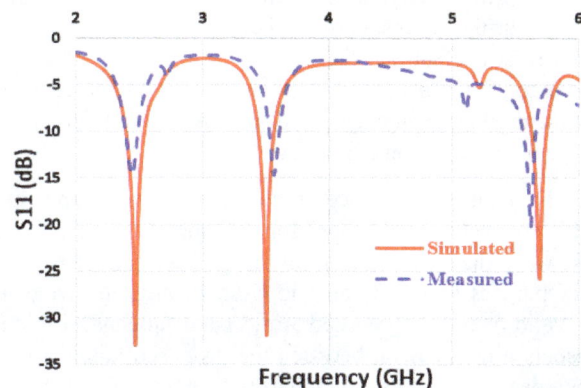

Fig. 18. Return loss comparison of the measured and simulated antenna.

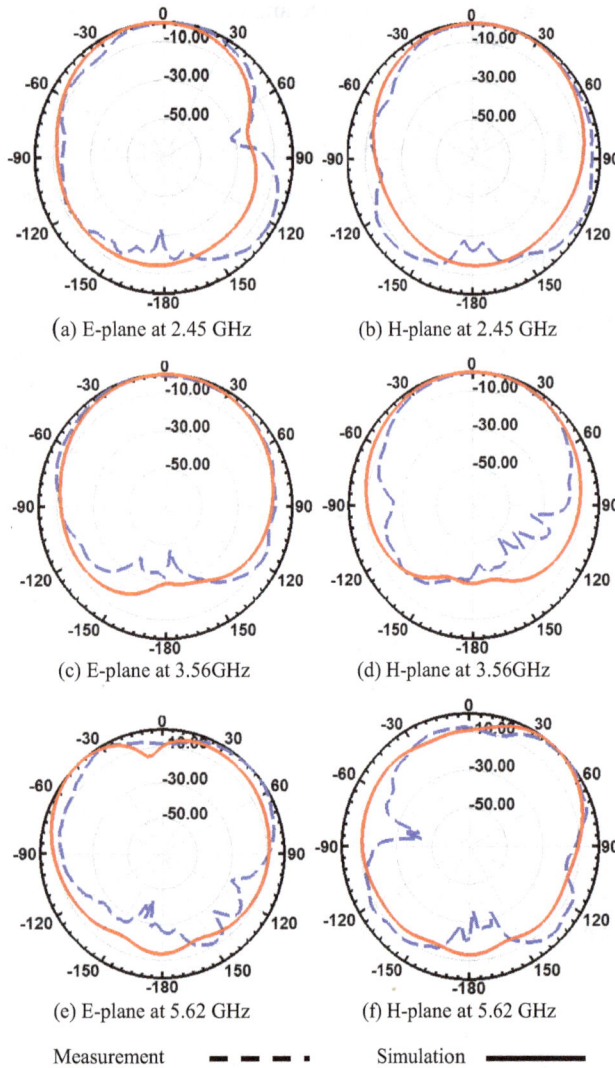

(a) E-plane at 2.45 GHz

(b) H-plane at 2.45 GHz

(c) E-plane at 3.56GHz

(d) H-plane at 3.56GHz

(e) E-plane at 5.62 GHz

(f) H-plane at 5.62 GHz

Measurement   ▬ ▬ ▬ ▪   Simulation ▬▬▬▬▬

Fig. 19. Simulated and measured E and H planes radiation patterns at the three frequencies.

beam. A good agreement between simulated and measured radiation patterns can be observed in the two planes.

The gain and radiation efficiency were measured using comparison method which required measuring the response of the standard antenna with a known gain (reference antenna). NSI-RF-RGP-10 probe with known parameters covering a range of frequencies 0.75–10 GHz is used for measuring the response of the standard antenna to get cable losses effect and the free space loss. Then the proposed antenna response was measured and compared with the standard antenna to get rid of the losses and to compute the gain and radiation efficiency.

Fabrication tolerance, welding problems, and the SMA connector that almost touches the rectangular unit cell, were the main reasons of return loss and gain little discrepancies especially at 2.45 GHz. In order to evaluate the validity of the proposed triple band antenna, Table 3 presents a comparison between the proposed antenna and other designs in terms of measured values of resonance frequency, peak gain, and radiation efficiency. It can be demonstrated from this comparison that the proposed an-

	Resonance frequencies(GHz)	Measured efficiency (%)	Peak gain (dBi)
[7]	2.4	43.7	0.27
	2.8	69.8	3.31
	3.4	77.6	4.45
[17]	2.17	57.6	0.9
	3.2	96.1	2.1
	5.25	80	2.85
This work	2.45	33.3	1.03
	3.56	29.3	5.1
	5.62	60.1	5.41

Tab. 3. Comparison between different metamaterial antenna designs.

tenna has higher gain when compared to the other antennas, even though it has lower efficiency due to fabrication tolerance as stated before.

6. Conclusion

A new design of triple band microstrip antenna printed on Rogers RT/duroid 5880 substrate was designed after studying the effect of etching metamaterial unit cells in the ground plane. Using two different types of meta-unit cells in the same design to generate multi-resonance response, was the main idea behind this work. Double circular CSRRs are loaded to the ground plane of microstrip antenna which operates at the middle Wi-MAX range (3.56 GHz) with 5.1 dBi gain, lead to add a new resonance for WLAN at 5.6 GHz with 5.41 dBi gain. The lower resonance of Wi-MAX at 2.45 GHz with 1.03 dBi gain is added by loading a rectangular CSRR in the ground plane. Intensive EM simulations using 3D FEM-based EM simulator to optimize the positioning of the unit cells for coupling effect reduction between incorporated metamaterial cells were introduced. An acceptable consistency between the fabricated and simulated results has been achieved which makes the proposed antenna suitable for WLAN/ WiMAX applications.

Acknowledgments

The author would like to thank Eng. Nagdy in the National Telecommunication Institute (NTI) in Egypt for fabricating the proposed models and Dr. Hadia Elhennawy and Eng. Abdel Hamid Hatem for their support in measuring the radiation characteristics of the fabricated models in the anechoic chamber of Ain Shams University's antenna laboratory in Cairo, Egypt.

References

[1] BALANIS, C. A. *Antenna Theory: Analysis and Design.* 3rd ed. Wiley Interscience, 2005. (Fundamental parameters of antennas, p. 27–132.) ISBN 978-0471667827

[2] PETRARIU, A. I., POPA, V. Analysis and design of a long range PTFE substrate UHF RFID tag for cargo container identification. *Journal of Electrical Engineering*, 2016, vol. 67, no. 1, p. 42–47. DOI: 10.1515/jee-2016-006

[3] ESHTIAGHI, R., SHAYESTEH, M. G., ZAD-SHAKOOIAN, N. Multi circular monopole antenna for multiband applications. *IEEE Antennas and Wireless Propagation Letters*, 2011, vol. 10, p. 1205–1207. DOI: 10.1109/LAWP.2011.217297

[4] CHAKRABORTY, U., KUNDU, A., CHOWDHURY, S. K., et al. Compact dual-band microstrip antenna for IEEE 802.11 a WLAN application. *IEEE Antennas and Wireless Propagation Letters*, 2014, vol. 13, p. 407–410. DOI: 10.1109/LAWP.2014.2307005

[5] DONG, Y., ITOH, T. Metamaterial-based antennas. *Proceedings of the IEEE*, 2012, vol. 100, no. 7, p. 2271–2285. DOI: 10.1109/JPROC.2012.2187631

[6] DONG, Y., TOYAO, H., ITOH, T. Compact circularly-polarized patch antenna loaded with metamaterial structures. *IEEE Transactions on Antennas and Propagation*, 2011, vol. 59, no. 11, p. 4329–4333. DOI: 10.1109/TAP.2011.2164223

[7] DONG, Y., TOYAO, H., ITOH, T. (2012). Design and characterization of miniaturized patch antennas loaded with complementary split-ring resonators. *IEEE Transactions on Antennas and Propagation*, 2012, vol. 60, no. 2, p. 772–785. DOI: 10.1109/TAP.2011.2173120

[8] SOLIMAN, A. M., ELSHEAKH, D. M., ABDALLAH, E. A., et al. Inspired metamaterial quad-band printed inverted-F (IFA) antenna for USB applications. *Applied Computational Electromagnetics Society Journal*, 2015, vol. 30, no. 5, p. 564–570.

[9] ANEESH, M., KUMAR, A., SINGH, A., et al. Design and analysis of microstrip line feed toppled T shaped microstrip patch antenna using radial basis function neural network. *Journal of Electrical Engineering and Technology*, 2015, vol. 10, no. 2, p. 634-640. DOI: 10.5370/JEET.2015.10.2.634

[10] WU, Z., LI, L., LI, Y., CHEN, X. Metasurface superstrate antenna with wideband circular polarization for satellite communication application. *IEEE Antennas and Wireless Propagation Letters*, 2016, vol. 15, p. 374–377. DOI: 10.1109/LAWP.2015.2446505

[11] KAHNG, S., JEON, J., PARK, T. An orthogonally polarized negative resonance CRLH patch antenna. *Journal of Electrical Engineering and Technology*, 2015, vol. 10, no. 1, p. 331–337. DOI: 10.5370/JEET.2015.10.1.331

[12] TAMANDANI, A., AHMADI-SHOKOUH, J., TAVAKOLI, S. Wideband planar split ring resonator based metamaterials. *Progress In Electromagnetics Research M*, 2013, vol. 28, p. 115 to 128. DOI: 10.2528/PIERM12120318

[13] BAGE, A., DAS, S. Studies of some non conventional split ring and complementary split ring resonators for waveguide band stop and band pass filter application. In *The IEEE International Conference on Microwave and Photonics (ICMAP)*. Dhanbad (India), 2013, p. 1–5. DOI: 10.1109/ICMAP.2013.6733474

[14] ZIOLKOWSKI, R. W. Design, fabrication, and testing of double negative metamaterials. *IEEE Transactions on Antennas and Propagation*, 2003, vol. 51, no. 7, p. 1516–1529. DOI: 10.1109/TAP.2003.813622

[15] FALCONE, F., LOPETEGI, T., LASO, M. A. G., et al. Babinet principle applied to the design of metasurfaces and metamaterials. *Physical Review Letters*, 2004, vol. 93, no. 19, 197401. DOI: 10.1103/PhysRevLett.93.197401

[16] XIE, Y., LI, L., ZHU, C., LIANG, C. H. A novel dual-band patch antenna with complementary split ring resonators embedded in the ground plane. *Progress In Electromagnetics Research Letters*, 2011, vol. 25, p. 117–126. DOI: 10.2528/PIERL11062802

[17] GUPTA, A., SHARMA, S. K., CHAUDHARY, R. K. A compact CPW-fed metamaterial antenna for high efficiency and wideband applications. In *The 21st National Conference on Communications (NCC)*. Bombay (India), 2015, p. 1–4. DOI: 10.1109/NCC.2015.7084825

About the Authors ...

Wael ALI received his B.Sc. and M.Sc. in Electronics and Communications Engineering from Arab Academy for Science, Technology and Maritime Transport (AASTMT), Alexandria, Egypt in 2004 and 2007, respectively. He obtained his Ph.D. in Electronics and Communications Engineering from Alexandria University, Alexandria, Egypt in 2012. He is a lecturer in AASTMT, Alexandria, Egypt. His research interests include smart antennas, microstrip antennas, microwave filters, and metamaterials.

Ehab HAMAD was born in Assiut, Egypt in 1970. He received his B.Sc. and M.Sc. in Electrical Engineering from Assiut University, Assiut, Egypt, in 1994 and 1999, respectively. He received his Ph.D. degree in Electrical Engineering from Otto-von-Guericke-University of Magdeburg, Magdeburg, Germany in 2006. He was a Teaching/ Research Assistant from 1996 to 2001 in Electrical Engineering Dep., South Valley University, Aswan, Egypt. From July 2001 to Dec. 2006, he was a Research Assistant in the Institute for Electronics, Signal Processing, and Communications, University of Magdeburg, Magdeburg, Germany. From July 2010 to April 2011, he joined the School of Computing & Engineering, University of Huddersfield, UK as a Post-doctoral Research Assistant. Mr. Hamad is working now as an Associate Professor for Antennas and Microwave Engineering at Aswan Faculty of Engineering, Aswan University, Aswan, Egypt. Dr. Hamad is member of the Institute of Electrical and Electronics Engineers (IEEE) and a member of the Egyptian Syndicate of Engineers. He has authored about 35 research articles and his current research interests include antenna design, UWB antennas and metamaterials, RFID, RF MEMS as applied to microwave and millimeter-wave circuits, and microwave passive planar structures.

Mohamed BASSIUNY received his B.Sc. in Electronics and Communications Engineering from the Military Technical College (MTC) in 1977 and M.Sc. from Alexandria Engineering College, Alexandria, Egypt in 1982, and Ph.D. in Electronics and Communications Engineering from MTC, in 1988. He is a lecturer in AASTMT, Alexandria, Egypt. His research interests include smart antennas, digital signal processing.

Mohamed HAMDALLAH received his B.Sc. and M.Sc. in Electronics and Communications Engineering from Arab Academy for Science, Technology and Maritime Transport (AASTMT), Alexandria, Egypt in 2012 and 2016, respectively. He is a part-time teaching assistant in AASTMT, Aswan, Egypt. His research interests include microstrip antennas, and metamaterials.

19

Interference Mitigation for the GPS Receiver Utilizing the Cyclic Spectral Analysis and RR-MSWF Algorithm

Yi HU[1], Maozhong SONG[2], Xiaoyu DANG[2], Hongli YAN[1]

[1] School of Electronics & Electrical Engineering, Chuzhou University, Chuzhou, China
[2] School of Electronics & Information Engineering, Nanjing University of Aeronautics & Astronautics, Nanjing, China

{hygps607, yhli81}@163.com, {smz106, dang}@nuaa.edu.cn

Abstract. *A method utilizing the cyclic spectral analysis (CSA) and reduced-rank multistage Wiener filtering (RR-MSWF) algorithm to mitigate the interference for the GPS receiver is proposed. In many cases, interference from adjacent channel or from cochannel overlaps on the weak global positioning system (GPS) signal in both time and frequency domains, and it is hard to mitigate this kind of strong interference with the conventional filtering techniques. While with the proposed method given in the paper, we can mitigate the interference effectively. The general process of the proposed method is that first we get the cyclic frequencies (CFs) of the strong interference by CSA of the received GPS signal. And then with the obtained CFs of the interference, we use the blind adaptive frequency shift (BA-FRESH) filter to get the principal process of mitigating the strong interference and separating the weak GPS signal. Finally by utilizing the efficient RR-MSWF algorithm to implement the BA-FRESH filtering, we can mitigate the strong interference effectively and hence improve the performance of the GPS receiver.*

Keywords

GPS, interference mitigation, cyclic spectral analysis, frequency shift filtering (FRESH), reduced-rank multistage Wiener filtering algorithm (RR-MSWF)

1. Introduction

The development of modern wireless communication systems and the emergence of many personal privacy devices (PPDs) make the working environment of the GPS receiver more severe and more complicated than ever [1–3]. The adjacent channel interference and/or the co-channel interference from these systems and/or from PPDs may cause the GPS receiver serious performance degradation when their strength is much stronger than that of the GPS signal. For the adjacent channel interference, it mainly arises from the communication satellite or from the wireless mobile communication system whose transmitting frequency or harmonics is close to the frequency of GPS signal [4–6]. While for the cochannel interference, it mainly arises from the other satellite navigation system such as Galileo or quasi-zenith satellite system (QZSS) which shares the same downlink frequency with the GPS signal [1]. For example, both GPS L1 C/A code signal and Galileo E1 OS signal have the downlink frequency of 1575.42 MHz. In addition, except these two types of unintentional interference, the GPS receiver may also be subject to the intentional interference from the hostile such as multi-tone interference [2], narrowband continuous wave interference [7], [8], linear frequency modulation interference [9] and so on. Though the nature of direct-sequence spread-spectrum (DSSS) of the GPS signal can reject the interference in some extent, when the strength of the interference becomes relatively stronger, the acquisition and tracking performance of the GPS receiver will degrade greatly, and this will correspondingly degrade the accuracy of positioning and ranging of the GPS receiver [10].

To mitigate the interference mentioned above, many filtering techniques such as matched filtering (MF) [11], finite impulse response (FIR) filtering [12], FFT-based interference excision method [13] and so on have been proposed. Though by these filtering techniques we can mitigate the interference overlapping on the GPS signal in time or frequency domain, in many cases the interference may overlap on the weak GPS signal in both time and frequency domains, and we cannot effectively mitigate them any more by these conventional filtering techniques. To solve this problem, methods such as short time Fourier transform (STFT) [4], wavelet transform (WT) and related methods [14] are proposed. For STFT, WT and related methods, their performance of interference mitigation often depends on the selection of transforming windows or wavelet basis functions, and this may bring some inconvenience in practical applications. On the other hand, for most man-made signals including the GPS signal and interference that we are to mitigate, they are spectral correlated in their corresponding cyclic spectral domains [15–17], and this feature can be used to mitigate the strong interference. Based on this, a new method utilizing CSA [18–20] of the received GPS signal and RR-MSWF algorithm [21], [22] to mitigate the strong interference for the GPS receiver is proposed. The general process of the proposed method is that first we get the CFs of strong interference by CSA of the received GPS signal. And then with the obtained CFs, we use the BA-FRESH filter [23–25] to get the principal process of mitigating the strong interference and separating the weak GPS signal. Finally by utilizing the efficient RR-MSWF algorithm to implement the BA-FRESH filtering, we can effectively mitigate the strong interference and hence improve the performance of the received GPS signal.

The rest of the paper is organized as follows. In Sec. 2, CSA of the received GPS signal overlapped with strong interference is given. Section 3 presents the detailed interference mitigation process with BA-FRESH filter and RR-MSWF algorithm based on the results given in Sec. 2. In Sec. 4, simulations on CSA of the received GPS signal, acquisition and tracking performance of the separated GPS signal are offered. Finally, the paper is concluded in Sec. 5.

2. CSA of the Received GPS Signal

2.1 Signal Model

Considering that single coarse acquisition (C/A) code signal is acquired and tracked in one channel of the GPS receiver, the received GPS L1 C/A code signal overlapped with interference after down-conversion can be given as [3]

$$x(t) = A_s s(t)\cos\left[2\pi(f_{IF} + f_d)t + \phi_s\right] + I(t) + n(t) \quad (1)$$

where A_S and Φ_S are the amplitude and initial phase of the carrier, respectively; f_{IF} is the intermediate frequency (IF) of the GPS receiver and f_d is the Doppler shift; $n(t)$ is assumed the additive white Gaussian noise; $I(t)$ is the interference, $s(t)$ is the baseband signal of GPS and

$$s(t) = \sum_{n=-\infty}^{+\infty} b_g(n)\sum_{m=0}^{M-1} c_m p_{T_c}(t - \tau_g - mT_c - nT_g) \quad (2)$$

where $b_g(n)$ is the data bit with value ±1 with equal probability; $\{c_m\}_{m=0}^{M-1}$ is one cycle of C/A code spreading sequence with length M, and $c_m \in \{\pm 1\}$; τ_g is the initial phase of GPS C/A code; T_g and T_c are the cycles of the data bit and C/A code chip, respectively; $p_{T_c}(\bullet)$ is a unit-height rectangular pulse whose support is on $[-T_c/2, T_c/2]$.

For convenience, we make two hypotheses on $I(t)$ of (1): (i) there has one strong interference in $I(t)$. In fact, if there has more than one strong interference, we can repeat the interference mitigation process with the proposed method till all the strong interference is cancelled; (ii) due that the paper is mainly focused on the interference mitigation by CSA method, the recognition of interference is out the scope of this paper (cf. [26], [27] if needed), and here we assume that the type of interference we are to mitigate is known. Based on these two hypotheses and without loss of generality, we choose a binary phase shift keying (BPSK) signal as $I(t)$ due that it has both the CFs related to

the carrier frequency and the CFs related to the bit/baud rate [22] which are the two key elements for the following BA-FRESH filter, thus $I(t)$ can be written as

$$I(t) = A_i i(t)\cos\left[2\pi(f_{IF} + \Delta f_i)t + \phi_i\right]$$
$$= A_i\left(\sum_{n=-\infty}^{+\infty} b_i(n)p_{T_i}(t - nT_i)\right)\cos\left[2\pi(f_{IF} + \Delta f_i)t + \phi_i\right] \quad (3)$$

where $i(t)$ is the baseband signal of $I(t)$; Δf_i is the frequency bias relative to f_{IF}; $b_i(n)$, T_i, ϕ_i and $p_{T_i}(\bullet)$ have the similar meanings as those given in (1) and (2).

2.2 CSA of the Received GPS Signal

For the received GPS signal $x(t)$ given in (1), its cyclic autocorrelation function (CAF) and the corresponding spectral correlation function (SCF) can be given as [23]

$$R_x^\alpha(\tau) = \left\langle x(t+\tau/2)x^*(t-\tau/2)\exp(-j2\pi\alpha t)\right\rangle, \quad (4)$$

$$S_x^\alpha(f) = \lim_{T\to\infty} T\left\langle X_T(t, f+\alpha/2)X_T^*(t, f-\alpha/2)\right\rangle \quad (5)$$

where the superscript "*" denotes complex conjugate, τ is the time lag, α is the CF of $x(t)$ and $\alpha \neq 0$, $\langle \bullet \rangle$ denotes the time average on t, and

$$X_T(t, v) = \frac{1}{T}\int_{t-T/2}^{t+T/2} x(u)\exp(-j2\pi vu)\mathrm{d}u . \quad (6)$$

For $R_x^\alpha(\tau)$ and $S_x^\alpha(f)$, they are actually a Fourier transform pair, i.e.,

$$S_x^\alpha(f) = \int_{-\infty}^{\infty} R_x^\alpha(\tau)\exp(-j2\pi f\tau)\mathrm{d}\tau . \quad (7)$$

For the conjugate CAF of $x(t)$ and its SCF, they have the similar relationships as (4) to (7). In addition, the discrete CAF and SCF corresponding to $x(t) = x(nT_S)$ are respectively as [15]

$$R_x^\alpha(k) = \lim_{N\to\infty} \frac{1}{2N+1}\sum_{n=-N}^{N} R_x\big((n+k), n\big)\exp\big(-j2\pi\alpha(n+k/2)T_S\big) , \quad (8)$$

$$S_x^\alpha(f) = \sum_{k=-\infty}^{+\infty} R_x^\alpha(k)\exp(-j2\pi kT_s f) \quad (9)$$

where $R_x(n,m)$ is the conventional autocorrelation function of $x(n)$, and $T_S = 1/f_S$ is the sampling time interval.

Substituting (1) and (3) into (4) to (7) or (8) to (9), we can get the CFs of $x(t)$ from $S_x^\alpha(f)$. But often the strength of interference is stronger than that of the GPS signal, the cyclic spectrum of GPS signal may be concealed by that of the interference, thus we can get the CFs of interference first, and then cancel the interference from $x(t)$ with the following BA-FRESH filter using the obtained CFs. Based on (1) the SCF of $I(t)$ can be given as [23]

$$S_I^\alpha(f) = \begin{cases} \dfrac{A_i^2}{4T_i}\left[\begin{array}{l} Q\left(f+f_i+\dfrac{\alpha}{2}\right)Q^*\left(f+f_i-\dfrac{\alpha}{2}\right)+ \\ Q\left(f-f_i+\dfrac{\alpha}{2}\right)Q^*\left(f-f_i-\dfrac{\alpha}{2}\right) \end{array} \right], & \alpha=\dfrac{k}{T_i} \\[4mm] \dfrac{A_i^2}{4T_i}Q\left(f\pm f_i+\dfrac{\alpha}{2}\right)Q^*\left(f\mp f_i-\dfrac{\alpha}{2}\right)e^{\pm j2\phi_i}, & \alpha=\mp 2f_i+\dfrac{k}{T_i} \\[3mm] 0, & \text{otherwise} \end{cases}$$

$$(10)$$

where α is the CF of $I(t)$, $k \in \mathbb{Z}$, $f_i \triangleq f_{IF}+\Delta f_i$ and $Q(f)=\sin(\pi f T_i)/(\pi f)$.

With (10) we can get the amplitude of $S_I^\alpha(f)$ at $f=0$ as

$$\left|S_I^\alpha(f)\right|_{f=0} = \begin{cases} \dfrac{A_i^2}{2T_i}\left|Q\left(f_i+\dfrac{\alpha}{2}\right)Q^*\left(f_i-\dfrac{\alpha}{2}\right)\right|, & \alpha=\dfrac{k}{T_i} \\[4mm] \dfrac{A_i^2}{4T_i}\left|Q\left(\dfrac{\alpha}{2}\pm f_i\right)\right|^2, & \alpha=\mp 2f_i+\dfrac{k}{T_i} \\[3mm] 0, & \text{otherwise.} \end{cases}$$

$$(11)$$

If we disregard $\alpha=0$ (in this case we cannot get any cyclic information from $S_I^\alpha(f)$), then from (11) we can find that $\left|S_I^\alpha(f)\right|_{f=0}$ takes the max values at $\alpha=2f_i$, and this can be used to get the estimates of f_i, i.e.,

$$\hat{f}_i = \frac{1}{2}\arg\max_{\alpha(\alpha>0)}\left(\left|S_I^\alpha(f)\right|_{f=0}\right). \tag{12}$$

Meanwhile, if we let $f=f_i$ in (10), we will have

$$\left|S_I^\alpha(f)\right|_{f=f_i} = \begin{cases} \dfrac{A_i^2}{4T_i}\left[\begin{array}{l}\left|Q\left(2f_i+\dfrac{\alpha}{2}\right)Q^*\left(2f_i-\dfrac{\alpha}{2}\right)\right| \\ +\left|Q\left(\dfrac{\alpha}{2}\right)\right|^2\end{array}\right], & \alpha=\dfrac{k}{T_i} \\[5mm] \dfrac{A_i^2}{4T_i}\left|Q\left(2f_i+\dfrac{\alpha}{2}\right)Q^*\left(\dfrac{\alpha}{2}\right)\right|, & \alpha=\mp 2f_i+\dfrac{k}{T_i} \\[3mm] 0, & \text{otherwise} \end{cases}$$

$$(13)$$

In (13), due that $Q(f)=\sin(\pi f T_i)/(\pi f)$, we can get that except the first right term takes the max value at $\alpha=1/T_i$, all the other values are equal or approximate equal to zero for $\alpha \neq 1/T_i$. Besides, from (11) we can also get that $\left|S_I^\alpha(f)\right|_{f=0}$ takes the submax values at $\alpha=2f_i\pm 1/T_i$ (or max values if we discard the point $\alpha=2f_i$). Combining these two results, we can get the estimate of data bit rate $R_i=1/T_i$ as

$$\hat{R}_i=\frac{1}{3}\left[\begin{array}{l}\arg\max_{\alpha,\,\alpha>0}\left(\left|S_I^\alpha(f)\right|_{f=\hat{f}_i}\right)+\arg\max_{\alpha,\,\alpha>2\hat{f}_i}\left(\left|S_I^\alpha(f)\right|_{f=0}\right) \\[3mm] +\arg\max_{\alpha,\,0<\alpha<2\hat{f}_i}\left(\left|S_I^\alpha(f)\right|_{f=0}\right)\end{array}\right]. \tag{14}$$

The results given in (12) and (14) indicate that by detecting the max value of SCF amplitude at $f=0$ (excluding point $\alpha=0$), we can easily get the estimate of carrier frequency $\hat{f}_i$, and by detecting the submax value of SCF amplitude at $f=0$ and max value at $f=\hat{f}_i$, we can get the estimate of data bit rate $\hat{R}_i$, thus we can get all the CFs of $i(t)$ [23], and they are critical to the following BA-FRESH filter and RR-MSWF algorithm given in Sec. 3. Note that because $S_x^{-\alpha}(f)=S_x^{\alpha}(f)$, the estimated results of $\hat{f}_i$ and $\hat{R}_i$ by $\left|S_x^{-\alpha}(f)\right|$ are the same as the results by $\left|S_x^{\alpha}(f)\right|$ ($\alpha>0$), and we have ignored them.

2.3 Obtaining of the Interference CFs from Cyclic Periodogram of the Received GPS Signal

From Sec. 2.2 we have known that by CSA of the received GPS signal, we can get the estimates of the interference CFs. But in practice it is often difficult to achieve this because of the finite processed data. To this problem, we can use the time smoothing or frequency smoothing cyclic periodogram [15] instead of the SCF to get the estimates of the interference CFs. Since the time smoothing method is often considered more computationally efficient than the frequency smoothing method [18], we will choose the former to get the CFs of the interference. The discrete time smoothing cyclic periodogram of $x(n)$ can be given as [15]

$$S_{X_{1/\Delta f}}^\alpha(n,f)=\frac{1}{KP}\sum_{k=0}^{KP-1}\left\{\begin{array}{l}\Delta f X_{1/\Delta f}(n-k/(\Delta fK),f+\alpha/2)\times \\ X_{1/\Delta f}^*(n-k/(\Delta fK),f-\alpha/2)\end{array}\right\} \tag{15}$$

where

$$X_{1/\Delta f}(k,f)=\sum_{i=0}^{N-1}a_{1/\Delta f}(i)x(k-i)\exp\left[-j2\pi f(k-i)T_s\right], \tag{16}$$

P is the number of non-overlap data blocks, K is a block-overlap parameter, $a_{1/\Delta f}(\bullet)$ is a data tapering window with the length of N samples, $\Delta f=1/[(N-1)T_s]$ is the spectral resolution. In (15), the time length of total data segment $\Delta t=[(1+P-1/K)N-1]T_s$, and $0\le n\le \Delta t/T_s$.

To get the reliable estimates of the interference CFs, the product of Δt and Δf should meet the condition [15]

$$\Delta t\Delta f=[(1+P-1/K)N-1]/(N-1)\approx P\gg 1. \tag{17}$$

In practical applications, the implementation of (15) can be fulfilled by FFT accumulation method (FAM) or by strip spectral correlation analyzer (SSCA) [19]. Based on (12), (14) to (16), and using FAM algorithm, the process of obtaining the interference CFs of $\hat{f}_i$ and $\hat{R}_i$ can be given as following:

Step 1 Suppose N is a power of 2 (if not, we can pad trailing zeros to the samples of $x(n)$ to realize this), based on (15) and (16), use FAM method to get $S_{X_{1/\Delta f}}^\alpha(n,f)$.

Step 2 Get the position corresponding to the max value of $\left|S_{X_{1/\Delta t}}^{\alpha}(n,f)\right|_{f=0}$, where $\alpha>0$. Suppose the result is n_f, we can get the estimated carrier frequency of interference as $\hat{f}_i=(n_f-n_m)/(N_tT_S)$, where $N_t=2N^2$ is the total number of processed data by FAM in each data segment, and $n_m=N_t/2$ is the middle position of N_t.

Step 3 For $\alpha>0$, find the positions corresponding to submax values of $\left|S_{X_{1/\Delta t}}^{\alpha}(n,f)\right|_{f=0}$ in the ranges of (n_m,n_f-3) and (n_f+3,N_t), and suppose the results are $n_b^{(l)}$ and $n_b^{(r)}$, where $n_b^{(l)}$ and $n_b^{(r)}$ are on the left and right of n_f, respectively. Here we have empirically neglected 6 spectral lines which are centered at n_f and may take the submax values due to cyclic spectral leakage [14].

Step 4 Find the positions corresponding to submax values of $\left|S_{X_{1/\Delta t}}^{\alpha}(t,f)\right|_{f=\hat{f}_i}$ in the range (n_m+3,N_t) and suppose the results is $n_b^{(f)}$, we can get the estimated data bit rate of interference as $\hat{R}_i=\left[(n_b^{(r)}-n_m)+(n_m-n_b^{(l)})+(n_b^{(f)}-n_m)\right]/(3N_tT_s)$.

3. Interference Mitigation with BA-FRESH Filtering and RR-MSWF Algorithm

3.1 Principles of BA-FRESH Filter

With the interference CFs obtained from the cyclic periodogram of the received GPS signal, we can effectively mitigate the interference and separate the weak GPS signal with the adaptive FRESH filter. Due that the reference signal often cannot be got beforehand, we can use the blind adaptive FRESH (BA-FRESH) filter to fulfill the filtering. The structure of BA-FRESH filter [24] based on $x(n)$ of (1) is shown as Fig. 1, in which $\hat{I}(n)$ and $\tilde{s}_g(n)$ are the estimated interference and the separated GPS signal, respectively.

In Fig. 1, $\alpha_\gamma\in\left\{k\hat{R}_i\right\}$ ($\gamma=1, 2, \cdots, \Gamma$) and $\beta_\varsigma\in\left\{2k\hat{f}_i\pm\hat{R}_i\right\}$ ($\varsigma=1, 2, \cdots, \Xi$) are the non-conjugate and conjugate CFs of the interference, respectively, $k\in\mathbb{Z}$. For ν of the reference signal $d(n)$, it meets $\nu\in\left\{k\hat{R}_i\right\}\cup\left\{2k\hat{f}_i\pm\hat{R}_i\right\}$ ($k\in\mathbb{Z}$) and $\nu\notin\left\{a_\gamma\right\}_{\gamma=1}^{\Gamma}\cup\left\{\beta_\varsigma\right\}_{\varsigma=1}^{\Xi}$. The results of $\hat{I}(n)$ and $\tilde{s}_g(n)$ can be given as [24], [25]

$$\hat{I}(n)=\mathbf{h}^H\tilde{\mathbf{x}}(n),\ \tilde{s}_g(n)=d(n)-\hat{I}(n)\qquad(18)$$

where the superscript H denotes transpose conjugate, and

$$\mathbf{h}=[\mathbf{h}^{\alpha_1},\cdots,\mathbf{h}^{\alpha_\Gamma},\mathbf{h}^{\beta_1},\cdots,\mathbf{h}^{\beta_\Xi}]^T,\qquad(19)$$

$$\tilde{\mathbf{x}}(n)=[\mathbf{x}^{\alpha_1}(n),\mathbf{x}^{\alpha_2}(n),\cdots,\mathbf{x}^{\alpha_\Gamma}(n),\mathbf{x}^{\beta_1}(n),\mathbf{x}^{\beta_2}(n),\cdots\mathbf{x}^{\beta_\Xi}(n)]^T\qquad(20)$$

where the superscript T denotes transpose, $\mathbf{h}^{\alpha_\gamma}$ and $\mathbf{x}^{\alpha_\gamma}(n)$ (or $\mathbf{h}^{\beta_\varsigma}$ and $\mathbf{x}^{\beta_\varsigma}(n)$) are respectively the coefficients and inputs of the FIR filter corresponding to α_γ with L_γ taps (or β_ς with L_ς taps),

$$\begin{cases}\mathbf{h}^{\alpha_\gamma}=[h^{\alpha_\gamma}(0),h^{\alpha_\gamma}(1),\cdots,h^{\alpha_\gamma}(L_\gamma-1)]^T,\ \gamma=1,2,\cdots\Gamma\\\mathbf{h}^{\beta_\varsigma}=[h^{\beta_\varsigma}(0),h^{\beta_\varsigma}(1),\cdots,h^{\beta_\varsigma}(L_\varsigma-1)]^T,\ \varsigma=1,2,\cdots\Xi\end{cases}\qquad(21)$$

and

$$\begin{cases}\mathbf{x}^{\alpha_\gamma}(n)=[x(n-L_\gamma+1)\exp\left(j2\pi\alpha_\gamma(n-L_\gamma+1)\right),\cdots,\\\qquad x(n)\exp\left(j2\pi\alpha_\gamma n\right)]^T,\\\qquad \gamma=1,2,\cdots\Gamma,\\\mathbf{x}^{\beta_\varsigma}(n)=[x^*(n-L_\varsigma+1)\exp\left(j2\pi\beta_\varsigma(n-L_\varsigma+1)\right),\cdots,\\\qquad x^*(n)\exp\left(j2\pi\beta_\varsigma n\right)]^T,\\\qquad \varsigma=1,2,\cdots\Xi.\end{cases}\qquad(22)$$

3.2 Implementation of BA-FRESH Filtering with RR-MSWF Algorithm

In practice, we can implement the BA-FRESH filtering with recursive least squares (RLS) algorithm, least mean square (LMS) algorithm and so on [24]. Considering the computation efficiency, in the following numerical simulations we have used the efficient RR-MSWF algorithm [21], [22] to mitigate the interference and separate the GPS signal. The schematic diagram of RR-MSWF algorithm based on Fig. 1 is shown as Fig. 2.

Based on Fig. 2 and equations (18) to (22), the detailed implementation of the interference mitigation with the batch adaptive RR-MSWF algorithm [21] is given as in Tab. 1.

4. Numerical Results

To show the effectiveness of the proposed method, simulations shown as Figs. 3 to 8 are offered for verification. Among them, Figure 3 offers CSA results of the received GPS C/A code signal overlapped with the strong interference, Figures 4 to 8 offer the acquisition, probability of detection, and tracking of the separated weak GPS signal with the proposed method, respectively, in which the CFs $\hat{f}_i$ and $\hat{R}_i$ of the BA-FRESH filter used in Figs. 4 to 6 are obtained from Figs. 3 and the CFs used in Figs. 7 to 8 are got in a similar way given as Fig. 3. Besides, in Figs. 4 to 8 simulations with the conventional MF filtering are offered for comparison, in Figs. 5 and 6 simulation results without the strong interference are offered for further comparison, and in Figs. 7 and 8 the cases in the presence of more strong interference are also offered for further verification.

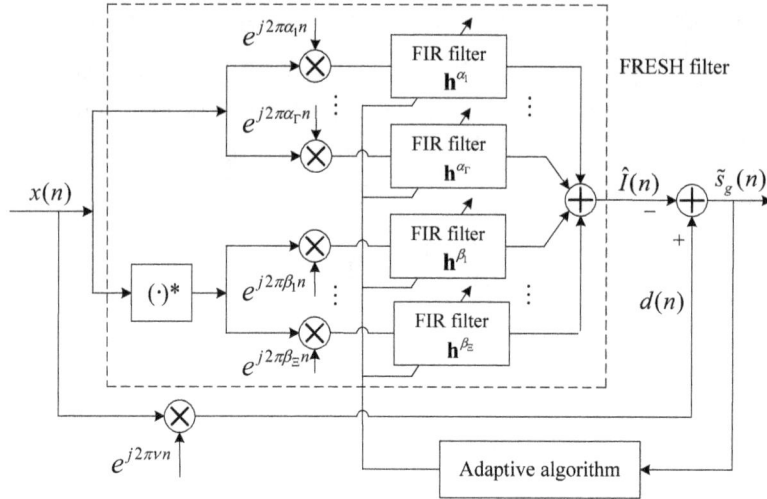

Fig. 1. Schematic diagram of BA-FRESH filter.

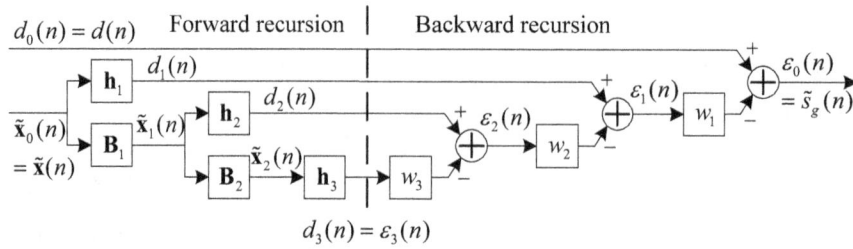

Fig. 2. Schematic diagram of RR-MSWF algorithm (stage $D = 3$).

Initialization:

$\mathbf{X}_0 = \left[\tilde{\mathbf{x}}(1),\ \tilde{\mathbf{x}}(2),\ \cdots,\ \tilde{\mathbf{x}}(k),\ \cdots,\ \tilde{\mathbf{x}}(N_{\Delta t}) \right]$,

$\mathbf{d}_0 = \left[d(1),\ d(2),\ \cdots, d(k),\ \cdots,\ d(N_{\Delta t}) \right]$.

Forward recursion:	*Backward recursion:*
for $r = 1,\ 2,\ \cdots,\ D$	$\boldsymbol{\varepsilon}_D = \mathbf{d}_D$;
$\qquad \boldsymbol{\lambda}_r = \mathbf{X}_{r-1} \mathbf{d}_{r-1}^{H}$;	for $r = D,\ D-1,\ \cdots,\ 1$
$\qquad \mathbf{h}_r = \boldsymbol{\lambda}_r \big/ \lVert \boldsymbol{\lambda}_r \rVert$;	$\qquad w_r = \boldsymbol{\varepsilon}_r \mathbf{d}_{r-1}^{H}$;
$\qquad \mathbf{d}_r = \mathbf{h}_r^{H} \mathbf{X}_{r-1}$;	$\qquad \boldsymbol{\varepsilon}_{r-1} = \mathbf{d}_{r-1} - w_r^{*} \boldsymbol{\varepsilon}_r$;
$\qquad \mathbf{B}_r = \mathbf{I} - \mathbf{h}_r \mathbf{h}_r^{H}$;	end
$\qquad \mathbf{X}_r = \mathbf{B}_r^{H} \mathbf{X}_{r-1}$;	
end	

Output of the separated GPS signal:

$\tilde{\mathbf{s}}_g = \boldsymbol{\varepsilon}_0 = \mathbf{d}_0 - w_1^{*} \boldsymbol{\varepsilon}_1$,

where

$\tilde{\mathbf{s}}_g = \left[\tilde{s}_g(1),\ \tilde{s}_g(2),\ \cdots,\ \tilde{s}_g(k),\ \cdots,\ \tilde{s}_g(N_{\Delta t}) \right]$.

Tab. 1. Implementation of BA-FRESH filtering with the batch adaptive RR-MSWF algorithm to mitigate the strong interference. In the table, $N_{\Delta t}$ is the total samples input the BA-FRESH filter in time interval Δt, $\mathbf{I}$ is a unit matrix, and $\lVert \bullet \rVert$ denotes 2-norm.

The common parameter setups used in Figs. 3 to 8 are given as following:

- Time length $\Delta t = 1$ ms except in Figs. 6 and 8 it is set to 40 ms, and sampling frequency $f_s = 1/T_s = 16$ MHz.

- Carrier IF, Doppler shift and initial C/A code phase of the GPS signal are respectively set as $f_{IF} =$

4.309 MHz, $f_d = 2$ kHz, and $\tau_g = 200$ chips or 3128 in samples.

- In Figs. 4 to 6, frequency bias and data bit rate of interference are respectively set as $\Delta f_i = 500$ kHz and $R_i = 1.544$ Mbit/s; while in Figs. 7 and 8, the setups of different type of interference are given in Tab. 2.

Other different parameter setups used in each simulation are given as following:

- In Fig. 4, while using the FAM algorithm to get the CFs of the interference, the cycle resolution $\Delta\alpha = 1/\Delta t$ is set to 1 kHz and the spectral resolution Δf is set to 120 kHz, K is chosen to be 4 according to [18];

- In Figs. 4 to 8, for the BA-FRESH filtering, $\alpha_\gamma \in \left\{\pm\hat{R}_i\right\}$, $\beta_\varsigma \in \left\{\pm 2\hat{f}_i, \ \pm 2\hat{f}_i \pm \hat{R}_i\right\}$, and $\nu = 0$ for the reference signal [24], $L_\gamma = L_\varsigma = 10$ ($\gamma = 1, 2$; $\varsigma = 1, 2, \cdots, 6$), and D of RR-MSWF algorithm is chosen to be 8 according to the result given in [21].

- In Figs. 5 and 7, the probability of detection is calculated by Monte Carlo method and the independent run number is 300, and P_{fa} in the figure is the false-alarm probability;

- In Figs. 6 and 8, the root mean square error (RMSE) of each tracking result is the average of 50 independent runs.

4.1 CSA of the Received GPS Signal

From Fig. 3(a) we can get that the estimated carrier frequency, and two estimated data bit rates respectively as $\hat{f}_i = 9.6182/2 = 4.8091$ MHz, $R_i^{(l)} = 9.6182 - 8.0742 = 1.544$ Mbit/s and $R_i^{(r)} = 11.1622 - 9.6182 = 1.544$ Mbit/s. From Fig. 3(b) we can get $R_i^{(0)} = 1.544$ Mbit/s. Then from these results we can get the final estimated data bit rate as $\hat{R}_i = (R_i^{(l)} + R_i^{(r)} + R_i^{(0)})/3 = 1.544$ Mbit/s. All these results show that they fit well with their preset values.

4.2 Acquisition and Detection of the Separated GPS Signal

From Fig. 4(a) we can see that when the strength of interference equals 5 dB, i.e., the interference is relatively weak, we may still acquire the weak GPS signal with the conventional MF filtering; but when the strength of interference increases to 15 dB as given in Fig. 4(b), i.e., the interference becomes relatively stronger, we almost cannot acquire the weak GPS signal. But on the other hand, in these two cases, we can well acquire both the weak GPS signals with the proposed method (denoted by "BAF-RR" filtering).

From Fig. 5 we can further learn the difference of two methods for the acquisition of the weak GPS signal from their probabilities of detection. From Fig. 5(a), we can see that when the strength of interference equals 5 dB, and under the condition that probability of detection $P_d = 0.8$, the proposed BAF-RR filtering outperforms the MF filtering about 4 dB for $P_{fa} = 10^{-3}$, and 6 dB for $P_{fa} = 10^{-6}$, and is only about 2 dB lower than the result without the strong interference for $P_{fa} = 10^{-3}$ or $P_{fa} = 10^{-6}$. While from Fig. 5(b) we also can see that when the strength of interference equals 12 dB, under the same condition that $P_d = 0.8$, the proposed BAF-RR filtering outperforms the MF filter-

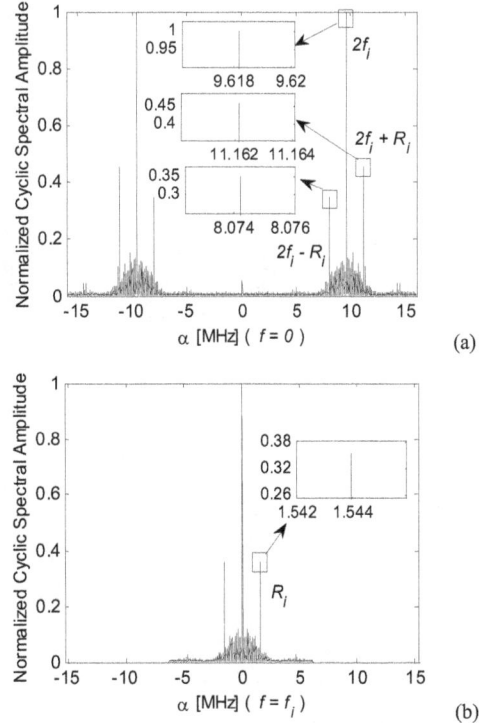

Fig. 3. Obtain the estimates of the strong interference CFs R_i and f_i from the special planes of cyclic periodogram, in which (a) is the plane at $f = 0$, and (b) is the plane at $f = f_i$. The signal to noise ratios (SNRs) of the GPS and the interference are −15 dB and 10 dB, respectively.

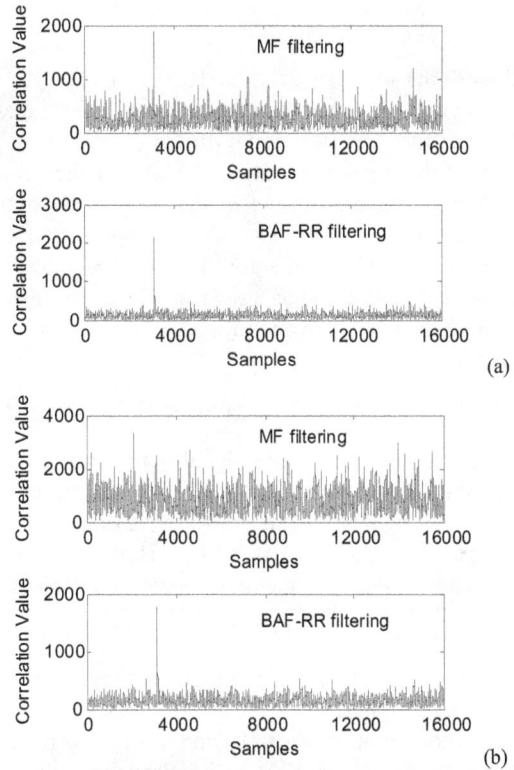

Fig. 4. Acquisition comparisons of the separated GPS signal filtered with the conventional MF method and with the proposed BAF-RR method, in which the SNR of the GPS signal is −15 dB and the SNRs of the strong interference are (a) 5 dB and (b) 15 dB, respectively.

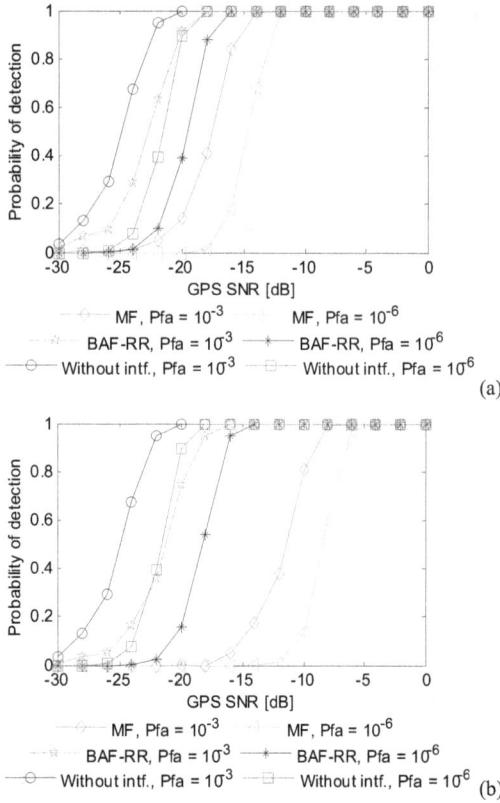

Fig. 5. Comparisons of detection probability of the separated GPS signal filtered with the conventional MF method and with the proposed BAF-RR method, in which the SNRs of the strong interference are (a) 5 dB and (b) 12 dB, respectively.

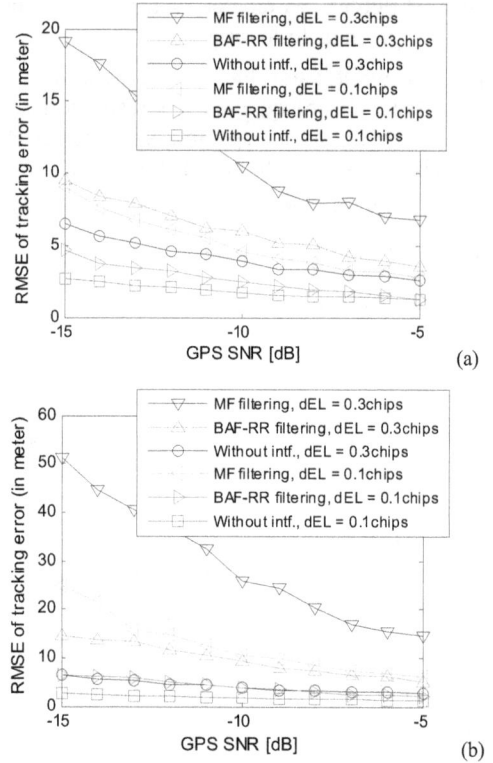

Fig. 6. RMSE comparisons of the tracking of the separated GPS signal filtered with the conventional MF method and with the proposed BAF-RR method, in which the SNRs of the strong interference are (a) 5 dB and (b) 12 dB, respectively.

ing about 7 dB for $P_{\mathrm{fa}} = 10^{-3}$ and 10 dB for $P_{\mathrm{fa}} = 10^{-6}$, and is only about 3 dB lower than the result without the strong interference for $P_{\mathrm{fa}} = 10^{-3}$, and 4 dB for $P_{\mathrm{fa}} = 10^{-6}$. In addition, under the condition that $P_{\mathrm{d}} = 0.8$, with the proposed BAF-RR filtering we can detect the weak GPS signal as low as -24 dB for the SNR of overlapped interference equal to 5 dB, and -20 dB for the SNR of interference equal to 12 dB. The results given in Fig. 5 are also consistent with the results given in Fig. 4.

4.3 Tracking of the Separated GPS Signal

The tracking of the GPS signal is the another critical stage of the GPS receiver, and it is mainly fulfilled by the discriminator of delay locked loop (DLL) aided with the carrier phase locked loop (PLL). Here we use the discriminator whose tracking error is given by

$$\Delta\tau = \left(\sqrt{I_{\mathrm{ES}}^2 + Q_{\mathrm{ES}}^2} - \sqrt{I_{\mathrm{LS}}^2 + Q_{\mathrm{LS}}^2}\right) \Big/ \left(\sqrt{I_{\mathrm{ES}}^2 + Q_{\mathrm{ES}}^2} + \sqrt{I_{\mathrm{LS}}^2 + Q_{\mathrm{LS}}^2}\right)$$

[28] to measure the DLL tracking performance of the separated GPS signal, where I_{ES} and I_{LS} are respectively the coherent integrations of early (E) and late (L) correlators of the in-phase, Q_{ES} and Q_{LS} are the similar results of the quadrature-phase. The parameter dEL in Fig. 6 is the spacing of E and L correlators.

From Fig. 6 we can see that with the BAF-RR filtering, we can also improve the tracking performance of the

separated GPS signal dramatically. For example, in Fig. 6(a) when the strength of the interference equals 5 dB, under the condition that GPS SNR = -13 dB and dEL = 0.3 chips, for the conventional MF filtering, RMSE of tracking error of the separated GPS signal is about 18 m; but after the interference mitigation with the proposed BAF-RR filtering, the RMSE of tracking error of the separated GPS signal will decrease to about 9 m. In Fig. 6(b) we can also get the tracking performance of the separated GPS signal filtered with the proposed BAF-RR filtering greatly outperforms that with the conventional MF filtering, and the tracking result with the proposed BAF-RR method is nearly close to the case without the strong interference.

In addition, from Fig. 6 we can also see that the dEL also has an important effect on the tracking accuracy of the GPS receiver. Generally speaking, the correlators of E and L with small dEL are preferable to those with large dEL in improving the tracking accuracy of the GPS receiver.

4.4 Acquisition and Tracking Performance of the Separated GPS Signal in the Presence of More Strong Interference

In this section, the acquisition and tracking performance of the separated GPS signal in the presence of more strong interference is also verified. The simulation results are shown as Figs. 7 and 8 respectively, in which the setups of different interference are given in Tab. 2.

Interference type	SNR (dB)	Carrier frequency (MHz)	Bit or chirp rate
BPSK	10	$f_i = f_{IF} + \Delta f_i = 4.809$	$R_i = 1.544$ Mbps
Binary amplitude shift keying (2ASK)	7	$f_i = 4.609$	$R_i = 1.2288$ Mbps
Binary frequency shift keying (2FSK)	5	$f_{i1} = 4.109$, $f_{i2} = 4.189$	$R_i = 1.2288$ Mbps
Linear frequency modulation (LFM or chirp)	5	$f_i = 3.909$ (initial frequency)	$k = 6.7 \times 10^7$ Hz/s $B = 0.2$ MHz

Tab. 2. The setups of different interference in which $f_{IF} = 4.309$ MHz, k and B of LFM are frequency change rate and bandwidth respectively.

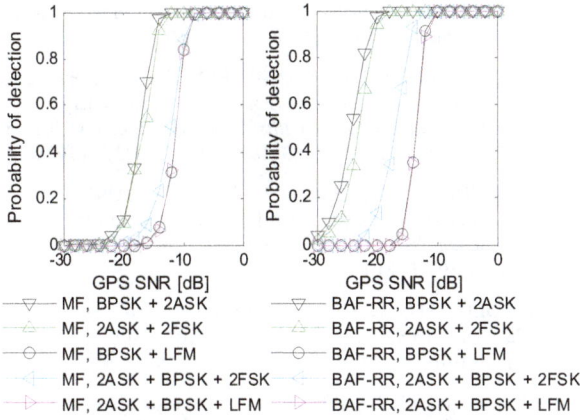

Fig. 7. Comparisons of detection probability of the separated GPS signal filtered with the conventional MF method and with the proposed BAF-RR method under more strong interference, in which $P_{fa} = 10^{-3}$ and the setups of the interference are given in Tab. 2.

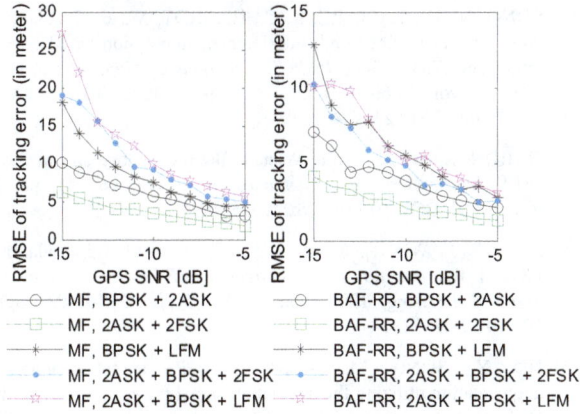

Fig. 8. RMSE comparisons of the tracking of the separated GPS signal filtered with the conventional MF method and with the proposed BAF-RR method under more strong interference, in which dEL = 0.1 chips and the setups of the interference are given in Tab. 2.

From Fig. 7 we can get that under more strong interference, for the same P_d of the separated GPS signal, the proposed BAF-RR algorithm still outperforms the MF algorithm greatly in GSP SNR. For example, for $P_d = 0.8$ and the case of two interference of BPSK + 2ASK, BAR-RR can outperform MF about 6 dB in GPS SNR. On the other hand, from Fig. 7 we also can get that for the two algorithms, when there exists the LFM interference, their acquisition performance will all degrade dramatically, and this can be explained by that the GPS signal and the given LFM interference almost cannot be distinguished in both correlation domain and cyclic spectral domain. But even for this worst case, the proposed BAR-RR algorithm still can filter some more LFM interference than the conventional MF algorithm, as can be seen from their comparisons of P_d in Fig. 7.

While from Fig. 8 we can see that when there is more strong interference, with the proposed BAR-RR algorithm we can also improve the tracking accuracy of the separated GPS signal a lot than with the conventional MF algorithm. For example, when GPS SNR = −15 dB and for the case of three interference of 2ASK + BPSK + 2FSK, RMSE of tracking error of the separated GPS signal is about 10 m, while for the conventional MF algorithm the result is about 20 m. In addition, from Fig. 8 we can also get that when there exists the LFM interference, the tracking performance of both MF algorithm and BAR-RR algorithm will degrade greatly, and this is just well consistent with the result given in Fig. 7.

5. Conclusions

A method based on CSA and RR-MSWF algorithm to mitigate the strong interference for the GPS receiver is proposed. Utilizing the CFs obtained by CSA of the received GPS signal and BA-FRESH filtering combined with the efficient RR-MSWF algorithm, we can effectively mitigate the strong interference overlapping on the weak GPS signal in both time and frequency domains, and hence improve the acquisition and tracking performance of the GPS receiver dramatically. On the other hand, due that the delayed signals input the FRESH filter are complex variables, the computation of BA-FRESH filtering is often more complicated than that of the conventional filtering. Concretely, based on (18)–(22) and RR-MSWF algorithm offered in Tab. 1, the computation complexity of BA-FRESH filtering can be given as about $O(N_{\Delta t}(\Gamma L_\gamma + \Xi L_\varsigma)D)$, while for the conventional FIR or MF filtering, the result is about $O(N_{\Delta t}\ell)$, where ℓ is the number of FIR filter taps or the Doppler shift searching step size of MF filtering, and often $\ell < (\Gamma L_\gamma + \Xi L_\varsigma)D$, correspondingly the computation time of the proposed method will be larger than that of the conventional one. For example, with the values of $\Gamma = 6$, $\Xi = 2$, L_γ, L_ς, D and f_s given in the beginning of Sec. 4, suppose CPU operation frequency is 1.0 GHz and $\Delta t = 1$ ms and $\ell = 50$, the time cost of MF is about 1.6 ms and the proposed method is about 10.2 ms. Often, to this problem of complexity or computation time, we can make a tradeoff between the number of CFs (= $\Gamma + \Xi$) and taps L_γ (or L_ς) used in the FRESH filter and the performance of the separated GPS signal.

Acknowledgments

This work was supported by the Initial Research Fund of Chuzhou University under grant 2015qd05 and NSFC under grant 61571224. The authors would also like to acknowledge the anonymous reviewers for their helpful comments.

References

[1] WILDEMEERSCH, M., FORTUNY GUASCH, J. Radio frequency interference impact assessment on global navigation satellite systems. *European Commission Joint Research Center, Science and Technique Report, JRC55767*, 2010. DOI: 10.2788/6033

[2] JAHROMI, A. J., BROUMANDAN, A., DANESHMAND, S., et al. Vulnerability analysis of civilian L1/E1 GNSS signals against different types of interference. In *Proceedings of the 28th International Technical Meeting of the ION Satellite Division.* Tampa, Florida (USA), 2015, p. 3262–3271.

[3] LI, J., NIE, J., LI, B., et al. Increase of carrier-to-noise ratio in GPS receivers caused by continuous-wave interference. *Radioengineering*, 2016, vol. 25, no. 3, p. 506–517. DOI: 10. 13164/re.2016.0506

[4] SAVASTA, S., PRESTI, L. L., RAO, M. Interference mitigation in GNSS receivers by a time-frequency approach. *IEEE Transactions on Aerospace and Electronic Systems*, 2013, vol. 49, no. 1, p. 415 to 438. DOI: 10.1109/taes.2013.6404112

[5] ANYAEGBU, E., BORDIN, G., COOPER, J., et al. An integrated pulsed interference mitigation for GNSS receivers. *The Journal of Navigation*, 2008, vol. 61, no. 2, p. 239–255. DOI: 10.1017/s0373463307004572

[6] RAGHAVAN, S., LAZAR, S., EDGAR, C., et al. Cochannel and adjacent channel interference to GPS use. In *17th AIAA International Communications Satellite Systems Conference and Exhibit*. 1998, p. 564–574. DOI: 10.2514/6.1998-1330

[7] LI, M., DEMPSTER, A. G., BALAEI, A. T., et al. Switchable beam steering/null steering algorithm for CW interference mitigation in GPS C/A code receivers. *IEEE Transactions on Aerospace and Electronics Systems,* 2011, vol. 47, no. 3, p. 1564–1579. DOI: 10.1109/taes.2011.5937250

[8] MAO, W. L. GPS interference mitigation using derivative-free Kalman filter-based RNN. *Radioengineering*, 2016, vol. 25, no. 3, p. 518–526. DOI: 10. 13164/re.2016.0518

[9] HOU, Y. G., GUO, W., LI, X. S. Design of a GPS receiver for the linear frequency modulation interference suppression. In *IEEE International Conference on Communications, Circuits and Systems*. 2009, p. 454–456. DOI: 10.1109/icccas.2009.5250504

[10] BHUIYAN, M. Z. H., KUUSNIEMI, H., SÖDERHOLM, S., et al. The impact of interference on GNSS receiver observables - a running digital sum based simple jammer detector. *Radioengineering*, 2014, vol. 23, no. 3, p. 898–906.

[11] LEE, Y. T., CHANG, C. M., MAO, W. L., et al. Matched-filter-based low-complexity correlator for simultaneously acquiring global positioning system satellites. *IET Radar, Sonar and Navigation*, 2010, vol. 4, no. 5, p. 712–723. DOI: 10.1049/iet-rsn.2009.0147

[12] FANTE, R., VACCARO, J. J. Wideband cancellation of interference in a GPS receiver array. *IEEE Transactions on Aerospace and Electronic Systems*, 2000, vol. 36, no. 2, p. 549–564. DOI: 10.1109/7.845241

[13] CAPPOZZA, P. T., HOLLAND, B. J., HOPKINSON, T. M., et al. A single-chip narrow-band frequency-domain excisor for a global positioning system (GPS) receiver. *IEEE Journal of Solid-State Circuits*, 2000, vol. 35, no. 3, p. 401–411. DOI: 10.1109/4.826823

[14] MUSUMECI, L., DOVIS, F. Performance assessment of wavelet based techniques in mitigating narrow-band interference In *IEEE International Conference on Localization and GNSS (ICL-GNSS)*. 2013, p. 1-6. DOI: 10.1109/icl-gnss.2013.6577264

[15] GARDNER, W. A. Measurement of spectral correlation. *IEEE Transactions on Acoustics, Speech and Signal Processing*, 1986, vol. 34, no. 5, p. 1111–1123. DOI: 10.1109/tassp.1986.1164951

[16] ŠEBESTA, V., MARŠÁLEK, R., FEDRA, Z. OFDM signal detector based on cyclic autocorrelation function and its properties. *Radioengineering*, 2011, vol. 20, no. 4, p. 926–931.

[17] DIMC, F., BALDINI, G., KANDEEPAN, S. Experimental detection of mobile satellite transmissions with cyclostationary features. *International Journal of Satellite Communications and Networking*, 2015, vol. 33, no. 2, p. 163–183. DOI: 10.1002 /sat.1081

[18] ROBERTS, R. S., BROWN, W. A., LOOMIS, H. H. Computationally efficient algorithms for cyclic spectral analysis. *IEEE Signal Processing Magazine*, 1991, vol. 8, no. 2, p. 38–49. DOI: 10.1109/79.81008

[19] NAPOLITANO, A., PERNA, I. Cyclic spectral analysis of the GPS signal. *Digital Signal Processing*, 2014, vol. 33, p. 13–33. DOI: 10.1016/j.dsp.2014.06.003

[20] HUANG, P., PI, Y., PROGRI, I. GPS signal detection under multiplicative and additive noise. *The Journal of Navigation*, 2013, vol. 66, no. 4, p. 479–500. DOI: 10.1017/s0373463312000550

[21] HONIG, M. L., GOLDSTEIN, J. S. Adaptive reduced-rank interference suppression based on the multistage Wiener filter. *IEEE Transactions on Communications*, 2002, vol. 50, no. 6, p. 986–994. DOI: 10.1109/tcomm.2002.1010618

[22] SONG, N., DE LAMARE, R. C., HAARDT, M., et al. Adaptive widely linear reduced-rank interference suppression based on the multistage Wiener filter. *IEEE Transactions on Signal Processing*, 2012, vol. 60, no. 8, p. 4003–4016. DOI: 10.1109/tsp.2012.2197747

[23] GARDNER, W. A. Cyclic Wiener filtering: theory and method. *IEEE Transactions on Communications*, 1993, vol. 41, no. 1, p. 151–163. DOI: 10.1109/26.212375

[24] ZHANG, J., WONG, K. M., LUO, Z. Q., et al. Blind adaptive FRESH filtering for signal extraction. *IEEE Transactions on Signal Processing*, 1999, vol. 47, no. 5, p. 1397–1402. DOI: 10.1109/78.757230

[25] HU, Y., SONG, M., MENG, B. GPS signal availability augmentation utilizing the navigation signal retransmission via the GEO comsat. *Wireless Personal Communications*, 2015, vol. 82, no. 4, p. 2655–2671. DOI: 10.1007/s11277-015-2371-9

[26] DOBRE, O. A., ABDI, A., BAR-NESS, Y., et al. Cyclostationarity based blind classification of analog and digital modulations. In *Proceedings of the 2006 IEEE Conference on Military Communications MILCOM 2006*. Washington (USA), 2006, p. 2176–2182. DOI: 10.1109/milcom.2006.302556

[27] RAMKUMAR, B. Automatic modulation classification for cognitive radios using cyclic feature detection. *IEEE Circuits and Systems Magazine*, 2009, vol. 9, no. 2, p. 27–45. DOI: 10.1109/mcas.2008.931739

[28] KAPLAN, E. D., HEGARTY, C. J. *Understanding GPS: Principles and Applications*. 2nd ed., Norwood (USA): Artech House, 2006. ISBN: 1580538940

About the Authors ...

Yi HU was born in Lu'an, China, in 1974. He is currently a lecturer and engineer in communication engineering in Chuzhou University. He received his Ph.D. degree from Nanjing University of Aeronautics & Astronautics (NUAA) in 2015. His research interests include satellite navigation signal processing and Sat-COM.

Maozhong SONG was born in Shexian, China, in 1962. He is a professor in the College of Electronic & Information Engineering, NUAA. His current research interests include satellite navigation signal processing, communication measurement and controlling techniques.

Xiaoyu DANG was born in Nanjing, China, in 1973. He is a professor in the College of Electronic & Information Engineering, NUAA. His current research interests include communication signal processing and Sat-COM.

Hongli YAN was born in Chuzhou, China, in 1981. She is a lecturer in Chuzhou University. She received her M.Sc. degree from Wuhan University of Science and Technology in 2008. Her current research interests include wireless communications and signal processing.

20

On the Equalization of an OFDM-Based Radio-over-Fiber System Using Neural Networks

Leila SAFARI [1], Gholamreza BAGHERSALIMI [1], Ali KARAMI [1], Abdolreza KIANI [2]

[1]Dept. of Electrical Engineering, University of Guilan, Rasht, Iran,
[2]Dept. of Electrical Engineering, Abadan Branch, Islamic Azad University, Abadan, Iran

leila_safari_81@yahoo.com, bsalimi@guilan.ac.ir, karami_s @guilan.ac.ir, ab2rezakiani@yahoo.com

Abstract. *In this study the impact of a Radio-over-Fiber (RoF) subsystem on the performance of Orthogonal Frequency Division Multiplexing (OFDM) system is evaluated. The study investigates the use of Multi-Layered Perceptron (MLP) and Radial Basis Function (RBF) neural networks to compensate for the optical subsystem nonlinearities in terms of bit error rate, error vector magnitude, and computational complexity. The Bit Error Rate (BER) and Error Vector Magnitude (EVM) results show that the performance of MLP neural network is superior to that of RBF neural network and time-multiplexed pilot-based equalizer especially in the case of highly nonlinear behavior of the RoF subsystem.*

Keywords

Radio-over-Fiber, OFDM, equalization, neural network, RBF, MLP, BER, EVM

1. Introduction

In recent years, with the growth of technologies, efficient high-speed data transmission techniques over wireless channels have become important topics for research. A key enabling technology supporting the provision of cellular communications is Radio-over-Fiber (RoF), alternatively known as Hybrid Fiber Radio (HFR). This technology combines two media: radio and optical. Typically, the optical part is used to interconnect a central radio processing facility with a remote radio antenna, the latter providing coverage to wireless broadband users. Some of the advantages offered by a RoF system include low signal attenuation (in the fiber), improved coverage and system performance, enhanced capacity, low Radio Frequency (RF) power dissipation, reduced complexity due to the centralized processing of RF signals and ultimately low system costs [1–3]. However, the performance of RoF systems can be severely affected by nonlinear effects in the transmission channel. In the RF part of a RoF system, the main source of nonlinearity is the power amplifier. In the optical part of a RoF system, the main sources of nonlinearity include the laser-diode (LD) light source, the optical fiber

and the PIN photo detector (PD). For short haul optical links, the fiber nonlinearity is usually small and can be neglected from system considerations. Also, the PD can be assumed to be linear for the same conditions.

One area of interest in modern communications is Orthogonal Frequency Division Multiplexing (OFDM) which is becoming widely used in wireless communication systems due to its high data rate transmission capability with high bandwidth efficiency and also its robustness to multipath fading without requiring complex equalization techniques [2, 4]. OFDM has been adopted in a number of wireless applications including Digital Audio Broadcast (DAB), Digital Video Broadcast (DVB), Wireless Local Area Network (WLAN) standards such as IEEE 802.11g and Long Term Evolution (LTE) [4, 5]. To mitigate ISI (inter-symbol interference) introduced by the channel, a Cyclic Prefix (CP) is used which in turn leads to spectral inefficiency [4]. In comparison to single carrier systems including code division multiple access (CDMA) systems, more vulnerability to nonlinearity, frequency offset due to fading etc., and phase noise are OFDM disadvantages [4], [6], [7].

Channel estimation is a subject which has received a great deal of attention by researchers in recent years. In wireless systems, channel estimation techniques are used for the estimation of RF channel impulse response (CIR) in order to compensate for the amplitude distortion and phase rotation introduced by the RF channel variations. In cellular communication, estimation is achieved by time-multiplexed pilot signals, code-multiplexed pilot signals or a combination of these signals [8]. In the wireless local area network (LAN) systems, on the other hand, estimation is carried out by frequency (subcarrier) multiplexed pilot signals, time (symbol) multiplexed pilot signals or a combination of these signals [4]. All such equalization techniques require extra hardware at either the transmit and/or the receive side of the communication link. To compensate for RF channel impairments, two basic one-dimensional (1-D) estimation techniques namely the block-type (time-multiplexed) and comb-type (frequency-multiplexed) along with some combinations of both schemes (scattered-type) are employed in OFDM systems [4]. Normally, the aforementioned techniques are used to estimate and compensate for fading channels.

Neural networks (NNs) have been evolved into a powerful tool in solving complex applications such as function approximation, nonlinear classification, nonlinear system identification, nonlinear mapping between the high-dimensional input and output spaces, and forming complex decision regions with nonlinear decision boundaries [9–16]. Further, because of the nonlinear characteristics of the NNs, these networks have been found successful in channel estimation and equalization in digital communication systems. In [13], a method for channel estimation and equalization based on the multi-layered perceptron (MLP) neural network was presented, which resulted in a robust solution. Radial basis function (RBF), as another NN scheme, was also used for channel equalization [10, 14].

For RoF-OFDM communications, there is a dearth of publications in the context of channel estimation and compensation [11, 17]. In particular, the effectiveness of all methods has not been substantially investigated before. Further, as will be shown, the system performance can be improved by using neural networks. It should be emphasized that both the amplitude and phase distortions of an optical channel affect the system performance. Also, due to the hysteresis-type memory of the optical subsystem, the entire RoF link suffers more from the phase impairments and frequency-dependent nonlinearities when compared to other nonlinear elements.

In this paper, the impact of the optical subsystem nonlinearities on the performance of an OFDM system (adopted from IEEE802.11g standards [18]) is studied with respect to the channel estimation and equalization functions using both the MLP and the RBF neural networks. The results for the OFDM system Bit Error Rate (BER), Error Vector Magnitude (EVM), and computational complexity performances are presented for different modulation schemes, neural networks, and Output Back-off (OBO) levels when two neural networks are used to compensate for the distortion introduced by both of the optical channel and the RF channel. Strictly speaking, the overall subsystem that is, the optical (nonlinear) subsystem and the RF channel (represented by AWGN – additive white Gaussian noise) is estimated and equalized using two different neural networks under different conditions. Further, the effectiveness of the proposed MLP NN and RBF NN-based approaches is compared with that of the (time-multiplexed) pilot-based equalizer. Also, as this research focuses only on the optical subsystem, other sources of nonlinearity, such as the RF High Power Amplifier (HPA), are not considered. Finally, for the sake of simplicity and focusing on the equalization of the nonlinear subsystem, operations such as coding and interleaving along with fading channel are not considered.

The rest of the paper is organized as follows. Section 2 introduces the theoretical background and system model for estimating and equalizing the optical subsystem. Section 3 presents the MLP and RBF NN structures for cannel equalization while the results of neural network-based equalization are presented in Sec. 4. Finally, Section 5 concludes the paper.

2. Description of the RoF-OFDM System

In this section, the mathematical description of the baseband RoF-OFDM system for transmission of one OFDM symbol is presented. Figure 1 shows the block diagram of the RoF-OFDM communication system employing a NN-based equalizer. Also, various data and signals are labeled properly. In general, this model consists of a transmitter, an optical subsystem, a RF channel and a receiver. These subsystems are explained in the following subsections.

2.1 Transmitter

RoF-OFDM transmitter consists of a random digital data generator and an OFDM modulator. This transmitter sends its data bit stream $\{a_k\}$ to the modulator where they are translated to a symbol train using the Quadrature Amplitude Modulation (QAM) mapping rule. Let the user signal over the OFDM block T be denoted by

$$x_D(t) = \sum_{n=0}^{N_s-1} D_n \delta(t - nT_s); \qquad 0 \le t \le T \quad (1)$$

where $D_n = b_n + jd_n$ is the n-th (baseband) data symbol, b_n is the in-phase component of the n-th data symbol, d_n is the quadrature component of the n-th data symbol, T_s is the data symbol duration, N_s is the number of data symbols being transmitted in the OFDM symbol (block) period $T = N_s \cdot T_s$, and $\delta(t)$ is the Dirac delta function. The serial stream of data symbols is first converted into N parallel streams. An inverse DFT (Discrete Fourier Transform) is taken on each set of data symbols, giving a set of complex time-domain values. Then, a Cyclic Prefix (CP) is prepended in order to compensate for the Inter-symbol Interference (ISI). Thus, the OFDM symbol $x(t)$ is given by [4]

$$x(t) = \sum_{k=-k_1}^{N_s} \sum_{n=0}^{N_s-1} D_n \exp\left(j2\pi \frac{n}{N_s}\right); \quad -\frac{k_1 T}{N_s + k_1} < t < \frac{N_s T}{N_s + k_1} \quad (2)$$

where k_1 is the CP length. Also, the OFDM signal can be described by

$$x(t) = r(t) \exp(j\theta(t)) \quad (3)$$

where $r(t)$ and $\theta(t)$ are the amplitude and phase of the transmit signal, respectively.

2.2 Optical Subsystem

It is customary in communication systems to analytically describe the system performance provided that the analytic model for all components/subsystems (i.e. input-output relationship) are available. Beside validity under certain circumstances and difficulty in mathematical operations, some physical problems cannot be described by mathematical models. In such cases behavioral modeling is used in such a way that the system is treated as a black box

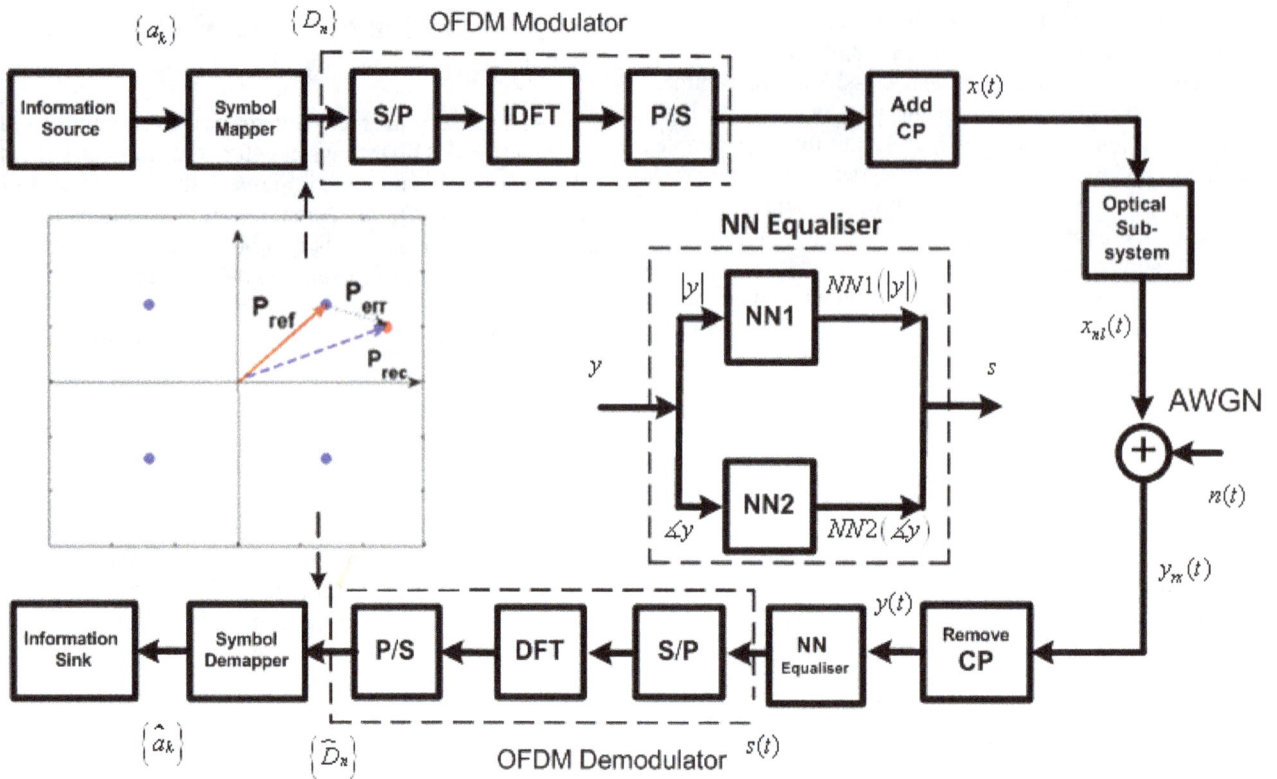

Fig. 1. Block diagram of a NN equalizer for a RoF-OFDM system.

whose characteristics are represented by the so called transfer functions or alternatively the large-signal response. These functions are expressed as AM-AM (amplitude-to-amplitude) and AM-PM (amplitude-to-phase) characteristics. It must be emphasized the all deficiencies in the system (such as optical noise and static and dynamic nonlinearities in the optical subsystem's case) are included in the black box model. The main advantage of behavioral modeling is that it can describe a complex system with no need to a thorough knowledge of its subsystems.

When the envelope of the signal is varying slowly, the large signal response can be used instead of the instantaneous value of the signal. The AM-AM/PM characteristics are valid provided that the device "memory time" (amplitude - and frequency-dependent time delay between the transmitted and received RF signal) is much smaller than the reciprocal of the input signal bandwidth [18]. As a rule of thumb, the AM-AM/PM models are used when the memory time of the nonlinear system is at least twice less than the reciprocal of the envelope frequency [19, 20]. This criterion is satisfied in this research.

In this study, the optical subsystem (shown in Fig. 1) includes a laser diode (LD), a 2.2 km long single-mode fiber and a photodiode (PD) [3]. The radio frequency is 1.8 GHz, however the frequency response of the optical link (centred at 1310 nm) is flat over frequencies $1.7 < f < 2.2$ GHz. The AM-AM/PM transfer characteristics are depicted in Fig. 2. Moreover, a pictorial description of OBO is shown in Fig. 2 and is defined (on a logarithmic scale) as the difference between the maximum output power and the output power at the operating point.

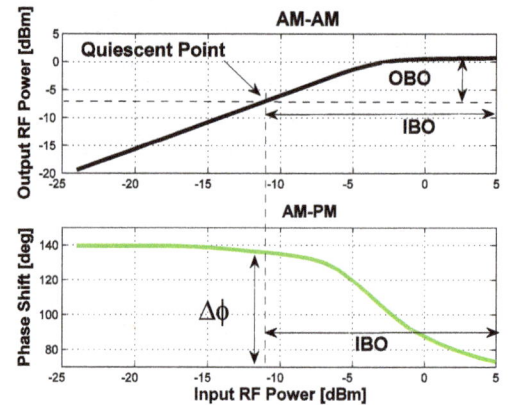

Fig. 2. AM-AM/PM characteristics (IBO: Input Back-off) [3].

The OFDM signal passes through the optical (nonlinear) subsystem whose output is described by

$$x_{nl}(t) = G_{out}\sqrt{\left(2R_{out}^{-1}\right)f_{AM\text{-}AM}(0.5R_{in}(G_{in}|x(t)|)^2)} \qquad (4)$$
$$\exp[jf_{AM\text{-}PM}(0.5R_{in}(G_{in}|x(t)|)^2 + j\psi_x(t)]$$

where *IBO* is the input back-off on a linear scale, $G_{in} = \sqrt{\max\left(P_{RF,i}\right)/\left(IBO \cdot P_m\right)}$ is the gain of the pre-amplifier which matches the output power of the transmit RRC filter to the input power of the optical subsystem [3], P_m is the maximum input RF power to the optical subsystem before the pre-amplifier, $\max(P_{RF,i})$ is the (measured) maximum input RF power before the pre-amplifier, $\psi_x(t)$ is the phase of the signal $x_t(t)$, $R_{in} = R_{out} = 50 \ \Omega$ is the input/output impedance of the optical subsystem, and G_{out} is a linear gain which sets the overall gain of the optical subsystem to unity. Also, $f_{\text{AM-AM}}(.)$ and $f_{\text{AM-PM}}(.)$ are the AM-AM and AM-PM transfer characteristics of the optical subsystem, respectively. To explicitly representing the amplitude and phase transfer characteristics of the optical subsystem, equation (4) can be written as

$$x_{\text{nl}}(t) = g\left[x_{\text{t}}(t)\right] = F\left[r(t)\right]\exp(j[\phi\left[r(t)\right]+\theta(t)]) \quad (5)$$

where g is a nonlinear function which represents the optical subsystem, and F and ϕ are known as AM-AM and AM-PM transfer functions, respectively.

2.3 RF Channel

The output signal of the optical subsystem is corrupted by a zero-mean complex Additive White Gaussian Noise (AWGN) process $n(t)$. So, the received signal is described by

$$y_{\text{rx}}(t) = x_{\text{nl}}(t) + n(t). \quad (6)$$

Also, equation (5) can be rewritten by using

$$x_{\text{nl}}(t) = x_{\text{d}}(t) + n_{\text{d}}(t) \quad (7)$$

where $x_{\text{d}}(t)$ is the amplified/attenuated version of signal $x(t)$, and $n_{\text{d}}(t)$ is the total distortion introduced by the optical subsystem and AWGN channel. In addition, E_{b} is defined as the average bit energy of the user. So, the received signal can be described as

$$y_{\text{rx}}(t) = x_{\text{d}}(t) + n_{\text{d}}(t) + n(t). \quad (8)$$

Then, the signal-to-interference-noise ratio (SINR) is given as follows

$$SINR = \frac{E[x_{\text{d}}^2(t)]}{E[n_{\text{d}}^2(t) + n^2(t)]}. \quad (9)$$

2.4 Receiver

The receiver consists of a NN equalizer to compensate for the channel impairments along with an OFDM demodulator. First, by removing the CP, the resultant signal is applied to the NN equalizer. The NN equalizer is composed of two sub-networks that compensate for the amplitude and phase of the transmitted signal. These sub-networks learn the knowledge of the channel by using the learning algorithm in the training phase. The output of the equalizer $s(t)$ is then compared with the transmit signal $x(t)$ to produce an error which is used to update the weights of the network in the training phase. Then, during system normal operation phase these sub-networks recover the amplitude and phase of signal $x(t)$. Following amplitude estimation and phase derotation of the received signal, the resultant serial data are converted into parallel, and a DFT is taken followed by a conversion to serial data whose output is denoted by $\{\hat{D}_n\}$. These serial data are then demodulated using 4/16QAM demapping rule. Finally the resultant bit stream $\{\hat{a}_k\}$ is compared with the transmit one i.e. $\{a_k\}$.

3. Neural Network Model for Channel Equalization

In this paper, two NN structures i.e. MLP and RBF are employed for channel estimation and equalization. These are described as follows.

Figure 3 shows the proposed MLP structure. In these configurations, I and J denote the number of neurons in the input and hidden layers, respectively. The proposed NN comprises two sub-networks NN1 and NN2 that estimate and compensate for the amplitude and phase of the transmitted signal x, respectively. The MLP NNs are usually trained in a supervised manner with a highly popular algorithm known as the error back-propagation (BP). However, the conventional back-propagation method is often too slow for many practical problems; thus, in this paper the *resilient* back-propagation technique, which is one of the fast training algorithms, is employed to accelerate the training process [21]. The system error is computed between the target outputs and the corresponding estimated outputs by the NN. The weights associated with the NN are then updated using the BP algorithm [13]. This procedure is continued till the Mean-Square Error (MSE) of the NN reaches a predetermined value.

Figure 4 shows the proposed RBF NN structure where m and n denote the number of neurons in the input and hidden layers, respectively. In the hidden layer, the input space is expanded onto a higher dimensional space by applying a nonlinear transformation using a set of RBFs. At the output layer, linear combinations of the outputs produced in the hidden layer neurons are calculated in order to obtain the RBF network final outputs [15]. The output of the network is represented by:

$$y(x) = \sum_{i=1}^{N} w_i \phi_i(x) \quad (10)$$

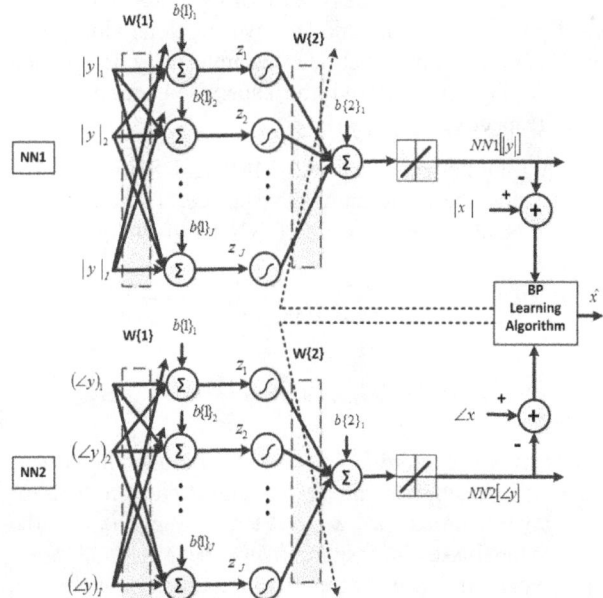

Fig. 3. The proposed MLP neural network to compensate for amplitude (top) and phase (bottom) distortions.

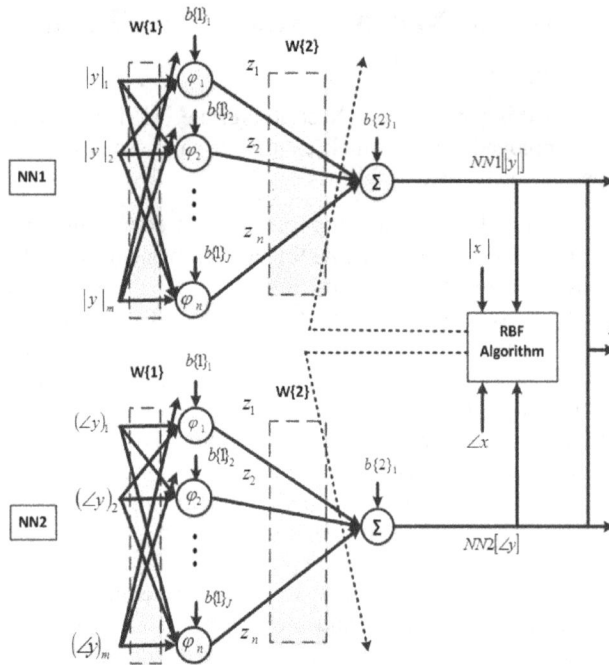

Fig. 4. The proposed RBF neural network to compensate for amplitude (top) and phase (bottom) distortions.

where x is the input vector and w_i's are the weight values corresponding to the output layer. A Gaussian basis function is usually used for the hidden neurons which is given by:

$$\phi(x) = \exp\left(-\frac{\|x-c\|^2}{\sigma^2}\right) \qquad (11)$$

where c is the center of the Gaussian function and σ is the spread parameter. Training of an RBF network is accomplished by obtaining suitable centers for the hidden layer neurons along with appropriate weight values for the output layer. In the training phase of an RBF-NN, the centers and spread parameters of the hidden layer neurons along with weights for the output layer neurons are usually determined by employing unsupervised and supervised training algorithms, respectively.

The sub-networks in both MLP and RBF NNs are trained using the same training sequence. The outputs of NN1 and NN2 in the case of MLP NN are given by

$$NN1\big[|y(t)|\big] = f_2 \sum_{i=1}^{J} c_{1i} f_1[w_{1i} \,|\, y(t) \,|\, + b_{1i}],$$

$$NN2\big[\angle y(t)\big] = f_2 \sum_{i=1}^{J} c_{2i} f_1[w_{2i}(\angle y(t)) + b_{2i}] \qquad (12)$$

where $f_1(.)$ is a sigmoid function used for hidden layer neurons, $f_2(.)$ represents the linear activation function used for output layer neurons, $\{w_{1i}, c_{1i}, b_{1i}\}$ and $\{w_{2i}, c_{2i}, b_{2i}\}$ are the weights and biases corresponding to the NN1 and NN2, respectively. Also, J is the number of neurons in the hidden layer. The outputs of NN1 and NN2 in the case of RBF NN are given by

$$NN1[|\,y(t)\,|] = a_{10} + \sum_{k=1}^{n} a_{1k}\varphi_k(|\,y(t)\,|),$$

$$NN2[\angle y(t)] = a_{20} + \sum_{k=1}^{n} a_{2k}\varphi_k(\angle y(t)) \qquad (13)$$

where φ_k represents the Gaussian function used for the k-th hidden layer neuron, a_{1k} and a_{2k} are the weights connecting the k-th hidden unit to the output unit in the NN1 and NN2 sub-networks. In addition, a_{10} and a_{20} are respectively the bias terms used in the NN1 and NN2 sub-networks, and n represents the number of hidden neurons in the two sub-networks.

After training, both trained NNs are ready to produce an estimate for the transmitted signal $\hat{x}$, which can be shown by (14)

$$\hat{x}(t) = NN1\big[|y(t)|\big] \exp\big(jNN2\big[|\angle y(t)|\big]\big). \qquad (14)$$

4. Results and Discussion

To assess the impacts of the optical subsystem on OFDM in the presence of nonlinearity, computer simulations were performed based on the system model shown in Fig. 1. A key performance metric for modern communication systems is EVM (in dB), which is defined as [5]

$$EVM(\text{dB}) = 10\log_{10}\left(\frac{P_{\text{err}}}{P_{\text{ref}}}\right) \qquad (15)$$

where P_{err} is the RMS (Root-Mean Square) power of the error (due to nonlinear time-variant channel) vector and P_{ref} is the reference constellation average power in the I (in-phase)-Q (quadrature) plane (see Fig. 1).

To enable statistically valid simulation results in reasonable simulation times, Monte-Carlo methods are used during the simulation. Also, the OFDM system parameters [18] and optical subsystem parameters [3] are summarized in Tab. 1.

DFT size (Number of subcarriers)	64
Data modulation	4/16QAM
OBO [dB]	0, 0.2, 0.5, 3
Number of input neurons	20
Number of hidden neurons in MLP-NN	65
Number of hidden neurons in RBF-NN(Number of epochs)	1 000
Number of epochs for training RBF-NN	1 000
Number of output neurons	1
Number of training bits	40 000
Total bits sent	10 000 000
Number of bits per block	800
Number of pilot bits per block (for the 1-D time-multiplexed case)	80 (10 % of the total power)

Tab. 1. Simulation parameters.

The estimation for both NNs by using 4/16QAM modulation schemes is achieved individually. In addition, the performance of a time-multiplexed (block type) pilot-aided equalizer is evaluated to compare with the results obtained by the NNs. Table 1 shows the simulation parameters used in this study.

Now, the performance of the proposed NN-based equalizers is discussed in terms of BER, EVM, and computational complexity.

4.1 BER Performance

Bit error rate was carried out for different values of OBO using 4QAM and 16QAM modulation schemes. The BER results obtained by the proposed NNs are compared with those of the pilot-aided estimation. BER curves for different E_b/N_0 values are shown at different OBO values. Figures 5 and 6 show the BER performance of the system in

the presence of AWGN for four OBO values, i.e. 0 dB (the worst case scenario), 0.2 dB, 0.5 dB, and 3 dB (the best case scenario) with 4QAM and 16QAM signal constellations, respectively. From the plots we observe that by increasing the nonlinearity of the optical subsystem (i.e., decreasing the OBO value) the MLP NN-based equalizer performs more efficiently than the pilot-aided equalizer. Upon a closer observation, we also note that the MLP NN performs better than the RBF NN and the pilot-based estimators. These observations confirm that the optical subsystem was estimated more efficiently by the trained MLP neural network in comparison to the pilot-based estimator.

4.2 EVM Performance

EVM is a measure used to quantify the performance of a digital radio transceiver. A signal sent by an ideal transmitter would have all constellation points precisely at the ideal locations (see Fig. 1 for the 4QAM case). However, various imperfections in the implementation such as carrier leakage, phase noise, nonlinearity etc. cause the actual constellation points to deviate from the ideal locations. Informally, EVM is a measure of how far the points are from the ideal locations. In this subsection, the EVM is carried out for different OBO values given in Tab. 1. Figures 7 and 8 show the EVM performance for different E_b/N_0 values and two extreme OBO values, i.e., 0 and 3 dB, respectively.

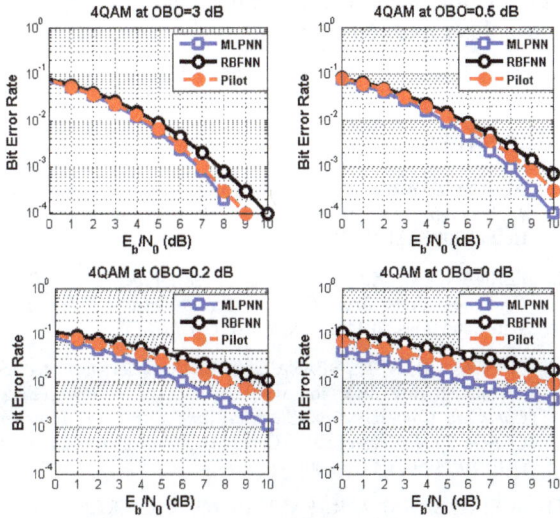

Fig. 5. BER for RoF-OFDM-4QAM at *OBO* = 3 dB (top left), *OBO* = 0.5 dB (top right), *OBO* = 0.2 dB (bottom left), and *OBO* = 0 dB (bottom right).

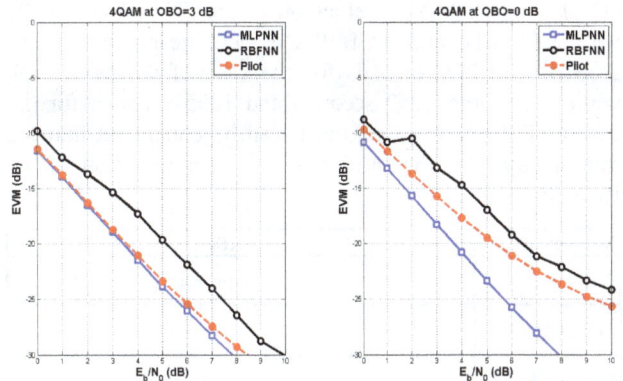

Fig. 7. EVM for RoF-OFDM-4QAM at *OBO* = 3 dB (left) and *OBO* = 0 dB (right).

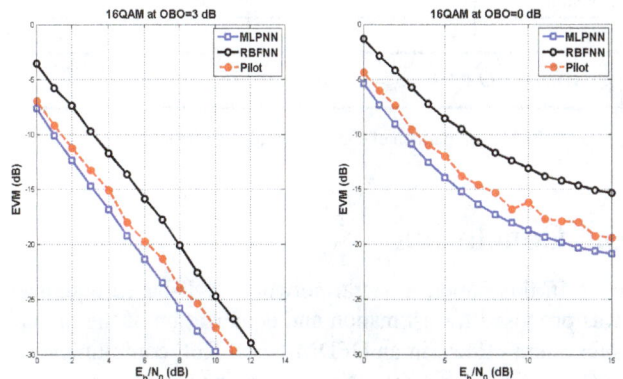

Fig. 6. BER for RoF-OFDM-16QAM at *OBO* = 3 dB (top left), *OBO* = 0.5 dB (top right), *OBO* = 0.2 dB (bottom left), and *OBO* = 0 dB (bottom right).

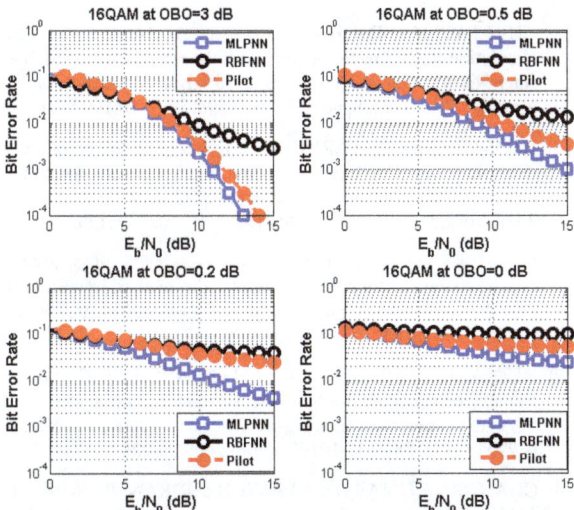

Fig. 8. EVM for RoF-OFDM-16QAM at *OBO* = 3 dB (left) and *OBO* = 0 dB (right).

The EVM performance of the NNs is compared with each other and with that of the pilot-aided estimator. It is clear that the EVM curve for the MLP NN is superior (lower) than EVM curves for the RBF and pilot-based equalizers. Therefore, the MLP NN has better performance in collecting the actual constellation points around their ideal locations in comparison to the other equalizers. It is also observed that the MLP NN performs more efficiently than the pilot-aided equalizer and RBF-based equalizer in the linear and nonlinear regions of the optical subsystem.

4.3 Computational Complexity

Table 2 compares the required computational parameters for 1 000 iterations of training in both types of NNs and modulation schemes. In MLP and RBF networks, the number of nodes in the input layer is 20, the number of neurons in the hidden layer in the MLP NN is 65, and one neuron in the output layer is considered for both NNs. To analyze the performance of NNs, each equalizer is trained for 1 000 epochs, and then the hidden neurons in the RBF NN are increased to 1 000. The number of training symbols is 40 000 for all NNs and it can be seen from Tab. 2 that the RBF NN requires an additional division and exp(.) operation because of using a Gaussian basis function as given by (11). In addition, since in the RBF NN the input space is mapped onto a high dimensional space, its computational complexity is higher than that of the MLP NN. Besides, the elapsed times for performing 1 000 epochs of training for 4QAM and 16QAM schemes on a core (TM) 2 Duo CPU with 2.67 GHZ and 4 GB RAM were, respectively, 156 seconds and 230 seconds for the case of the MLP NN, while they were 1052 seconds and 1095 seconds for the case of RBF NN. Therefore, it can be concluded that the training time for the MLP NN is shorter than that of the RBF NN.

Number of	MLP		RBF	
Weights	φ_1	1431	$n(m + k)$	21000
Additions	$3\varphi_2 + 3K - IJ$	2798	$2mn + m + n + k$	41021
Multiplications	$4\varphi_2 + 3\varphi_3 + 2K - IJ$	4420	$m n + m + 2n + k$	22021
tanh(.)	φ_3	86	----	----
Division	----	----	$m + n$	1020
Exp(.)	----	----	n	1000
$\varphi_1 = (I + 1)J + (J + 1)K,\quad \varphi_2 = IJ + JK,\quad \varphi_3 = I + J + K$				

Tab. 2. Comparison of computational complexity.

5. Conclusion

In this paper, a neural network (NN)-based equalizer was proposed for estimation and equalization of the optical subsystem effects in an OFDM-based RoF communication system. Two different NNs were trained considering two modulation schemes, and their testing results were compared. It was shown that the MLP NN could estimate and

compensate for the subsystem effects more efficiently than the RBF NN. The comparison was achieved in terms of BER and EVM for a range of E_b/N_0 values. The simulation results corresponding to different linear and nonlinear behaviors of the subsystem confirmed the superiority of the MLP NN-based equalizer in estimating the (optical) channel behavior over the RBF NN-based equalizer in terms of EVM, BER, and computational complexity. It was also shown that an MLP-NN with a proper architecture could be trained in a shorter time compared to the RBF-NN. Finally, a comparison of the proposed NN-based equalizer with a pilot-based estimator revealed that the MLP-based equalizer had a better performance compared to the pilot-based equalizer.

Acknowledgments

The authors would like to thank the anonymous reviewers for their helpful comments that helped to improve this paper considerably.

References

[1] AL-RAWESHIDY, H., KOMAKI, S. *Radio-over-Fibre Technologies for Mobile Communications Networks.* Artech House, 2002. ISBN: 978-1580531481

[2] FERNANDO, X. N. *Radio over Fiber for Wireless Communications: From Fundamentals to Advanced Topics.* John Wiley and Sons, 2014. ISBN: 978-1-118-79706-8

[3] FERNANDO, X. N., SEASAY, A. B. Characteristics of directly modulated RoF link for wireless access. In *Proceeding of Canadian Conference on Electrical and Computer Engineering CCECE 2004.* Niagara Falls (Canada), 2004, vol. 4, p. 2167–2170. DOI: 10.1109/CCECE.2004.1347673

[4] SHEIKH BAHAI, A. R., SALTZBERG, B. R., ERGEN, M. *Multicarrier Digital Communications.* Kluwer/Plenum, 2002. ISBN: 978-0387225753

[5] GOLDSMITH, A. *Wireless Communications.* Artech House, 2005. ISBN: 978-0521837163

[6] CHETTAT, H., SIMOHAMED, L. M., ALGANI, C., et al. Co-simulation-based modeling and performance analysis of hybrid fiber-wireless links. *International Journal of Communication Systems*, 2013, vol. 26, no. 5, p. 583–596. DOI: 10.1002/dac.1361

[7] HOSSAIN, MD. A., TARIQUE, M. Effect of multipath fading and multiple access interference on broadband code division multiple access systems. *International Journal of Communication Systems*, 2012, vol. 25, no. 7, p. 874–886. DOI: 10.1002/dac.1293

[8] BAGHERSALIMI, G. Performance assessment of a wideband code-division multiple access-based radio-over-fibre system with near–far effect: downlink scenario. *IET COM*, 2014, vol. 8, no. 7, p. 1056–1064. DOI: 10.1049/iet-com.2013.0805

[9] PATRA, J. C., MEHER, P. K., CHAKRABORTY, G. Nonlinear channel equalization for wireless communication systems using Legendre neural networks. *Signal Processing*, 2009, vol. 89, no. 5, p. 2251–2262. DOI: 10.1016/j.sigpro.2009.05.004

[10] CHANDRA KUMAR, P., SARATCHANDRAN, P., SUNDARA-JAN, N. Minimal radial basis function neural networks for nonlinear channel equalization, *IEEE Proceedings - Vision, Image and Signal Processing*, vol. 147, no. 5, Oct. 2000, p. 428–435. DOI: 10.1049/ip-vis:20000459

[11] NAWAZ, S. J., MOHSIN, S., IKRAM, A. A. Neural network based MIMO-OFDM channel equalizer using comb-type pilot arrangement. In *International Conference on Future Computer and Communication*. Kuala Lumpur (Malaysia), 2009, p. 36–48. DOI: 10.1109/ICFCC.2009.136

[12] PATRA, J. C., POH, W. B., CHAUDHARI, N. S., DAS, A. Nonlinear channel equalization with QAM signal using Chebyshev artificial neural network. In *Proceedings of International Joint Conference on Neural Networks*. Montreal (Canada), 2005, p. 3214–3219. DOI: 10.1109/IJCNN.2005.1556442

[13] RAHMAN, Q. M., IBNKAHLA, M., BAYOUMI, M. Neural network based channel estimation and performance evaluation of a time varying multipath satellite channel. In *Proceedings of the 3rd Annual Communication Networks and Services Research Conference (CNSRC'05)*. Halifax (Canada), May 2005, p. 74–79. DOI: 10.1109/CNSR.2005.44

[14] DENG JIANPING, SUNDARAJAN, N., SARATCHANDRAN, P. Communication channel equalization using complex-valued minimal radial basis function neural networks. *IEEE Transactions on Neural Networks*, 2002, vol. 13, p. 687–696. DOI: 10.1109/TNN.2002.1000133

[15] PATRA, J. C., CHIN, W. C., MEHER, P. K., CHAKRABORTY, G. Legendre-FLANN-based nonlinear channel equalization in wireless communication system. In *IEEE International Conference on Systems, Man and Cybernetic*. Singapore, 2008, p. 1820–1831. DOI: 10.1109/ICSMC.2008.4811554

[16] DEVELI, I. Application of multilayer perceptron networks to laser diode nonlinearity determination for radio-over-fibre mobile communications. *Microwave and Optical Technology Letters*, 2004, vol. 42, no. 5, p. 425–427. DOI: 10.1002/mop.20325

[17] BAGHERSALIMI, G. A comparative performance study of optical subsystem equalization in OFDM-based and OWDM-based radio-over-fiber systems. In *Proceedings of the 11th International Conference on Telecommunications ConTEL2011*. Graz (Austria), 2011, p. 315–320. ISBN: 978-3-85125-161-6

[18] *IEEE std 802.11g*; Supplement to IEEE standard for information technology telecommunications and information exchange between systems-local and metropolitan area networks specific requirements-part 11: Wireless LAN medium access control (MAC) and physical layer (PHY) specifications; amendment 4: Further higher data rate extension in the 2.4 GHz band. Tech. rep., IEEE, 2003. ISBN : 978-0-7381-6324-6

[19] BOSCH, W., GATTI, G. Measurement and simulation of memory effects in pre-distortion linearizers. *IEEE Transactions on Microwave Theory and Techniques*, 1989, vol. 37, no. 12, p. 1885 to 1890. DOI: 10.1109/22.44098

[20] WAY, W. I., AFRASHTEH, A. Linearity characterization of connectorized laser diodes under microwave intensity modulation by am/am and am/pm measurements. In *IEEE MTT-S International Microwave Symposium Digest*. Baltimore (USA), 1986, vol. 86, no. 1, p. 659–662. DOI: 10.1109/MWSYM.1986.1132274

[21] DEMUTH, H., BEALE, M., HAGAN, M. Neural networks toolbox user's guide for use with matlab 7. The Mathworks Inc, 2006. Available at: http://www.mathworks.com

About the Authors ...

Leila SAFARI received M.Sc. degree in Electronics Engineering from the University of Guilan, Rasht, Iran in 2011. Her research interests include neural networks and radio-over-fiber systems.

Gholamreza BAGHERSALIMI (corresponding author) is with the Department of Electrical Engineering, the University of Guilan, Rasht, Iran. He received his B.Sc. degree from Tehran University, Tehran, Iran in 1990, M.Sc. degree from Tarbiat Modarres University, Tehran, Iran in 1994, and Ph.D. degree from the University of Leeds, Leeds, UK in 2006, all in Communication Engineering. His research interests include equalization and synchronization of multicarrier systems, radio-over-fiber networks, free-space optical communications, visible light communications and spread spectrum techniques.

Ali KARAMI is an Associate Professor in the University of Guilan in Iran. He received B.Sc. at Sharif University of Technology, Tehran, in 1992, M.Sc. at Iran University of Science and Technology, Tehran, in 1994 and Ph.D. at Amirkabir University of Technology in 1999, all in Electrical Engineering. His research interests include transient stability, transient energy function method and neural network applications.

Abdolreza KIANI received his M.Sc. degree in Electronics Engineering from the University of Guilan, Rasht, Iran in 2011. His research interests include equalization and radio-over-fiber systems. Now, he is with the Islamic Azad University, Abadan branch.

21

Miniaturized Substrate Integrated Waveguide Diplexer Using Open Complementary Split Ring Resonators

Mostafa DANAEIAN, Kambiz AFROOZ, Ahmad HAKIMI

Department of Electrical Engineering, Shahid Bahonar University of Kerman, Kerman, Iran

mdanaeian@eng.uk.ac.ir, {afrooz, hakimi}@uk.ac.ir

Abstract. *In this paper, two miniaturized planar diplexers based on the substrate integrated waveguide (SIW) structure loaded by open complementary split-ring resonators (OCSRRs) are proposed. The working principle is based on the theory of evanescent mode propagation. The proposed SIW diplexers operate below the cutoff frequency of the waveguide. Both the complementary split-ring resonators (CSRRs) and the OCSRRs behave as electric dipoles however, the resonance frequency of the OCSRRs is approximately half of the resonance frequency of the CSRRs. Therefore, the electrical size of the OCSRRs is larger than the CSRRs. Accordingly, the OCSRRs are more appropriate for the SIW miniaturization. At first, the filtering response of the SIW structure loaded by OCSRR unit cells is investigated. Then, two miniaturized SIW diplexers which consist of two cascaded OCSRR unit cells with different orientations are designed. For the first diplexer (Type I), the fractional bandwidths of operation for the up and down channels are 9.52 % and 2.59 % at 4.2 GHz and 5.8 GHz, respectively. For the second diplexer (Type II), the fractional bandwidths of operation for the up and down channels are 5.95 % and 2.51 % at 4.7 GHz and 5.6 GHz, respectively. Finally, in order to validate the ability of the proposed OCSRR unit cells in the size reduction, two designed diplexers are fabricated and experimental verifications are provided. A good agreement between the results of measurement and simulation is achieved. The proposed diplexers show significant advantages in terms of size reduction, low loss, high isolation, and integration with other planar circuits.*

Keywords

Open Complementary Split Ring Resonators (OCSRRs), electric dipoles, substrate integrated waveguide (SIW), evanescent mode, miniaturization

1. Introduction

Recently, substrate integrated waveguide (SIW) has already attracted an extreme interest in the design of mi-crowave and millimeter-wave integrated circuits [1–4]. The SIW structure is synthesized on a planar dielectric substrate with linear arrays of metallized via holes or posts connected with two metal plates on the top and bottom sides. This structure has been quickly developed because of its advantages such as high quality factor, low insertion loss, high power-handling capability, low cost, and more importantly, easy integration with planar circuits [1–4].

Diplexers are one of essential components in transceivers for microwave multi-service and multi-band communication systems. Diplexer is a three ports device which is used to separate transmit/receive channels connected to a common antenna. This component provides a high band-isolation between two different output ports to avoid interference between each other. In order to achieve a few desired specifications for a high-performance diplexer in terms of low insertion losses, strong channel rejection and high isolation, this device can be realized by different filter configurations [5–10]. The SIW structures are suitable to realize essential filter elements of the diplexers because of the advantage of these structures. However, these structures suffer from large sizes.

On the other hand, there has been a rising attention in the field of the double negative (DNG) structures which can be realized by resonance structures or non-resonance structures [11–13]. It has been shown by Marques et al. that the resonant-type approach is useful for the synthesis of bandpass filters (BPFs). The essential elements for implementations of the resonant-type approach are the split ring resonators (SRRs) and complementary split ring resonators (CSRRs). These metamaterial resonant elements are frequency-selective structures and the electrical size of them is significantly small. The SRRs provide a negative effective permeability while the CSRRs exhibit an effective negative permittivity in a narrow band at the resonate frequencies [12], [13]. Other types of these resonant particles are the open split ring resonator (OSRR) and the open complementary split ring resonator (OCSRR). The electrical size of the OCSRR is larger than the electrical size of the CSRR. Therefore, this particle can be more appropriate to miniaturize the physical size of microwave components which are implemented by the resonant-type approach [12–13].

In this paper, the evanescent-mode technique is used to miniaturize the SIW diplexers. Two compact planar SIW diplexers loaded by the OCSRR unit cells with different orientations are designed. These proposed diplexers are realized by a three-port device with two cascaded SIW-OCSRR filters whose working principle is based on the theory of evanescent mode propagation. Therefore, more miniaturization is achieved by using OCSRRs instead of CSRRs. Finally, two proposed SIW diplexers are fabricated and tested. A good agreement between the simulation and measurement's results is obtained. The designed diplexers exhibit some useful characteristics, such as high isolation, low loss and more important, high compactness.

2. Realization of the SIW-OCSRR Filters

Several methods and topologies are reported [5–6] in order to miniaturize the SIW filters. Among them, the evanescent-mode technique attracts an extreme attention because of easy etching some components on the top or bottom metal surface of the SIW structure [7–10]. According to this theory, an additional forward passband at the frequency range below the waveguide cutoff could be achieved by loading the electric dipoles on the metal cover of the SIW structure. Since the CSRRs behave as electrical dipoles, so that some miniaturized SIW filters based on this technique have been reported [14]. The passband of the miniaturized filter is located below the cutoff frequency of the SIW structure [14].

The structures and the equivalent-circuit models of the CSRRs and the OCSRRs are shown in Fig. 1(a) and Fig. 1(b), respectively. Both the CSRR and the OCSRR are modeled by means of the shunt-connected resonant tank formed by one capacitance and one inductance. As shown in Fig. 1, the inductance of the CSRR is a quarter of the OCSRR, so the resonance frequency of the CSRR is twice the resonance frequency of the OCSRR with the same footprint. Therefore, the electrical size of the OCSRR is larger than that of the CSRR and more miniaturization could be achieved by loading the OCSRR on the SIW surface [15]. To achieve further miniaturization, the SIW bandpass filters loaded by the OCSRRs have been proposed in [15] where two prototypes of the SIW-OCSRR filters have been designed. The configurations of these filters are depicted in Fig. 2. In order to increase the coupling between the two OCSRRs as well as suppress the propagation of the dominant mode of the SIW structure, the metal strip in the middle of two OCSRRs is removed.

To meet the given specification in the filter design, it is important to follow the classic methodology. In this method, the essential parameters could be defined in terms of the circuit elements of a low-pass prototype filter. These design parameters include the coupling coefficient and external quality factor. To specify the physical configuration of the filters based on this method, the relationship

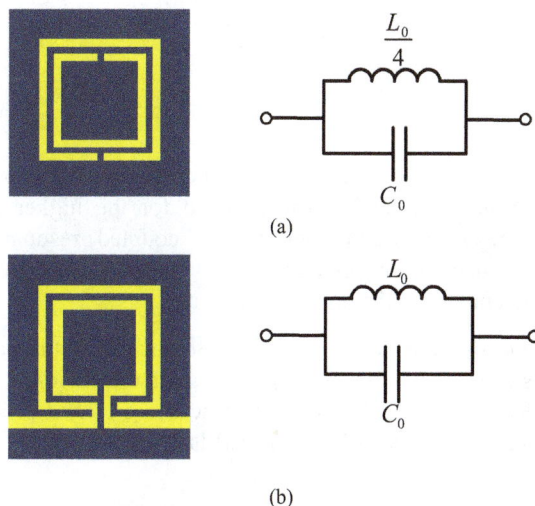

Fig. 1. Topology and equivalent circuit model of (a) the CSRR and (b) the OCSRR.

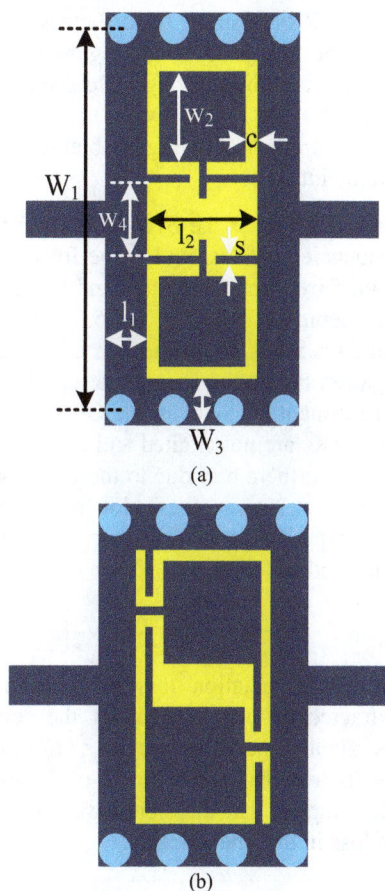

Fig. 2. Configuration of the SIW-OCSRR filters: (a) face-to-face oriented (Type I), (b) side-by-side reversely oriented (Type II).

between coupling coefficient, external quality factor and physical structures should be determined. Consequently, the distances w_4, l_1, and l_2 can be achieved by using their relationship with the coupling coefficient and extracted external quality factor. The external quality factor and coupling coefficient can be evaluated by using the following relations [15]:

$$Q_e = \frac{2f_0}{BW}, \tag{1}$$

$$K = \frac{f_1^2 - f_2^2}{f_1^2 + f_2^2} \tag{2}$$

where f_0 refers to the resonance frequency, BW refers to the 3 dB bandwidth, and f_1 and f_2 stand for the higher and lower resonant frequencies of the coupled resonators. Fig. 3 illustrates the external quality factor and coupling coefficient versus the distances l_1, l_2, and w_4.

On the other hand, the center frequency of the pass-bands can be easily moved by resizing the dimension of the OCSRRs which includes the distances w_2, c, and s. The resonance frequency of the OCSRR is achieved from:

$$f = \frac{1}{2\pi\sqrt{L_0 C_0}}. \tag{3}$$

Also, by resizing the length and width of the distances between two OCSRRs (resizing the distances l_1 and w_4) the bandwidth can be easily adjusted. Figure 4 presents the results from simulation by resizing the distances l_1 and w_4. As can be seen in these figures, when the distances l_1 and w_4 are smaller, a stronger coupling is obtained which leads to a larger bandwidth.

A current diagram of the SIW-OCSRR filter for two different frequencies (that is, within the forward passband below the cutoff frequency and stopband below the cutoff frequency) is depicted in Fig. 5. At 5.5 GHz, the signal propagates and OCSRRs are excited. In this case, forward-wave propagation below the waveguide cutoff is obtained based on the resonant behavior of the OCSRRs. However, at 3 GHz, OCSRRs are not excited and the injected power is retuned back. Furthermore, due to the use of slots in the proposed OCSRR structures, these slots may produce radiation in the upper half space. Therefore, the radiation loss could be quantified as follows:

$$R_r = 1 - |S_{11}|^2 - |S_{12}|^2. \tag{4}$$

The simulated radiation loss is depicted in Fig. 6 which is extracted by using (4) under the conditions of lossless substrate and metal. As we can see from Fig. 6 the radiation loss is less than 0.026 for Type I and 0.032 for Type II which implies that the slots have a little effect on the radiation loss in the filters design.

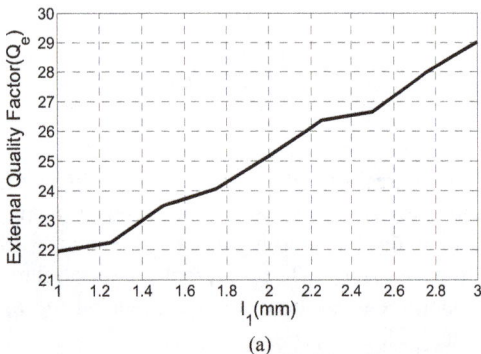

(a)

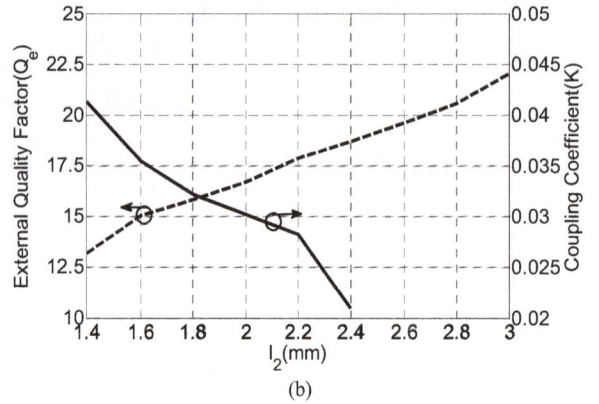

(b)

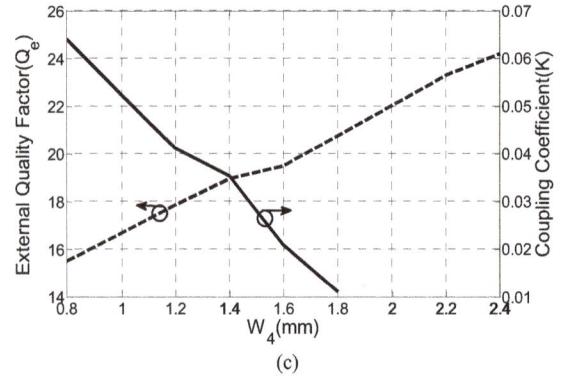

(c)

Fig. 3. (a) External quality factor and coupling coefficient as a function of l_1. (b) External quality factor as a function of l_2. (c) External quality factor and coupling coefficient as a function of W_4.

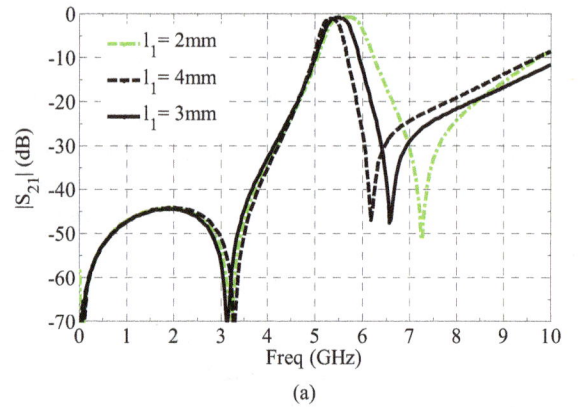

(a)

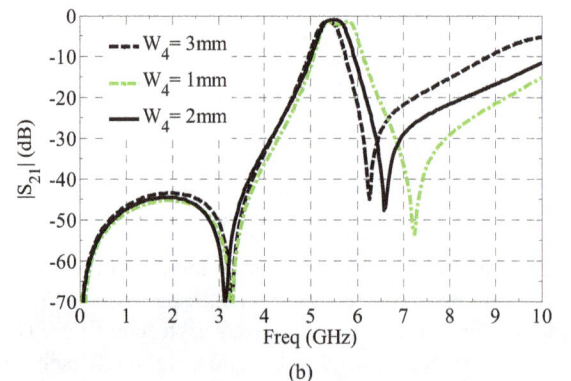

(b)

Fig. 4. (a) Simulated $|S_{21}|$ in dB with different l_1 and (b) simulated $|S_{21}|$ in dB with different W_4.

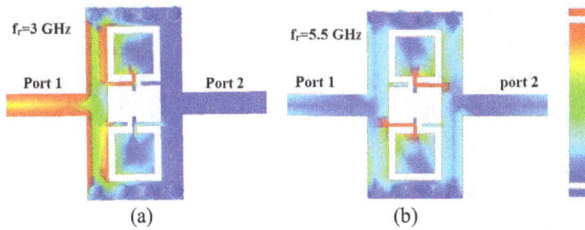

$f_r = 3$ GHz — Port 1 / Port 2

$f_r = 5.5$ GHz — Port 1 / port 2

(a) (b)

Fig. 5. Current diagrams for the SIW-OCSRR filter at two different frequencies.

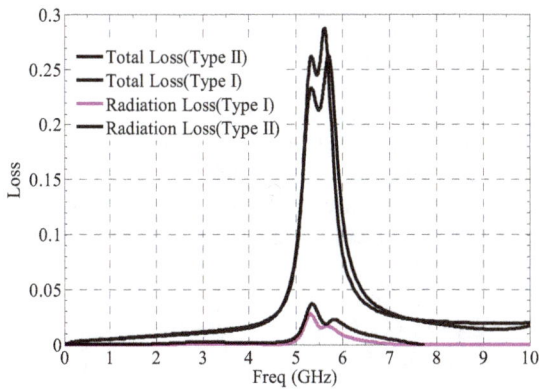

Fig. 6. Calculated radiation loss and total transmission losses (radiation loss + dielectric loss + conductor loss) of the proposed SIW-OCSRR.

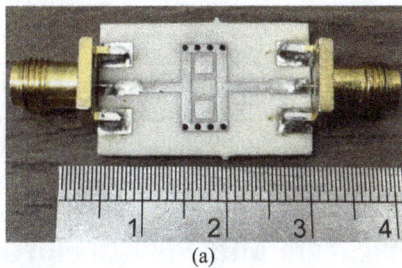

(a)

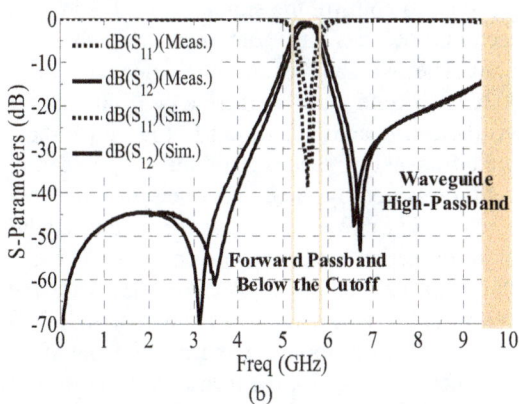

(b)

(c)

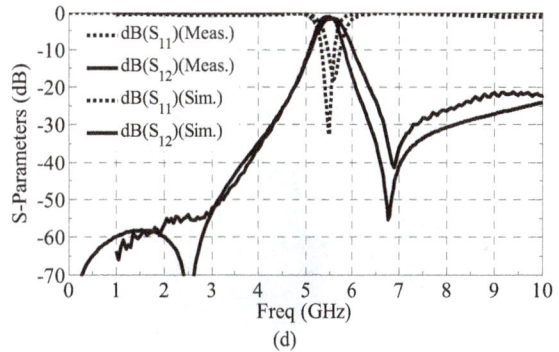

(d)

Fig. 7. Photographs, simulated and measured frequency responses of the fabricated SIW-OCSRR filters: (a) Type I, (b) Type II [15].

Also, in Fig. 6, a comparison by including all losses (which contain radiation loss + dielectric loss + conductor loss) for these resonator is presented which finally leads to 1.2 dB insertion loss for Type I and 1.6 dB insertion loss for Type II.

Figure 7 shows the photograph, simulated and measured transmission responses of the SIW-OCSRR filters which demonstrates that the designed filters have very small size compared with similar filters. As can be seen in Fig. 7 (b) and (d), a forward passband below the waveguide cutoff at center frequency of 5.5 GHz is achieved which is good for WLAN applications [15].

3. Design Procedure of the Proposed SIW Diplexers

Configurations of the proposed diplexers with their physical parameters are depicted in Fig. 8. Figure 8(a) shows the first layout of the proposed diplexer with a pair of OCSRRs in the face-to-face orientation referred to as Type I. The second configuration of the proposed diplexer with a pair of OCSRRs in the side-by-side reverse orientation is referred to as Type II and is shown in Fig. 8(b). The proposed diplexers consist of two SIW channel filters loaded by OCSRRs and three 50 Ω SIW-microstrip transitions. Three 50 Ω microstrip feed lines have been used here for the purpose of measurement where these microstrip feed lines provide the pads for the attachment of the edge launch SMA connectors.

The channel SIW-OCSRR filters play an essential role in designing the proposed diplexers in which a transmitter and a receiver filter are cascaded together. These proposed diplexers have been achieved after designing two SIW filters with appropriate features. The frequency responses of the proposed diplexers are shown in Fig. 9. The frequency responses of the proposed diplexers which are realized based on SIW-OCSRR filters are simulated by using 3-D electromagnetic simulator. In the simulations, the metallic and dielectric loss have been taken into account by using the conductivity of copper $\sigma = 5.8 \times 10^7$ S/m and the loss tangent $\tan\delta = 0.0027$ for the substrate. As shown in Fig. 9, the simulated insertion losses

(a)

(b)

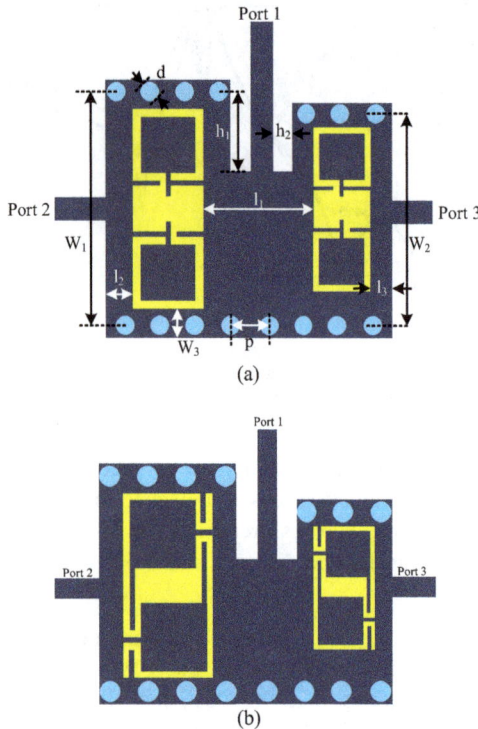

Fig. 8. Configuration of the proposed SIW-OCSRR diplexers: (a) Type I, (b) Type II.

(a)

(b)

Fig. 9. Simulated frequency responses and isolation of the proposed SIW diplexers: (a) Type I, (b) Type II.

at the two bands for diplexer Type I are approximately 1.5 dB and 2.8 dB, respectively and the simulated insertion losses at the two bands for diplexer Type II are approximately 1.7 dB and 2.1 dB, respectively. The simulated return losses at the lower and higher channels of the proposed diplexer (Type I) are better than 16 dB and for the

proposed diplexer (Type II) arre better than 17 dB. The simulated out-of-band rejection of each channel filter is better than 23 dB for both diplexers. It can be seen that low in-band insertion loss, high isolation, and high stopband rejection with high selectivity are obtained.

By resizing the dimension of the OCSRRs in the SIW-OCSRR filters and consequently, by changing the center frequency of the channel filters, the two passbands could be very close to each other or away. In order to get a low in-band reflection, we can tune the inset length h_1, slot width h_2 and the distance between the feeding and resonators (l_2 and l_3).

The obtained design dimensions of the proposed diplexer (Type I), which are shown in Fig. 8(a), and the dimensions of the proposed diplexer (Type II), which are shown in Fig. 8(b) are exhibited in Tab. 1 and Tab. 2, respectively. The overall dimensions of the diplexer Type I is less than $0.15\,\lambda_0 \times 0.14\,\lambda_0$ (11.4 mm × 10.2 mm) and the total sizes of the diplexer Type II is less than $0.22\,\lambda_0 \times 0.13\,\lambda_0$ (14.2 mm × 8.5 mm) where λ_0 is the free space wavelength of the first channel center frequency.

$W_1 = 10.2$ mm	$W_2 = 9.2$ mm	$W_3 = 1$ mm	$l_1 = 3.2$ mm
$l_2 = 1$ mm	$l_3 = 1$ mm	$h_1 = 3.4$ mm	$h_2 = 0.7$ mm
$p = 1.5$ mm	$d = 0.8$ mm	$c = 0.3$ mm	$s = 0.2$ mm

Tab. 1. Dimensions of the proposed diplexer (Type I).

$W_1 = 8.5$ mm	$W_2 = 7.8$ mm	$W_3 = 1$ mm	$l_1 = 3.8$ mm
$l_2 = 1$ mm	$l_3 = 0.8$ mm	$h_1 = 3.4$ mm	$h_2 = 0.8$ mm
$p = 1.5$ mm	$d = 0.8$ mm	$c = 0.4$ mm	$s = 0.2$ mm

Tab. 2. Dimensions of the proposed diplexer (Type II).

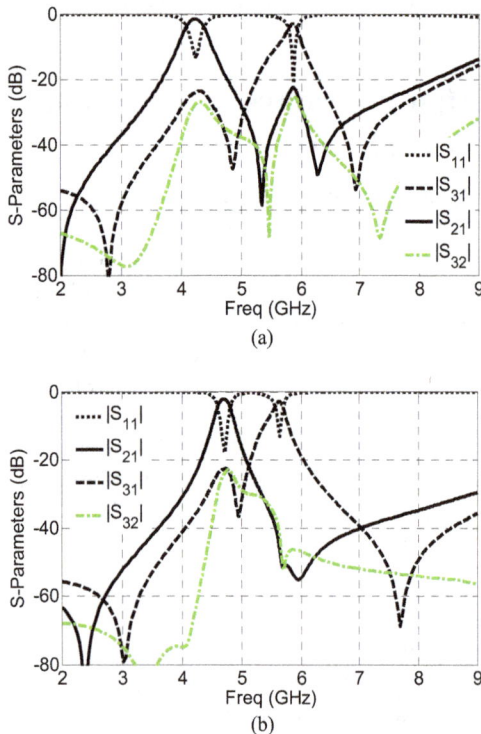

4. Fabrication and Measurements

In order to confirm the simulated results in the previous section, two designed diplexers are fabricated and measured. The two samples are implemented on a Rogers RO4003C substrate with a thickness of 0.508 mm and relative dielectric constant of 3.55. Figure 10 shows the measured frequency responses of the proposed diplexers which are measured by the employment of a network analyzer Rohde & Schwarz, zvk. These results contain the measured reflection and transmission coefficients ($|S_{11}|$, $|S_{21}|$ and $|S_{31}|$), as well as the isolation ($|S_{32}|$) for the both diplexers which are measured. It is obvious that the measured results are in a good agreement with the simulation ones. As shown in Fig. 10, the measured insertion losses at 4.33 GHz and 5.99 GHz for diplexer Type I are approximately 1.6 dB and 3.2 dB, respectively and the measured insertion losses at 4.78 GHz and 5.77 GHz for diplexer Type II are approximately 1.8 dB and 2.3 dB, respectively.

The measured return losses in the passbands are achieved less than 22 dB and 15 dB at lower (4.33 GHz) and higher (5.99 GHz) bands of the Type I diplexer, respectively, and less than 20 dB and 14 dB at lower (4.78 GHz) and higher (5.77 GHz) bands for the Type II diplexer, respectively. Similarly, the measured out-of-band

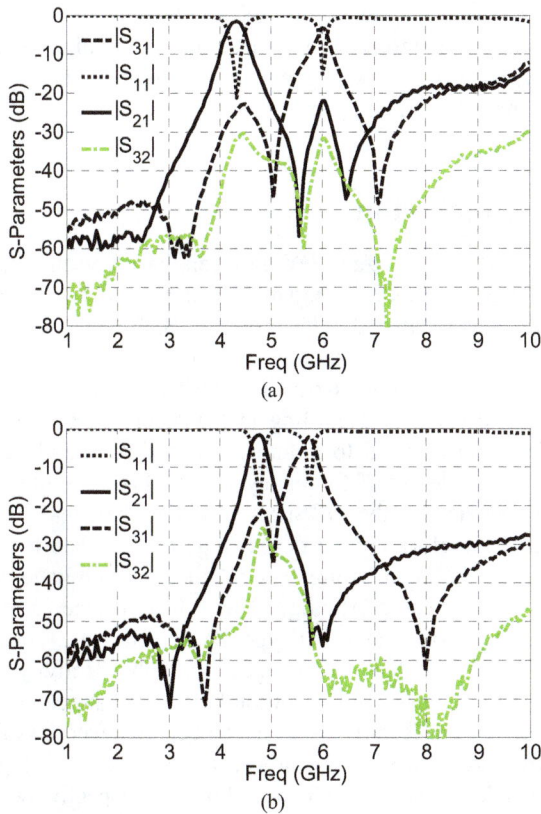

Fig. 10. Measured frequency responses and isolation of the proposed SIW diplexers: (a) Type I, (b) Type II.

rejection of each channel filter is better than 23 dB for Type I diplexer and better than 22 dB for Type II diplexer. Some minor differences between simulation and measurement results may be caused by the extra loss from the SMA connectors, by the limited accuracy of fabrication and measurement. The simulated channel return losses at Port 2 and Port 3 are also plotted in Fig. 11. As shown in these figures, the simulated return losses in both channels for both diplexers are better than 18.2 dB. Here both proposed diplexers work in C-band based on two different channel filters. However, by changing the center frequency of the channel filters which is discussed above, the proposed diplexers can operate in different band like S-band and C-band or C-band and X-band. Generally, the wave propagation at these frequencies for the SIW structure loaded by OCSRRs still keeps the manner of TE_{10} mode. These particles at the cutoff frequency of the SIW structure are not resonant and do not affect the initial propagation of the SIW structures [14].

Figure 12 shows the group delay for both channels of the Type I and Type II diplexers, respectively, which are flat. In the passband, the group delay from input port 1 to output port 2 with the center frequency 4.33 GHz is less than 1.65 ns and the group delay from input port 1 to output port 3 with the center frequency 5.99 GHz is less than 1.5 ns for Type I diplexer. In the passband, the group delay from input port 1 to output port 2 with the center frequency 4.78 GHz is less than 2.1 ns and the group delay from input port 1 to output port 3 with the center frequency 5.77 GHz is less than 2.1 ns for Type II diplexer.

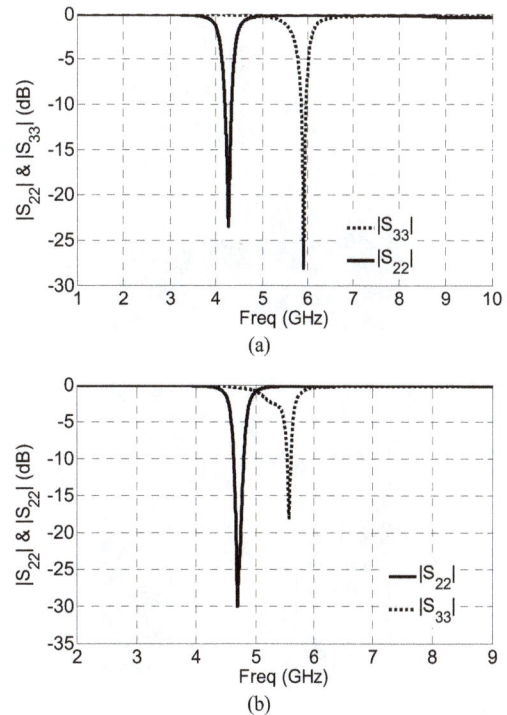

Fig. 11. Simulated channel filter return losses for ports 2 and 3: (a) Type I, (b) Type II.

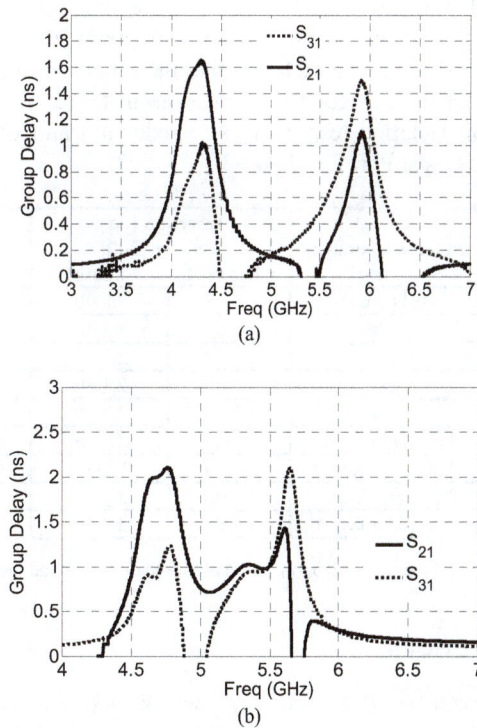

Fig. 12. (a) Simulated group delays of the Type I at two passbands (maximum group delay variation 0.5 ns). (b) Simulated group delays of the Type II at two passbands (maximum group delay variation 0.8 ns).

The photograph of the fabricated diplexers are shown in Fig. 13 which demonstrates that the proposed diplexers have very compact size compared with similar diplexers and miniaturization is achieved. Finally, a performance comparison between the proposed diplexers with other

(a)

(b)

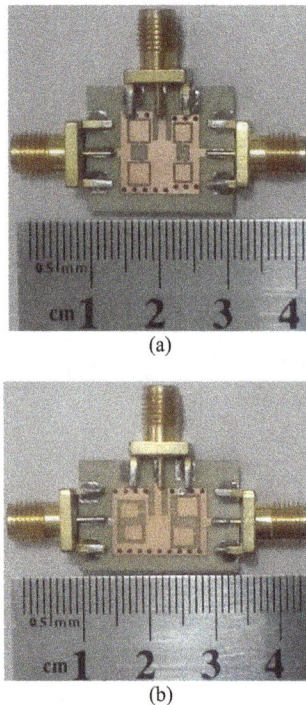

Fig. 13. Photograph of the fabricated diplexers:
(a) Type I, (b) Type II.

reported diplexers which include some miniaturized designs using the microstrip and SIW diplexers are summarizes in Tab. 3. It can be found that the proposed diplexers in this paper have good performances in terms of losses, rejection, isolation, especially size reduction and channel return losses at Port 2 and Port 3.

Generally, the proposed diplexers have many advantages in terms of low insertion loss, high quality factor, easy integration, good passbands performance and most importantly, compact size.

5. Conclusion

Two miniaturized SIW diplexers based on the theory of evanescent mode propagation have been designed, fabricated and measured. The proposed SIW diplexers have been realized by cascading two compact SIW-OCSRR filters with different sizes. The OCSRRs behave as an electric dipole and according to the theory of evanescent mode could be used to obtain a forward passband below the waveguide cutoff frequency of the SIW structure. The electrical size of the OCSRRs is larger than that of the CSRRs, therefore this particle is a good candidate to miniaturize the SIW structure. The OCSRRs which have been used in these cases are aligned with face-to-face (Type I) and side-by-side reverse (Type II) orientations. The designed diplexers have advantages like high selectivity, low loss and low cost. High isolation between channels is presented as well. Return losses have been achieved less than 22 dB and 15 dB at the lower (4.33 GHz) and higher (5.99 GHz) bands of the Type I diplexer, respectively, and less than 20 dB and 14 dB at the lower (4.78 GHz) and higher (5.77 GHz) bands of the Type II diplexer, respectively. The total size of the proposed Type I is less than $0.15 \lambda_0 \times 0.14 \lambda_0$ and the proposed Type II is less than $0.22 \lambda_0 \times 0.13 \lambda_0$ which illustrate the ability of the OCSRR unit cells on the size reduction.

Reference number	Insertion Loss ($|S_{21}|$ or $|S_{31}|$)		Suppression ($|S_{21}|$ or $|S_{31}|$)		Size ($\lambda_0 \times \lambda_0 \times \lambda_0$)	Isolation ($|S_{32}|$)
	Channel 1	Channel 2	Channel 1	Channel 2		
[6] (part1)	2.7 dB	2.8 dB	40 dB @ 1.5 GHz	48 dB @ 2.0 GHz	0.17×0.272×0.003	42 dB
[6] (part2)	2.8 dB	3.2 dB	38 dB @ 1.5 GHz	39 dB @ 1.76 GHz	0.187×0.30×0.003	30 dB
[7]	1.6 dB	2.3 dB	43 dB @ 4.66 GHz	28 dB @ 5.80 GHz	0.27×0.217×0.008	32 dB
[8]	1.6 dB	2.1 dB	42 dB @ 9.5 GHz	37 dB @ 10.5 GHz	2.04×0.65×0.053	35 dB
[9]	2.2 dB	2.4 dB	22 dB @ 7.75 GHz	22 dB @ 8.25 GHz	1.44×0.98×0.022	–
[10] (Diplexer 1)	1.92 dB	2.14 dB	33 dB @ 7.4 GHz	42 dB @ 8.2 GHz	3.06×1.38×0.019	–
[10] (Diplexer 2)	1.83 dB	2.13 dB	34 dB @ 7.4 GHz	43 dB @ 8.2 GHz	2.96×1.38×0.019	–
This work (Type I)	1.6 dB	3.2 dB	24 dB @ 4.33 GHz	23 dB @ 5.99 GHz	0.15×0.14×0.007	30 dB
This work (Type II)	1.8 dB	2.3 dB	22 dB @ 4.78 GHz	52 dB @ 5.77 GHz	0.22×0.13×0.007	26 dB

Tab. 3. Comparison between the proposed diplexers and the references. λ_0 is the free space wavelength of the first channel center frequency.

References

[1] BOZZI, M., GEORGIADIS, A., WU, K. Review of substrate-integrated waveguide circuits and antennas. *IET Microwaves, Antennas & Propagation,* 2011, vol. 5, no. 8, p. 909–920, DOI: 10.1049/iet-map.2010.0463

[2] MOSCATO, S., TOMASSONI, C., BOZZI, M., PERREGRINI, L. Quarter-mode cavity filters in substrate integrated waveguide technology. *IEEE Transactions on Microwave Theory and Techniques,* 2016, vol. 64, no. 8, p. 2538–2547. DOI: 10.1109/TMTT.2016.2577690

[3] CHEN, X., WU, K. Substrate integrated waveguide filters: Design techniques and structure innovations. *IEEE Microwave Magazine,* 2014, vol. 15, no. 6, p. 121–133, DOI: 10.1109/MMM.2014.2332886

[4] GARG, R., BAHL, I., BOZZI, M. *Microstrip Lines and Slotlines.* 3rd ed. Artech House, 2013. ISBN: 9781608075355

[5] BONACHE, J., GIL, I., GARCA-GARCIA, J., MARTIN, F. Complementary split ring resonators for microstrip diplexer design. *Electronics Letters,* 2005, vol. 41, no. 14, p. 810–811. DOI: 10.1049/el: 20050895

[6] CHEN, C.-H., HUANG, T.Y., CHOU, C.P., WU, R.B., Microstrip diplexers design with common resonator sections for compact size but high isolation. *IEEE Transactions on Microwave Theory and Techniques,* 2006, vol. 54, no. 5, p. 1945–1952. DOI: 10.1109/TMTT.2006.873613

[7] DONG, Y., ITOH, T. Substrate integrated waveguide loaded by complementary split-ring resonators for miniaturized diplexer design. *IEEE Microwave Wireless Component Letters*, 2011, vol. 21, no. 1, p. 10–12. DOI: 10.1109/LMWC.2010.2091263

[8] SIRCI, S., MARTINEZ, J. D., VAGUE, J., BORIA, V. E. Substrate integrated waveguide diplexer based on circular triplet combline filters. *IEEE Microwave and Wireless Components Letters*, 2015, vol. 25, no. 7, p. 430–432. DOI: 10.1109/LMWC.2015.2427516

[9] GARCIA-LAMPEREZ, A., SALAZAR-PALMA, M., YEUNG, S.-H. SIW compact diplexer. In *IEEE MTT-S International Microwave Symposium Digest*. Tampa Bay (FL, USA), Jun. 2014, p. 1–4. DOI: 10.1109/MWSYM.2014.6848514

[10] ZHAO, C., Z., FUMEAUX, C., LIM, C-C. Substrate-integrated waveguide diplexers with improved Y-junctions. *Microwave and Optical Technology Letters*, 2016, vol. 58, no. 6, p. 1384–1388. DOI: 10.1002/mop.29807

[11] CALOZ, C., ITOH, T. *Electromagnetic Metamaterials: Transmission Line Theory and Microwave Applications.* New York (USA): John Wiley & Sons, 2006. DOI: 10.1002/0471754323

[12] MARQUES, R., MARTIN, F., SOROLLA, M. *Metamaterials with Negative Parameters: Theory, Design and Microwave Applications.* New York (USA): John Wiley & Sons, 2011. ISBN: 9781118211564

[13] FERRAN, M. *Artificial Transmission Lines for RF and Microwave Applications.* John Wiley, 2015. ISBN: 978-1-118-48760-0

[14] DONG, Y., YANG, T., ITOH, T. Substrate integrated waveguide loaded by complementary split-ring resonators and its applications to miniaturized waveguide filters. *IEEE Transactions on Microwave Theory and Techniques*, 2009, vol. 57, no. 9, p. 2211–2223. DOI: 10.1109/TMTT.2009.2027156

[15] DANAEIAN, M., AFROOZ, K., HAKIMI, A., MOZNEBI, A.-R., Compact bandpass filter based on SIW loaded by open complementary split-ring resonators (OCSRRs). *International Journal of RF and Microwave Computer-Aided Engineering*, 2016, vol. 26, no. 8, p. 674–682. DOI: 10.1002/mmce.21017

About the Authors ...

Mostafa DANAEIAN was born in Yazd, Iran, in 1985. He received the B.Sc. degree in Electrical Engineering from Yazd University, Yazd, Iran, in 2008, the M.S. degree in Electrical Engineering from Shahid Bahonar University of Kerman, Kerman, Iran, in 2011, and is currently working toward the Ph.D. degree at the Shahid Bahonar University of Kerman, Kerman, Iran. His research interests are MTM, RF/microwave circuits design. He is currently involved with microwave devices based on SIW structures.

Kambiz AFROOZ was born in Baft, Iran, in 1983. He received the B.Sc. degree from the Shahid Bahonar University of Kerman, Kerman, Iran, in 2005, the M.Sc. and Ph.D. degrees from the Amirkabir University of Technology (Tehran Polytechnic), Tehran, Iran, in 2007 and 2012, all in Electrical Engineering. In May 2011, he joined the CIMITEC group, University Autonoma de Barcelona (UAB), Barcelona, Spain, as a Visiting Student. He is currently an Associate professor with the Electrical Engineering Department, Shahid Bahonar University of Kerman, Kerman, Iran. His research interests are in the areas of computer-aided design of active and passive microwave devices and circuits, computational electromagnetic, metamaterial transmission lines, and substrate integrated waveguide structures.

Ahmad HAKIMI was born in Rafsanjan, Iran, in 1961. He received his B.Sc. degree in Electrical Engineering from the Technical College of Shahid Bahonar University of Kerman, Kerman, Iran, in 1986. Using the scholarship granted by the Ministry of Higher Education of Iran and Istanbul Technical University (ITU) in 1987, he has been studying for his M.Sc. and Ph.D. degrees in the Faculty of Electrical and Electronic Engineering at the ITU. He received his M.Sc. and Ph.D. degrees from ITU in 1996 and 1995 in the field of high frequency electronics. His research interests include design and analysis of nonlinear RF circuits, numerical analysis and advanced engineering mathematics, analog filter design, and linear integrated circuits.

22

Supercontinuum Source for Dense Wavelength Division Multiplexing in Square Photonic Crystal Fiber via Fluidic Infiltration Approach

Hamed SAGHAEI

Faculty of Engineering, Shahrekord Branch, Islamic Azad University, Shahrekord, Iran

h.saghaei@iaushk.ac.ir

Abstract. *In this paper, a square-lattice photonic crystal fiber based on optofluidic infiltration technique is proposed for supercontinuum generation. Using this approach, without nano-scale variation in the geometry of the photonic crystal fiber, ultra-flattened near zero dispersion centered about 1500 nm will be achieved. By choosing the suitable refractive index of the liquid to infiltrate into the air-holes of the fiber, the supercontinuum will be generated for 50 fs input optical pulse of 1550 nm central wavelength with 20 kW peak power. We numerically demonstrate that this approach allows one to obtain more than two-octave spanning of supercontinuum from 800 to 2000 nm. The spectral slicing of this spectrum has also been proposed as a simple way to create multi-wavelength optical sources for dense wavelength division multiplexing.*

Keywords

Supercontinuum generation, photonic crystal fiber, optofluidic, dispersion, dense wavelength division multiplexing

1. Introduction

Optical supercontinuum (SC) is a coherent and broadband light that is generated when a short laser pulse causes nonlinear effects in nonlinear materials including crystals, glasses, noble gasses and organic liquids [1]. It is unique because of possessing both the bandwidth of a white light source and the coherence properties of a laser [2, 3]. SC generation naturally requires high intensities which are achieved by confining the energy in space and time [2]. The SC generation mechanism has been recently recognized as a sequence of nonlinear physical processes, each occurring consecutively along the propagation axis such as self-phase modulation (SPM), self-steepening (SS), stimulated Raman scattering (SRS), and four-wave mixing (FWM) [4]. The first SC generation experiments in silica were obtained by standard optical fiber with zero group velocity dispersion (GVD) wavelength around 1.3 μm and

several hundred meters of the fiber length [5]. The advent of photonic crystal fibers (PCFs) in the form of triangle and square lattices at the late 1990s attracted widespread interest throughout the scientific community, and has led to a revolution in the generation of ultra-broadband high brightness spectra through SC generation [6–8]. The single-mode propagation over broad wavelength ranges, the enhanced modal confinement, the elevated nonlinearity, and the ability to engineer group velocity dispersion are the unique and excellent characteristics of PCFs [8]. Low dispersion is the key parameter for spectral broadening and other nonlinear phenomena such as FWM [9]. It enables single-mode phase matching of the nonlinear processes that broaden the spectrum. Optimizing dispersion relaxes the need for high intensity, allowing SC generation for larger cores, longer pulses and shorter length of the fiber [10]. It should be noted that the reduced effective area of the propagating mode in PCF enhances the Kerr nonlinearity effect relative to standard fiber [11]. SC generation in PCFs has many applications in optical frequency metrology [12], optical coherence tomography (OCT) [13], pulse compression [14], and the design of tunable ultrafast femtosecond laser sources [15]. In a telecommunication context, the spectral slicing of broadband SC spectra has also been proposed as a simple way to create multi-wavelength optical sources for dense wavelength division multiplexing (DWDM) applications [16, 17]. There are various techniques that the dispersion of PCF can be engineered and customized for a particular application. However, most of the proposed techniques, so far, are based on varying the PCF geometry; such as varying the circular air-hole diameter (d) [18, 19], the pitch size (Λ) of the periodic lattice [9], as well as the number of air-holes rings (N) surrounding the PCF core or even using ring-shaped air-holes [20]. These techniques depend on the technological capability to realize a specific design with high precision, and also are limited by the size of the PCF cross-sectional area. In particular, it is quite difficult to control the accurate positions and radii of the air-holes within the PCF both triangle and square lattices. Precision in control of these two geometrical parameters during the fabrication process in addition to the maximum number of the rings of air-holes within a PCF are the most critical constraints in acquiring the desired dispersion. To overcome these topological limitations, optofluid approach is proposed and our focus in this paper for dispersion engineering is based on the optofluidic: A new branch in photonic that attempts to merge fluidic and optic [21]. In a meticulous manner, due to their intrinsic porous nature photonic crystal (PhC) devices infiltrated with optofluid have demonstrated tunable and reconfigurable optical properties [22]. Furthermore, selective liquid infiltration of individual air pores of a planar PhC lattice has shown to extend the number of opportunities associated with this optofluidic platform [23–25]. This offers the potential for realizing integrated microphotonic devices and circuits, which could be (re)configured by simply changing the liquid and/or the pattern of the infiltrated area within the PhC lattice.

Our research group numerically investigated the potential of selective optofluidic infiltration of air-holes within the triangular-lattice photonic crystal fiber (TPCF) in [24] in which different properties of TPCF composed of silica background were studied. The aim of that study was to investigate the effects of fluid infiltration on the dispersion profile and its nonlinear parameter to obtain optimal structure of the PCF at the desired zero dispersion wavelength depending on the application. In this paper, a square-lattice photonic crystal fiber (SPCF) based on optofluidic infiltration technique is proposed for supercontinuum (SC) generation. Using this approach, without nano-scale variation in the geometry of the SPCF, ultra-flattened near zero dispersion centered about 1500 nm will be achieved. By choosing the suitable refractive index of the liquid to infiltrate into the air-holes of the SPCF, the SC will be generated for 50 fs input optical pulse of 1550 nm central wavelength with 20 kW peak power. We numerically demonstrate that this approach allows one to obtain more than two-octave spanning of SC from 800 to 2000 nm. The spectral slicing of this spectrum has also been proposed as a simple way to create multi-wavelength optical sources for dense wavelength division multiplexing.

The rest of this paper is organized as follows. In Sec. 2, we propose the procedure for dispersion engineering of solid core SPCF, by means of selective optofluidic infiltration. The mathematical analysis of SC generation is presented in Sec. 3. Section 4 is dedicated to the discussion of the simulation results. Finally, we will conclude the paper in Sec. 5.

2. Dispersion Engineering of SPCF

The goal of this section is to investigate the possibility of tailoring the dispersion profile of a SPCF by means of selective optofluidic infiltration of SPCF's air-holes to achieve SC generation. In doing so, we have we have considered a solid core SPCF that consists of circular air-holes of diameter $d = 0.6\Lambda$ arranged in a square-lattice of constant $\Lambda = 2\ \mu m$, as depicted in Fig. 1(a).

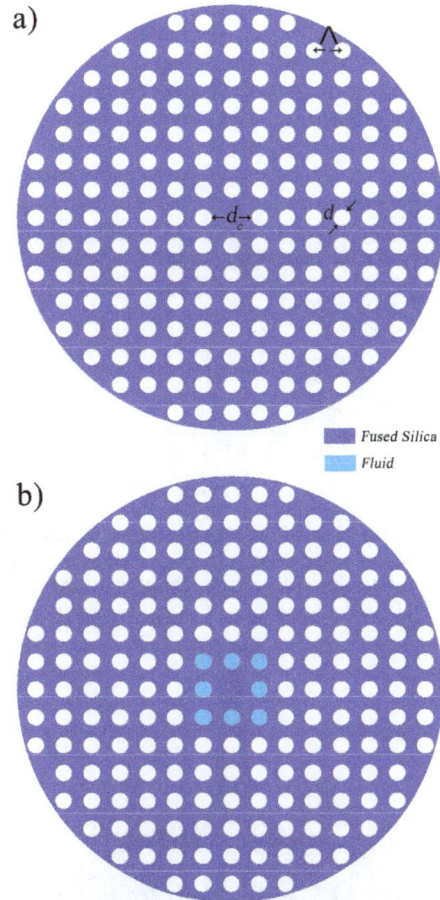

a)

b)

Fig. 1. A cross-sectional view of (a) un-infiltrated SPCF and (b) the inner most ring of SPCF to be selectively infiltrated with an optofluidic of index n_f, two have the same the lattice constant of 2 μm and air-hole diameters of 0.6Λ.

The air-holes lattice forms seven square rings co-centered with the solid core. Figure 1(b) illustrates the case for which the air-holes form the inner most ring which is infiltrated by an optofluidic of refractive index n_f. The white and dark circles in Fig. 1 represent the un-infiltrated and selectively infiltrated air-holes, respectively. In order to investigate the dispersion properties of the SPCF, we have employed a full vectorial finite element method (FV-FEM) [26]. The refractive index of silica as a background material can be expressed by the Sellmeier expansion as

$$n(\lambda) = 1 + \sum_{n=1}^{3} \frac{A_n \lambda^2}{\lambda^2 - B_n^2} \qquad (1)$$

where A_n and B_n are the Sellmeier coefficients. For silica, their values are as follows [27]: $A_1 = 0.6961663$, $A_2 = 0.407942$, $A_3 = 0.897479$, $B_1 = 0.068404\ \mu m^2$, $B_2 = 0.116241\ \mu m^2$, and $B_3 = 9.89616\ \mu m^2$.

Eigenvalues and the propagation constants k_z are obtained by solving the previous equations. Effective refractive index n_{eff} is calculated as

$$n_{eff} = \frac{k_z}{k_0}. \qquad (2)$$

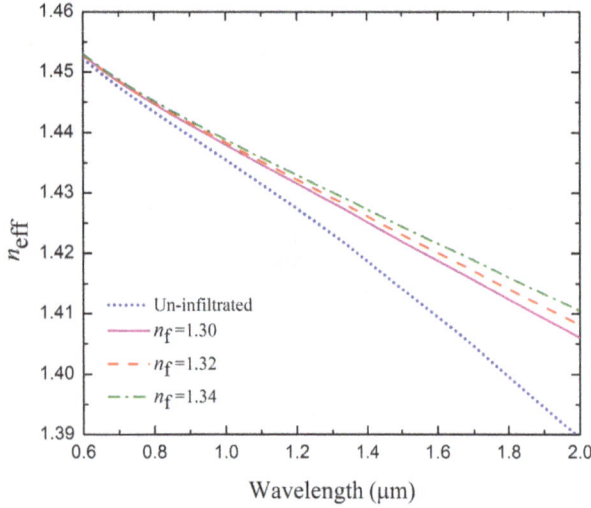

Fig. 2. Comparison of the effective refractive index for the un-infiltrated SPCF of Fig. 1(a) with Fig. 1(b) infiltrated with optical fluids of various indices ($1.3 \leq n_f \leq 1.34$).

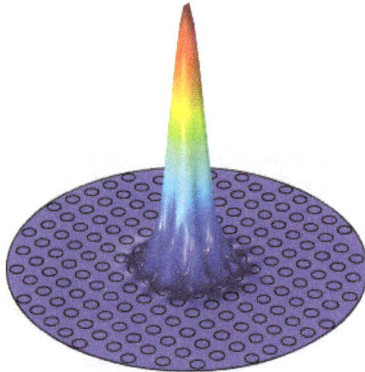

Fig. 3. 3D schematic view of fundamental mode distribution for the SPCF with the inner most ring to be selectively infiltrated with an optofluidic for $n_f = 1.30$.

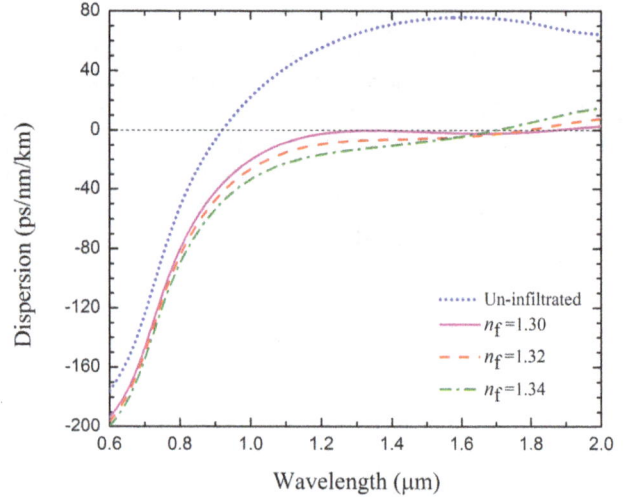

Fig. 4. Comparison of the dispersion for the un-infiltrated SPCF of Fig. 1(a) with those of Fig. 1(b) infiltrated with optical fluids of various indices ($1.3 \leq n_f \leq 1.34$).

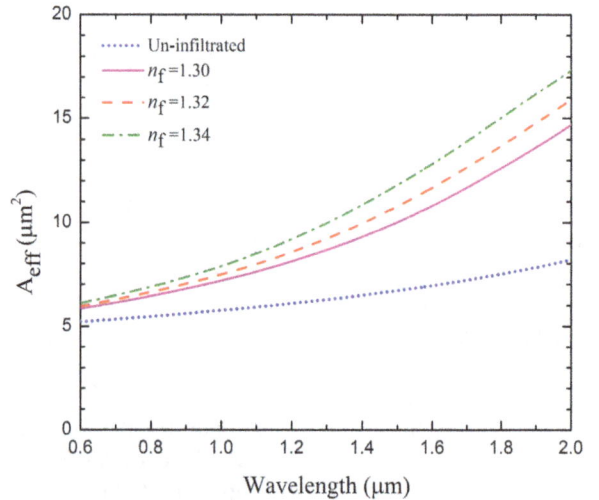

Fig. 5. Comparison of effective mode for the un-infiltrated SPCF of Fig. 1(a) with Fig. 1(b) infiltrated with optical fluids of various indices ($1.3 \leq n_f \leq 1.34$).

Figure 2 compares effective refractive index for the un-infiltrated SPCF of Fig. 1(a) with Fig. 1(b) infiltrated with optical fluids such as water, methanol, and ethanol with refractive indices of 1.3, 1.32 and 1.34 in 1550 nm, respectively. The refractive indices versus wavelength of methanol and ethanol are given by:

$$n(\lambda) = \begin{cases} \sqrt{1.74 - 0.005\lambda^2 + 0.004\lambda^{-2}}, & \text{methanol} \\ \sqrt{1 + \dfrac{0.83\lambda^2}{\lambda^2 - 0.009} - \dfrac{0.156\lambda^2}{\lambda^2 - 0.494}}, & \text{ethanol} \end{cases} \quad (3)$$

In the following simulations, the dependencies of refractive indices of fluids on wavelength are taken into account. The total dispersion D is computed as

$$D = \left(\frac{-\lambda}{c}\right)\frac{d^2 n_{\text{eff}}}{d\lambda^2}. \quad (4)$$

The 3D schematic view of the fundamental mode distribution is shown in Fig. 3 for the case of $n_f = 1.32$. Fundamental mode assumption is used for the rest of the paper.

In order to investigate the effect of the infiltration on the SPCF dispersion and effective mode area profiles, we compare these profiles for the un-infiltrated and infiltrated cases. Figures 4 and 5 compare the numerical results of dispersion profiles and effective mode area for $1.3 \leq n_f \leq 1.34$, respectively. The comparison shown in Fig. 4 reveals that infiltrating the most inner ring of the SPCF with various optical fluids reduces the SPCF dispersion value significantly and also enables one to design SPCFs suitable for various applications such as dispersion flattened PCF (DF-PCF), dispersion shifted PCF (DS-PCF), and dispersion compensated PCF (DC-PCF). For example, this figure shows that the infiltration with $n_f = 1.3$ has reduced the dispersion profile, in the wavelength range of 1.2 μm $<\lambda<$ 2 μm, to insignificant values of $0 \leq D \leq 2$ (ps/km.nm). This nearly flattened profile is an example of DF-PCF, over the given range of the wavelengths that makes this particular infiltrated PCF suitable for DWDM applications that can be used in the optical telecommunication systems.

Furthermore, the comparison of curves of Fig. 5 illustrates that by increasing the value of refractive index infiltration, the effective mode area increases and consequently the value of nonlinear coefficient, γ decreases as

$$\gamma = \frac{n_2 \omega_0}{c A_{\text{eff}}(\omega_0)} \quad (5)$$

where n_2 is the nonlinear index which is assumed by 2.7×10^{-20} m^2/W for silica. c is the speed of light. Results show that there is a trade-off between the dispersion and nonlinear coefficient by infiltrating the fluid. For this reason, we restrict our simulations only to the case of most inner ring infiltration.

3. Mathematical Background

Nonlinear Schrödinger equation (NLSE) can truly simulate the pulse propagation in single mode SPCF and can be derived under the slowly varying envelope approximation as [4]

$$\frac{\partial A}{\partial z} + \frac{\alpha}{2} A + \sum_{n=1}^{4} i^{(n-1)} \frac{\beta_n}{n!} \frac{\partial^n A}{\partial t^n} = \\ i \left(\gamma(\omega_0) + i\gamma_1 \frac{\partial}{\partial t} \right) \cdot A(z,t) \int_{-\infty}^{t} R(t') \left| A(z, t-t') \right|^2 dt \quad (6)$$

where $A(z,t)$ is the intensity temporal profile of the pulse, α is the fiber loss, β-terms correspond to the chromatic dispersion of the fiber. The mode propagation constant is:

$$\beta(\omega) = n(\omega)\frac{\omega}{c} = \sum \frac{\beta_m}{m!}(\omega - \omega_0)^m, \\ \beta_m = \left(\frac{d^m \beta}{d\omega^m} \right)_{\omega = \omega_0} \quad (7)$$

where β_m describes the wave-vector of the light in the fiber. The term proportional to β_2 is responsible for group velocity dispersion (GVD). It causes temporal pulse broadening when a pulse propagates in a single-mode fiber. Higher orders of dispersion become dominant when the input pulse central wavelength is near the zero dispersion wavelength or when the bandwidth of the pulse becomes a significant fraction of the central frequency. Figure 6 compares dispersion characteristics of the un-infiltrated SPCF of Fig. 1(a) with those of Fig. 1(b). The nonlinear parameter γ is a very important parameter that determines the magnitude of the optical nonlinearity and leads to SPM, self-steepening (SS), stimulated Raman scattering (SRS), and FWM. The right-hand side of (6) accounts for the nonlinear response of the fiber. The response function $R(t)$ including instantaneous Kerr and delayed Raman response effects can be written as

$$R(t) = (1 - f_R)\delta(t) + f_R h_R(t) \quad (8)$$

where $f_R = 0.18$ represents the fractional contribution of the delayed Raman response. The Raman response function $h_R(t)$ takes an approximate analytic form as ($\tau_1 = 0.12$ fs and $\tau_2 = 0.32$ fs for silica)

$$h_R(t) = \frac{\tau_1^2 + \tau_2^2}{\tau_1 \tau_2} \exp(-t / \tau_2) \sin(t / \tau_1) \quad (9)$$

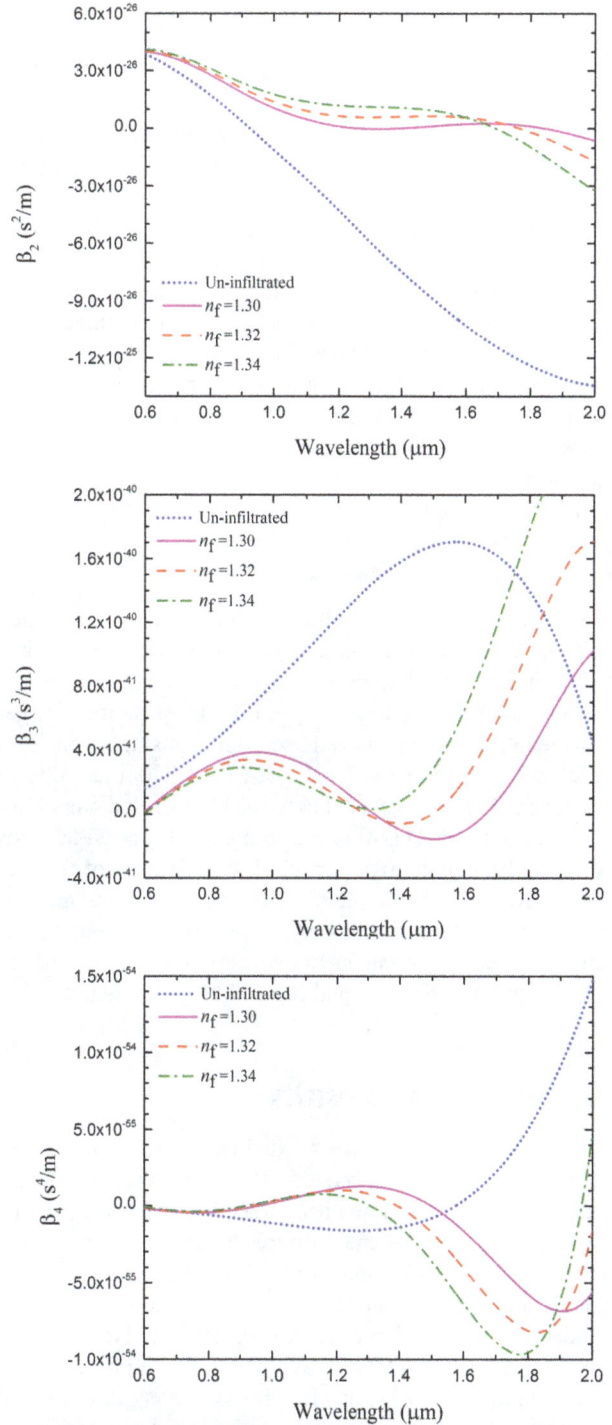

Fig. 6. Cross-sectional view of (a) un-infiltrated SPCF and (b) the inner most ring of SPCF to be selectively infiltrated with an optofluidic of index n_f, two have the same the lattice pitch ($\Lambda = 2$ μm) and air-hole diameters ($d = 0.6\Lambda$).

The Split-Step Fourier Method (SSFM) is used to solve the NLSE which has proven to be a well-suited numerical technique. The input pulse is given by

$$A(0,T) = \sqrt{P_0}\,\text{sech}\left(\frac{T}{T_0}\right).$$ (10)

At $z = 0$, and define soliton order N is written as

$$N^2 = \frac{T_0^2\gamma P_0}{|\beta_2|}.$$ (11)

The soliton order N is determined by the characteristics of input pulse (T_0 and P_0), and the properties of the fiber (γ and β_2). In the mathematically ideal case, where A is a perfectly hyperbolic-secant-shaped pulse, N is an integer, and only self-phase modulation and anomalous group velocity dispersion are present, higher-order solitons propagate in a periodic evolution pattern with original shape recurring at multiples of the soliton period. This is called as ideal soliton conditions. Without any perturbation, these higher order solitons will propagate in a periodic evolution pattern and SC generation would never occur. SC generation is the result of deviations from ideal soliton conditions experienced by a higher order soliton. It means that a small perturbation results in a higher order soliton with the order of N breaks up to N fundamental solitons [25, 28, 29].

These small perturbations include higher order dispersion, SS and SRS. As a result of self-frequency shift (SFS) induced by intrapulse Raman scattering, the fundamental solitons continuously shift toward the longer wavelength of the broadened spectrum, causing a considerable spectral expansion on the red side. The blue side of the spectrum is developed due to the blue shifted nonsolitonic radiation (NSR) [28] or phase-matched. The SPM leads to initially symmetrical spectral broadening of optical pulses after which the phase matching conditions are fulfilled. Then soliton fission occurs and finally SFS and NSR shift the spectral components toward longer and shorter wavelength, respectively and SC will be generated.

4. Simulation Results

Section 2 gives us a qualitative idea about the engineering of the dispersion which is suitable for SC generation in SPCF, now, it is crucial to use the extracted parameters of the presented figures to take into account the influence of dispersion and fluid injection on SC generation. We calculate the SC generation using the model of (6), for input source of 50 fs pulse duration, 20 kW peak power (50 mW average power) and the center wavelength of 1550 nm. In this wavelength different commercial Er-doped femtosecond fiber lasers, such as Femtolite Ultra Bx-60, the CF1550-HP, and Buccaneer are available [30]. The simulations are run for the 100-mm long engineered SPCF with the dispersion and effective mode area calculated in Figs. 4, and 5, respectively. Figure 7 shows spectrum evaluation of SC generation for

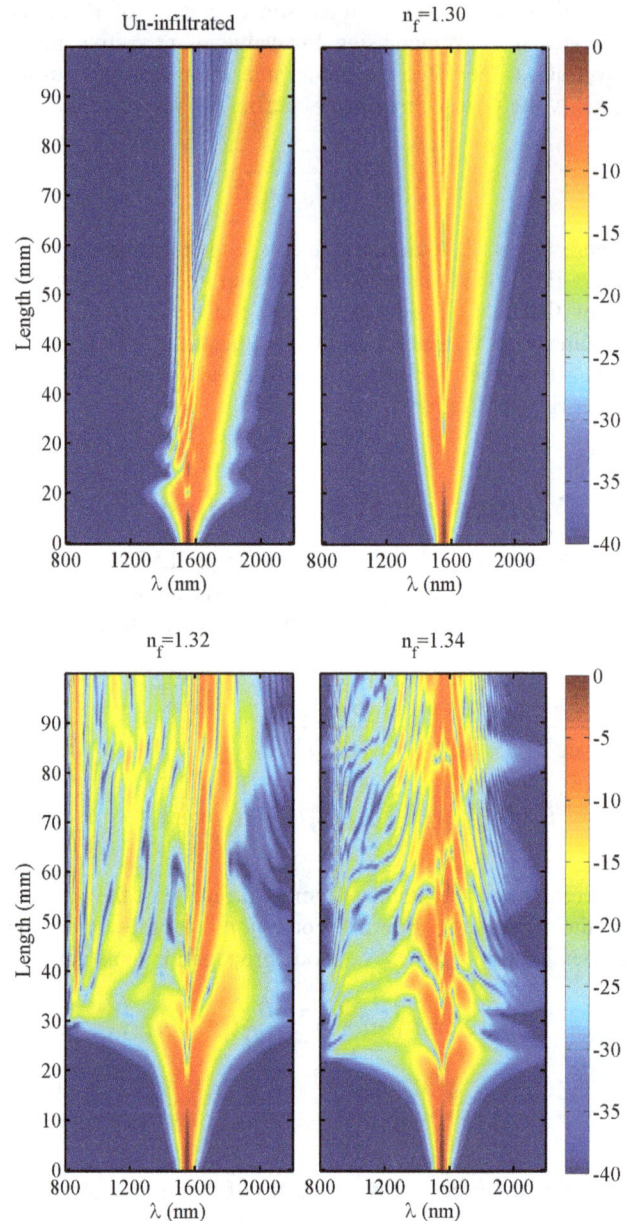

Fig. 7. Simulated evolution of the SC spectrum over propagation distance through SPCF for the uninfiltrated case, 1.3, 1.32, and 1.34 the infiltration case when pumped at 1550 nm.

infiltrations with $1.3 \leq n_f \leq 1.34$ (shown in Fig. 1(b)) with the uninfiltrated case when setting the pump wavelength in the 1550 nm. The simulation results reveal that we can obtain an optimum SC generation simply by changing the fluid refractive index through the SPCF holes. For input pulse with a central wavelength corresponding to 1550 nm, the suitable fluid has a refractive index of 1.32 to have ultra-broadband SC as wide as 1200 nm from 800 to 2000 nm.

Aiming to study the coherence degradation, the analysis of the coherence properties of the generated supercontinua from the presented PCF infiltrated by fluids is performed. It considers the addition of one photon per mode noise with a random phase and it is evaluated within the first order degree of coherence given by

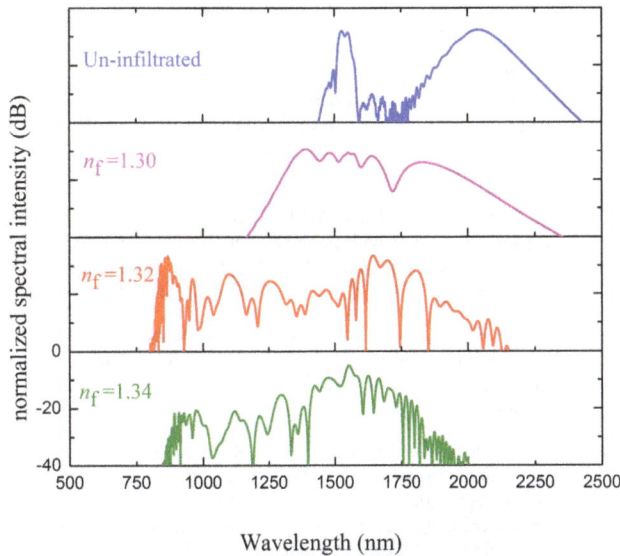

Fig. 8. SC coherence at the output of the fiber for the uninfiltrated case, 1.30, 1.32, and 1.34 the infiltration case when pumped at 1550 nm.

$$\left| g_{12}^{(1)} \left(\lambda, t_1 - t_2 = 0 \right) \right| = \frac{\left| \left\langle E_1^* \left(\lambda, t_1 \right) E_2 \left(\lambda, t_2 \right) \right\rangle \right|}{\sqrt{\left\langle \left| E \left(\lambda, t_1 \right) \right|^2 \right\rangle \left\langle \left| E \left(\lambda, t_1 \right) \right|^2 \right\rangle}} \quad (12)$$

where $E_1(\lambda)$ and $E_2(\lambda)$ are two electric fields. Angular brackets indicate an ensemble average over independently generated pairs of SC spectra, and t is the time measured at the scale of the temporal resolution of the spectrometer used to resolve these spectra. g_{12} is considered at $t_1 - t_2 = 0$ in order to focus on the wavelength dependence of the coherence. It is known that the coherence property of SC light is greatly affected by the quantum noise of the pump pulse. From an ensemble average of 50 independent simulations, we find that the generated supercontinuum in the 100 mm long PCF infiltrated by methanol with an input peak power of 20 kW is coherent over the entire generated bandwidth that is shown in Fig. 8 in which methanol has the flattest coherence among them. It shows that SPCF infiltrated by the fluid of $n_f = 1.32$ is a suitable source for dense wavelength division multiplexing systems. Extracting the desired bandwidth of high intensity and coherence optical signal can be done using some optical filters such as optical fiber Bragg grating that is constructed in a short segment of optical fiber [31].

5. Conclusion

In summary, we investigated and discussed some nonlinear phenomena that result in SC generation in SPCF and proposed optofluidic infilteration as an alternative approach. We numerically investigated the generation of SC with and without optofluidic infilteration and our results confirmed that using the proposed approach, without nano-scale variation in the geometry of the SPCF, its dispersion was simply engineered as desired for fluid with $n_f = 1.32$. Also, for 1550 nm central wavelength of an optical

pulse with 50 fs duration and 20 kW peak power, SC with 1200 nm spectral spanning was generated in 100-mm length of the fiber. The generated SC is a suitable source for wavelength division multiplexing system.

References

[1] DUDLEY, J. M., COEN, S. Coherence properties of supercontinuum spectra generated in photonic crystal and tapered optical fibers. *Optics Letters,* 2002, vol. 27, no. 13, p. 1180–1182. ISSN: 1539-4794. DOI: 10.1364/OL.27.001180

[2] DUDLEY, J. M., GENTY, G., COEN, S. Supercontinuum generation in photonic crystal fiber. *Reviews of Modern Physics,* 2006, vol. 78, no. 4, p. 1135–1184. ISSN: 0034-6861. DOI: 10.1103/RevModPhys.78.1135

[3] DUDLEY, J. M., TAYLOR, J. R. *Supercontinuum Generation in Optical Fibers.* Cambridge University Press, 2010. ISBN: 1139486187

[4] AGRAWAL, G. P. *Nonlinear Fiber Optics.* Academic Press, 2007. ISBN: 0123695163

[5] BIRKS, T., BAHLOUL, D., MAN, T., et al. Supercontinuum generation in tapered fibres. In *Proceedings of Lasers and Electro-Optics, CLEO'02.* USA, 2002, p. 486–487. isbn: 1557527067. DOI: 10.1109/CLEO.2002.1034235

[6] RANKA, J. K., WINDELER, R. S., STENTZ, A. J. Visible continuum generation in air–silica microstructure optical fibers with anomalous dispersion at 800 nm. *Optics Letters,* 2000, vol. 25, no. 1, p. 25–27. ISSN: 1539-4794. DOI: 10.1364/OL.25.000025

[7] WADSWORTH, W. J., ORTIGOSA-BLANCH, A., KNIGHT, J. C., et al. Supercontinuum generation in photonic crystal fibers and optical fiber tapers: a novel light source. *Journal of the Optical Society of America B,* 2002, vol. 19, no. 9, p. 2148–2155. ISSN: 1520-8540. DOI: 10.1364/JOSAB.19.002148

[8] REEVES, W., SKRYABIN, D. V., BIANCALANA, F., et al. Transformation and control of ultra-short pulses in dispersion-engineered photonic crystal fibres. *Nature,* 2003, vol. 424, no. 6948, p. 511–515. ISSN: 0028-0836. DOI: 10.1038/nature01798

[9] MONAT, C., EBNALI-HEIDARI, M., GRILLET, C., et al. Four-wave mixing in slow light engineered silicon photonic crystal waveguides. *Optics Express,* 2010, vol. 18, no. 22, p. 22915 to 22927. ISSN: 1094-4087. DOI: 10.1364/OE.18.022915

[10] SHEN, L., HUANG, W.-P., CHEN, G., et al. Design and optimization of photonic crystal fibers for broad-band dispersion compensation. *IEEE Photonics Technology Letters,* 2003, vol. 15, no. 4, p. 540–542. ISSN: 1041-1135. DOI: 10.1109/LPT.2003.809322

[11] MORTENSEN, N. A. Effective area of photonic crystal fibers. *Optics Express,* 2002, vol. 10, no. 7, p. 341–348. ISSN: 1094-4087. DOI: 10.1364/OE.10.000341

[12] UDEM, T., HOLZWARTH, R., HÄNSCH, T. W. Optical frequency metrology. *Nature,* 2002, vol. 416, no. 6877, p. 233 to 237. ISSN: 0028-0836. DOI: 10.1038/416233a

[13] MOON, S., KIM, D. Y. Ultra-high-speed optical coherence tomography with a stretched pulse supercontinuum source. *Optics Express,* 2006, vol. 14, no. 24, p. 11575–11584. ISSN: 1094-4087. DOI: 10.1364/OE.14.011575

[14] SCHENKEL, B., PASCHOTTA, R., KELLER, U. Pulse compression with supercontinuum generation in microstructure

fibers. *Journal of the Optical Society of America B,* 2005, vol. 22, no. 3, p. 687–693. ISSN: 1520-8540. DOI: 10.1364/JOSAB.22.000687

[15] LIU, B., HU, M., FANG, X., et al. High-power wavelength-tunable photonic-crystal-fiber-based oscillator-amplifier-frequency-shifter femtosecond laser system and its applications for material microprocessing. *Laser Physics Letters,* 2008, vol. 6, no. 1, p. 44. ISSN: 1612-202X. DOI: 10.1002/lapl.200810084

[16] NAKASYOTANI, T., TODA, H., KURI, T., et al. Wavelength-division-multiplexed millimeter-waveband radio-on-fiber system using a supercontinuum light source. *Journal of Lightwave Technology,* 2006, vol. 24, no. 1, p. 404–410. ISSN: 0733-8724. DOI: 10.1109/JLT.2005.859854

[17] SAGHAEI, H., SEYFE, B., BAKHSHI, H., et al. Novel approach to adjust the step size for closed-loop power control in wireless cellular code division multiple access systems under flat fading. *IET Communications,* 2011, vol. 5, no. 11, p. 1469–1483. ISSN: 1751-8636. DOI: 10.1049/iet-com.2010.0029

[18] HANSEN, K. P., FOLKENBERG, J. R., PEUCHERET, C., et al. Fully dispersion controlled triangular-core nonlinear photonic crystal fiber. In *Proceedings of Optical Fiber Communication Conference.* Atlanta (GA, USA), 2003, p. 505–509. ISBN: 1-55752-731-8.

[19] WU, T.-L., CHAO, C.-H. A novel ultra flattened dispersion photonic crystal fiber. *IEEE Photonics Technology Letters,* 2005, vol. 17, no. 1, p. 67–69. ISSN: 1041-1135. DOI: 10.1109/LPT.2004.837475

[20] SAGHAEI, H., HEIDARI, V., EBNALI-HEIDARI, M., et al. A systematic study of linear and nonlinear properties of photonic crystal fibers. *Optik-International Journal for Light and Electron Optics,* 2015, vol. 127, no. 24, p. 11938–11947. ISSN: 0030-4026. DOI: 10.1016/j.ijleo.2016.09.111

[21] MONAT, C., DOMACHUK, P., EGGLETON, B. Integrated optofluidics: A new river of light. *Nature Photonics,* 2007, vol. 1, no. 2, p. 106–114. ISSN: 1749-4885. DOI: 10.1038/nphoton.2006.96

[22] MONAT, C., DOMACHUK, P., GRILLET, C., et al. Optofluidics: a novel generation of reconfigurable and adaptive compact architectures. *Microfluidics and Nanofluidics,* 2008, vol. 4, no. 1-2, p. 81–95. ISSN: 1613-4982. DOI: 10.1007/s10404-007-0222-z

[23] EBNALI-HEIDARI, M., DEHGHAN, F., SAGHAEI, H., et al. Dispersion engineering of photonic crystal fibers by means of fluidic infiltration. *Journal of Modern Optics,* 2012, vol. 59, no. 16, p. 1384–1390. ISSN: 0950-0340. DOI: 10.1080/09500340.2012.715690

[24] EBNALI-HEIDARI, M., SAGHAEI, H., KOOHI-KAMALI, F., et al. Proposal for supercontinuum generation by optofluidic infiltrated photonic crystal fibers. *IEEE Journal of Selected Topics*

in Quantum Electronics, 2014, vol. 20, no. 5, p. 582–589. ISSN: 1077-260X. DOI: 10.1109/JSTQE.2014.2307313

[25] SAGHAEI, H., EBNALI-HEIDARI, M., MORAVVEJ-FARSHI, M. K. Midinfrared supercontinuum generation via As_2Se_3 chalcogenide photonic crystal fibers. *Applied Optics,* 2015, vol. 54, no. 8, p. 2072–2079. ISSN: 1539-4522. DOI: 10.1364/AO.54.002072

[26] SAITOH, K., KOSHIBA, M. Full-vectorial finite element beam propagation method with perfectly matched layers for anisotropic optical waveguides. *Journal of Lightwave Technology,* 2001, vol. 19, no. 3, p. 405. ISSN: 0733-8724. DOI: 10.1109/50.918895

[27] MALITSON, I. Interspecimen comparison of the refractive index of fused silica. *Journal of the Optical Society of America,* 1965, vol. 55, no. 10, p. 1205–1208. ISSN: 0030-3941. DOI: 10.1364/JOSA.55.001205

[28] DIOUF, M., SALEM. A., CHERIF, R., et al. Super-flat coherent supercontinuum source in $As_{38.8}$ $Se_{61.2}$ chalcogenide photonic crystal fiber with all-normal dispersion engineering at a very low input energy. *Applied Optics,* 2017, vol. 56, no. 2, p. 163–169. ISSN: 1539-4522. DOI: 10.1364/AO.56.000163

[29] SAGHAEI, H., MORAVVEJ-FARSHI, M. K., EBNALI-HEIDARI, M., et al. Ultra-wide mid-infrared supercontinuum generation in $As_{40}Se_{60}$ chalcogenide fibers: Solid core PCF versus SIF. *IEEE Journal of Selected Topics in Quantum Electronics,* 2016, vol. 22, no. 2, p. 1–8. ISSN: 1077-260X. DOI: 10.1109/JSTQE.2015.2477048

[30] MIRET, J., SILVESTRE, E., ANDRÚS, P. Octave-spanning ultraflat supercontinuum with soft-glass photonic crystal fibers. *Optics Express,* 2009, vol. 17, no. 11, p. 9197–9203. ISSN: 1094-4087. DOI: 10.1364/OE.17.009197

[31] KASHYAP, R. *Fiber Bragg Gratings.* Academic Press, 1999. ISBN: 0080506275

About the Author ...

Hamed SAGHAEI was born in Shahrekord, Iran, in 1982. He received the B.Sc., M.Sc. and Ph.D. degrees from Amirkabir University of Technology, Shahed University, and Science and Research branch of IAU, Tehran, Iran, all in Electrical Engineering in 2004, 2007, and 2015, respectively. He joined Islamic Azad University, Shahrekord, Iran, in 2011, where he is currently an assistant professor of electronics and the head of the Faculty of Electrical and Electronic Engineering. His research interests are photonic crystal fibers and waveguides, supercontinuum generation, and optical communication.

Permissions

List of Contributors

Vaclav Prajzler and Milos Neruda
Dept. of Microelectronics, Faculty of Electrical Engineering, Czech Technical University, Technická 2, 166 27 Prague, Czech Republic

Pavla Nekvindova
Institute of Chemical Technology, Technická 5, 166 27 Prague, Czech Republic

Petr Mikulik
Dept. of Condensed Matter Physics, Masaryk University, Kotlářska 2, 611 37 Brno, Czech Republic

Mohammad T. Kawser, Mohammad R. Islam, Muhammad R. Rahim and Muhammad A. Masud
Dept. of Electrical and Electronic Engineering, Islamic University of Technology, Board Bazar, Gazipur-1704, Bangladesh

Saulius Japertas and Vitas Grimaila
Dept. of Telecommunications, Kaunas University of Technology, K. Donelaičio str. 73, 44249 Kaunas, Lithuania

Alberto Tekovic
Access and Transport Network Engineering, VIPnet Ltd., Vrtni put 1, 10 000 Zagreb, Croatia

Davor Bonefacic, Gordan Sisul and Robert Nad
Dept. of Radiocommunications, Faculty of Electrical Engineering and Computing, Unska 3, 10 000 Zagreb, Croatia

Mohd Hidir Mohd Salleh, Norhudah Seman and Akaa Agbaeze Eteng
Wireless Communication Centre (WCC), Universiti Teknologi Malaysia, 81310, Johor, Malaysia

Dyg Norkhairunnisa Abang Zaidel
Dept. of Electrical and Electronics Engineering, Faculty of Engineering, Universiti Malaysia Sarawak, 94300 Kota Samarahan, Sarawak, Malaysia

Hristo Zhivomirov
Dept. of Theory of Electrical Engg. and Measurements, Technical Univ. of Varna, Sudentska Street 1, Varna, Bulgaria

Nikolay Kostov
Dept. of Telecommunications, Technical University of Varna, Sudentska Street 1, Varna, Bulgaria

Hongshu Lu, Weiwei Wu, Jingjian Huang, Xiaofa Zhang and Naichang Yuan
College of Electronic Science and Engineering, National University of Defense Technology, Changsha, Hunan, 410073, China

Kaibo Cui, Weiwei Wu, Xi Chen, Jingjian Huang and Naichang Yuan
College of Electronic Science and Engineering, National University of Defense Technology, Changsha, Hunan, 410073, China

Slawomir Gajewski
Dept. of Radio Communication Systems and Networks, Faculty of Electronics, Telecommunications and Informatics, Gdansk University of Technology, Narutowicza 11/12, 80233 Gdansk, Poland

Sandra Costanzo
DIMES, University of Calabria, Via P. Bucci cubo 42C, 87036 Rende (CS), Italy

Martin Kenyeres, Vladislav Skorpil and Radim Burget
Dept. of Telecommunications, Brno University of Technology, Technická 12, 612 00 Brno, Czech Republic

Jozef Kenyeres
Sipwise GmbH, Europaring F15, 2345 Brunn am Gebirge, Austria

Khachen Khaw-Ngam and Montree Kumngern
Faculty of Engineering, King Mongkut's Institute of Technology Ladkrabang, Bangkok 10520, Thailand

Fabian Khateb
Department of Microelectronics, Brno University of Technology, Technická 10, Brno, Czech Republic
Faculty of Biomedical Engineering, Czech Technical University in Prague, nám. Sítná 3105, Kladno, Czech Republic

Manuel Cedillo-Hernandez, Antonio Cedillo-Hernandez, Mariko Nakano-Miyatake and Hector Perez-Meana
Instituto Politecnico Nacional SEPI ESIME Culhuacan, Avenida Santa Ana 1000, San Francisco Culhuacan Coyoacan, Ciudad de Mexico, Mexico

Francisco Garcia-Ugalde
Universidad Nacional Autonoma de Mexico, Facultad de Ingenieria, Avenida Universidad 3000 Ciudad Universitaria Coyoacan, Ciudad de Mexico, Mexico

Peng Li, Hongwei Ge and Jinlong Yang
Key Laboratory of Advanced Process Control for Light Industry (Jiangnan University), Ministry of Education, Wuxi, 214122, China
School of Internet of Things Engineering, Jiangnan University, Wuxi, 214122, China

Zeynep Hasirci, Ismail Hakki Cavdar and Mehmet Ozturk
Dept. of Electrical and Electronics Engineering, Karadeniz Technical University, 61080, Trabzon, Turkey

Zhao Zhang, Xiangyu Cao, Jun Gao and Sijia Li
Information and Navigation College, Air Force Engineering University, Xi'an 710077, China

Liming Xu
Science and Technology on Electronic Information Control Laboratory, Chengdu 610036, China

Kai-Da XU
Inst. of Electromagnetics and Acoustics & Dept. of Electronic Science, Xiamen University, Xiamen 361005, China
Shenzhen Research Inst. of Xiamen University, Shenzhen 518057, China

Meng-Ze LI, Yan-Hui LIU and Ye-Cheng BAI
Inst. of Electromagnetics and Acoustics & Dept. of Electronic Science, Xiamen University, Xiamen 361005, China

Jing AI
EHF Key Lab of Science, University of Electronic Science and Technology of China, Chengdu 611731, China

Wael Ali
Dept. of Electronics & Comm. Engineering, College of Engineering, Arab Academy for Science, Technology and Maritime Transport (AASTMT), Alexandria, Egypt

Ehab Hamad
Dept. of Electrical Engineering, Aswan Faculty of Engineering, Aswan University, Aswan 81542, Egypt

Mohamed Bassiuny and Mohamed Hamdallah
Dept. of Electronics & Comm. Engineering, College of Engineering, Arab Academy for Science, Technology and Maritime Transport (AASTMT), Aswan, Egypt

Yi Hu and Hongli Yan
School of Electronics & Electrical Engineering, Chuzhou University, Chuzhou, China

Maozhong Song and Xiaoyu Dang
School of Electronics & Information Engineering, Nanjing University of Aeronautics & Astronautics, Nanjing, China

Leila Safari, Gholamreza Baghersalimi and Ali Karami
Dept. of Electrical Engineering, University of Guilan, Rasht, Iran

Abdolreza Kiani
Dept. of Electrical Engineering, Abadan Branch, Islamic Azad University, Abadan, Iran

Mostafa Danaeian, Kambiz Afrooz and Ahmad Hakimi
Department of Electrical Engineering, Shahid Bahonar University of Kerman, Kerman, Iran

Hamed Saghaei
Faculty of Engineering, Shahrekord Branch, Islamic Azad University, Shahrekord, Iran

Index

www.ingramcontent.com/pod-product-compliance
Lightning Source LLC
Chambersburg PA
CBHW080704200326
41458CB00013B/4954